Conservation Science

Heritage Materials
2nd Edition

Conservation Science

Heritage Materials
2nd Edition

Edited by

Paul Garside

British Library, UK
Email: dr.paul.garside@gmail.com

and

Emma Richardson

Image Permanence Institute, USA
Email: ejrpph@rit.edu

ROYAL SOCIETY
OF **CHEMISTRY**

Print ISBN: 978-1-78801-093-1

EPUB ISBN: 978-1-78801-934-7

A catalogue record for this book is available from the British Library

The Royal Society of Chemistry is a charity, registered in England and Wales, Number 207890, and a company incorporated in England by Royal Charter (Registered No. RC000524), registered office: Burlington House, Piccadilly, London W1J 0BA, UK, Telephone: +44 (0) 20 7437 8656.

Visit our website at www.rsc.org/books

Printed in the United Kingdom by CPI Group (UK) Ltd, Croydon, CR0 4YY, UK

Preface to the Second Edition

Since the first edition of *Conservation Science: Heritage Materials* there have been a number of challenges within the museum and heritage profession, from the financial crisis of 2008 that saw sweeping cuts in funding to arts and humanities, to the current global pandemic that has seen institutions close their doors to visitors. Through each of these changes conservation professionals have demonstrated their capacity to problem solve, maintain rigour and draw on the breadth of knowledge and skills from across this cross-disciplinary field.

Conservation science as a sub-discipline provides the underpinning knowledge that allows conservation professionals to make informed decisions around treatment, storage and display. To achieve greatest beneficial impact, conservation science necessitates collaboration with key stakeholders such as conservators, curators, artists and the public, providing context for scientific analysis and ensuring critical questions are addressed. The aim of this book is to provide a starting point for dialogue between the disciplines and a common language of understanding for such collaboration. It draws together materials science with conservation science case studies, demonstrating how both tried-and-tested methods and the latest technologies can be applied to materials characterisation and technical art history, and to improve our understanding of degradation and means of condition assessment, thus informing preservation decisions.

It is worth noting that over the past decade the term heritage science has moved into the vernacular of the profession. Whilst this term can be helpful in providing distinction from the field of ecological conservation science, heritage science is such a broad term that it does not

Conservation Science: Heritage Materials, 2nd Edition
Edited by Paul Garside and Emma Richardson

Published by the Royal Society of Chemistry, www.rsc.org

provide definition. In this second edition of *Conservation Science: Heritage Materials* we are primarily focused on science within conservation and how technologies may support preservation, and therefore in this case we believe conservation science is a more appropriate term.

Following a thoughtful Introduction by Dr Cordelia Rogerson of the British Library, reflecting on the multidisciplinary nature of the profession, there is an introductory chapter on the use of analytical techniques in conservation science, which is aimed at providing a foundation for discussions later in the book. The later chapters are divided into broad material types and focus on practical, evidence-based approaches to the analysis and characterisation of heritage materials, reflecting the increasing use of these techniques in mainstream conservation decision making. Each of these chapters follows a similar format, covering chemistry and structure, analytical techniques, impact on conservation and preservation, and case studies appropriate to the material. Whilst the format is aimed at ensuring accessibility for the reader, the level of detail within each chapter varies and is a reflection on the differing complexities and priorities within both the science and the particular conservation requirements of the materials.

Although not explicitly addressed by this book, the continuously evolving nature of the heritage sector and the conservation profession means that factors such as environmental sustainability, inclusivity and cultural awareness will play an increasingly important role in decision making, and an informed understanding of the composition, properties and analysis of heritage materials will be an important factor in ensuring the best possible outcomes for collections and communities. Similarly, this knowledge will enable resources, which are likely to be placed under increasing demands, to be used in the most efficient and appropriate manner.

This book would not have been possible without the many years of research and development of expertise by the contributing authors. We are grateful for their willingness to share their knowledge and also their enduring patience in getting this edited volume over the finishing line. Each editor and author has balanced a great many professional and personal demands along the way, so it is with a great sigh of relief that we finally have a volume to which everyone can be proud to have contributed. We are eternally grateful to the publications team at the Royal Society of Chemistry for their gentle nudging and unwavering belief that the book would eventually materialise (no pun intended!).

Paul Garside
Emma Richardson

Preface to the First Edition

Conservation science is a broad church and a rapidly developing discipline. The aim of this book is to provide, in one volume, an account of the composition and conservation of historic materials. It brings together recent information on decay and degradation that are unusually scattered in journals and not readily available to many practising conservators. By doing this, it is hoped that the potential of materials conservation, together with the processes that are necessary to maximise it, can be made more understandable to all those interested in historic objects. We believe that a clear understanding of the deterioration mechanisms of historic and artistic works will help conservators make informed decisions about their care and treatment.

At the time of writing, there is concern throughout Europe about the fragmented nature of research funding in conservation science, despite the fact that heritage generates considerable income through tourism for national economies. There is parliamentary activity at national and EU levels as we move to the 7th European Research Framework. This book demonstrates that a very wide range of accumulated deterioration problems affect materials that are part of our movable and immovable heritage. The threat of environmental climate change now compounds those problems and seriously extends the scale and nature of the challenge that conservators and conservation scientists face. The wealth of knowledge offered by specialists in this book illustrates the commitment to good science and reinforces the need for on-going research to meet the new challenges.

The book is intended for graduates and professionals who are involved in the understanding and care of historic materials,

Conservation Science: Heritage Materials, 2nd Edition
Edited by Paul Garside and Emma Richardson

Published by the Royal Society of Chemistry, www.rsc.org

whether conservators, finds specialists, museum curators or administrators. It is also of interest to undergraduates studying archaeological conservation at university, or scientists taking subsidiary courses in conservation. It should also prove useful to those involved in the training of students in material conservation. Inevitably, when describing deterioration (decay), material composition and treatment methods, science is essential for the explanation. However, the general reader with basic science knowledge should have no difficulty in following the text. It should provide a good introduction to the subject and we hope that it will be a valuable reference work and precious reading companion for both students and professionals.

The text is divided into chapters covering analytical aspects, organic materials, inorganic materials, wall paintings and *in situ* preservation. Contributions are from a wide range of practising international experts from leading research institutes who outline the problems and provide many interesting and exciting case studies. The text offers a basic introduction to the science of heritage materials, their deterioration and restoration processes. The complexities and problems faced by conservators and scientists are addressed together with the processes of decay and degradation that affect ancient materials. It describes in detail the processes of decay and degradation for each major category of material (paper, wood, stone, glass, metals, leather, plastics *etc.*). The chapters on plastics and *in situ* conservation provide insights into the rapidly developing new areas of conservation science.

A word here about topic coverage and our policy on referencing. These are generally full with topics where there exists no adequate books or review articles – metals and wood are cases in point – and, as a consequence, there is no necessary correlation between length of a chapter and the practical importance of the subject. In relation to the references, we have adopted a nonreferenced style and provided reference sources and further reading at the end.

This book is clearly the product of many specialists and the editors are conscious that without their input it would not exist. It has had a very long gestation period. There is never a good time to edit a book; it is a labour of love that surpasses all understanding and common sense, measured against the increasing pressures of academic life. The contributions have come in over a long period and in that sense we are grateful for the understanding of not only the authors, but the RSC, particularly Janet Freshwater and Annie Jacob, who have cajoled, but mostly supported, us throughout a prolonged delivery.

We are also grateful for assistance and help from our organisations, namely, the Mary Rose Trust (MJ) (notably Andy Elkerton with the index) and the University of Portsmouth (EM). This could not have happened without the support of our families, especially Helen and Jeannette. We hope you enjoy the final product.

Eric May
Mark Jones

Glossary

Acicular – Needle-like or whisker-like crystalline structures.
Aliphatic Molecule – An organic compound in which the constituent atoms are joined in straight chains, branched chains or non-aromatic rings; they may contain single, double or triple bonds (*i.e.*, saturated or unsaturated).
Annealing – A heat treatment that alters the properties of a material (typically a metal), increasing its ductility and reducing its hardness, to improve its ability to be worked.
Aromatic Molecule – An organic compound which contains one or more ring structure with fully conjugated bonds; this allows certain electrons to disassociate across the ring structure (π bonding), granting increased stability compared to similar non-aromatic structures. Benzene is the most common aromatic compound.
Auxochrome – A functional group which modifies the observed colour of a chromophore.
Azo Compound – A molecule containing the diazenyl group ($R{-}N{=}N{-}R'$); many of these materials are strongly coloured and so are important in dye chemistry.
Barkcloth – A non-woven fabric made from the inner bark of certain trees (usually of the *Moraceae*, or mulberry, family), produced through retting and pounding.
Biopolymer – A polymer that is produced by (or derived from) a living organism.
Chelating Agent – An organic molecule capable of forming multiple bonds to a metal ion, often used in conservation to remove soiling through the formation of water-soluble ring-like complexes that can be washed away.

Conservation Science: Heritage Materials, 2nd Edition
Edited by Paul Garside and Emma Richardson

Published by the Royal Society of Chemistry, www.rsc.org

Chromophore – The region in a molecular structure which leads to its colour, due to its electronic structure, usually resulting from extended series of conjugated bonds in aliphatic and aromatic compounds.
CIE – An organisation (Commission International de l'éclairage) that establishes methods of colour measurement. The 1976 CIELab system relates mainly to a colour space but the more recent CIE 2000 formula is a calculation that uses CIELab values but brings colour differences closer to the perception of the human eye.
Complex – A molecular entity, formed of a central atom or ion (frequently a metal ion) surrounded by a specific number of loosely bonded ions or molecules in a particular geometry.
Conjugation – A sequence of alternating single and double bonds in aliphatic and aromatic organic compounds.
Copolymer – A polymer containing more than one species of monomer.
Degree of Polymerisation (DP) – The average number of monomer/repeat units in a polymer molecule.
Dendrite – Branchlike structures of crystals which form within metal as it cools and solidifies from the molten phase; in multicomponent systems the composition of the dendrites may differ from that of the bulk.
Detrital – Sedimentary rocks composed of rock debris or fragments.
Earlywood – Wood formed early in the growing season of a tree, comprised of larger cells with thinner cell walls, and forming the lighter part of the annual growth ring (*cf.* latewood); also called 'spring wood'.
Efflorescence – The formation of crystals on the surface of an object, as the result of the migration of a mobile species (such as a soluble salt or a plasticiser) from the bulk to the surface.
Emulsion – A mixture of two mutually insoluble liquids, in which one liquid is found as a dispersion of fine droplets within the other.
Flavonoids – Naturally occurring vegetable pigments named after the compound flavone, commonly used to produce yellow dyestuffs.
Fluorophore – A chemical species that exhibits fluorescence due to its electronic structure (*cf.* chromophore).
Flux – In the context of glass and ceramics, a substance that helps lower the melting point of the main constituents.
Functional Group – Specific chemical structures or regions in a molecule that are responsible for its chemical reactivity.
Glass Transition – A physical process that occurs in amorphous materials as temperature increases, characterised by a transition from a hard and brittle (glass-like) state to a soft, viscous (rubbery) state. The glass transition temperature (T_g) is the temperature range over which this process occurs.

Green Body – A mass of weakly bound clay material (often with organic or inorganic additives), before it has been fired, sintered or vitrified.
Hygroexpansivity – Change in dimensions due to variations in moisture content.
IR – Infrared, referring to electromagnetic radiation in the region of the spectrum adjacent to visible light, and of a lower frequency (longer wavelength): wavelength 700 nm–1 mm, frequency 430 THz–300 GHz.
Isomer – Compounds containing the same atomic composition but displaying different atomic arrangements, resulting in different chemical and physical properties.
Lamellar – Alternating layers of different materials producing inhomogeneous microstructures, formed naturally or through material processing.
Latewood – Wood formed late in the growing season of a tree, comprised of smaller cells with thicker cell walls, and forming the darker part of the annual growth ring (*cf.* earlywood); also called 'summer wood'.
Lees – Dead yeast cells, left over from the fermentation process.
Libriform Fibres – Highly elongated cells, with a high proportion of lignin in the cell walls, found in xylem, which contribute to the hardness and resilience of wood.
Lixiviation – The separation of a mixture of insoluble and soluble materials by dissolving the latter in water or another suitable solvent.
Macromolecule – A large molecule formed from linked monomeric units. The term is often used interchangeably with 'polymer', although it applies specifically to the molecular species rather than the bulk material.
Masstone – The full colour pigment or coating, when applied thickly.
Minimal Intervention – A conservation ideal in which the minimum of treatment is performed to maintain the stability of an object.
Monomer – A molecule capable of bonding with other monomers to form large chain-like or networked molecular structures (polymers), through the process of polymerisation; the basic building block of a polymer.
Mordant – A substance (typically a polyvalent metal ion) which enhances the affinity of a dye for a substrate and improving its fastness; the use of a mordant may also alter the colour of the dyed material.
Mullite – A needle-like silicate mineral, formed during various melting and firing processes, and found in porcelain. Also known as porcelainite.
Mucilage – Hydrophilic polysaccharide secretions from certain plants, often gel forming, such as those originating from aloes.

Native Metal – A metal that is found in its pure, metallic form in nature, rather than as a mineral.
Oligomer – A molecule comprised of a few monomers; dimers and trimers (two and three monomeric units, respectively) are examples of oligomers.
Passivation – A process leading to the reduction of the reactivity of a material (often a metal) to its environment, though the formation of an inert or minimally reactive surface layer.
Perspex – A transparent polymer made of poly(methyl methacrylate).
Plasticiser – Low molecular weight species, which can be added to polymers or other bulk materials to modify their properties, typically to increase plasticity and decrease viscosity (thus enhancing flexibility and durability).
Plasticiser Migration – The tendency of plasticisers in a bulk material to travel within that material over time, often forming sticky or crystalline deposits on the surface. The process generally leads to the embrittlement of the original object.
Polymer – A large molecule (macromolecule) comprised of many monomeric units linked together by a polymerisation reaction. It can also refer to the bulk material formed from polymer molecules.
Polysaccharide – A class of polymers including cellulose, starch and chitin, which are formed from simple sugars (monosaccharides, such as glucose, fructose and arabinose).
Quenching – The rapid cooling of a material to prevent the formation of large crystals, by which desirable physical properties are promoted, typically hardness.
Radical – An atom, molecule or ion with an unpaired valence electron; these species are normally highly reactive.
R Group – An abbreviation used to denote the part of a molecule outside the immediate region of interest.
Retting – A process in which vegetation is left in water for an extended period during which microbial and chemical processes break down substances holding cellulose fibres together, rendering them easy to separate and subsequently use. The term is most frequently encountered in the treatment of flax to obtain linen fibres.
Saturated Compound – An organic molecule which contains only single bonds.
Scission – The breaking of a chemical bond.
Sinter – To fuse a powder or particulate material to into a cohesive mass, through the use of heat or pressure, without causing it to fully melt; the resulting structure has a network of small cavities and is often porous.

Size – A substance applied to a material to improve its absorption or handling characteristics. Sizing of paper reduces the ability of the material to absorb liquid, allowing inks or pigments to remain on the surface rather than bleeding or spreading; it may also alter other properties. Sizing of textiles or yarns improves strength and abrasion resistance, reducing breakages during manufacture.
Stereochemistry – The aspect of chemistry concerned with three-dimensional (spatial) arrangement of atoms and molecules, and the way in which this geometry influences their interactions and reactivity.
Stoichiometry – The ratio of quantities of species involved in a chemical reaction, allowing the calculation of reagents and products in such reactions.
Surrogate Material – Substitute or replicate materials that are often employed when original objects cannot be sampled or tested, and whose properties and behaviours are chosen or modified to mimic those of such objects.
Tempering – A slow heat treatment performed after quenching to increase toughness.
Tracheids – Highly elongated cells found in the xylem of plants, which allow the transport of water and soluble material such as salts.
Tristimulus Colour Measurement – A method of measuring and representing colour using three values, such as the CIE $L^*a^*b^*$ system.
Unsaturated Compound – An organic molecule which contains at least one double or triple bond.
UV – Ultraviolet, referring to electromagnetic radiation in the region of the spectrum adjacent to visible light, and of a higher frequency (shorter wavelength): wavelength 10 nm–400 nm, frequency 30 PHz–790 THz.
Valence – The number of electrons that an atom uses in bonding; a measure of the ability of an atom to form bonds.
Viscoelasticity – The mechanical characteristic of materials that exhibit both instantaneous, elastic behaviour and time-dependant, viscous behaviour when placed under stress.
Vitrification – The transformation, through heat, of a material into a glassy solid.
Wavenumber – A measure of frequency, normally defined as the reciprocal of the wavelength in centimetres, with the units cm^{-1}; frequently used in vibrational spectroscopy.

Abbreviations

ABS	Acrylonitrile Butadiene Styrene
AFM	Atomic Force Microscopy
AM	Additive Manufacturing
ATR	Attenuated Total Reflectance
BSE	Back-Scattered Electron
CAD	Computer Aided Design
CD	Cross Direction
CIE	Commission Internationale de l'Éclairage (International Commission on Illumination)
CPMG	Carr–Purcell–Meiboom–Gill
CT	Computed Tomography
DR	Draw Ratio
DFT	Density Functional Theory
DEP	Diethyl Phthalate
DMA	Dynamic Mechanical Thermal Analysis
DNA	Deoxyribonucleic Acid
DP	Degree of Polymerisation
DRIFTS	Diffuse Reflectance Infrared Fourier Transform Spectroscopy
DSC	Differential Scanning Calorimetry
DT	Direct Temperature
DTMS	Direct Temperature-resolved Mass Spectrometry
DTPA	Diethylenetriaminepentaacetic Acid
E'	Storage Modulus
E''	Loss Modulus
EB	Binding Energy
EDDHMA	Ethylenediamine-*N*,*N'*-bis(2-hydroxyphenylacetic acid)

Conservation Science: Heritage Materials, 2nd Edition
Edited by Paul Garside and Emma Richardson

Published by the Royal Society of Chemistry, www.rsc.org

EDS	Energy Disperse X-Ray Spectroscopy (also EDX)
EDTA	Ethylenediaminetetraacetic Acid
EDX	Energy Disperse X-Ray Spectroscopy (also EDS)
ELISA	Enzyme-Linked Immunosorbent Assay
EIS	Electrochemical Impedance Spectroscopy
EMF	Electromotive Force
ESEM	Environmental Scanning Electron Microscopy
EXAFS	Extended X-ray Absorption Fine Structure
FDM	Fused Deposition Modelling
FFF	Fused Filament Fabrication
FIB	Focused Ion Beam
FID	Flame Ionisation Detection
FORS	Fibre Optics Reflectance Spectroscopy
FTICR	Fourier Transform Ion Cyclotron Resonance
FTIR	Fourier Transform Infrared Spectroscopy
FTMIR	Fourier Transform Mid-Infrared Spectroscopy
FTNIR	Fourier Transform Near-Infrared Spectroscopy
FTMS	Fourier Transform Mass Spectrometry
GC	Gas Chromatography
GC–MS	Gas Chromatography–Mass Spectroscopy
HPLC	High Performance Liquid Chromatography (or High-Pressure Liquid Chromatography)
HSQC	Heteronuclear Single Quantum Coherence
ICP-MS	Inductively Coupled Plasma Mass Spectroscopy
IR	Infrared
LA	Laser Ablation
LC	Liquid Chromatography
LIBS	Laser Induced Breakdown Spectroscopy
LM	Light Microscopy
MALDI	Matrix-Assisted Laser Desorption/Ionization
MAS NMR	Magic Angle Spinning Nuclear Magnetic Resonance
MD	Machine Direction
MHT	Micro Hot Table
MIC	Microbiologically Induced Corrosion
MIR	Mid Infrared
Mn	Number Average Molar Mass
MRI	Magnetic Resonance Imaging
MS	Mass Spectroscopy
MSTBFA	*N*-methyl-*N*-(*t*-Butyldimethylsylil)trifluoroacetamide
Mw	Weight Average Molar Mass
NAA	Neutron Activation Analysis
NIR	Near Infrared

NLO	Nonlinear Optical
NMR	Nuclear Magnetic Resonance
NRA	Nuclear Reaction Analyses
OM	Optical Microscopy
PA	Polyamide
PC	Principal Component
PCA	Principal Component Analysis
PCR	Polymerization Chain Reaction
PEF	Photon Excitation Fluorescence
PEG	Polyethylene Glycol
PET	Polyethylene Terephthalate
PIGE	Particle-Induced Gamma-ray Emission
PIXE	Particle Induced X-Ray Emission Spectroscopy
PLA	Biopolymer Polylactic Acid
PLM	Polarized Light Microscopy
PMF	Peptide Mass Fingerprinting
PY	Pyrolysis
PY-GC–MS	Pyrolysis Gas Chromatography–Mass Spectroscopy
PUR	Polyurethane
PVAc	Polyvinyl Acetate
PVC	Polyvinyl Chloride
RBS	Rutherford Backscattering Spectroscopy
RF	Radio Frequency
R_f	Retardation Factor
RH	Relative Humidity
RP	Rapid Prototyping
RTI	Reflectance Transformation Imaging
SAN	Styrene Acrylonitrile
SAXS	Small Angle X-ray Scattering
SE	Secondary Electron
SEC	Size Exclusion Chromatography
SEM	Scanning Electron Microscopy
SERS	Surface Enhanced Raman Spectroscopy
SHE	Standard Hydrogen Electrode
SHG	Second Harmonic Generation
SLS	Selective Laser Sintering
SS NMR	Solid State Nuclear Magnetic Resonance
T	Temperature
TCD	Thermal Conductivity Detection
TEY	Total Electron Yield
TEM	Transmission Electron Microscopy
T_g	Glass Transition Temperature

TGA	Thermogravimetric Analysis
THM	Thermally Assisted Methylation
TLC	Thin Layer Chromatography
T_m	Melting Temperature
TMA	Thermomechanical Analysis
TMAH	Tetramethylammonium Hydroxide
TOF	Time of Flight
TPE	Thermoplastic Elastomers
TPP	Triphenyl Phosphate
TPU	Thermoplastic Polyurethane
T_s	Shrinkage Temperature
UV	Ultraviolet
Vis	Visible
VOC	Volatile Organic Compound
WD	Wavelength Dispersion
WDS	Wavelength Dispersive Spectroscopy
XANES	X-ray Absorption Near Edge Structure
XAS	X-ray Absorption Spectroscopy
XRD	X-ray Diffraction
XRF	X-Ray Fluorescence
ZD	Depth Direction

Contents

Conservation Science: Heritage Materials, 2nd Edition
Edited by Paul Garside and Emma Richardson

Published by the Royal Society of Chemistry, www.rsc.org

1 Introduction

C. Rogerson

British Library, 96 Euston Road, London, NW1 2DB, UK
Email: cordelia.rogerson@bl.uk

1.1 Conservation as a Multidisciplinary Profession

Perhaps the stand-out feature of the conservation profession is its multidisciplinary nature. A conservator assimilates a widespread range of information in order to consider the stability, care and interpretation of cultural materials. Typically, a conservator draws information from the broad areas of humanities and science, and then combines the data by means of practical decision making and often hand skills. Resultantly these arenas of knowledge end up cohabiting the same space when applied to an object since separating the scientific data from the cultural significance is artificial, if not impossible.

The main purpose of this book is to reinforce and enhance the understanding and interpretation of material culture by elucidating the science required for conservation purposes. By addressing the science aspect, and more specifically the analytical methods available to assemble information about the materials in hand, the book ultimately aims to support the entire multidisciplinary process of conservation.

Not that all conservation processes are identical or that conservators have matching skill sets or roles. An increasingly wide range of specialisms exists in the profession. Whilst some conservators will work directly with objects daily, others may do this only as part of

Conservation Science: Heritage Materials, 2nd Edition
Edited by Paul Garside and Emma Richardson

Published by the Royal Society of Chemistry, www.rsc.org

their role and others again may be concerned with processes of risk management to care for collections more holistically. Individual conservators may specialise quite closely on one material type or group of related materials, whilst others may be required to apply knowledge across a range of materials as multicomponent objects or collections are repeatedly encountered. What underpins each strand of the conservation profession, however, is a desire to prolong the life cycle of cultural materials to enable interpretation and use for the enrichment of society now and in the future.

At its root, the process of achieving the longevity of heritage material is a series of interrelated investigations enabling decisions to establish a course of action or actions for a single item or a collection. In this respect, further similarities exist between seemingly disparate roles across the conservation profession. Ultimately, this sequence of investigation and decision reduces risk to cultural materials through the understanding of composition and structure, identifying the nature, cause and extent of deterioration pathways and enabling judgements to be made on how stability may be achieved. Conservation activities may be interventive in nature but not in every case, since altering the surrounding environment can often be effective enough. Additionally, the investigation and treatment process can purposefully, or potentially unexpectedly, reveal more information to enhance interpretation and historic appreciation of the material. Unseen detail, underdrawings, atypical material, hitherto unrealised alteration or a maker's mark, to name a few, can potentially be discovered in the course of conservation work.

A conservator customarily has ready access to sophisticated investigation techniques. Eyes and hands provide observation and touch. Combined with taught and accumulated knowledge and the ability to interpret the evidence presented by collection material, an enormous quantity of information is gleaned and posited by these methods. Oftentimes this may be sufficient. Equally, there will be occasions where it is not and key questions remain.

The analytical techniques presented in this volume therefore endeavour to reveal additional information or sometimes hidden evidence that cannot be detected or discerned reliably by eyes or touch alone.

1.2 By Chapter

The book augments an existing corpus of conservation science publishing and brings an understanding of materials and application of analyses and treatments up to date. The previous edition (2006) provided an emphasis on archaeological materials and contexts in particular.[1] This newer version includes additional materials and emphasises the application of currently available methods of analysis. Other conservation science texts complement the information presented. Lister, for example, specifically discusses the chemical composition of materials to understand their behaviour in cultural materials.[2] Shashoua promotes the understanding of the vast subject area of plastics in the framework of heritage by considering their chemical composition, manufacture and deterioration pathways.[3] Timár-Balázsy and Eastop present textiles within the context of science as a means to promote better decision making for treatment and empathy with differing contexts. Other texts centre specifically the on the choice of materials for treatments.[4] Horie accentuates the requirement to understand the behaviour of resins and other polymers for treating items, particularly since once applied they cannot be easily removed or reversed.[5] Varella meanwhile, introduces the theoretical aspects of instrumental analysis within the conservation discipline and gives examples of applications on several materials.[6]

What is distinctive in the present volume is that it is designed as a single point of reference and starting point for conservators seeking to understand a range of materials within the context of science and analysis. Whilst the text does not seek to cover any single material in infinite detail, the reader will gain sufficient knowledge of the materials to appreciate, select and apply appropriate investigative methods.

Bearing in mind that this volume is intended as a starting point for conservators seeking to understand analytical techniques, Chapter 2 opens up the subject area by describing the main analytical techniques currently available. The reader is encouraged to understand that the most appropriate technique is selected by bringing together the material type and research question to be answered and through understanding what differing techniques can achieve. The practicalities of sample requirements and the nature of the results are equal considerations. Matching the most appropriate technique or techniques to the scenario in hand is imperative to operative and cost-effective analysis.

Each subsequent chapter examines a different material or group of related materials. Organic and inorganic materials are included in this volume and both moveable and non-moveable heritage addressed in contexts ranging from archaeological to the modern day. This in itself demonstrates the wide-ranging circumstances where analysis can enhance the preservation, use and interpretation of material culture. Broadly, each chapter introduces the material composition, comments on deterioration pathways before addressing the main analytical approaches available to characterise the material and to help understand its vulnerability and condition.

Not all analytical techniques are complex or expensive. Whilst many may require equipment that is intricate and potentially costly this is not always the case. Chapter 3 focuses on paper, and highlights two major deterioration mechanisms, which can alter and disrupt its stability. Spot tests, perhaps the simplest tests available, are described alongside more complex possibilities with insight as to the most useful.

Wood is encountered in cultural heritage in a very wide range of environments from archaeological contexts to fine and applied art and buildings. Chapter 4 addresses the challenges of comprehending and detecting the deterioration mechanisms that can result in such varied applications.

Differences are also noted between perhaps superficially similar materials in Chapter 5. Leather and parchment have differing properties as a result of their manufacturing processes despite both originating from animal skin. Divergent deteriorating mechanisms are addressed alongside means to determine the animal species and ways to understand the degradation state.

Metal and associated alloys are introduced in Chapter 6 and continue the theme of variability within a material group. The occurrence and severity of corrosion on metal are related not only to its environment and application but also heavily depend on the composition of the material itself. Techniques to comprehend both the composition and deterioration of metals are described.

Additional inorganic materials are explored in Chapter 7, focussing on glass, ceramics and enamelled objects. Here the technology to create these materials is emphasised, because understanding manufacturing methods helps elucidate the subsequent behaviour and deterioration. Techniques for both laboratory based and mobile set ups are considered addressing the fact that conservation is undertaken both by bringing material to the conservator and, *in situ*, where the conservator is the moveable asset.

Chapters 8 and 9 explore colouring materials and their variety. Chapter 8 focuses on the similarities and differences between dyes and pigments. Various contexts are presented from works of art to a case study of Le Corbusier's use of colouring materials to present classifications and differing perspectives. Paint and binding media are explored in Chapter 9 to suggest analytical strategies for protein and oil-based media using intricate and intimate instrumental analysis with necessary samples, albeit small. Case studies address both more established and newer 20th century artist materials.

Textiles are the focus of Chapter 10. Ubiquitous in society but also sometimes overlooked because the multiple layers of historic evidence and significance are lost in their overall familiarity, the most commonly met types are placed centre stage.

The final Chapters 11 and 12 explore materials encountered within history that is more recent or through predominantly contemporaneous acquisitions; plastics and 3D printed material. Plastics have a growing familiarity within heritage collections and their deterioration is also becoming a customary element of conservation knowledge. Chapter 11 emphasises how the mechanical and thermal characteristics of plastics are assessed and investigated *via* analytical strategies for investigation. Items created *via* 3D printing may be less familiar generally. Chapter 12 emphasises their structural possibilities and material compositions to bring these possibly lesser-known pieces to attention.

1.3 Applied Science and Emphasising Practical Application

Applied science is the use of scientific processes and knowledge as the means to achieve a particular practical or useful result. Conservation science is a branch of applied science since conservators are inherently practical problem solvers who apply knowledge to a scenario.

Whilst the aim of the book is to position and strengthen conservation science within the professional discipline, a prime objective is to draw the fields of practical conservation and conservation science closer together *via* a common purpose. The techniques and applications described throughout the volume are concerned with aiming for and obtaining meaningful and useful results. Unapologetically so. Current economics within the heritage industry rarely allow for analysis as merely an academic exercise. Equally, well-intentioned research into processes and deterioration pathways that ultimately lack useful application for conservators 'in the field' seems profligate given the extent of cultural heritage needing attention. The contributors to this volume emphasise the application of science for conservation to bridge the gap between pure science and applied practical analysis to aid care and treatment decisions.

Accordingly, the volume is not describing how to develop new techniques, research, or pure science but rather how to apply established knowledge and analytical techniques to situations and materials encountered. That is not to say the conservation profession and allied conservation science professionals do not or cannot develop new scientific knowledge, far from it. Nevertheless, the intended focus of this volume is to inform and guide conservators in a hands-on sense using existing technologies and applying up to date understanding and processes, rather than generating new methods. Ultimately, new ideas in applying techniques to additional scenarios may result.

1.4 The Enduring Importance of Communication

Despite holding a diverse skill set, a conservator does not work in isolation. A key aspect of successful conservation decisions is collaboration with other specialists. To appreciate historic and current significance and determine future uses for cultural materials that are inclusive and empathetic, teamwork with the creators and user communities can be imperative. Equally, in facilitating conservation science and interpreting the resulting data, partnership and discussion with science professionals can be required.

The second objective for this work, connected to the first is, therefore, to enhance communication. A long established but growing feature of the conservation profession is the ability to communicate with others. As noted above, drawing science and conservation closer together requires a clear understanding of the reasons for and output of the analysis. Successful analysis is achieved through being able to establish and communicate the question to be answered and then understanding how it may be resolved. Appreciating the process, advantages and limitations of a technique for a particular material, therefore, seems fundamental to suggesting scientific analysis and communicating with a scientific specialist to achieve this. Not least in understanding the resultant data sets.

Furthermore, communication within conservation science is not just an internally focussed activity for the process of enhancing treatment decisions. Communication is an increasingly externally focussed endeavour. Behind the scenes tours of conservation studios, museum stores and heritage sites are enduringly popular for visitors to these institutions. The British Library has been hosting twice-monthly sell-out visits to the Centre for Conservation since its opening in 2007. Demand is not waning. The Victoria & Albert Museum, Birmingham Museums, Fishbourne Roman Palace, Chichester, and the Peoples History Museum, Manchester are just a mere snapshot of the institutions where tours, viewing areas and open days allow access to the normally unseen realm of the heritage professional and conservation in particular. Visitors seek to learn more about the process of conserving heritage and conservation science is a component of this story.

The growth of social media has triggered a steadily increasing expectation to communicate more widely. Despite the prolific output of social media platforms, the quality of the information is imperative to represent the heritage industry successfully. Conservators are progressively contributing to this newer web based agenda and learning the necessary skills to connect with clarity and conciseness. Arguably, sharing knowledge and experience of conserving cultural heritage can only enhance the relevance and significance of the work done and any means to enrich this narrative is positive.

1.5 A Reality Check

Finally, a dose of pragmatism is required. Scientific analysis, even when appropriately applied with knowledge and acumen, has limitations. A result or results may be ambiguous, suggestive of material components and deterioration mechanisms rather than a definitive answer. Analysis may only tell us what a material is not, rather than what it is. A process of elimination and judgement will ensue.

Moreover, a data set is not necessarily an end in itself but requires further interpretation and application to the original question and circumstances. Unlike a factory production line where all components and outcomes typically strive to be identical, conservation science does not have this luxury. Every item and its analysis has its own qualities and peculiarities, meaning the adaptation of technique and judgement of the outcome may become necessary elements of the end-to-end process. The results of appropriately chosen and interpreted analysis will, however, inform and guide decisions, adding to the body of knowledge surrounding a collection or individual item. Whilst this volume is quite specific in promoting these activities, the reader should always remember that conservation is a holistic and connected process.

References

1. *Conservation Science: Heritage Materials*, ed. E. May and M. Jones, Royal Society of Chemistry Publishing, Cambridge, 2006.
2. T. Lister, *Conservation Chemistry: An Introduction*, Royal Society of Chemistry Publishing, Cambridge, 2005.
3. Y. Shashoua, *Conservation of Plastics*, Routledge, Abingdon, 2008.
4. *Chemical Principles of Textile Conservation*, ed. Á. Timár-Balázsy and D. Eastop, Butterworth Heinemann, Oxford, 1998.
5. C. V. Horie, *Materials for Conservation*, Routledge, Abingdon, 2nd edn, 2010.
6. *Conservation Science for Cultural Heritage – Application of Instrumental Analysis*, ed. E. Varella, Springer-Verlag, Berlin, 2013.

2 Analytical Techniques in Conservation Science

P. Garside*[a] and E. Richardson[b]

[a]The British Library, 96 Euston Road, London, NW1 2DB, UK; [b]Image Permanence Institute, Rochester Institute of Technology, 70 Lomb Memorial Drive, Rochester, NY 14623-5604, USA
*Email: dr.paul.garside@gmail.com

2.1 Introduction

Modern instrumental analysis can reveal a wide range of information about historic artefacts and can be of great value in the study and preservation of such items. However, the first question to ask should always be 'why carry out an analysis?'. Without a specific question about a material or object in mind, performing any sort of instrumental analysis represents a waste of time, money and resources, as well as an unnecessary intervention to the object itself. Nonetheless, there are many reasons why it may be appropriate to study artefacts using these techniques. The information may help to: determine the provenance, history, use and adaptation of an object; characterise its composition, construction and state; identify the presence and effect of previous treatments; and inform future decisions about conservation, use and storage. This can be achieved through the identification of the materials used to construct the item, the nature of structural and microstructural features, and the presence of soiling, degradation products, wear and other signs of use.

Conservation Science: Heritage Materials, 2nd Edition
Edited by Paul Garside and Emma Richardson

Published by the Royal Society of Chemistry, www.rsc.org

Not all techniques are appropriate to all items, due to their specificities or sampling requirements, so it is necessary to have a general understanding of the different methods available to make informed decisions about analytical choices. This enables productive conversations between conservators and conservation scientists, helping choose methods which are most sympathetic to the object. It also ensures that the best use can be made of any samples that are taken, that analytical results will provide useful and usable information, and that time and resources are not wasted. Therefore, the aim of this chapter is to provide this broad overview of the main analytical methods which may be of use when studying historic and archaeological materials. For each technique, a summary of its method of operation will be given, along with the kinds of materials with which is it best suited to work, the type of sample preparation that is required, the nature of the results or data it generates and its limitations. The chapter is not intended to give an exhaustive examination of such techniques, nor to discuss all of the variations and options that are available, but instead aims to provide a starting point by which these approaches may be better understood, and to give examples of the typical kinds of results generated by them. Analytical methods appropriate to particular material types will be discussed in more detail within the individual chapters, and more specific or specialist techniques may also be explored.

2.2 Microscopy

2.2.1 Light Microscopy (and Hand Lenses)

Light microscopy is probably the most widely used method of investigating cultural heritage artefacts, due to its availability, low costs, ease of use and straightforward sample preparation. The most common forms of the technique are stereomicroscopy and transmitted light microscopy, as well as the use of hand lenses, although there are many variations.

Light microscopy can be carried out with either reflected or transmitted light. In the former, light falls on the sample and is either reflected or absorbed, providing an image that is principally dependent on its surface. In the latter, light passes through the sample to the observer, allowing internal structures to be seen.

Hand or mounted lenses, loupes and thread counters provide a simple and convenient method of examining objects in detail.

They rarely provide magnifications greater than ×20, and at higher levels it is difficult to keep such devices steady enough to allow a good view.

Stereomicroscopes allow higher levels of magnification (typically up to about ×100) and possess a variety of other advantages. They are designed to allow different light paths to reach each eye, giving a stereoscopic view of the specimen, which aids the examination of topographical features. They are often mounted on robust stands which are both extendible and positionable, allowing large, bulky or awkward objects readily examined, potentially *in situ*, and without requiring samples to be taken. Stereomicroscopy is a reflectance technique.

Conventional light microscopes may be designed to work in reflectance, transmission, or both. Although in theory magnifications of up to ×1000 may be achieved, practical considerations may limit this. At higher magnifications, the depth of focus (the thickness of the sample which is in focus) becomes increasing limited, so for samples that are highly three-dimensional at these sorts of scales, such as textile fibres, the useful maximum magnification may be limited to around ×400. Higher levels are achievable with flat or polished samples, where the narrow depth of focus does not present a problem. If used in transmission, a sample must normally be taken from the object and prepared on a slide; it is particularly useful for samples which have characteristic internal morphologies, such as plant fibres. Reflected light microscopy, on the other hand, principally allows the surface of samples to be examined; samples may be prepared in a similar way to those for transmitted light microscopy, and it is also useful for resin-set cross-sections, paint fragments, fracture surfaces, and the like.

All of these techniques rely on the specimen having distinct and observable visual differences. Auxiliary methods may assist if this visual contrast is limited: raking light helps emphasise and distinguish surface relief; polarised light creates apparent contrast based on the chemistry, thickness, orientation and long-range ordering of the sample; lighting which will induce fluorescence can highlight fluoroscopic components; dyes and stains can selectively colour regions based on chemistry and structure.

Samples and Preparation: Light microscopy is applicable to a wide range of heritage artefacts and materials. Some versions of the technique (such as stereomicroscopy) potentially allow *in situ* examination, but generally, a sample of some sort must be taken, although these can be small. These may then be prepared as slides or cross-sections, stained or etched to reveal particular features, treated with solvents or reagents to separate out components, *etc.*

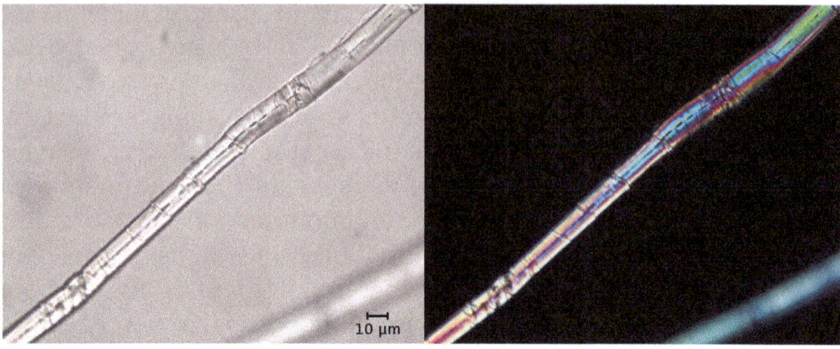

Figure 2.1 Transmitted light and polarised light micrographs of a flax fibre.

Results: In addition to the direct examination of features of interest, most microscopes can now be used with digital cameras which allow micrographs to be readily recorded. Typical transmitted light and polarised light micrographs of a flax fibre are shown in Figure 2.1.

Limitations: In many cases, sampling is required. Care must be taken that samples or areas observed are representative of the whole. Deterioration can obscure or alter features of interest.

2.2.2 Electron Microscopy

Scanning electron microscopy (SEM) uses an electron beam, rather than light, to generate an image of the surface of a specimen. The very short wavelength of the electrons, compared to visible light, allows far higher levels of magnification and better depths of field. The electron beam interacts with the sample in a variety of different ways that, with appropriate detection, permits different imaging techniques. The most common use secondary electrons, backscattered electrons or X-rays (Figure 2.2), although other imaging and analytical methods are possible with more specialist equipment.

Secondary electrons (SE) are generated when an incident electron interacts with an outer shell electron in an atom of the sample, which is then ejected; they are relatively low energy, so only those generated in the upper layer (several nanometres) of the sample can escape to be detected, meaning that they can be used to generate a detailed image of the topography of the material. Backscattered electrons (BSE) result from an incident electron being elastically scattered by an atomic nucleus; they are relatively high energy and so can escape from deeper within the sample. The back-scattering effect increases rapidly with increasing atomic number, so regions containing heavier elements

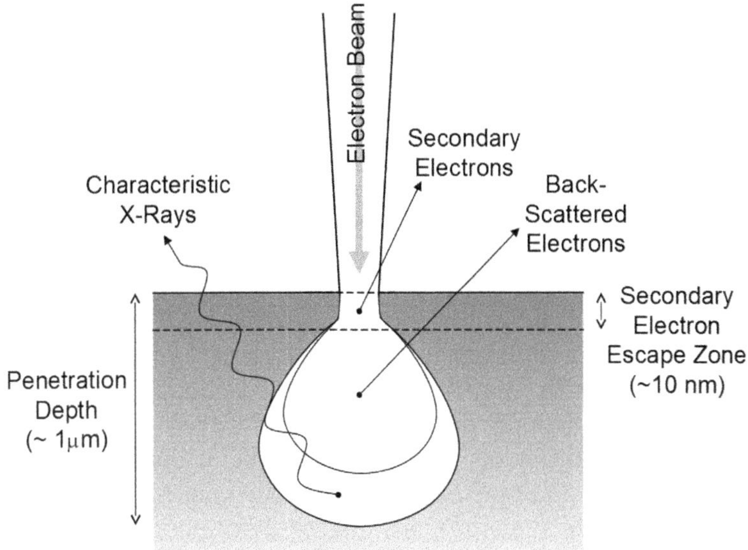

Figure 2.2 The generation of secondary electrons, backscattered electrons and characteristic X-rays in scanning electron microscopy.

will appear brighter; as a result, BSE images provide information on the elemental composition of the surface, as well as data on phase boundaries and topography (though of a lower resolution than SE images). The interaction of incident electrons can also lead to the emission of X-rays, which is exploited by energy dispersive X-ray spectroscopy (EDS or EDX; see elemental analysis, Section 2.7.2). This allows elemental compositions and distributions to be assessed; EDS mapping allows such data to be linked to structural features observed in SE or BSE imaging.

Electron microscopy must be carried out under significantly reduced pressures (typically a vacuum), as an atmosphere would impede the incident and emitted electrons.

Samples and Preparation: The technique typically requires the specimen to be placed on a conductive stub in the chamber of the microscope, which means that samples must generally be taken unless the item as a whole is small enough that it can be placed in the device. To prevent a build-up of charge and heat on the material, it is conventionally coated with a conductive layer of carbon or gold, unless it is inherently conductive in its own right. This can be avoided in low vacuum or environmental SEMs, in which sufficient gas is introduced into the chamber to dissipate charge and heat, whilst still allowing the electron beam to operate; this can also allow dynamic phenomena such as wetting and heating to be observed. Specimens can be prepared in much the same way as those for examination by optical microscopy – for example, by resin mounting and sectioning to reveal desired aspects of the sample.

Results: Electron micrographs can be recorded from the various imaging methods (*e.g.*, SE and BSE), and different methods will often provide complementary information. A typical scanning electron micrograph, of triphenyl phosphate crystals from a cellulose acetate object, is shown in Figure 2.3.

Limitations: Unlike light microscopy, electron microscopy only provides surface information, so internal structures cannot be observed without sectioning. No information is provided about colour or visual contrast. Generally, samples must be taken and if these samples require coating before examination, they will also be irreparably changed. The low-pressure conditions in the chamber can result in the dehydration of organic materials, leading to deformation, dimensional changes and damage, which can yield misleading results. Particularly sensitive materials may also be directly damaged by the primary (incident) electron beam itself. Fibrous, filamentary or friable materials can

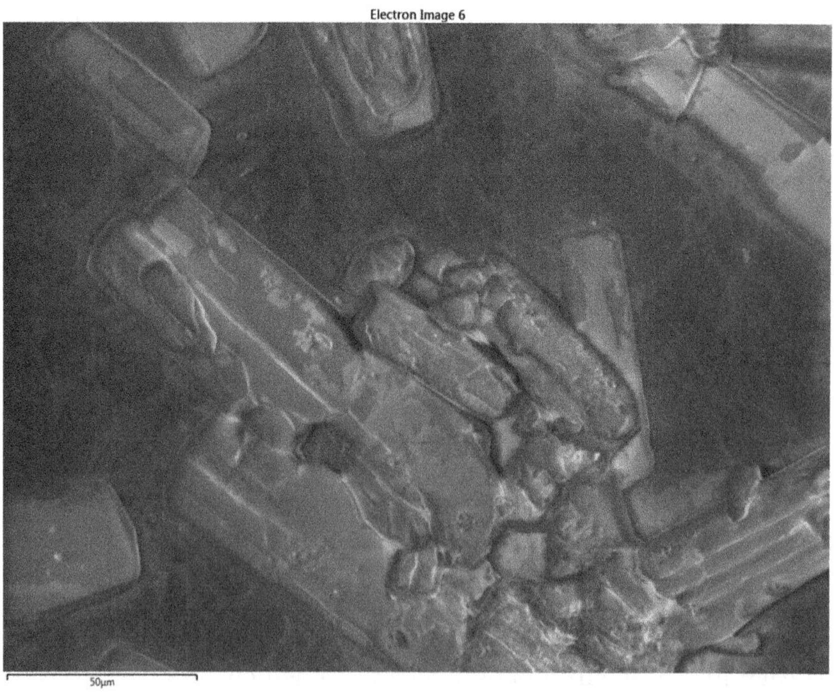

Figure 2.3 Scanning electron micrograph of triphenyl phosphate crystals on cellulose acetate film.

be subject to charging effects, even if coated, causing parts of the specimen to move during imaging, leading to a blurred or overexposed image.

2.2.3 Other Forms of Microscopy

Other specialist forms of microscopy which may be found at larger institutions include techniques such as confocal light microscopy (which allows effective magnifications greater than ×1000), transmission electron microscopy (TEM) and atomic force microscopy (AFM).

2.3 Analytical Imaging

2.3.1 Multispectral and Hyperspectral Imaging

Multispectral and hyperspectral imaging are techniques which capture image data at numerous specific wavelength bands in the electromagnetic spectrum (typically the visible region, and often extending into either or both the infrared and ultraviolet regions too), and so provide a range of information which would not be visible to the unaided eye. The two terms are often used somewhat interchangeably, but multispectral imaging typically relies on tens of discrete bands at separate and specific wavelengths, whereas hyperspectral imaging uses hundreds or thousands of bands over a continuous wavelength range. (By contrast, the human eye and the majority of conventional digital photography rely on three broad wavelength bands, denoted as 'red', 'green' and 'blue'). Images can be captured by a variety of different methods, specific to the particular equipment used.

These techniques are particularly useful for objects with detailed, complex, layered or indistinct surface features comprised of a variety of different materials, including paintings, pictures, printed and written texts, as well as items with repairs, modifications or alterations. They can reveal and clarify aspects such as deliberately or accidentally damaged, obscured or faded features, highlight the presence of alterations, additions or different but superficially similar materials, detect underdrawings, *etc.*

Samples and Preparation: As a form of digital photography, both of these imaging techniques can ordinarily be applied to any object which can be suitably positioned in front of the camera, allowing for the typically complex nature of the camera equipment and lighting. Sampling is not required, nor is specific preparation of the object

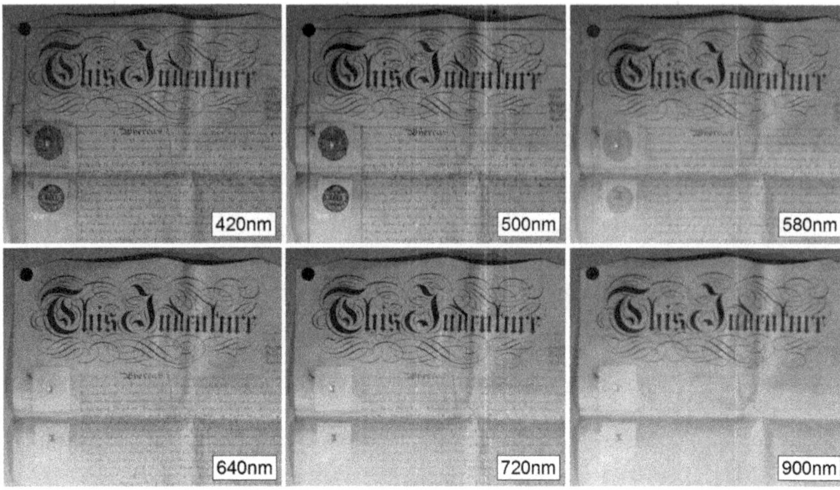

Figure 2.4 Multispectral image of a parchment document (at 420, 500, 580, 640, 720 and 900 nm).

beyond ensuring that it can be presented to the camera in such a way as to give a representative image.

Results: The techniques produce a multispectral or hyperspectral image, a file containing wavelength data over multiple bands spatially mapped as image pixels. Figure 2.4 shows six images (at wavelengths of 420, 500, 580, 640, 720 and 900 nm) extracted from a multispectral image of a parchment document.

Limitations: The image sets are complex and large. They often need considerable processing in order to derive useful information from them, which requires time and expertise. The very large file sizes may necessitate special considerations for storing, transferring and using the data.

2.3.2 X-radiography

X-radiography uses X-rays to create images which can reveal the internal structures of an object; the technique has a long history of medical use. X-rays penetrate solid materials to differing extents depending on their density, and so images can be generated by positioning a photographic plate or a suitable detector on the opposite side of the subject to the source. These images can be used to provide information on the way in which an object is constructed, including structural elements, fillers and voids, the presence and distribution of different materials and components, and the location of joins, repairs or internal damage.

Samples and Preparation: Due to the use of X-rays, the technique must usually be carried out in a designated area or somewhere to which both the equipment can be transported and suitable temporary shielding and restrictions on access can be established. Beyond these requirements, it can typically be applied to any object which can be suitably positioned with respect to the source and detector. Ideal samples contain a variety of materials with varying atomic numbers, thus providing high radiographic contrast.

Results: The technique provides an X-radiograph. For earlier versions of the technique, this will be as a physical image on a photographic film or plate. More modern versions capture a digital image using a suitable detector.

Limitations: As this technique requires the use of an X-ray source, health and safety considerations will apply to its application. Predominantly organic materials, such as textiles, may be difficult to image as they provide a low radiographic contrast, resulting in images with 'soft' edges and limited detail. If the technique uses photographic plate or film, these are bulky to handle and store, and generally require further digitisation for analysis or publication.

2.3.3 Computed Tomography

Computed tomography (CT) scanning is an imaging method that utilises numerous individual radiographic images of an object, captured at different angles, from which a three-dimensional dataset is created through computational analysis. This dataset can then be interrogated to generate virtual cross-sections (tomographic images) of the original object, allowing its interior structure and composition to be explored without further intervention. The most common form of the technique is based around X-radiography, and so the term 'CT scan' is often used without further elaboration to refer to this method, but other forms are possible, including positron emission tomography (PET) and single-photon emission computed tomography (SPECT).

In the context of cultural heritage, this method provides a non-invasive way of investigating the inner structures of objects which could not otherwise be readily examined due to physical limitations, fragility or ethical considerations – some examples include: archaeological materials which may be intractably encased in corrosion, degradation products or other deposits; tightly rolled, folded or packaged textiles or documents, especially if these objects are brittle or otherwise difficult to handle; and items with internal structures or components which cannot be directly observed without causing damage to the outer layers. The technique will only provide useful information with objects which have good radiographic contrast between the components of interest.

Samples and Preparation: Little sample preparation is required, but the object must be of an appropriate size to position in the imaging chamber.

Results: The technique generates a three-dimensional dataset of the object, from which tomographic images can be extracted.

Limitations: The technique is expensive and requires specialist operators, so access is limited to larger heritage institutions or requires collaboration with off-site third parties. Typically, the equipment is not mobile, so items must be taken to it. The datasets are large and can be difficult to interpret. As with all radiographic and similar techniques, there are health and safety considerations and radiological protection issues which must be taken into account.

2.3.4 Other Imaging Techniques

A range of other imaging techniques may also be found. There are a variety of different forms of radiography, using gamma rays and other species of ionising and non-ionising radiation. There are also simpler methods, such as reflectance transformation imaging (RTI), a computational imaging technique which uses a conventional digital camera to record multiple images of an object from a fixed position under varying lighting conditions, and combines these images to reveal subtleties of surface structures and topography.

2.4 Colourspace Measurements

Colourspace measurement is a method of representing colour using numerical values; the name refers to the fact that if three values are used to specify a colour, plotting these as three axes then defines a three-dimensional space in which each colour is represented by a particular point. The most familiar of these are the red-green-blue (RGB) and cyan-magenta-yellow (CMY) systems, but the method used most widely in cultural heritage is probably the International Commission on Illumination (CIE) 'L*a*b*' system. One particular advantage of this system is that it is based on the way in which the human eye and brain observe and determine differences between colours. 'L' represents lightness, from 0 (pure black) to 100 (pure white), 'a' measures the green-red axis (negative values are green and positive values, red) and 'b', the blue-yellow axis (negative values are blue and positive values, yellow). The difference between two colours can be determined as a ΔE^* value, representing the distance between colour '1' and colour '2' within the colourspace, as:

$$\Delta E^* = \sqrt{((L_1 - L_2)^2 + (a_1 - a_2)^2 + (b_1 - b_2)^2)}$$

In theory, a ΔE^* of 1 represents the smallest change detectable by a human observer, but in practice most people can only typically detect a ΔE^* of about 2 or 2.5.

The measurements can be made with a colourimeter and allow accurate and reproducible colour matching, as well as permitting changes over time, such as light induced fading, to be monitored.

Samples and Preparation: Provided the instrument can be correctly positioned with respect to the object, no special preparation is required. Most of the instruments are small and portable allowing *in situ* measurements to be made.

Results: The technique produces a numerical profile of the measured colour.

Limitations: Depending on the instrument, it may be difficult to position it on three-dimensional objects. Stray light may interfere with the readings. In general, the sampling area is relatively large (several millimetres in diameter), which means that colours on detailed or intricate surfaces cannot accurately be measured, due to overlapping contributions.

2.5 Vibrational Spectroscopy

Vibrational spectroscopy allows the chemistry of a specimen to be investigated and, in the context of conservation science, it is of great value in identifying unknown materials, assessing the effect and extent of degradation, and potentially probing microstructural features. There are various forms of vibrational spectroscopy, but they all rely on illuminating the specimen with electromagnetic radiation and measuring the radiation that is consequently transmitted, absorbed, reflected or emitted by it, according to particular selection rules governed by the interaction of specific chemical bonds with corresponding frequencies of radiation. The vibrational frequencies of bonds depend on factors such as the elements involved (the O–H bond typically has a higher vibrational frequency than C–H), the nature of the bonding (the C≡C bond has a higher frequency than C=C) and the local environment (factors like hydrogen bonding, long-range ordering or the presence of plasticisers will all influence the vibrational frequency); in general, stronger bonds vibrate at higher frequencies. As a result, measuring these vibrational frequencies allows an understanding of the types of bonds present in a sample, their local environments and their proportions. The data are presented as spectra of intensity against frequency (wavenumbers, with the unit cm^{-1}) or wavelength (in nanometres, nm), with specific interactions giving rise to peaks or bands at particular positions. (Frequency (f) and wavelength (λ) are linked by the formula $f = c/\lambda$, where 'c' is the speed of light; wavenumbers are measured as the reciprocal of the wavelength in centimetres.)

The two most common forms of this technique are infrared and Raman spectroscopy, though each has many variations. At its simplest, the data provided by these methods can be used as a 'fingerprint' to identify a sample by comparison of its spectrum with a library of spectra of known materials. However, as the peaks or bands within a spectrum relate to the presence and local environment of specific types of chemical bond, it is also possible to use this data to investigate the way in which the chemistry of a material changes over time (through degradation, modification or other chemical reactions), and how the bonds respond to changes in their immediate surroundings (such as the presence of plasticisers or absorbed water, the nature of long-range ordering including crystallinity, polymer orientation and protein secondary structures, *etc.*).

2.5.1 Infrared Spectroscopy (Mid and Near Infrared)

Infrared (IR) spectroscopy provides a simple, rapid and reproducible technique for investigating the chemistry of materials. It can be used to characterise and identify the composition of a sample, assess degradation and other chemical changes, explore microstructural properties such as crystallinity and degree of polymerisation, and probe dynamic phenomena such as moisture sorption. It is particularly well suited to organic materials, though it is also widely applicable to inorganic species. In this form of spectroscopy, bands are observed for those vibrational modes in which the dipolar character of a chemical bond changes (Figure 2.5). Most infrared spectroscopy is carried out in the mid infrared region (typically 4000–400 cm^{-1}, or 2500–25 000 nm); for clarity, this is sometimes explicitly referred to as MIR spectroscopy, but frequently is simply given as IR spectroscopy. It is also possible to analyse a sample using the near infrared region of the spectrum, and this is referred to as NIR spectroscopy; this is discussed in more detail later in this section.

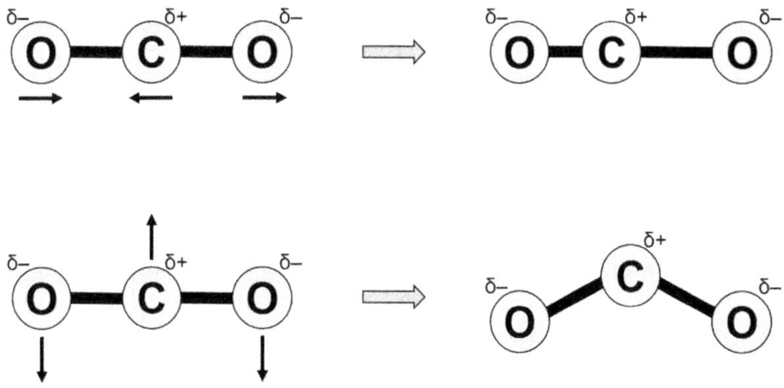

Figure 2.5 Vibrational modes in which the dipolar character of a chemical bond changes.

Modern spectrometers typically generate data using Fourier transform processing, which allows high quality data to be gathered rapidly; as a result, the technique is often referred to as Fourier transform infrared (FTIR) spectroscopy.

Historically, IR spectroscopy required a sample through which radiation could pass, which was typically achieved with a powdered sample suspended in a transparent solid or liquid medium, a film of liquid or solution, a very fine cross-section or a gas in a suitable cell. As this requires samples to be taken and heavily processed it is not ideal for many historic materials, and more modern applications of the technique can avoid this. With suitable equipment, reflectance spectra can be recorded without sampling (or without destroying an existing sample), but the quality of the spectra is highly dependent on the specimen's topography, uniformity and orientation, so reproducibility can be problematic, the strength of certain peaks may be distorted and only the surface is analysed. Diffuse reflectance infrared Fourier transform spectroscopy (DRIFTS) allows reflectance spectroscopy to be carried out on small quantities of a powdered specimen with minimal preparation, or on material gathered onto a finely abrasive surface, minimising sampling requirements. Attenuated total reflectance (ATR) spectroscopy uses a crystal 'window' (typically diamond or germanium) against which the sample is pressed; particular optical phenomena allow a spectrum to be recorded from the surface of the sample. High quality, reproducible data can be gathered from small samples and, in some cases, it may be possible to position objects to record spectra without sampling providing that good contact with the crystal can be achieved. The pressure necessary for this contact can damage fragile materials and, in all cases, the spectra only provide surface information. Microspectroscopy and micro-ATR spectroscopy provide further refinement to these techniques, in which a spectrometer is attached to a microscope with infrared optics, allowing spectra to be recorded from precise areas (of the order of square micrometres) of very small samples; with motorised stages, infrared spectral mapping of a specimen is possible, allowing both the presence and location of specific individual components to be investigated.

Near infrared spectroscopy exploits similar properties of the material, but spectra are generated from combinations and overtones of particular bands seen in mid infrared spectra; as a result, the spectral range covers higher frequencies (or shorter wavelengths), typically 12 000–4000 cm^{-1} (833–2500 nm). The peaks themselves tend to be much broader and less distinctive than those of the mid infrared. As a result, it is often more difficult to derive information directly from the spectra, but they can be used for 'fingerprinting' in a similar way. Computational analysis using the properties of known and well-characterised reference materials can reveal much more about the properties of the sample (multivariate or chemometric analysis). A significant advantage of the technique is that it can be used with conventional fibre-optic probes, potentially allowing objects to be investigated *in situ*. It is also simple to operate, rapid and has good reproducibility.

Samples and Preparation: Sampling requirements and preparation vary significantly between the different implementations of this technique, so generalisations are difficult. If sampling is required (*e.g.*, for ATR or DRIFTS), these samples are generally small and can often be used subsequently for further analysis. Reflectance spectroscopy can obviate the need for any sampling, providing the object can be suitably positioned.

Results: The data is presented as a spectrum of intensity against frequency (or wavelength), with each peak or band representing a type of chemical bond in a particular environment. Transmission and absorption spectra represent equivalent information and can be converted through an appropriate algorithm. Transmission and absorption mid infrared spectra of cellulose can be seen in Figure 2.6, and a typical near infrared spectrum, of polyvinyl chloride film, in Figure 2.7.

Limitations: Depending on the exact application of the technique, sampling may be required. Accurate identification of materials often

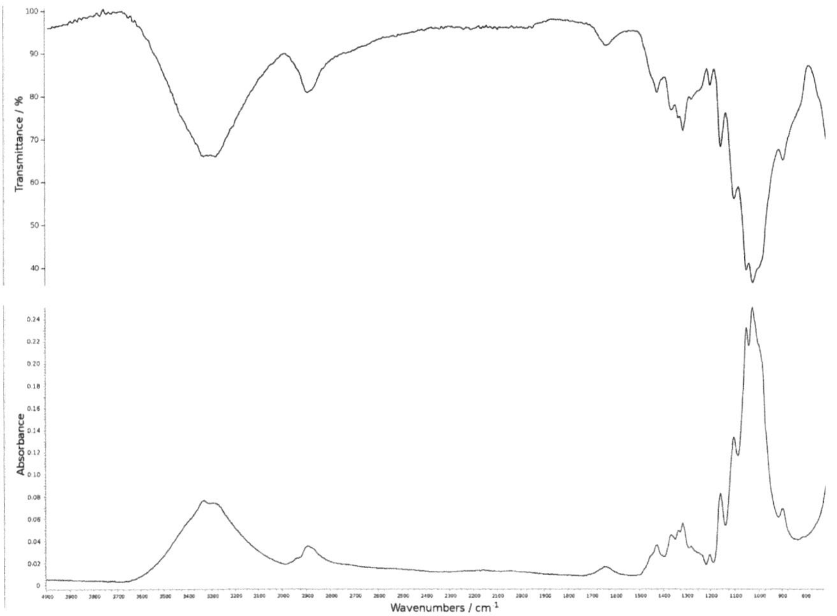

Figure 2.6 FTMIR transmission and absorption spectra of cellulose.

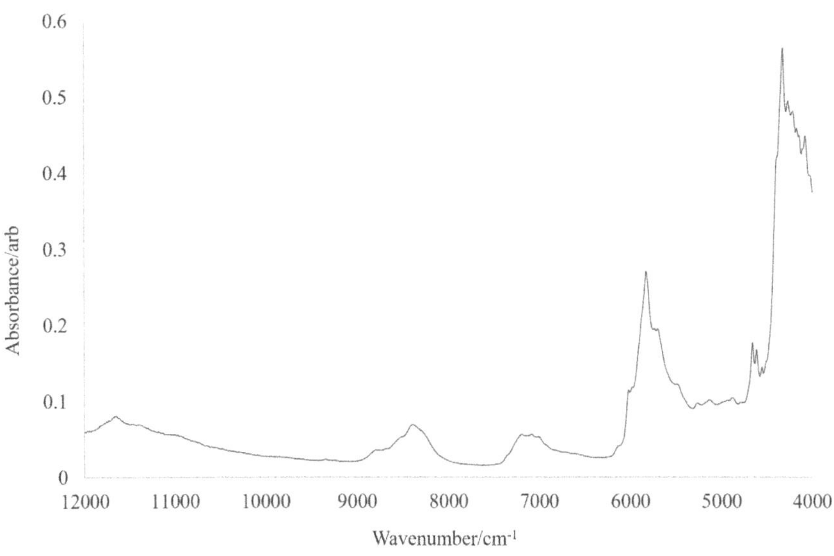

Figure 2.7 FTNIR spectrum of polyvinyl chloride film.

requires a comprehensive library of reference spectra and mixtures of materials can be difficult to differentiate. Inorganic materials may yield spectra with broad, poorly defined bands.

2.5.2 Raman Spectroscopy

Raman spectroscopy provides information complementary to that derived from infrared spectroscopy. Raman bands are observed for vibrational modes that result in a change in the polarisability of a bond, rather than changes in dipolar character (Figure 2.8). Absorbed energy causes bond vibration and subsequently some of the absorbed energy is re-emitted. This is Raman scattered light and is detected and plotted as a Raman spectrum of intensity against the shift of scattered energy with respect to that of the original beam.

Raman spectroscopy gives good results from both organic and inorganic materials, so it will also provide information on substances that often cannot readily be investigated with infrared spectroscopy, such as mineral pigments. As the technique relies on illuminating the surface of a specimen and observing the scattered radiation, it may be possible to carry out the analysis without requiring samples to be taken. As with the infrared technique, it is possible to carry out Raman microspectroscopy, allowing specific points of interest of samples to be investigated in detail and, with the right equipment, to create Raman maps of the surface.

Samples and Preparation: With the correct equipment, this technique can be applied *in situ*, thus avoiding the need for sampling, and will give results from a wide range of organic and inorganic materials. The information is complementary to that provided by infrared spectroscopy.

Results: Data is presented as a spectrum of intensity against frequency (or wavelength), with each peak or band representing a type of chemical bond within a particular environment (Figure 2.9), similar to those obtained by IR spectroscopy.

Limitations: A potential limitation of the technique is the risk of causing damage to the sample from the incident radiation – as the Raman signal is very weak relatively intense illumination is required, which can result in the sample surface suffering localised damage. This is principally a risk if the technique is performed by an inexperienced operator, or if standardised instrument settings are employed rather than tailoring the settings to the particular requirements of the

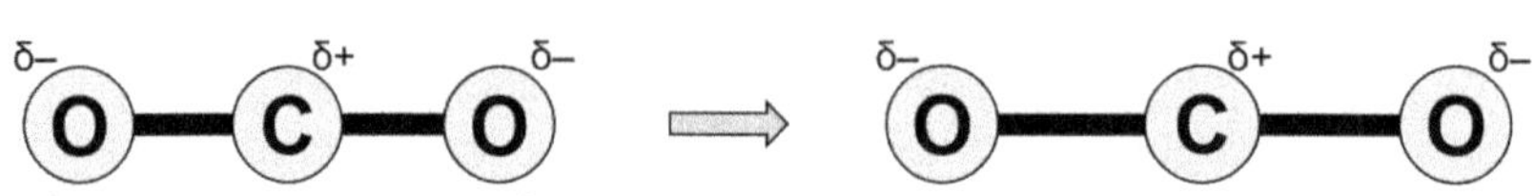

Figure 2.8 Vibrational modes that result in a change in the polarisability of a bond.

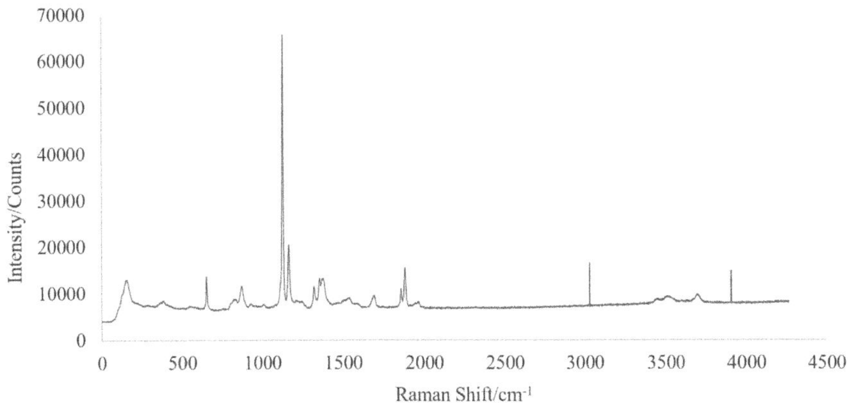

Figure 2.9 Raman spectrum of polystyrene.

material. Reducing the laser power can mitigate the risk of damage, but this may come at the cost of significantly increased accumulation times.

2.6 UV-visible Spectrophotometry

UV–visible (UV–Vis) spectrophotometry relies on the absorption or reflectance of light in the visible and near ultraviolet region of the electromagnetic spectrum. It is inherently linked to the perceived colours of the material under examination, and so can provide a 'fingerprint' of coloured specimens that contains greater information than simple colourspace measurements. The technique can be used in transmission to study samples in solution, which is valuable for extractable materials such as dyestuffs, and also allows concentrations to be calculated on the basis of the Beer–Lambert law. If the technique is used with reflectance measurements, coloured surfaces such as pigments and paints can be studied. Some forms of the technique also extend into the near infrared to provide a broader range of spectral information (UV–Vis-NIR). The instruments may be single beam or double bean. In the former, a separate reference spectrum is required but allows for simpler, more compact devices with a greater dynamic range. In the latter case, double beam instruments record the sample and a reference simultaneously and tend to be easier to use and more stable.

UV–Vis spectroscopy may be used as the detection method for techniques such as high-performance liquid chromatography (HPLC; see Section 2.10.2).

Samples and Preparation: The technique will provide information about coloured materials. In reflectance mode, it can be carried out *in situ*, but

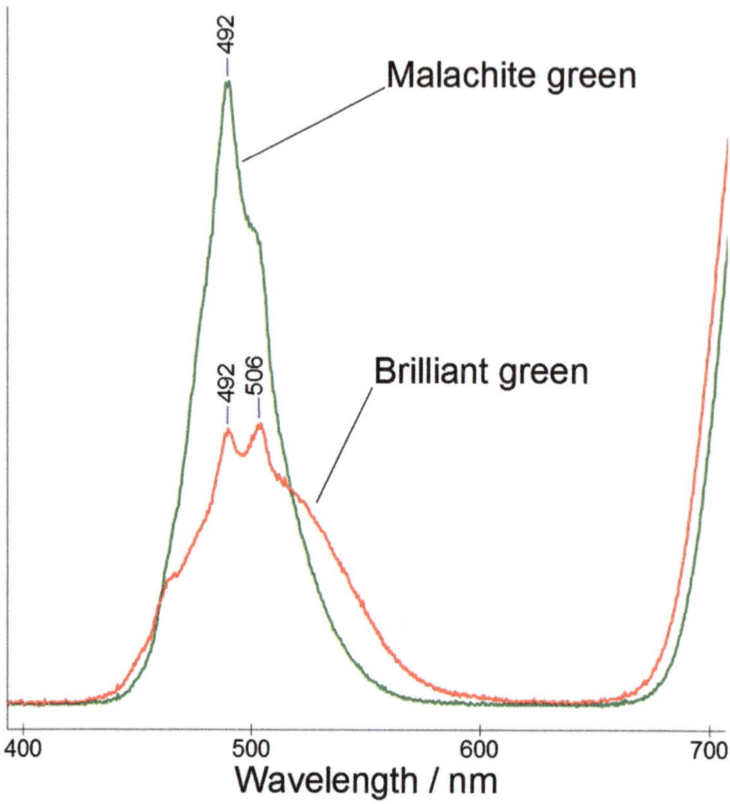

Figure 2.10 Visible reflectance spectra of two green dyes, malachite green and brilliant green.

this provides little information about concentration. In transmission, a sample must be taken and dissolved in a suitable solvent.

Results: The technique provides a plot of intensity against wavelength (Figure 2.10).

Limitations: The technique will only give information with materials that absorb light in the UV, visible and IR regions of the spectrum (depending on the spectral range observed). To use the transmission technique, sampling is required.

2.7 Elemental Analysis

2.7.1 X-ray Fluorescence

X-ray fluorescence (XRF) analysis permits the study of the elemental composition of an object. In the field of conservation science, it is particularly useful for the study of inorganic materials such as metals, stone, minerals (including mineral-based pigments), ceramics, glass and the like; the technique is not sensitive to low atomic weight elements, however, so it is less useful for the study of organic materials. It relies on excitation by high-energy X-rays or gamma rays to cause the emission of secondary or fluorescent X-rays. If an atom is exposed to radiation with sufficiently high energy it may be ionised by the ejection of an electron. X-rays and gamma rays have enough energy to eject electrons from the tightly bound inner atomic orbitals, and when this happens a higher orbital electron 'falls' into the gap, releasing an X-ray photon with energy equal to the difference between those of the two orbitals. As these energy differences are characteristic of particular elements, measuring the emitted X-rays allows the element to be identified, and this enables the content and proportion of elements within the specimen to be determined.

Bench-top XRF devices can have very high spatial resolutions and high sensitivities (up to the order of parts per million); the technique can be used without sampling if items can be adequately positioned to allow measurements, but otherwise specimens must be taken. Portable XRF devices have the advantage that they can generally be used *in situ*, but they usually have relatively large lateral measurement areas (typically of the order of 5 mm in diameter) and unless they have imaging capabilities it can be difficult to precisely localise the sampling position; furthermore, their sensitivity tends to be lower than that of bench-top instruments.

Samples and Preparation: Many objects can be examined without taking samples, and often *in situ*, providing they can be suitably positioned with respect to the instrument. The technique is non-destructive, so if samples are taken, they can subsequently be used for additional analyses. Fully quantitative analysis of element compositions is best achieved with flat, polished surfaces, which may necessitate sampling, as well as reference to calibration materials.

Results: The data is usually presented as a plot of intensity against energy, with individual peaks representing specific X-ray emissions (Figure 2.11). Individual elements generally have multiple such emissions.

Limitations: Although measurements can be taken without any particular sample preparation, the topography of the specimen will influence the quality of the results (with flat, polished samples providing the best data). If accurate elemental proportions are required, then calibration with similar samples of known composition is typically necessary.

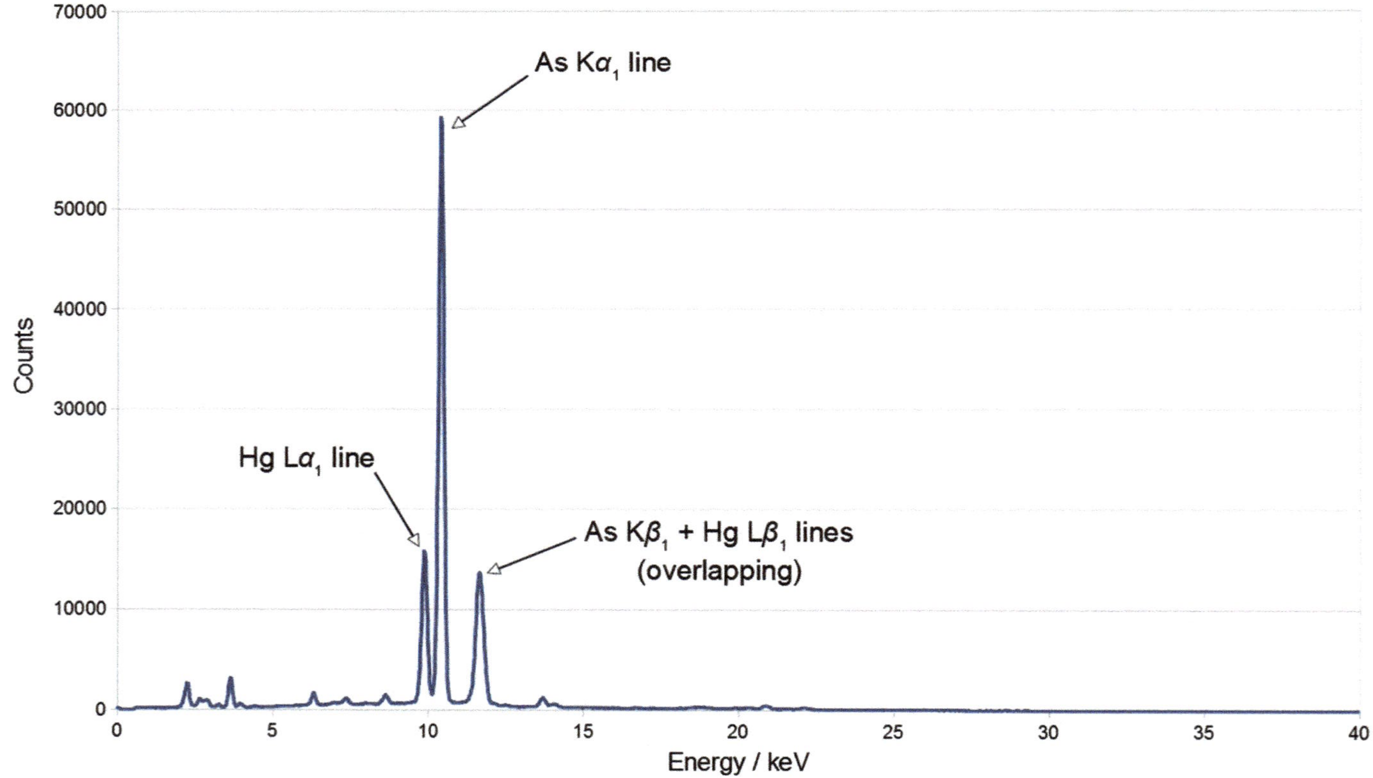

Figure 2.11 XRF data from a mixture of red (vermilion, HgS) and yellow (orpiment, As_2S_3) pigments.

In theory, the lightest element that can be detected by this technique is beryllium (atomic number of 4), but in practice the combination of instrumental limitations and low X-ray yields of lighter elements means that detection of elements lighter than sodium (atomic number of 11) is generally impractical.

A significant practical limitation of this technique is the fact that it involves X-ray emission, which in most countries will be governed by specific health and safety legislation and will often require special controls or safety measures to be put in place when using the equipment.

2.7.2 Energy Dispersive X-ray Spectroscopy

Energy dispersive X-ray spectroscopy (EDX or EDS) is closely related to XRF, but rather than using radiation to induce the emission of an X-ray it instead relies on the energy imparted by an incident high-energy electron to do so. As a result, it is often combined with SEM as the electron beam used by the technique for imaging can also generate these X-rays. Particular advantages of the technique include the ability to precisely localise the region from which the measurement is taken, resulting in accurate spot analysis. It can also readily be used for line scans, giving linear elemental profiles, or mapping in which the elemental composition across a region can be plotted. The use of this technique in combination with SEM imaging allows such information to be associated with specific structural and topographical features.

Samples and Preparation: The sampling and preparation requirements for this technique are similar to those of SEM, although coating the sample with a conductive layer may interfere with the results. It is suitable for a similar range of material types as XRF.

Results: Analysis at a single point produces a plot of intensity against energy, similar to those generated by the XRF technique. The technique can also be used to record line scans or maps of elemental data, in which each point of pixel is comprised of an EDS spectrum (Figure 2.12).

Limitations: The limitations of this technique are similar to those of SEM and XRF. However, the chamber of the SEM, in which the analysis is performed, acts as an effective X-ray barrier, which minimises radiation control issues.

2.7.3 Other Elemental Analysis Techniques

Other related methods of elemental analysis include techniques like PIXE and NAA, but these are more specialist approaches and

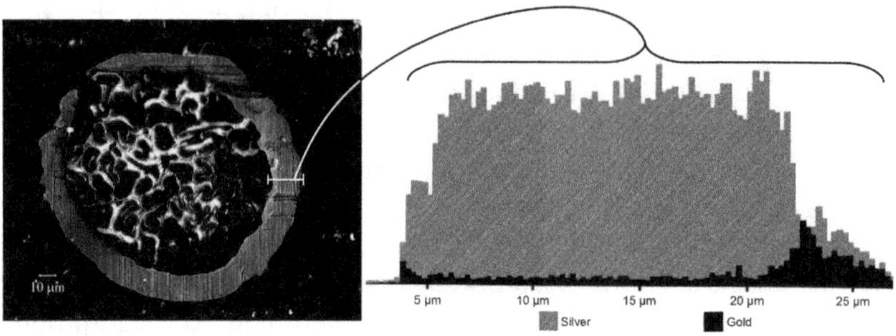

Figure 2.12 Gold and silver distribution, derived from an EDS line-scan of a cross-section of a silver gilt metal thread (with a silk core) set in resin.

availability are more limited. Particle induced X-ray emission spectroscopy (PIXE) generates characteristic X-rays by bombarding the sample with high-energy charged particles. The method has the advantage of being very sensitive and thus able to detect trace elements (of the order of parts per million). In neutron activation analysis (NAA), the bulk sample is bombarded with neutrons, a process which causes the emission of characteristic (prompt) gamma rays and also converts some of the atoms in the specimen to their radioactive counterparts. As these radioisotopes decay, over the period of up to several months, they also emit (delayed) gamma rays. Both of these emissions are at energies characteristic of the elements involved and are identified by either prompt or delayed gamma ray detection. The technique has the advantage that it is suitable for bulk specimens, so entire artefacts, rather than samples, can be analysed.

2.8 Crystallographic Analysis – X-ray Diffraction

X-ray diffraction (XRD) can be used to explore the structure of crystalline materials. X-rays scatter when they interact with atoms, and if these atoms are found in regular arrays (as is the case with crystals) they will act as a diffraction grating, causing the scattered X-rays to form regular patterns through constructive and destructive interference. These patterns are dependent on the arrangement and spacing of the atoms, so can be used to build up a picture of the structure of a crystal. The technique can be used to characterise materials that possess a crystalline or partially crystalline nature such as minerals, pigments, metals or textile fibres. It can also be used to investigate crystalline degradation products, such as the surface crystallisation of plasticiser from polymers or the efflorescence of absorbed salts in archaeological materials.

Single-crystal XRD is the oldest form of the technique, and gives the most precise information, but can be difficult to carry out, not least due to the requirement to acquire or grow a single crystal of the material to be studied, without impurities or significant imperfections. The crystal is placed in a monochromatic X-ray beam, and the diffraction pattern is measured; the crystal is slowly rotated in the beam, allowing its three-dimensional structure to be explored.

Powder XRD is generally more useful for the kind of materials encountered in conservation science. The diffraction of an X-ray beam by a powdered sample is measured, and it is assumed that the random arrangement of the particles will ensure that all crystal planes are observed in statistically significant proportions. In practice, it may also be necessary to rotate the sample to generate good data.

Samples and Preparation: In the case of powder XRD, a sample must be taken and finely ground. For single-crystal XRD, a crystal with minimal faults and of appropriate size is required, which is often grown from solution specifically for the purpose.

Results: The technique generates an XRD pattern or plot; the former is a visual representation of the diffraction of the X-rays, presented as a series of spots of specific intensities at particular distances and angles from a central point, and the latter a graph of intensity against angle (given as 2θ).

Limitations: The technique generally requires a sample to be taken, which in the case of powder XRD must also be pulverised. The interpretation of the diffraction pattern will be complicated by the presence of mixtures and the technique will reveal little useful information about largely amorphous materials. Growing crystals of appropriate size and purity for single-crystal XRD is time-consuming and difficult.

2.9 Nuclear Magnetic Resonance

Nuclear magnetic resonance (NMR) spectroscopy encompasses a range of related techniques that can be used to provide detailed information on the composition and structure of a sample, study the dynamics of atomic and molecular motion and provide three-dimensional images of objects and structures. NMR exploits the specific magnetic properties of particular atomic nuclei. Subatomic particles have a property called ‘spin’ and the total spin of an atomic nucleus determines whether it is detectable by the technique. Nuclei with an odd number of protons and/or neutrons have a non-zero spin, and are detectable, whereas nuclei with even numbers of both have a total spin of zero and are not. It is important to note that this means that NMR is sensitive to specific isotopes, not merely certain elements. If these nuclei are placed in a strong constant magnetic field, which is then perturbed by a weak oscillating field, they respond by generating an electromagnetic signal with a characteristic frequency as they return to equilibrium. This can provide information on the presence, proportion and distribution of particular nuclei, their local environment (that is, chemical bonding and structure) and dynamic processes (linked to relaxation times).

Originally, NMR was restricted to nuclei with high natural abundance (^{1}H, ^{19}F, ^{31}P, *etc.*) and was mainly limited to liquid samples. Superconducting magnets and refinements to the techniques allow a much greater sensitivity and the detection of nuclei with low natural abundance. This has allowed the development of ^{13}C NMR, for example, which is particularly valuable in the study of organic materials. Solid phase NMR spectra normally have broad, poorly defined features, but the introduction of magic angle spinning (MAS) NMR allows the technique to be applied to solids in a way which generates narrower, higher resolution features. Magnetic resonance imaging (MRI) allows large objects to be 'mapped'; this technique is now commonly used in medical diagnostics, and also provides a means by which large, complex cultural heritage objects may be studied in detail and their internal structures explored. In recent years, conservation studies have investigated the use of single-sided (SS) NMR for *in situ* analysis of objects, where non-invasive depth profiles in the region of 5000–25 000 μm are possible. In the case of SS-NMR, Carr–Purcell–Meiboom–Gill (CPMG) radiofrequency pulse sequences are used to obtain spin echoes at different depths through the substrate, enabling spatially resolved profiles of the effective spin–spin relaxation time ($T_{2\mathrm{eff}}$) and proton density.

Samples and Preparation: For conventional NMR, only very small sample sizes are required, but it must be possible to dissolve them in a suitable solvent; deuterated solvents are used in order to prevent a signal from the solvent itself interfering with that of the sample, and, in modern NMR systems, to provide a reference which allows the device to correct the field strength ('frequency lock').

One of the advantages of SS-NMR in conservation applications is that the excitation region of the instrument is positioned above the magnet, meaning analysis is both non-invasive and non-destructive. However, due to the instrument providing lateral depth profiles objects or samples do need to be flat to enable resolution of different stratigraphies; SS-NMR is not suitable for curved objects.

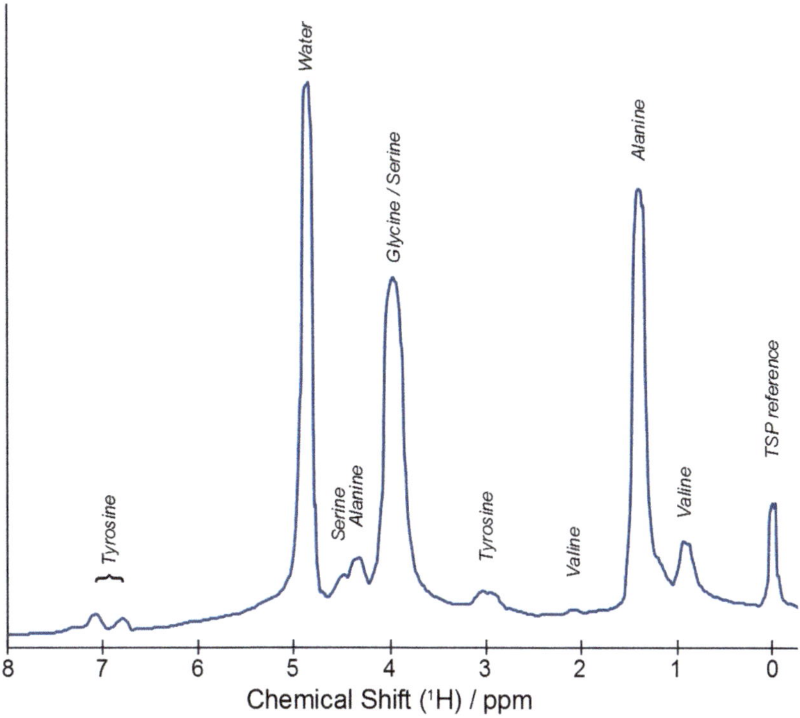

Figure 2.13 ^{1}H NMR spectrum of silk.

Results: A typical ^{1}H NMR spectrum of dissolved silk can be seen in Figure 2.13. Figure 2.14 shows the CPMG echo relaxation decay of a cellulose diacetate acetate film containing 12.5% triphenyl phosphate and 12.5% diethyl phthalate. Fitting a double exponent to the data gives a $T_{2\text{eff}}$ relaxation time of 0.12 ms and 0.44 ms attributed to the hard and soft components, respectively. Figure 2.15 shows a 2D contour map compiled from consecutive CPMG echo relaxation decays acquired at 50 µm cross-sections through the cellulose diacetate film sandwiched between glass plates. The signal originating from the film is seen between steps 7 to 9, which equates to a thickness of 150 µm. There is a reduced signal visible at analysis step 6 and this lower signal indicates that the lateral resolution of 50 µm is sampling both sample and glass.

2.10 Chromatography

Chromatography allows the separation of mixtures of materials. The sample must be in a form that allows it to travel, which is normally

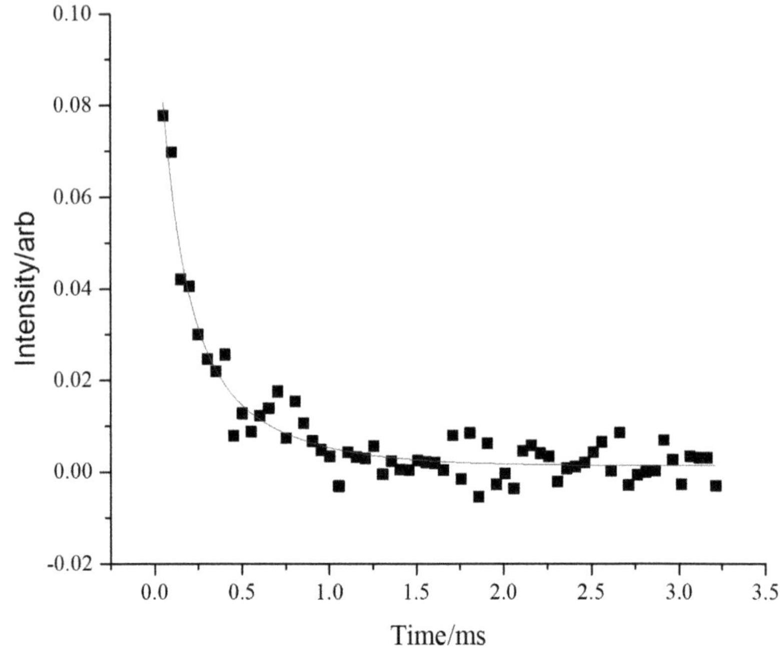

Figure 2.14 NMR CPMG T_2 relaxation curve of cellulose acetate film.

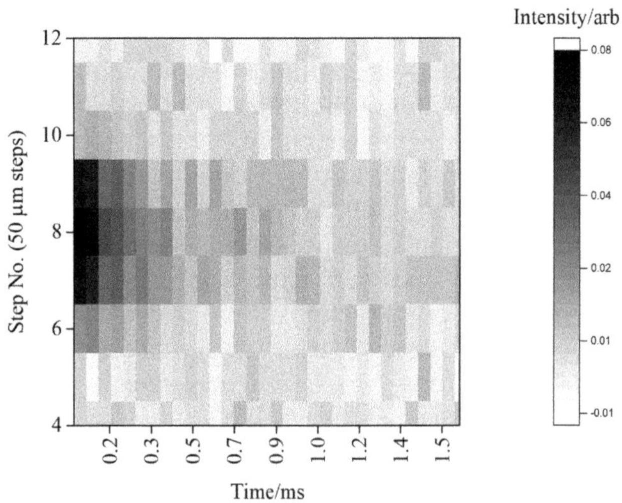

Figure 2.15 A 2D contour map compiled from consecutive CPMG echo relaxation decays.

achieved by dissolving it in a suitable solvent or using it in a volatilised or gaseous form. This is carried by a liquid or gas stream, the mobile phase, through another material, the stationary phase, which causes the components of the mixture to travel at different speeds and so separate out (known as partitioning), due to interactions based on molecular size, charge or chemical affinity. After partitioning, the components are detected using a variety of different techniques and characterised by their separation or retention time, often by comparison to those of known reference materials. It may be possible to determine the proportions of components in the original sample and to collect each component for additional analysis either through linked systems such as gas chromatography–mass spectroscopy (often referred to as 'hyphenated techniques') or for investigation at a later date. The specific features of the most common forms of chromatography are discussed in Sections 2.10.1–2.10.3.

Samples and Preparation: Chromatography is useful in the identification of organic compounds, degradation products, dyes and similar material. It requires samples to be taken, although these are generally small. The samples must be soluble in an appropriate solvent (for liquid chromatography), volatile (for gas chromatography) or capable of being derivatised or otherwise treated to impart such properties.

2.10.1 Liquid Chromatography - Planar Chromatography

In planar chromatography, the stationary phase is found as a two-dimensional plane (*i.e.*, a sheet of suitable material). These methods are simple, rapid and readily available, requiring little specialist equipment. The sample is applied as a dot to the stationary phase, which is then placed in a shallow trough of solvent, the mobile phase. As the solvent is drawn up, it encounters the sample, the components of which then begin to separate as they are transported at different rates depending on their affinity with the solvent and substrate. Components are characterised by their *retardation factor*, R_f, measured as the ratio of the distances travelled by the component and solvent front. In many cases, the position of the components can be observed visually, but this may be enhanced through the use of ultraviolet illumination or through selective staining. A further advantage of these methods is that several samples can be run in parallel on a single sheet of material. Paper chromatography and thin layer chromatography (TLC) are examples of this approach.

The simplest form of this technique, paper chromatography, is one that most people will be familiar with from the school experiment of separating the components of an ink dot using blotting paper dipped in water (Figure 2.16). In this example, the ink is the sample, the water the mobile phase and the blotting paper the stationary phase. The effectiveness and applicability of this method can be improved by using high quality chromatography paper and selecting a solvent which will enhance the separation of the components.

A more sophisticated approach is afforded by thin layer chromatography (TLC), in which the stationary phase consists of a thin layer of adsorbent material, such as alumina, silica gel or cellulose, bonded to a flat, inert substrate. The variety of possible stationary phases makes this method more versatile than paper chromatography, as the affinity of this phase can be selected to suit the sample, and it is possible to use a wider range of solvents.

Samples and Preparation: A sample must be taken which is then dissolved in an appropriate solvent. It is also necessary to identify an

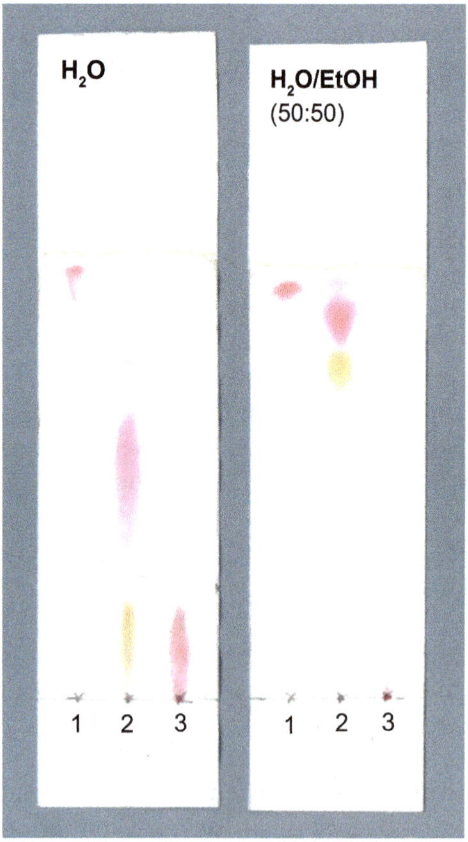

Figure 2.16 Planar chromatography of three different samples of red ink, in water and water/ethanol.

appropriate mobile phase which will give the required separation of components.

Results: The separation of the components is observed visually on the chromatography paper or plate (Figure 2.16). If the components are not themselves strongly coloured, a stain or other visualisation method (such as UV induced fluorescence) may be required in order to see them.

Limitations: The technique requires a sample to be taken, which must also be soluble in a suitable solvent and the components must be capable of appropriate separation in the mobile phase. The components must either be visible or capable of being visualised.

2.10.2 Liquid Chromatography – HPLC

High-performance liquid chromatography (HPLC) employs a stationary phase packed in a column and operates under pressure rather than relying on gravity or capillary action. The stationary phase is composed of irregularly or spherically shaped particles, a porous monolithic layer or a porous membrane. The sample is forced by a high-pressure liquid (the mobile phase) through this column, and separation is achieved through physical, chemical or ionic interactions. The liquid is often a mixture of solvents, and the ratio of these components and their temperature may be varied over the course of the analysis to enhance separation. As the separated components leave the column, they are detected using a suitable detector (often UV–visible spectroscopy at a specific wavelength); the factions can also be collected for further analysis or processing or sent directly to another linked analytical technique. HPLC is capable of dealing with very small sample sizes and can achieve a high degree of separation between components of a sample (resolving power).

Samples and Preparation: As with planar chromatography, a sample must be taken, and suitable solvents and mobiles phases are needed; however, the sample size can be very small.

Results: The technique generates a chromatogram, a plot of intensity (as derived from the detection method) against time (retention times), as shown in Figure 2.17; more sophisticated detection methods may generate a series of spectra linked to specific retention times.

Limitations: Samples must be taken. It can be difficult to find a suitable mixture or sequence of mobile phase solvents, as well as column materials, to achieve the desired separation. The technique requires relatively large quantities of solvents which cannot be reused.

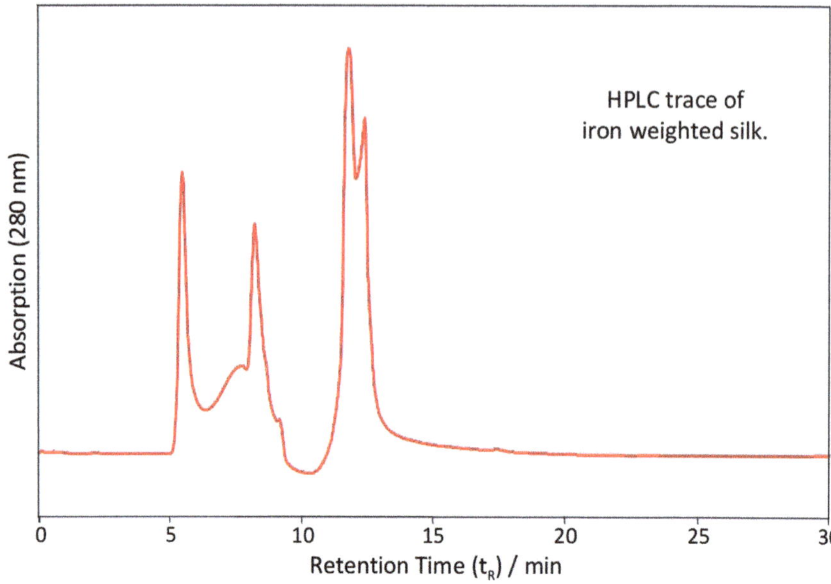

Figure 2.17 HPLC trace of iron weighted silk, dissolved in sodium thiocyanate.

2.10.3 Gas Chromatography

Gas chromatography (GC) requires a material that can be volatilised without decomposition, which is carried by an inert gas (mobile phase), with separation achieved *via* a column packed with a suitable stationary phase, usually a polymer or liquid layer on an inert support. In addition to the chemical or physical interaction with the stationary phase, separation can also be influenced by the rate of gas flow, the length of the column and the temperature.

A variety of different detectors may be used to analyse the partitioned components as they leave the column, but the two most common are flame ionisation detection (FID) and thermal conductivity detection (TCD). FID pyrolyses material as it leaves the column, forming cations that generate a detectable current between the electrodes; this method has a low detection limit, but only works for organic or hydrocarbon-based materials. TCD uses a heated filament cooled by the carrier gas stream, with the presence of sample molecules in this gas reducing its thermal conductivity and so causing a measurable increase in the temperature of the filament; it is non-destructive, so can be used in conjunction with other detection methods to improve versatility and sensitivity. There are many other types of detectors that can be used, with differing specificities, concentration ranges and compatibilities with carrier gases, allowing the technique to be optimised for the material under analysis.

Gas Chromatography–Mass Spectroscopy: Gas chromatography–mass spectroscopy (GC–MS) combines the advantages of each of the individual techniques to separate and characterise substances, including complex mixtures, with a very high degree of specificity. The sample is separated into components as it passes along the GC column, and then these components undergo ionisation before they enter the linked MS device. These ionised fragments are then detected as described in the section on the MS technique, 2.11.

Pyrolysis Chromatography: Most chromatographic techniques rely on the material being volatile or soluble in a benign solvent, and so are not suited to many of the materials found in archaeological or heritage collections, such as wood, paints and fabrics, all of which contain cross-linked high molecular weight compounds. Pyrolysis GC–MS (py-GC–MS or pyGCMS) provides a method of studying these materials. The sample (typically less than 1 mg) is introduced into a hot zone created by heated filaments, lasers or micro-furnaces, where it is rapidly volatilised and broken down into characteristic fragments. This material is then carried to the GC–MS system. The spectra of the fragments produced by the mass spectrometer can be employed as a fingerprint to compare with those of known materials or used to 'reconstruct' the molecules of the original polymer (Figure 2.18).

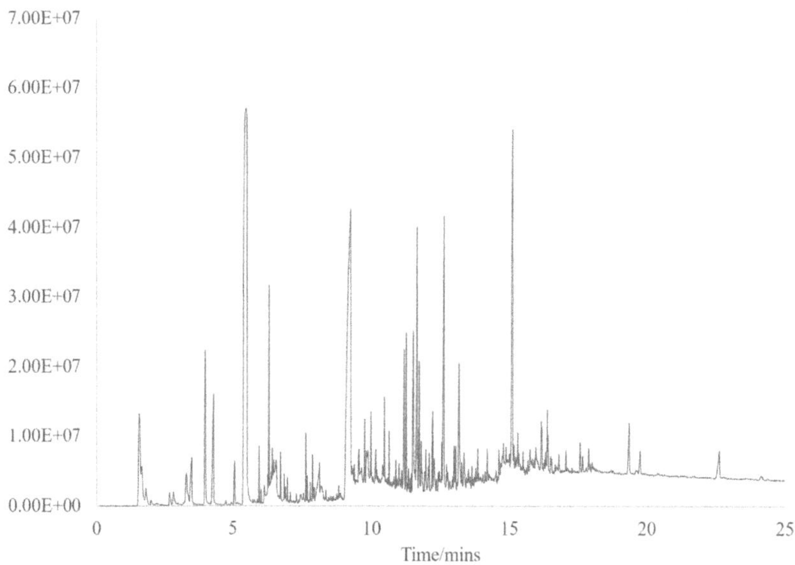

Figure 2.18 py-GC–MS chromatogram of polyester thermoplastic resin.

2.11 Mass Spectroscopy

Mass spectroscopy (MS) allows the determination of mass, chemical characteristics and structure of molecules, and as well elemental and isotopic signatures. The sample is ionised, and the components are then separated by accelerating them through a strong electric or magnetic field that interacts with their charge; the acceleration and deflection of the ions by these fields is determined by both their mass and their charge, and they are detected dependent upon these properties. This allows the mass-to-charge (m/z) ratio of the component ions to be measured, and the data is presented as a mass spectrum.

A wide range of different ionisation, separation (mass selection) and detection methods are possible. Ionisation can be achieved with bombardment by electrons or atomic or molecular ions, exposure to a plasma or irradiation by a laser. Separation can be achieved by deflection of the flight path of the ions (sector MS), by measuring the time taken for ions to reach the detector [time-of-flight (TOF) MS] or by selectively destabilising flight paths (quadrupole MS); in all cases, the separation is ultimately dependent on the m/z ratios of the ions. The detectors typically record either an induced charge or a current produced when an ion interacts with a surface; an electron multiplier is often used to boost the signal.

Sampling and Preparation: Depending on the set-up used, the technique can be applied to a variety of archaeological and historic materials including biological materials, degradation products, binders, adhesives and the like. It can also be used to characterise materials separated by gas chromatography (see GC–MS, Section 2.10.3).

Results: The data is used to create a mass spectrum, a plot of mass-to-charge ratio against intensity (Figure 2.19).

Limitations: The technique may provide limited information about materials which produce similar ionic fragments (such as closely related organic materials) and cannot distinguish geometric and optical isomers; these limitations can be dealt with using 'hyphenated' techniques such as GC–MS, however. The technique is also expensive and highly specialised, requiring an expert operator.

2.12 Physical Analysis

Physical characteristics of a material can include strength, tenacity, elasticity, extensibility, viscosity, thermogravimetric and

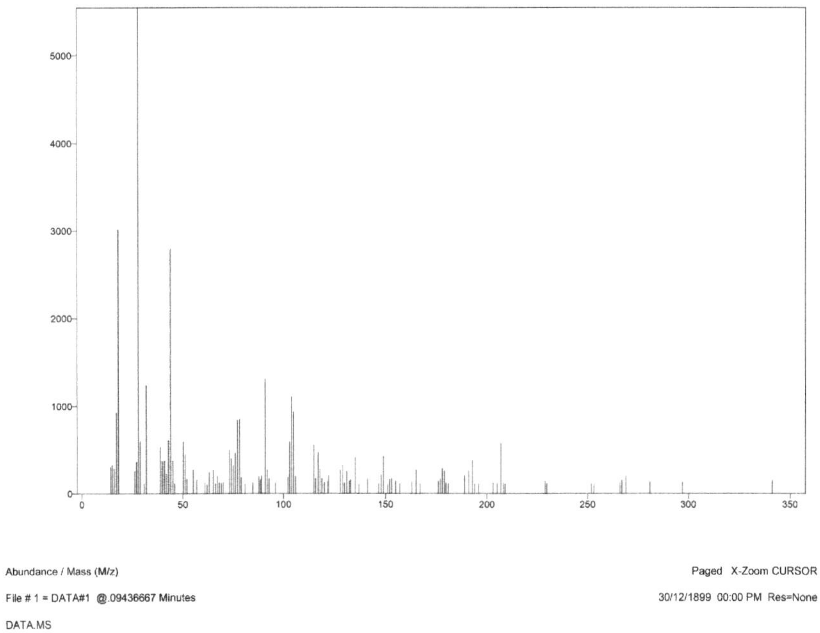

Figure 2.19 Mass spectrum of polyester thermoplastic resin.

thermomechanical properties, transition temperatures and the like, which can be measured using various different techniques.

In general, these tests are destructive and require sampling; these samples may be relatively large, especially for techniques such as mechanical testing, and replicate tests are almost always necessary for reliable results.

2.12.1 Mechanical Testing

All materials will deform to some extent on the application of a high enough load. Deformation is defined as *a change in the shape or dimensions of a body resulting from stress* and it is a material's response to that deformation that will govern its application. Stress–strain analysis provides information on the bulk physical properties of a sample, such as the resistance to deform, and the extensibility of a material. In conservation it is often helpful to monitor changes in physical properties during ageing and degradation processes or to assess a material's strength prior to its application as a conservation treatment, such as an adhesive. Although there are many modes of mechanical deformation, tension is the most commonly applied within conservation and refers to the application of a force at opposing ends of a sample. The sample is clamped at each end and subjected to a constant extension, whilst the force required to maintain this extension rate is measured.

For simple tension, if the length of a rod (L_0) is subjected to a tensile force (F_t) there will be a change in length (δ). The elongation of the sample is known as the strain (ε) and is defined as the extension per unit length.

$$\varepsilon = \delta / L_0$$

The stress exerted on an object is defined as the (measured) force divided by the unit area; force is proportional to the cross-sectional area (A) of the sample, calculated as F/A. This is termed the stress, σ.

$$\sigma = F_t / A_0 = \mathrm{N\,m^{-2}}$$

An extension of (L_0) will cause a corresponding contraction in the perpendicular coordinates of the sample. However, in simple uniaxial experiments it is often assumed that the cross-sectional area (A) of the sample will remain constant, A_0; this measure of stress is known as engineering stress.

Sampling and Preparation: Sampling requirements and preparation vary significantly between the different geometries of this technique, so generalisations are difficult. Samples might range from a single fibre length to a steel girder. Depending on how the material has been produced, the orientation of testing can significantly impact the results. If the samples are highly oriented (anisotropic) then the physical properties will differ if samples are prepared in different directions.

A critical factor to keep in mind during testing is the free length of the sample, also known as the gauge length. For instance, when tension testing a yarn, the longer the free length of the yarn is the lower the breaking strength. This is due to an increase in yarn ends within the gauge length. It is therefore important to have a constant gauge length for comparative studies.

Results: The data is usually presented as load *versus* extension, but is often converted to stress *versus* strain (Figure 2.20) when the cross-sectional area is known and comparisons between materials are desired.

Limitations: Analysis is always invasive but does not always result in the destruction of the sample. Due to the need to extend, compress or bend the sample testing will require the removal of a sample from an object. If elastic, reversible properties are being studied, it may be possible to retain the sample for further analysis. However, if the

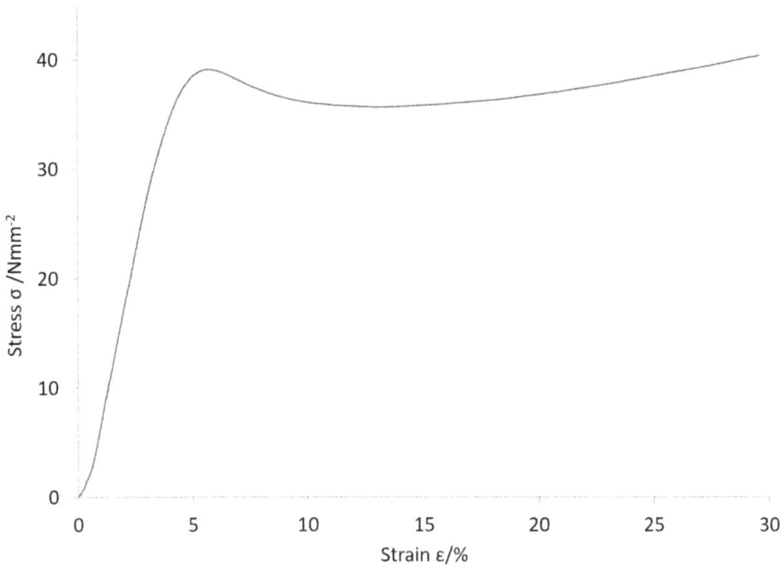

Figure 2.20 Tensile test of cellulose acetate film.

sample is taken beyond its elastic limit then permanent changes will be induced in the sample and any further analysis will reflect these changes.

Due to the variability in bulk physical properties along the length of a material replicate measurements are always required.

2.12.2 Thermogravimetric Analysis

Thermogravimetric analysis (TGA) determines the change in weight of a sample in relation to a change in temperature. Accurately weighed samples are introduced into the furnace and the weight of the sample measured as a function of temperature. TGA monitors the release of volatiles from the sample being formed through simple evaporation or by decomposition of the material into low molar mass fragments. It may be used for quantitative estimation of volatile organic compounds (VOCs), water content or other low molar mass additives present in an object. It also allows for an assessment to be made of the point of thermal decomposition. TGA is often coupled with MS (discussed in Section 2.11), which allows for the identification of volatile and decomposition products.

TGA can be undertaken in an oxygen or oxygen-free nitrogen atmosphere, depending on whether oxidative or thermal reactions are

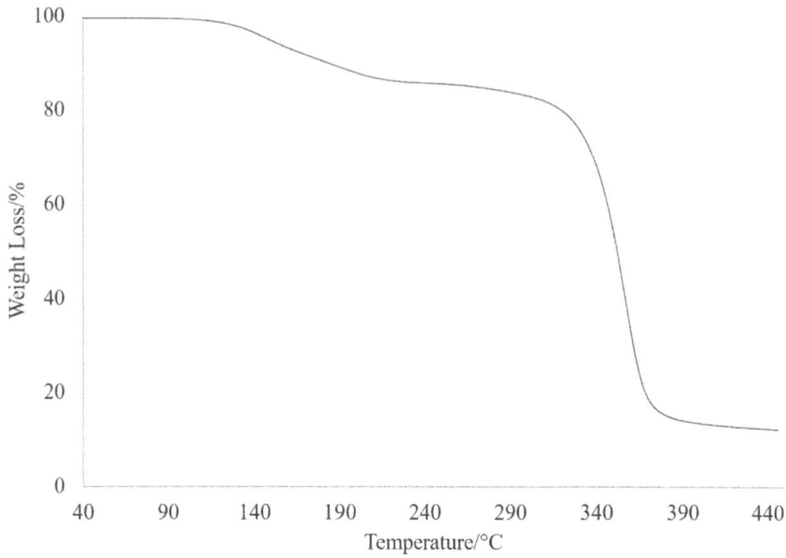

Figure 2.21 Weight loss curve cellulose acetate film.

of interest. Samples are placed into an inert crucible, both being accurately weighed prior to analysis and throughout the heating regime.

Sampling and Preparation: Although TGA is a destructive test, sample masses are low and often in the region of 2–5 mg. For best result samples are analysed in powdered form to enable the rapid evolution of volatile reaction products from the sample; reactions will be quicker when there is a greater surface area.

Results: Data is usually presented as mass *versus* temperature, enabling the temperature of any weight loss event to be identified (Figure 2.21). Data are either presented as absolute weight loss or as a percentage of the original mass of the sample. It is not uncommon to see derivatives calculated for TGA heating curves, which not only show the temperature of an event but also indicate the rate of change.

Limitations: TGA is both an invasive and destructive analytical technique and the sample is not retained after analysis. As such, if sample size is limited it is recommended that any non-destructive tests are undertaken first.

Where comparative studies are being undertaken sample masses should, as far as practicable, be constant to ensure equal heat transfer within each crucible. However, this is not always possible given the challenges of sampling artefacts and therefore should be borne in mind during interpretation.

2.12.3 Thermomechanical Analysis

Thermal mechanical analysis (TMA) monitors changes in mechanical properties whilst subjecting a sample to a temperature regime. TMA is used to determine the glass transition temperature and the percentage elongation of the sample as a function of temperature. Different types of mechanical deformation can be determined by employing different sample geometries during analysis, such as bending, compression and tension, which allows the temperature dependant responses of a material to be assessed under the conditions of use. Similarly to TGA, the sample furnace is purged with an oxygen or oxygen-free nitrogen atmosphere, depending on whether oxidative reactions are being monitored.

TMA can be run in static or dynamic mode. The former is used to measure dimensional changes and derive coefficients of expansion, whilst the latter allows for the determination of the glass transition temperature by monitoring changes in elastic properties with respect to temperature.

Sampling and Preparation: As is the case with mechanical analysis, sampling requirements and preparation vary significantly between the different geometries of this technique. However, the sample furnace is a limiting factor in terms of sample size, with sample lengths typically in the order of 2–20 mm, depending on the analysis geometry and sample stiffness.

Results: Data is presented as modulus *versus* temperature or elongation as a function of temperature or time (Figure 2.22).

Limitations: Limitations are similar to the mechanical analysis discussed in Section 2.12.2. However, in the case of TMA deforming the sample at raised temperatures induces plastic, non-reversible changes and therefore the retention of the sample for further analysis is not possible.

2.12.4 Hydrogravimetry (Dynamic Vapour Sorption)

The water sorption properties of materials are a critical factor in determining storage conditions of objects, due to the relationship with both dimensional changes and, for certain materials, an increased risk of hydrolysis. Sorption experiments are carried out to determine degrees of moisture absorption as a function of relative humidity, calculated as a percentage weight change when compared to the desiccated sample. The degree of absorption is dependent on a number of factors such as porosity, the presence of hydrophobic species, and the degree of crystallinity.

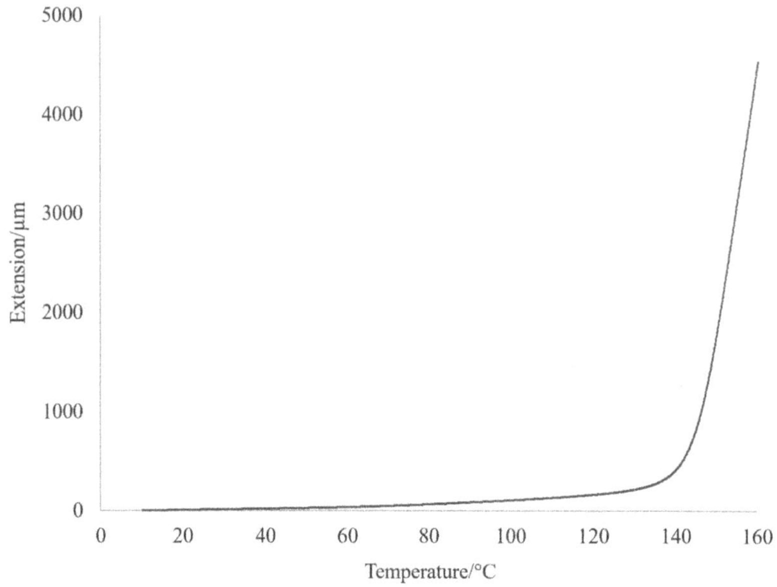

Figure 2.22 Heating curve of cellulose acetate film.

During such experiments it is common to monitor the desorption behaviour to determine whether the process of moisture absorption is reversible. It is often the case that physical processes are dependent on the history of the material, and therefore altering the environment can cause a non-reversible state known as hysteresis.

In its simplest form, gravimetry can be carried out by subjecting samples to incremental changes in relative humidity (RH) controlled by saturated salt solutions and weighing samples following equilibrium at each RH interval. This requires samples to be equilibrated in rubber-sealed RH chambers and weighted at each interval using a microbalance. This approach is limited to change in mass at set RH levels and does not provide dynamic information relating to the absorption and desorption of moisture. An alternative approach is to combine a humidity generator with an enclosed microbalance so that dynamic humidity programmes can be used. This not only allows for changes in mass to be assessed, but also provides information on the rates of change and the time it takes for a material to equilibrate at a particular RH level. This latter piece of information can be derived by comparing the RH data from the humidity generator with that of the microbalance at a given time.

Sampling and Preparation: Sample size and shape will be limited by the size of the RH chamber required to maintain the setpoint humidity surrounding the sample and the working range of the balance. For comparative purposes, gravimetry data are presented as the percentage change in mass rather than absolute change in mass, which means that the different samples can be compared. However, it must be borne in mind that different sized samples will have different surface areas that will have an impact on the moisture absorption profile, so samples should be kept as consistent as possible. The percentage change in mass is calculated relative to the starting humidity, so each sample should be allowed to equilibrate at the initial RH prior to recording their mass.

Results: If coupling an RH generator with a microbalance two graphs will be generated that need to be superimposed. The data from the generator will be presented as RH against time and that of the balance presented as mass against time. So long as the *x*-axes are in the same units of time and the start of the experiments can be aligned then the RH information will enable the mass of the sample at each RH point to be determined and any lag in equilibrium to be determined.

Figure 2.23 shows a gravimetric experiment for a cellulose acetate film, with percentage RH *versus* time (upper curve) overlaid with the percentage change in mass *versus* time (lower curve). In the upper RH curve, the data is gathered as the humidity is increased, highlighted here in dark grey, after which the generator maintains the set RH level. It is apparent here that at the end of each set increase the RH continues

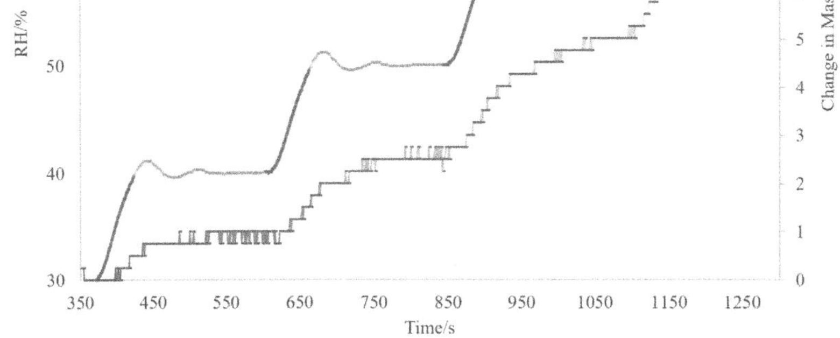

Figure 2.23 Gravimetric experiment of cellulose acetate film.

to rise, seen as a hump in the curve. This is owing to a lag between the equilibrium of the chamber and the RH sensor, meaning that equilibrium is not immediately reached within the enclosure and can take many minutes to reach the setpoint. This needs to be born in mind when assessing the absorption profile of the sample (lower curve).

Due to fluctuations in the mass of the sample during sorption, the gravimetry curve (lower curve) is discontinuous, but the general shape of the curve allows the point of sample equilibrium to be determined, seen as a plateau at each RH setpoint. The sample reaches equilibrium and plateaus more readily at lower RH levels than at high RH, with the sample not having enough time during this experiment to reach equilibrium at 70%.

Limitations: Depending on the sensitivity of the balance and the type of material being studied weight changes might be negligible. As mentioned previously, differences in the surface area of a sample will also affect the results, therefore differences in the physical form of the material, such as fibres, films and foams, will be determining factors for moisture sorption and need to be considered when making comparisons.

2.13 Chemical Tests

Chemical tests exploit the particular chemistry and reactivity of a sample to characterise it. This may be through a distinctive colour change on exposure of the material to a certain reagent, its solubility or patterns of dissolution in specific solvents, or its flammability and combustion properties (flame colour, smell, *etc.*). These tests were once commonplace, as they require little in the way of analytical equipment, can be carried out quickly, easily and cheaply, have a high throughput, and in theory produce distinctive results. However, they are also destructive, limited in their scope and may involve the use of toxic chemicals or be otherwise hazardous. Furthermore, although the results may be presented as unambiguous when performed on pristine materials, in practice they can be misleading or difficult to interpret, due to a variety of factors including: the presence of dyestuffs or discolouration which can obscure colour changes; the presence of additives, soiling or other components that may also react to the test; physical and chemical changes due to degradation which can alter the material's response; the subjectivity of the observer. As a result, these tests have generally fallen out of favour, and have been replaced by instrumental analyses. However, their ready accessibility and ease of use mean that certain tests of this kind are still used, and the results will be found widely in older literature. As the application and interpretation of these tests are specific to particular material types, they will be discussed in individual chapters, where appropriate.

2.14 Sampling

When dealing with historic artefacts, it is obviously important to intervene as little as possible and to ensure that any intervention yields enough knowledge or benefits to justify the action. This is of particular importance when considering analytical techniques, as many require samples to be taken and these samples may be destroyed or altered in the process of the analysis. Any such decision would normally be made in conjunction with conservators, curators and other stakeholders. If it is necessary and appropriate to take samples, in some cases it is possible to do this in a way that minimises changes to the object. For example, paint or pigment flakes which have already become loosened or detached may provide useful samples; similarly, it may be possible to take fibre samples from loose threads or frayed edges, and areas of pre-existing damage can often yield specimens without causing further losses. If multiple forms of analysis are to be carried out, it is important to plan these in such a way as to allow the maximum information to be derived from the minimum number of samples, bearing in mind which techniques allow samples to be recovered unchanged, which alter them and which destroy them.

When considering the most appropriate techniques to use, a balance must be struck between the techniques most capable of giving the best range of information and those with sampling requirements that are best suited to the object in question. Various terms are used to describe techniques which minimise sampling requirements, including 'non-sampling', 'non-invasive', 'non-contact' and 'non-destructive', but the precise definitions of these can vary between equipment manufacturers or literature sources, and to a certain extent are used interchangeably, which can lead to confusion. Non-sampling techniques do not require a sample to be taken from the object, but may still require contact with it; for example, the ATR FTIR technique can be regarded as non-sampling if the object can be suitably positioned on the instrument, but it requires pressure which may prove damaging to fragile materials. Non-invasive and non-contact techniques can record data without direct physical contact (usually through interaction with some form of electromagnetic radiation); this minimises the risk of damage to the object, providing it can be adequately presented to the instrument to allow data of sufficient quality to be recorded. Non-destructive techniques allow a sample to be recovered unchanged at the end of the analysis, but nonetheless still often require a sample to be taken initially, although this does then allow further analyses to be performed on a single specimen.

2.15 Artificial Ageing

Artificial (accelerated) ageing is often used to create surrogate materials whose properties are intended to mimic those of naturally aged samples. This may be done for a variety of reasons. It helps understand degradation processes. It reduces the need to sample historic objects. It can be used to create large amounts of samples which can be destructively tested without ethical issues. It can provide samples to test, refine and calibrate analytical techniques before valuable historic samples are analysed. The ageing itself may be achieved by a variety of processes, the most common of which are thermal ageing and light ageing. Thermal ageing involves exposing a sample to elevated temperatures (often with raised humidity), to increase rates of reaction in the material. Light ageing exposes a sample to intense visible light (UV may be present or absent, depending on the result required), to promote various photodegradative reactions. Other methods of ageing may also be encountered, including exposure to acid or alkaline conditions to induce hydrolysis, incubation with mould or other microorganisms, treatment with pollutants and other reagents, or subjecting the sample to prolonged vibration or other physical stress. These factors may be used in combination, either simultaneously or sequentially, and may also be applied in a cyclic manner. The particular choice of ageing regime will depend on both the nature of the material and the type of history that is being replicated, and numerous examples can be found in the literature. In archaeological contexts, ageing of surrogate samples may be achieved by burying them under specific conditions, although this is a slightly different form of artificial ageing as it normally relies on reactions and other processes which occur at ambient rates.

Although it can be a very useful technique, artificial ageing must also be treated with caution. The properties of artificially aged materials are not likely to perfectly mimic those of naturally aged materials, for a variety of reasons. Firstly, it is difficult to ensure that unaged modern materials are identical to their historical counterparts, in terms of composition, impurities, structure and the like, so the 'starting point' for creating surrogates may well be different. Secondly, it is impossible to know exactly what conditions an object will have been subjected to over its history, so the choice of artificial ageing mechanism(s) will always be speculative to a greater or lesser extent. And thirdly, the accelerated conditions employed by artificial ageing do not necessarily produce the same reactions as those experienced by similar material under natural conditions. This last point is particularly important in the case of thermal ageing: raising the temperature will not only increase the rates of reactions seen under normal conditions, but it will also start to enable reactions that would not normally be observed (as their activation energies are too high for them to occur except at elevated temperatures); furthermore, physical effects are likely, such as dehydration, state changes (glass transitions, softening, melting, fusing, *etc.*) and the migration of mobile species (such as plasticisers).

With those caveats in mind, however, artificial ageing is a useful source of surrogate specimens, which can allow the deterioration of historic materials to be better explored and understood, and permit the development of analytical approaches to investigate such objects.

Recommended Reading

S. Bradbury and B. Bracegirdle, *Introduction to Light Microscopy*, BIOS Scientific Publishers, Oxford, 1998.

J. M. Brown, *Molecular Spectroscopy*, Oxford University Press, Oxford, 1998.

Single-Sided NMR, ed. F. Casanova, J. Perlo and B. Blumich, Springer, Dordrecht, 2011.

M. Derrick, D. Stulik and J. M. Landry, *Infrared Spectroscopy in Conservation Science*, Getty Publications, Los Angeles, 1999.

J. P. Eberhart, *Structural and Chemical Analysis of Materials*, John Wiley & Sons Ltd, Chichester, 1991.

C. B. Faust, *Modern Chemical Techniques*, Royal Society of Chemistry, London, 1995.

R. L. Feller, *Accelerated Aging: Photochemical and Thermal Aspects*, Getty Publications, Los Angeles, 1994.

Principles and Applications of Thermal Analysis, ed. P. Gabbott, Blackwell Publishing, Hoboken, 2007.

P. J. Goodhew, J. Humphreys and R. Beanland, *Electron Microscopy and Analysis*, 3rd edn, Taylor & Francis Inc., New York, 2001.

R. W. G. Hunt and M. R. Pointer, *Measuring Colour*, 4th edn, Wiley, Hoboken, 2011.

C. Jones, C. Duffy, A. Gibson and M. Terras, Understanding Multispectral Imaging of Cultural Heritage: Determining Best Practice in MSI Analysis of Historical Artefacts, *J. Cult. Herit.*, 2020, **45**, 339.

L. MacDonald, *Digital Heritage – Applying Digital Imaging to Cultural Heritage*, Butterworth-Heinemann, Oxford, 2006.

S. O'Connor and M. M. Brookes, *X-radiography of Textiles, Dress and Releated Objects*, Routledge, Abingdon, 2007.

K. A. Rubinson and J. F. Rubinson, *Contemporary Instrumental Analysis*, Prentice-Hall Inc., Upper Saddle River, 2000.

H. W. Siesler, Y. Ozaki, S. Kawata and H. M. Heise, *Near Infrared Spectroscopy*, Wiley-VCH, Weinheim, 2002.

D. Skoog, F. Holler and C. Stanley, *Principles of Instrumental Analysis*, 7th edn, Cengage Learning, Boston, 2017.

M. F. Striegel and J. Hill, *Thin-Layer Chromatography for Binding Media Analysis*, Getty Publications, Los Angeles, 1996.

N. Odegaard, S. Carroll and W. S. Zimmt, *Material Characterization Tests for Objects of Art and Archaeology*, 2nd edn, Archetype Publications, London, 2005.

B. H. Stuart, *Analytical Techniques in Materials Conservation*, John Wiley & Sons Ltd., Chichester, 2008.

3 Paper

V. Daniels

The British Museum, Department of Scientific Research, London WC1B 3DG, United Kingdom
Email: vdaniels@britishmuseum.org

3.1 Introduction

Paper is a very common material in everyday life. When indoors we may be literally surrounded by paper either on the walls as wallpaper or as pictures or volumes in bookcases. Paper has been in common use for hundreds of years all over the world and is used as a support for works of art and record-keeping, as well as for cleaning, ephemera and packaging to protect useful things made of other materials. Packaging accounts for 55% of all paper and paperboard production, and this figure looks likely to increase. Most of us own far more objects made of paper than any other man-made material.

Libraries and record offices hold vast amounts of paper and even if the quantity of information held digitally may exceed that held on paper, the physical bulk of the paper generally exceeds that of the apparatus for the stored digital information. As far as works of art on paper are concerned, the original work is held in high esteem while an electronic copy is of small value. However, the mass storage of digitally stored images of printed paper has meant the destruction of some newspapers and other paper records. There are several reasons for this: firstly, some paper records have been and still are being produced in such a high volume that physical storage has become too expensive; secondly, the deterioration of some types of paper means

Conservation Science: Heritage Materials, 2nd Edition
Edited by Paul Garside and Emma Richardson

Published by the Royal Society of Chemistry, www.rsc.org

that many newspapers and books have become too brittle and discoloured to be used by readers, and the information therein is better copied, often being transferred to digital form, where it is easily and inexpensively copied, transferred to other users and searched.

The aim of this chapter is to outline the chemistry of paper, its manufacture, subsequent mechanisms of ageing and principal conservation methods.

3.2 Chemistry, Structure and Microstructure of Paper

A consequence of the ubiquity of paper is that everyone can recognise it when they see it. The vast majority of paper is, and has been, made from cellulose fibres. The fibrous nature of paper may be seen if it is torn, when the tear edge may show the individual fibres. Plants contain fibres principally made of the carbohydrate called cellulose. Not all parts of plants contain cellulose but it still makes up about a third of the mass of the higher plants, and it is the most abundant biopolymer on Earth. Nowadays, the vast majority of books such as this will contain pages made of cellulose fibres extracted from wood. Paper is believed to have first been made in China in 105 AD. Traditional Chinese and Japanese papers are made from indigenous plants, which yield long fibres of relatively pure cellulose by processes involving a combination of prolonged soaking in water or solutions followed by mechanical beating. Soaking in water for several weeks encourages microbial decomposition and softening of the plant stems (retting). The earliest European papers were made in about 1200 AD from fibres from textile rags, which in turn were made from the fibres from cotton or linen plants. Cotton occurs as the fluffy seed heads of the cotton plant and, being almost pure cellulose, can be used for papermaking or weaving with little preparation. Linen fibres are obtained from the stems of the flax plant by similar techniques to those just described for traditional Chinese and Japanese papers. Eventually, demand for paper became so large that other sources of fibres had to be found and for a relatively short period in the 19th century grasses became popular, but today wood cellulose is used for the vast majority of papers, being both inexpensive and readily renewable. Wood cellulose is intimately associated with unwanted impurities and purification stages are normally part of papermaking processes.

Both cellulose and starch can be thought of as polymers of glucose, however the cellulose molecule is different from starch in that the glucose molecules are linked together in a slightly different way, and the repeating unit in cellulose can be considered to be the dimer cellobiose. Two glucose molecules link together by losing a molecule of water and forming an oxygen bridge between the 1-carbon on one ring and the 4-carbon on the other, this is called a 1,4-glucosidic bond. A representation of the cellulose molecule is shown in Figure 3.1. Note the large numbers of hydroxyl groups (–OH).

The average number of glucose monomer units in the cellulose chains is called the degree of polymerisation (DP) (see Chapter 11). In bleached, new, wood pulp-derived paper the DP is in the order of 1000–2000 while that of cotton and flax (linen) cellulose is 8000–10 000. The cellulose molecules tend to lie parallel to one another and form strong hydrogen bonds between the partial negative charges on the oxygen atoms and the partial positive charges on the hydrogen atoms of the hydroxyl groups. Hydrogen bonding is sufficiently dominant that ordered regions are formed and the cellulose is sufficiently crystalline that it diffracts X-rays and will resist chemical attack by limiting access of reactive chemical species to the more ordered regions. These slender bundles of cellulose molecules are called microfibrils and are the smallest building blocks of the fibre cell wall; they aggregate further to form fibrils. The fibre cell wall is made of layers (lamellae) of fibrils (see Chapter 10). All dry fibres contain a central empty core (the lumen) surrounded by several lamellae. On drying, the cells collapse, often flattening, here seen as flattened ribbon shapes (see Figure 3.2). When undamaged, paper fibres are long and thin with little surface roughness.

Wood and grasses contain organic materials other than cellulose, in particular lignin which is a large three-dimensional polymer made of building blocks containing rings of six aromatic carbon atoms called phenyl-propanes. The typical structures of phenyl-propanes in

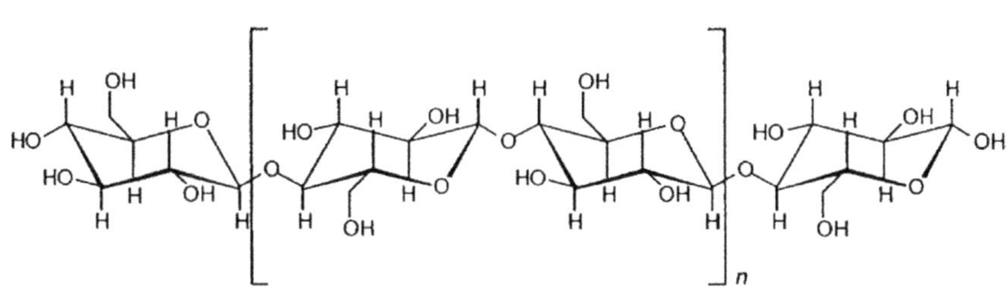

Figure 3.1 The cellulose molecule.
Reproduced from ref. 1 with permission from the Royal Society of Chemistry.

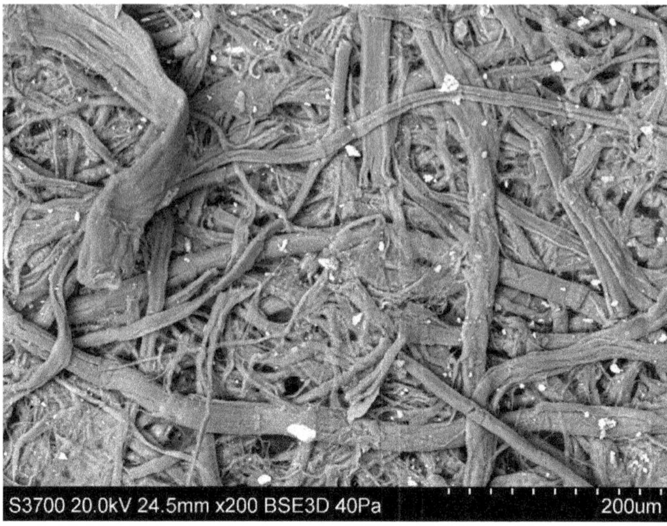

Figure 3.2 A scanning electron micrograph of the paper on a 17th century print. The white spots are probably gypsum, a few are composed of iron or calcium without sulfur. Each small division represents 20 microns.
Reproduced with permission from Nigel Meeks, Department of Scientific Research of the British Museum.

grasses and coniferous and deciduous woods are slightly different from one another.

Lignin is the stiffening material that gives wood its characteristic mechanical properties. Individual fibres are stiffened by lignin which occurs mostly in the outer cell wall. Lignin is hydrophobic and, if still present in the paper fibres, inhibits them from interacting with water and forming strong hydrogen bonds with adjacent fibres. Lignin is easily oxidised compared with cellulose, forming yellow oxidation products and acidic compounds which accelerate cellulose degradation. Wood cellulose is intimately associated with lignin in natural products, so vigorous physical and chemical treatments are needed to remove the lignin which can cause considerable damage to the cellulose and reduce the initial DP.

Wood can be ground to a powder and with little extra treatment can be used to make a type of paper. These papers have low strength as there are relatively few free fibres and the opportunities for hydrogen bonding between them is limited. These "groundwood" or "mechanical wood pulp" papers are inexpensive to make and are used in newspaper and cheap paper-back books. Such papers used to be made with acidic alum-rosin size and have long had a reputation for becoming discoloured and weak on ageing; however, changes in production methods have meant that even these cheapest papers are now more permanent.

Hydrogen bonding between cellulose fibres can be increased by beating, which is a process in which the fibres undergo mechanical forces so that they delaminate and the newly separated fibrils still remain attached to the fibre; this enhances the flexibility of the paper and provides opportunities for enhanced hydrogen bonding and physical entanglement between the fibres. The beating machinery can be as simple as mallets which hit the fibres or, on a larger scale, industrial machines which subject the fibres to strong shearing forces between moving metal blades.

Papermaking fibres may contain carbohydrate polymers other than those which are polymers of glucose. There are many other sugars in plants which can polymerise to form carbohydrates and these can become included in plant cells from which fibres are made. Carbohydrates similar to cellulose, but made from sugars other than glucose, are called hemicelluloses. Often the hemicellulose is named after the principal sugar that they contain, *e.g.* xylan, a hemicellulose in wood, is named after the sugar xylose. These polymers may be branched, unlike cellulose molecules. With their branching and their mixture of sugar monomers, hemicellulose molecules cannot pack closely together and unlike cellulose do not form ordered regions in fibres. Hemicelluloses are more reactive than cellulose because of the reactivity of some of the sugars they contain and because they have no crystallinity; if desired they can be dissolved from fibres using alkalis.

3.3 Paper Forming

Single sheets of paper are made by producing a suspension of cellulose-containing fibres in aqueous solution which is then drained through a fine mesh (the mould). Some definitions of paper demand that the fibres be dispersed in water before the paper sheet is formed. This definition excludes materials such as papyrus and barkcloth. The damp paper, formed of retained fibres and any other particulates, is next removed from the mould and allowed to dry. On an industrial scale, machine-made paper is made by allowing the fibre suspension to flow onto a moving mesh (the web) thus forming a continuous strip of paper which may be coated, dried and pressed on the machine before being made into a roll for storage. Two of the skills of papermaking are to form the suspension of fibres in the first place and then stop the fibres from becoming entangled before they settle evenly on the web. When clumps of fibres do occur on paper, they are called knots. Modern papermakers can use synthetic suspending agents to avoid knots, but papermakers who wish to use more traditional methods may use alum; the aluminium ions are attracted to the fibres which acquire a positive charge and become mutually repulsive. Mucilages made from seaweed are used in traditional Oriental papermaking.

In most processes for the manufacture of paper, the fibres may align themselves in a preferred direction on the web during the separation of fibres from the water. In machine-made paper, the orientation is particularly marked, as the fibres become oriented in the direction in which the web is moving, the machine direction (MD). The MD of a piece of paper is of particular importance in determining its working properties. The direction across the paper at right angles to the MD is called the cross direction (CD). Additionally, a paper usually has differing properties in the thickness of the paper (ZD). When paper is made from pulp the smallest particles will penetrate furthest into the paper as the water drains through it; thus the upper and lower surfaces may have particles of different size distributions. The lower surface may also pick up the texture of the web. Paper conservators and artists are particularly aware of the surface textures which occur on the different sides of a sheet of paper. Artists' papers often have a preferred front side.

A simple experiment can be performed to determine MD. Take a blank sheet of paper of little value and cut it into a square, keeping the sides of the square parallel to the original edges of the paper sheet. Bending the paper into a U-shape first along then across the sheet will usually show that the paper bends more easily one way than the other. The direction made by the "valley" of easiest bending is the MD. Paper will also tear more easily in the MD. Gently bending the corner of the pages of this book will probably show the MD is parallel to the spine; this enables the pages to be turned more readily. There are some other MD-related properties of paper which will be described later.

As previously indicated, there are two types of cellulose-based writing surface which may not be included in everyone's definition of paper. Although paper was first made in China in about 105 AD, the Ancient Egyptians made a type of cellulose-based writing surface from about 3000 BCE. The papyrus plant (*Cyperus papyrus*) has tall stems approximately triangular in cross-section in which there is a high concentration of cellulose fibres in a cream-coloured spongy matrix. Once the tough green outside layer is removed, thin strips of papyrus a few centimetres wide are cut and laid side-by-side to form a rectangular sheet. More strips are laid at right angles to these and the sheet is then pressed and dried. Another material which is similar to, but not paper, is barkcloth (also known as *tapa* or *kapa*) originating in the Pacific islands. Barkcloth is made from the cellulose fibres that are found beneath the bark of certain trees and shrubs (inner bark). The fibres are first retted and then beaten with grooved mallets to further soften them and form sheets of matted fibres. While still wet the sheets can be folded and beaten further to spread and reform the sheets. Barkcloth can be made into sheets for writing on or made into clothing.

3.4 Paper Size

The filter paper typically found in the chemical laboratory is made of pure cellulose fibres, but it is of little use as a writing or printing paper as it is too absorbent for inks. Generally, paper is sized to make it more hydrophobic (water repellent). Sizing also changes the surface texture and increases both the wet and dry strengths of the paper. Traditionally, a hot solution of gelatine was brushed onto the paper in its original unsized state (water leaf); this is called tub sizing. Papermaker's alum (aluminium sulphate) was sometimes added to the gelatine. Chemists will realise that this is a misnomer as aluminium sulphate is not a true alum; however, its chemical properties are similar to aluminium potassium sulphate, which is a true alum and also used by papermakers. Papermakers say that the use of alum is to "harden the size" but it is possibly of as much value as a biocide to help prevent the gelatine solution from developing microbial colonies while it is being stored. A popular sizing agent from the mid-19th to the mid-20th century was alum-rosin size. Rosin is a natural product derived from coniferous trees and is chemically a mixture of rosin acids. It was mainly used industrially in conjunction with alum to produce a size which produces optimum results on mechanical wood pulp papers. In water the positively charged trivalent aluminium ion is attracted to the negatively charged paper fibre surface and there forms a link to the negatively charged carboxylic acid groups of the rosin acid. There are many ways in which rosin can be used as a paper size and several theories on how they work.

Rosin-based sizes are not so effective on alkaline papers and are prone to yellowing. Modern papers are often sized with an alkyl ketene dimer that reacts with hydroxyl groups on the cellulose, and the large alkyl group imparts water repellency (see Figure 3.3). The product is a white solid that is emulsified in water for application to the paper. This type of sizing agent is mixed with the cellulose pulp rather than coated onto the paper and is called an internal size (as opposed to a surface size). Another type of internal size is based on alkyl succinic acids.

Papers often contain internal fillers, which may be polymers or minerals, or have coatings to provide a good surface for printing, to texture them or provide a layer of adhesive. Thick sheets of paper may be formed by casting a thick layer of fibres on the web or by laminating thinner sheets together. Some papers are dyed, waterproofed, fireproofed or perfumed. Paper is the most common support for photographs and can be made into books and magazines or three-dimensional structures such as boxes. Paper with painting, drawing or printing on it is much more complicated than a plain sheet of paper and any ageing or conservation process performed on a piece of paper will be influenced by the media on the surface and *vice versa*. These examples illustrate that the subject of paper conservation is very wide-ranging and that it is beyond the scope of a single chapter to attempt a comprehensive overview of all the materials of interest to the paper conservator/scientist.

3.5 Analytical Techniques Appropriate to Paper

Paper items in collections may have had chemical conservation treatments applied to them and applied dyes, pigments and adhesives may slow down or accelerate deterioration; these are of great interest to the paper conservation scientist who may wish to identify them. However, this section will concentrate on types of analysis of interest for the paper itself. Sometimes large amounts of expendable material are available for analysis, while at other times the researcher will be studying valuable objects which cannot be contaminated or sampled. In the latter case, the requirement is that analysis should be non-destructive or at least use the very minimum of a sample. The average

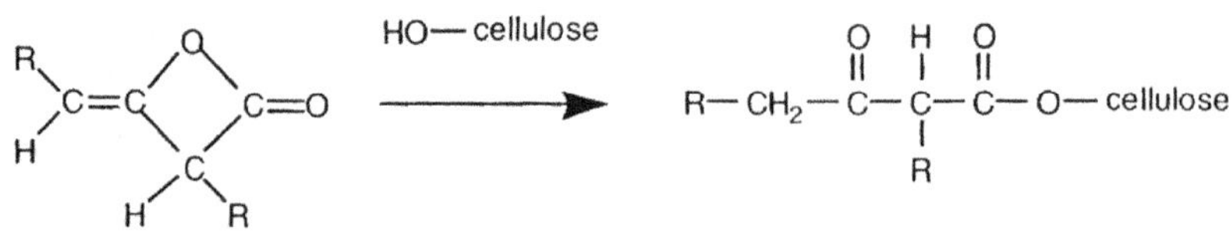

Figure 3.3 An alkyl ketene dimer.
Reproduced from ref. 1 with permission from the Royal Society of Chemistry.

paper conservator will not have access to sophisticated analytical instrumentation; however, there are still some types of analysis which can be performed, and both physical and chemical characterisation can be attempted. In all analytical work, sampling is important as a paper object is rarely homogenous. In an old book, for example, the edges of a book block are often more degraded than the centre because the edge has absorbed air pollutants and is often more degraded and of higher acidity; it is thus both physically and chemically inhomogeneous. A picture in a frame may have had its front surface exposed to light for many years while the back will have been subjected to less light exposure but had greater proximity to compounds off-gassing from the mounting materials.

"The pH of paper" is of interest as it is a major influence on the stability of paper on ageing. Although it is a phrase in common usage, it is a trifle misleading to talk about the "pH of paper", as the pH of a dry piece of paper can only be determined by extracting soluble acid and measuring the pH of the water extract (generally produced by macerating 1 g of paper in 50 ml of pure water).

The pH is defined as $pH = -\log_{10}[H^+]$ where $[H^+]$ is the hydrogen ion concentration. On this scale, pH 7 is neutral, pH 1 is extremely acidic and pH 14 extremely alkaline, with each pH decrease of 1.0 representing a 10-fold increase in hydrogen ion concentration. For example, there is ten times more free acid at pH 4 than pH 5. The pH is generally measured using a glass electrode coupled to a pH meter, the electrode being designed to dip into the water extract. With the aim of minimising the damage to an object, special electrodes have been designed for measuring very small amounts of extract, or that have a flat sensing area designed to be placed on the dampened surface of the paper. A pH lower than about 4.5 is generally considered too low for the long-term safety of the paper.

Many national or international standards describing paper to be used for conservation purposes specify that the papers should be free of alum and lignin. Several spot tests can be used to detect these and other additives of interest to be found in paper. One test for alum uses a solution of aluminon. The pale pink solution is applied to the surface of the paper and allowed to dry. A pale pink spot is produced when no aluminium salts are present; a darker pink or red is seen when such salts are present. Lignin can be detected using a freshly prepared solution of phloroglucinol in hydrochloric acid. A drop of the colourless solution is applied to the paper and a red colouration indicates the presence of lignin. There are several tests for rosin. In one of these, the Raspail test, a drop of concentrated sugar solution is added to the paper, blotted off after 1–3 minutes and a drop of concentrated sulfuric acid is added. Rosin gives a red colour. Paper strips impregnated with 1,10-bathophrenantholein are commercially available to test for ferrous ions in the ink or the adjacent paper; a red stain is produced when the ferrous ions react with the paper.

Two types of traditional sizes, namely starch and glue (protein), can be tested for, and the tests are also used for the identification of adhesives on paper artefacts. Starch can be detected by placing a drop of iodine solution onto the area to be tested; a blue-black colouration reveals the presence of starch. There are many tests for protein, but in one of these a solution of ninhydrin in methyl cellosolve (2-methoxyethanol) with a trace of surfactant is added to the paper which is then heated in an oven at 100 °C. A purplish stain indicates the presence of protein.

The polarised light microscope is an extremely useful piece of equipment for a paper conservator as it is not only possible to identify many inorganic pigments on and in the paper but also the fibre types used for paper manufacture. For paper fibre identification a few fibres can be scraped off a piece of paper and dispersed in water, although special treatments may be needed if the paper has been treated to enhance wet strength or it is coated. If placed on a microscope slide the dispersed fibres in water can be examined under a coverslip, or they may be dried and mounted in a permanent mounting medium. Partial crossing of the polarising filters can enhance surface or internal details. It is relatively easy to identify some fibres by their morphology and the presence of associated cells but others are very difficult indeed, especially if exotic plants are a possible source of fibre. Identification can be aided by stains which colour the fibres of different plant species differently. Pigment identification may be similarly performed by mounting the dispersed pigment particles in a medium of refractive index 1.66. This refractive index is chosen as roughly half of the common mineral particles have a refractive index greater than 1.66 and the other half less than 1.66. A simple optical test using a phenomenon called the Becke line is used to determine this. Other optical crystallographic properties are then used to further characterise the particles.

In the well-equipped laboratory, elemental analysis of paper may be performed by X-ray fluorescence. Such laboratories may have access to fixed or portable machines and analyses may aid in the identification of inorganic components of coatings, paints and fillers in the paper. It is important to remember that many pieces of paper in a collection may have been conserved in the past, maybe more than once, introducing elements not originally present, *e.g.* calcium and magnesium compounds may have been applied as deacidification treatments. It is possible to identify iron gall ink by the presence of iron. Generally, the concentrations of metals are just too low for the analysis of metal-point drawing (*e.g.* gold or silver); however, detection limits are improving.

The scanning electron microscope can also be used for elemental analysis of pieces of paper with the advantage over X-ray fluorescence in that elements of lower atomic numbers can be detected. In the recent past objects had to be coated with a conducting layer of carbon or gold but with the advent of environmental SEMs this is no longer always necessary and as a large sample chamber can accommodate a piece of paper up to 200 mm diameter; this opens up the theoretical possibility of entire objects being studied without the need for them to become contaminated. The effects of placing paper in a near-vacuum have not been studied but may not be serious especially if an acceptable relative humidity can be maintained. However, it is probable that the electron bombardment necessary for imaging and analysis could be damaging to the small area involved.

It is possible for X-ray diffraction to produce a pattern produced by the crystallinity of paper cellulose but unfortunately the information has little diagnostic value. However, the identity of fillers and other deposited crystalline materials can be determined by this method. Samples need to be removed from the objects unless the objects themselves are small (*e.g.* postage stamps), when the entire object may be analysed, but this depends on the type of equipment used.

Fourier transform infrared (FTIR) spectroscopy instruments are frequently found in the major heritage science laboratories; paper produces a spectrum with many peaks, each peak corresponding to vibrations in a particular chemical bond. The frequency of infrared peaks is usually measured in wavenumbers, with the unit cm^{-1}, the number of wavelengths which can be fitted into a centimetre. Peaks can often be attributed to types of chemical bond; for example, a peak at 2900 cm^{-1} may correspond to the stretching of C–H bonds, while a peak at 1740 cm^{-1} may correspond to a stretching of a carbonyl group. As the strength of the C–H peak remains more or less constant during the ageing of paper, the ratio of the strength of the carbonyl peak with the C–H peak might be used to follow the oxidation of the paper. The ratio of the peaks at 1372 cm^{-1} (C–H) with that at 2900 cm^{-1} has been used to monitor the crystallinity of the cellulose.

The FTIR spectrum of a piece of paper can be measured non-destructively using an FTIR-microscope or, probably more reproducibly, by taking a sample and either placing it directly in a diamond cell or placing it in an attenuated total reflectance (ATR) accessory. In addition to the cellulose in the paper, it is also possible to identify organic or inorganic fillers and coatings.

The remainder of the section is devoted to methods used principally for the evaluation of the state of preservation of paper. The strength of paper decreases when paper ages and it may discolour. The weakening of paper is caused by oxidative damage and acid-catalysed hydrolysis of the cellulose; both of these degradation mechanisms can cause cellulose chain scission and reduction of DP. These tests are used in industrial paper laboratories for assessing paper pulp during the manufacture of paper and in museum laboratories for evaluating the short and long-term properties of paper conservation techniques.

The solubility of paper carbohydrates in sodium hydroxide (NaOH) of various strengths is a useful method for characterising new or old papers. Pure cellulose of long chain length will dissolve very little in strong alkalis but as the amount of alteration by ageing increases so does the solubility. One method uses a 17.5% solution of NaOH to treat paper so that, of the carbohydrate component, only cellulose of $DP > 200$ remains undissolved, and lignin and mineral fillers remain largely unaffected. The insoluble carbohydrate is called α-cellulose which will decrease on ageing and degradation. After the 17.5% NaOH treatment the solution contains β- and γ-cellulose. The β-cellulose, which has a DP of 14–200, is soluble until the alkali is neutralised, while the γ-cellulose remains in solution after neutralisation; it has a DP of less than 10. Materials other than paper carbohydrates, *e.g.* lignin, interfere with the results.

A related method involves measuring the solubility of the paper in 1% NaOH at 100 °C. The solubility is proportional to the copper number (see below) that is itself proportional to the number of carbonyl groups in the paper, a measure of the oxidation and hydrolysis of the paper. This method has also been used on lignin-containing fibres of various kinds with some success.

The copper number of paper may be obtained by reacting it with an alkaline copper(II) solution, some of which is reduced to Cu(I) oxide as the carbonyl groups are oxidised to carboxylic acid. There are several variants of the test, one of which involves treating the formed Cu(I) oxide with phosphomolybdic acid. The reduced molybdenum is then titrated with potassium permanganate.

Phenylhydrazine is able to react with carbonyl groups to give coloured compounds. Paper sheets can be reacted with this reagent or modifications of it (*e.g.* 4-nitrophenylhydrazine) and the colour produced can be measured by reflectance spectroscopy. The colour may also be measured by dissolving the paper and measuring the absorbance of the solution. Lignin interferes badly with this test.

A test that measures the quantity of carboxyl groups uses methylene blue absorption. A sample of the paper is immersed in a solution of methylene blue and the absorbance of the dye solution at 620 nm is measured before and after immersion of the paper. The amount of sample needed to produce 50% exhaustion of the dye solution is used to calculate the carboxyl content.

Viscometry can be used to assess the DP of cellulose. The most commonly used solutions for dissolving cellulose contain copper ethylenediamine or cuprammonium hydroxide. These reagents all cause some damage to the cellulose by alkaline degradation during dissolution. Some workers perform a pre-treatment in which oxidised groups are reduced, which lessens the amount of this degradation. Measurements of cellulose solutions in the copper-containing reagents should ideally be performed in the absence of air as oxygen can cause degradation of the cellulose. The higher the DP of the cellulose, the greater is the viscosity of the solution (all other things being equal).

An alternative method for the determination of DP is size exclusion chromatography (SEC). LiCl/dimethyl acetamide can be used to dissolve the cellulose. The size exclusion column holds back larger molecules and when the cellulose emerges from the column, the smallest molecules come out first. A detector that measures the amount of material emerging from the column enables the user to see not only the average molecular weight but also the way it is distributed about the mean. Instruments need to be calibrated with solutions containing carbohydrates or other polymers of known DP.

The strength properties of paper are very important as some papers need to be handled and can easily become damaged. Some papers become so brittle that they cannot be handled at all and digitisation may be the only practical course. The measurement of strength properties is thus very important for the paper conservation scientist. Many conservation treatments produce an immediate effect on the strength of the paper, *e.g.* aqueous treatments, consolidations or lining techniques, but the researcher may also be interested in the long-term effects. More than one mechanical property may need to be measured, *e.g.* consolidation may produce a desirable increase in tensile strength but an undesirable decrease in flexibility. Recently the concept of "damage functions" has been proposed in which several properties are combined to reflect the use of the object. If the item is a book then the ability to turn the pages without the paper breaking may be combined with discolouration, with the former given more importance; however, in a work of art the discolouration would be more important than folding strength. The best single mechanical test would be the one in which the properties most important to the use of the paper were tested.

If folding is of paramount importance, the MIT folding endurance test is of great value. In this test, the test strip is held vertically under tension. The upper jaw that holds the paper is fixed but the lower jaw repeatedly rotates back and forth by a predetermined angle, with the centre of rotation being the point where the paper strip enters the jaws. The tips of the jaws are curved in a defined manner. A counter measures the number of folds required before the paper breaks. A strong paper will endure over a hundred folds but a very weak paper will only endure a few folds, if any, and for these the test is less suitable. For extremely brittle papers there is a test which bends papers over a set of rods of decreasing diameter.

Tensile strength is one of the commonly measured mechanical properties. One of the reasons for this is that tensile testing machines are very versatile and used for several types of conservation research work; they can be used to pull apart, and thereby measure the strength, of yarns, adhesive joints, plastics, *etc.* and also be adapted to crush materials such as stone. In the tensile testing of paper, a strip is clamped between two jaws and is pulled apart as the separation between the jaws increases. A load cell measures the stress on the paper. As the paper stretches, the stress increases until the paper ceases to stretch elastically and fibres start to slip over one another, or the paper starts to tear, in which case the rate of increase of stress will decrease. The graph of stress against elongation or strain can be used to calculate several useful parameters relating to the strength and elasticity of the paper.

When paper breaks under tension, the fibres themselves and the bonds between the fibres may break. In strong papers, the paper breaks mostly by failure of the fibre-to-fibre bonds, while in brittle papers the break occurs by failure of the fibres themselves. If the jaws holding the paper strip are very close together, the fibre strength dominates the results obtained as the contribution from fibre-to-fibre bonding is minimised. This is called zero span tensile testing.

In all the mechanical tests described in this section, relative humidity (RH) must be controlled carefully as the properties measured change significantly. International and national standards will specify a required RH and RH history for samples to be tested. As there is a hysteresis effect with water content and RH of paper, it may be important that the desired RH for testing is approached from one of either the dry side or the damp side [see Section 3.7]. Different results may be obtained if properties are measured along the CD or MD, but also some properties, *e.g.* gloss, colour, roughness, *etc.* may differ from one side of the sheet to the other. However, a major problem with all the mechanical tests is the variability of the results. Not only is there non-uniformity in the paper, but test results which involve breakage of the sample have a scatter in values which is larger than desirable because of the failure mechanism. Because of this scatter, a large number of samples have to be tested to obtain a result that is statistically useful. The smaller the difference needed to be detected the larger the sample size needs to be. Testing using 15–30 strips is usual. This consumes a large amount of sample and this type of test is not so useful for inherently irregular papers or those in short supply such as those from valuable objects.

Colour change caused by natural ageing in the absence of light may produce discolouration of initially "white" paper to noticeably yellow/brown. Ageing principally caused by light exposure may also produce this type of discolouration but may instead produce bleaching (see also Section 3.6). Dyes and pigments in the paper usually bleach when exposed to light. When measuring colour change, spectroscopy can be used to measure the entire visible reflectance spectrum of the paper, but often the reflectance at a single wavelength near the blue/violet end of the spectrum is used; 400 and 457 nm have been used in recent research. Another frequently used method for recording discolouration is to use a tristimulus colour measuring meter which records the hue as three parameters. Many systems for colour measurement have been devised, at present the most used is the CIELab (1976) system. Total colour change (ΔE) or a change in b^* (which represents yellowness) have been used when studying discolouration (see Chapter 2). There is also a newer CIE 2000 system.

3.6 Accelerated Ageing Tests

An important part of a scientist's work is not only assessing whether the paper is in a worse or better state after a conservation treatment, but also predicting whether this will still be the case after many years. As well as predicting the effects of ageing on paper, conservation materials (*e.g.* adhesives, packaging, *etc.*) need to be aged to see if they will develop undesirable characteristics on ageing (*e.g.* discolouration, brittleness, shrinkage, acidity, *etc.*).

Accelerated ageing tests work partly by increasing the energy in the material, and this can be done by either increasing the amount of light falling on an object or by increasing its temperature. Accelerated ageing using light should only be used to anticipate the effects of light on paper and not for the effects of ageing generally. The energy of photons in light transfers energy to paper in a different manner and intensity to the thermal energy of heat ageing and the results of one type of testing may be entirely different from the other. A paper that yellows in thermal ageing may bleach in light ageing. The vast majority of paper in libraries, archives and museums is kept in conditions of total darkness or in low incident light that contains little or no ultraviolet light. In such conditions light degradation is of little or no importance. However, objects on display, or which form part of a historical room, may receive sufficient light to cause serious concerns about degradation. A few dyes and pigments are extremely sensitive to light and objects containing such materials may need to have a ban on being exhibited at all.

Accelerated ageing may be carried out at various temperatures and relative humidities which may either be static or fluctuating. Barrow (see Section 3.8) used ageing in an oven for 72 hours at 100 °C to replicate the effects of natural ageing of paper for approximately 25 years. Nowadays, these ageing conditions are still sometimes used but the 25 year equivalence would probably not be claimed and certainly not for all types of paper. Many laboratories prefer to perform their paper ageing in an oven at a controlled RH and/or at a lower temperature. As a general rule, the closer the conditions of accelerated ageing are to ambient storage conditions, the better the results will be. The only drawback in ageing at lower temperatures is that the time needed for accelerated ageing is increased. There is a remarkable lack of conformity in the ageing conditions used by individual laboratories.

Fluctuating RH can be bad for paper because of loss of dimensional stability of the paper, *i.e.* production of cockling. However, recent work has also shown that loss of tensile strength can also result from large fluctuations in RH; this promises to be an interesting field of research in the future. Elevated temperatures combined with fluctuating RH have been used in the accelerated ageing of iron gall ink on paper.

Recent research from several independent laboratories has shown that the volatile degradation products from paper ageing appear to accelerate ageing. Accordingly, some scientists age paper in sealed glass tubes which not only retain degradation products but keep the water content of the paper approximately constant.

3.7 How Paper Interacts with Water

The interaction of paper with water is the single most important property of paper for conservators, so much so that an entire book has recently been published on this subject, and an example is given in the case study (Section 3.17).

Cellulose will reversibly absorb or release water vapour as the RH is increased or decreased. Figure 3.4 shows a typical absorption/desorption curve for cellulose. Note that the two curves are not the same, as the desorption curve lags relative to the absorption curve; this is called hysteresis.

When water is absorbed, the dimensions of the fibres change. As the general orientation of the cellulose molecules is along the fibre, most of the expansion occurs sideways. The increase in size is proportional to the water content, thus a graph of the fibre diameter *versus* RH has the same shape as the graph in Figure 3.4 and hysteresis is again exhibited. Individual fibres can expand in width up to 20% when going from 0 to 100% RH but lengthen perhaps about only one-tenth of that percentage. The paper does not expand up to 20% because the fibre-expansion effect is attenuated by the microstructure of the paper. Typical maximum expansion of paper is about 2–3%. As most fibres are aligned along the MD, the hygroexpansivity of paper (produced by the RH changes) is mostly in the CD. This may be demonstrated on a piece

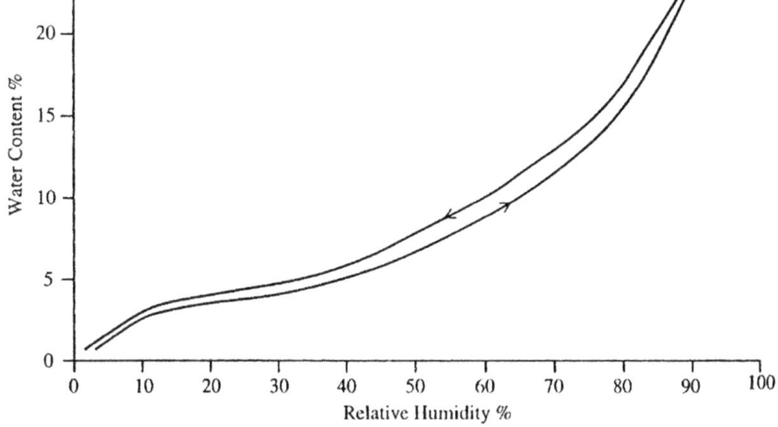

Figure 3.4 The absorption and desorption of water vapour by cellulose fibres. Reproduced from ref. 1 with permission from the Royal Society of Chemistry.

of paper; when wetting one side of a small square, the paper will form a U-shape along the MD as the fibres on the wetted side expand sideways while those on the dry side will not.

Recently-made paper may contain dried-in stress introduced during the drying of the paper. The stress will be removed if the paper goes to a sufficiently high RH during storage, when the paper may shrink irreversibly. Old papers do not appear to have dried-in stress – possibly this has been relieved during repeated small RH fluctuations over many years.

Uneven dimensional changes have serious consequences for the paper conservator. If a piece of paper expands on one edge while an adjacent area remains unchanged, then there will be a connecting area where there is stress. In areas of sufficiently high stress, the paper may undergo irreversible deformation with resulting loss of flatness. The production of such resulting wavy surfaces is called "cockling". The phenomenon is often seen on damp-damaged books where water has gained local access and the paper cannot expand freely as it is confined by other pages and the binding. The paper conservator has only to join two pieces of paper together to generate the potential for cockling, as the two pieces will have differing hygroexpansivities and one or both may have dried-in stress. The conservator may take care to use similar papers that have been stored for some time, align the MDs and avoid getting the paper too wet if using a water-based adhesive. Figure 3.5 shows an example of cockling; raking light illumination emphasises the irregularities in the paper surface.

3.8 Deterioration of Paper

Paper objects may suffer from damage by vermin, insect and microbial attack, theft, fire and flood. However, from a chemist's point of view there are two important mechanisms by which paper deteriorates, both of which may change its hue and make it weaker; these are acid-catalysed hydrolysis and oxidation. Papers vary a lot in their stability and the worst can become so brittle that they become unusable.

3.8.1 Acid-catalysed Hydrolysis of Cellulose

The first comprehensive and systematic work on the reasons for the deterioration of paper was performed by W. J. Barrow in the 1960s who firmly established the link between the acidity of paper and its permanence.

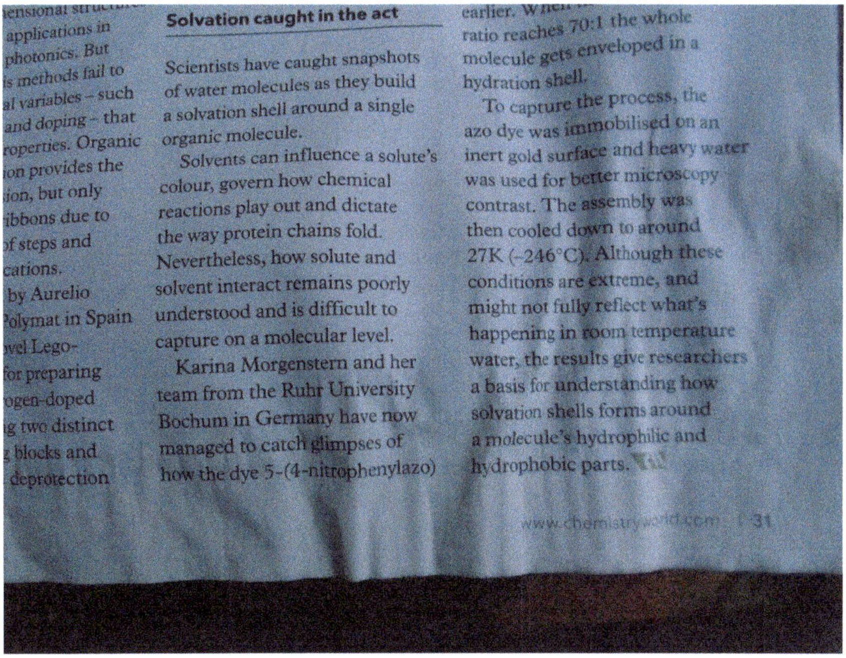

ensional structu…
applications in
photonics. But
is methods fail to
al variables – such
and doping – that
roperties. Organic
ion provides the
ion, but only
ibbons due to
of steps and
cations.
by Aurelio
Polymat in Spain
vel Lego-
for preparing
ogen-doped
g two distinct
g blocks and
deprotection

Solvation caught in the act

Scientists have caught snapshots of water molecules as they build a solvation shell around a single organic molecule.

Solvents can influence a solute's colour, govern how chemical reactions play out and dictate the way protein chains fold. Nevertheless, how solute and solvent interact remains poorly understood and is difficult to capture on a molecular level.

Karina Morgenstern and her team from the Ruhr University Bochum in Germany have now managed to catch glimpses of how the dye 5-(4-nitrophenylazo)

earlier. When … ratio reaches 70:1 the whole molecule gets enveloped in a hydration shell.

To capture the process, the azo dye was immobilised on an inert gold surface and heavy water was used for better microscopy contrast. The assembly was then cooled down to around 27K (–246°C). Although these conditions are extreme, and might not fully reflect what's happening in room temperature water, the results give researchers a basis for understanding how solvation shells forms around a molecule's hydrophilic and hydrophobic parts.

www.chemistryworld.com | 31

Figure 3.5 Cockling.

Barrow measured the pH and strength of many old pieces of paper and found that alkaline papers were generally strong but acidic papers were generally weak. There are many ways of measuring paper strength but Barrow used a folding endurance tester which measured how many times a strip of paper could be folded under tension before it broke, as mentioned in Section 3.5.

Barrow's other approach to the research was to obtain pieces of paper of measured pH and age them artificially, allowing decreases in strength to be followed. It was considered that storage at 100 °C for 3 days reproduced natural ageing of 25 years. The results reinforced the relationship between initial acidity and loss of strength on ageing.

The loss of strength of paper produced by acidity results from acid-catalysed hydrolysis (scission) of the cellulose chains. At ambient RH, cellulose contains several per cent of water that may participate in hydrolysis reactions. The point of scission is the link between the glucose (pyranose) rings. Loss of DP does not produce an effect on the loss of strength immediately. It has been proposed that the DP has to reach a critical point before a loss of strength is manifested. Cellulose fibres of an initially high DP, such as natural cellulose from cotton or linen, thus have a strength advantage over other types with an initially lower DP such as chemically purified wood celluloses. Loss of strength occurs mostly in amorphous areas as water and acids have difficulty in gaining access to crystalline areas. Paper can start life acidic because of the method of manufacture and its ingredients, or it may absorb acidic air pollutants such as sulphur dioxide and nitrogen oxides. Sulphur dioxide is believed to form sulphuric acid in the paper. Papermaker's alum produces an acidic solution; it can be thought to be a salt of a weak base (aluminium hydroxide) and a strong acid (sulphuric acid). Recent research has demonstrated that the deleterious effect of alum can be attenuated by the use of a gelatine size. Rosin, used in some kinds of size, contains rosin acids and can oxidise to form coloured compounds.

Alkaline degradation is possible but occurs much less frequently than acid degradation. Degradation in alkaline conditions is mostly of interest because of the deterioration which might occur during papermaking, conservation or analytical procedures, rather than during natural ageing.

3.8.2 Oxidation of Cellulose

Paper exists in air that contains about 20% oxygen. Organic materials are capable of reacting slowly with oxygen in ambient conditions (autoxidation). Such oxidation reactions involve intermediates called free radicals. In a simple compound A–B, the single bond between atoms A and B contains two shared electrons, if both go to B then two ions are produced: A^+ and B^-; however, if the electrons in the bond are shared between A and B then two free radicals, $A^\bullet$ and $B^\bullet$, are produced because each now has a free unpaired electron. Such free radicals can participate in chain reactions in which the reaction of a free radical with a neutral molecule creates a product and another free radical (propagation). Such autoxidation reactions could include:

Initiation:

$$RH + O_2 \rightarrow R^\bullet + HOO^\bullet$$

Propagation:

$$R^\bullet + O_2 \rightarrow ROO^\bullet$$

Propagation:

$$ROO^\bullet + RH \rightarrow ROOH + R^\bullet$$

(where RH represents an organic compound)

Autoxidation is slow in pure cellulose and it may be considered to take place at a negligible rate; however, certain conditions will initiate and accelerate free radical reactions. If light is absorbed by cellulose or a material associated with cellulose (*e.g.* a dye or lignin), the absorbing molecule's energy increases greatly. Many things may happen to this energy but one is that the energy can be transferred to another molecule, *e.g.* when lignin transfers its energy to cellulose; this is called photosensitisation of the cellulose by the lignin. The energetic cellulose could split open one of its chemical bonds producing two free radicals, or the energy may be transferred to an oxygen atom which, then being in a highly energetic state (*e.g.* singlet oxygen), can oxidise the cellulose. Ultraviolet light is generally more harmful as its photons are more energetic than those of longer wavelengths.

Metal ions that are capable of changing their valency; in particular, ions of copper and iron are powerful oxidation catalysts as they can initiate oxidation reactions. During a valency change, negatively charged electrons (e^-) are donated or accepted which then initiate a free radical chain reaction.

$$Cu^+ \rightarrow Cu^{2+} + e^-$$

$$Fe^{3+} + e^- \rightarrow Fe^{2+}$$

Certain pigments and inks, *e.g.* copper acetate pigments and iron gall ink, are notorious for their ability to induce metal-catalysed degradation of paper, and damage can be so severe that the area covered by colourant falls out of the sheet and is lost. During industrial oxidative bleaching of paper pulp, trace amounts of iron and copper can cause extensive damage to the cellulose and measures are taken to minimise this.

3.9 Deacidification

Much of the acidity in paper can be washed out using water; however, not all the acidity can be washed out of oxidised paper as some of the cellulose will have produced carboxylic acid groups on the cellulose chain. During deacidification treatments, many conservators will wish that some of the introduced alkaline material to neutralise acids is not washed out by water and remains behind to neutralise any acidity that may develop in the future. Deacidification is a process which was once keenly practised by most conservators but in recent decades it has fallen out of favour as conservators practise "minimal intervention", relying on good storage conditions to prolong the life of the paper.

Barrow's examinations of stable, old papers found that most had a significant calcium or magnesium carbonate content. Today, some of the best aqueous deacidification treatments introduce these compounds. There are several popular materials used for aqueous deacidification, one of these is calcium hydroxide solution (lime water). Calcium hydroxide is not very soluble in water; a saturated solution contains 1.69 $g L^{-1}$ at 20 °C and has a pH of 12.4 which is quite high. Severely degraded papers are suspected to undergo alkaline hydrolysis and some inks can bleed or change hue at such a high pH, thus the saturated solution is often diluted before use. While the paper is in the solution, the acids are neutralised. When treated paper is removed from the solution, the imbibed calcium hydroxide reacts with carbon dioxide in the air to form calcium carbonate. If the object has a very black area, *e.g.* as in some mezzotint prints, the dried deposited carbonate can leave a white bloom, however, usually it is not seen. The eventual pH of the paper is about 8.0–9.0.

If carbon dioxide is bubbled through a calcium hydroxide solution the carbonate is precipitated at first but eventually it dissolves again with the formation of calcium bicarbonate. This solution varies in pH depending on the quantity of carbon dioxide in solution but the pH is generally about 7. It can be used as a deacidifying solution when a high pH needs to be avoided.

Of historic interest is the Barrow two-stage treatment which deposits twice as much $CaCO_3$ by first immersing the paper in calcium hydroxide, then calcium bicarbonate:

$$Ca(OH)_2 + Ca(HCO_3)_2 \rightarrow 2CaCO_3 + 2H_2O$$

Barrow's method is seldom used now as the quantity of calcium carbonate deposited with the single-step processes is considered adequate.

Magnesium hydrogen carbonate may be prepared by passing carbon dioxide through a suspension of magnesium carbonate (light) in water. Many conservators make this by placing the suspension in a soda siphon and screwing on a bulb of compressed carbon dioxide. This has a neutral, if not slightly acidic, pH. The eventual product, on drying, is a form of basic magnesium carbonate.

Currently, the calcium-based treatments are preferred over those containing magnesium as some research has shown that the former may induce less yellowing on accelerated ageing, particularly on papers containing lignin. Alkaline compounds of Group I have never found general favour in deacidifying treatments as they cause yellowing on ageing, are too alkaline or are hygroscopic; however, a few workers have recommended slightly alkaline sodium salts, *e.g.* borates, as buffers.

If a paper becomes too fragile when immersed in water, or has dyes or image materials that are water soluble or otherwise fugitive, immersion in aqueous solutions cannot be used for deacidification. One option is to use one of a range of non-aqueous treatments. Calcium and magnesium are both in Group II of the periodic table. The next heaviest element in the group is barium that has a hydroxide which, unlike those for Ca and Mg, is soluble in methanol. $Ba(OH)_2$ solution was used for deacidification in the period 1970–1985 (approximately) but fell out of use as health and safety concerns about the toxicity of barium compounds became more important.

Historically, the next successful agent to be used was magnesium methoxide, $Mg(OMe)_2$, which can also be dissolved in methanol. The solution could be diluted in fluorocarbons, which have the advantage of being poor solvents, thus reducing the tendency for dyes and inks to bleed from the paper that might have occurred if pure methanol had been used as a diluent. However, the solution proved too water-sensitive for many purposes as the $Mg(OMe)_2$ reacts readily with water and precipitates out as $Mg(OH)_2$. This problem was overcome by saturating the solution with carbon dioxide. The result was a popular product called methyl magnesium carbonate, although its precise composition is uncertain. The fluorocarbon diluent has now been replaced by environmentally safer solvents.

A different approach is to use an inert organic solvent that contains micro-particles of an alkaline material such as calcium hydroxide or magnesium oxide. These treatments are suitable for single sheets where the suspension can be sprayed on or the sheet immersed. Another process uses micro-particles directed at the paper in a stream of air. Not that long ago, it was thought that this type of treatment could not work, as without liquid water the base and acid do not have an easy time making contact to bring about a neutralisation reaction. However, the processes perform well in accelerated ageing tests and we can only assume the paper contains sufficient water in ambient conditions for acid to move through fibres until it makes contact with one of the neutralising particles.

Many libraries have large holdings of books printed on very acidic paper. Deacidifying books individually is a very time-consuming and expensive process. A process in which large numbers of books could be deacidified at the same time (mass deacidification) is a desirable invention. The problem has produced much research and some novel deacidifying agents have been devised. One historically important and effective method was the use of diethyl zinc vapour but this is no longer in use. Three current treatments are propoxy-magnesium carbonate in an alcohol/fluorocarbon solvent, magnesium titanium alkoxide in a silane solvent and methyl and ethyl magnesium ethoxy carbonates in silane/alcohol solvent.

Cyclohexylamine carbonate, morpholine, amines in general and ammonia itself have all been considered as deacidifying treatments. Many are liquids with a significant vapour pressure or are gases, which makes them potentially useful for mass processes. However, all suffer from the problem that the deacidification produced is not permanent and some cause yellowing of the paper. None are currently in favour.

3.10 Bleaching and Washing of Paper

As paper ages it may become discoloured; some of this discolouration is water soluble and some is not. That which is soluble may be washed out with water but that which is not must either be tolerated and admired as part of the "patina" of the object or bleached away with an oxidising or reducing agent. Uniform discolouration is nowadays considered much more acceptable on paper than it was a few decades ago and consequently, in line with the "minimal intervention" approach less washing and very much less bleaching is now performed. If a piece of paper becomes locally wet *e.g.* during flooding, on drying the soluble discolouration tends to concentrate at the boundary where the wet area met the dry paper. Paper stored in high RH may develop foxing spots (see Section 3.11). These localised areas of discolouration are aesthetically less acceptable and prime candidates for bleaching.

There are ways to maximise the efficiency of water washing. Higher temperatures produce faster and more complete washing. The addition of a surfactant can also help, especially with a well-sized (hydrophobic) paper. Many different washing processes are performed by conservators because the image materials can be fugitive in water, *e.g.* some watercolour paints or inks. In such a case it is desirable to carry out washing while exposing the paint to the minimum possible amount of water or for the minimum time. When fugitive pigment is present, simple immersion of the paper in a bath of water may be unsuitable but a process where the paper is placed on a suction table and exposed to ultrasonically generated water mist might be much better as the principal mechanisms for removal of discolouration and the movement of particles through liquid water are different.

Paper ageing may produce double bonds in the constituent molecules; when they are present as alternating double and single bonds they are called conjugated double bonds. When there is a sufficiently long series of conjugated double bonds, they form chromophores which absorb light in the blue/violet region of the spectrum causing the paper to appear yellow/brown. Organic dyes and most discolouration in paper owe their colour to this mechanism. If some or even one of the double bonds can be destroyed during bleaching the chromophore may stop absorbing visible light as the conjugation is reduced. In paper conservation bleaching, double bonds in discoloured paper are usually eliminated by oxidising agents, or less frequently by reducing agents. If discolouration returns several months or more after bleaching, this is known as reversion. Much evidence of reversion occurring in paper that has been treated in conservation studios is anecdotal and reversion produced in the laboratory is not reliably reproducible, however the presence of carbonyl groups, produced during oxidative bleaching, is believed to be a cause of reversion.

In medieval times, undyed cellulose textiles were bleached by laying them out on fields of grass ('grassing'). The combined effect of morning dew and the sun being allowed to shine on them for several months produced bleaching. The invention of cheap ways of producing chlorine, *circa* 1774, made grassing redundant. An adaptation of the slow but effective method of grassing has been used for the last few decades for bleaching discoloured paper. The paper is immersed in water in a shallow tray and the dish is covered with a sheet of UV-absorbing Perspex and laid in strong sunlight or under a bright artificial light for several hours. Many conservators use a slightly alkaline solution containing magnesium carbonate instead of pure water. The mechanism for this type of bleaching is not fully understood. Probably the discoloured areas or other organic chromophores present absorb light and transfer the energy to the oxygenated water which then produces oxidising species which in turn destroy double bonds and hence the colour. Recent work on the effect of UVA radiation (long wave ultraviolet light) on wet Chinese papers has shown that hydrogen peroxide, singlet oxygen, hydroxyl and superoxide radicals are produced while the paper either bleaches or yellows. A problem with bleaching treatments that are based on the use of oxidising reagents is that the clean paper fibres are oxidised as well as the discoloured ones, resulting in unwanted chain scission and oxidation. The advantage of light bleaching is that as the discoloured areas are the only ones that absorb light, they are preferentially oxidised.

Many of the other conservation bleaching processes are derived from the paper industry as paper pulps often need to be bleached before being formed into paper products. Much more research has been carried out on bleaching than has ever been done by conservation scientists and the paper science literature is a good place for any reader wanting more information.

When chlorine comes into contact with either sodium, potassium or calcium hydroxide, the corresponding hypochlorite is formed. The hypochlorite ion (OCl^-) can oxidise double bonds in discoloured paper producing bleaching. Sodium and potassium hypochlorites can be purchased in an aqueous solution and are highly alkaline because excess hydroxide is present. Undiluted, the commercially available solutions are so alkaline that swelling of the cellulose and unwanted alkaline degradation may occur. Of the hypochlorites, it is usually safest to use calcium hypochlorite, also known as bleaching powder, as, in practice, it produces less alkaline conditions. Bleaching powder has to be dissolved in water prior to use.

In hypochlorite solution, three species are present which can produce bleaching: chlorine, hypochlorite ions (OCl^-) and hypochlorous acid (HOCl). In pH neutral conditions HOCl is the dominant oxidising species. This is considered the least desirable species for bleaching as it produces aldehyde groups in oxidised cellulose that may later produce yellowing on ageing. The safest bleaching occurs above pH 9.5–10.0. When the hypochlorite ion is the dominant bleaching species, the damage to cellulose is least and the product of oxidation is the carboxyl group that does not cause colour reversion on ageing. When bleaching an initially acidic paper, care should be taken that the paper does not alter the pH of the bleaching solution outside the desired range. Burgess gives details of bleaching procedures. Chloramine-T is a compound which produces small concentrations of hypochlorites when added to water. It was fashionable at one time for paper bleaching but fell out of favour when it was discovered that aluminium ions in the paper could cause it to be retained by the paper during subsequent washing. This is not a great problem as the bleach can be neutralised with a reducing agent. Chloramine-T is domestically available in aquarium shops. It is sometimes seen being sold for localised treatment of foxing spots on paper, however, if it is not washed out it can cause extensive degradation of the paper and holes can appear. Residual chloramine-T eventually turns into *p*-toluene sulphonamide which can sometimes be seen on unwashed, treated objects as white crystals.

Hydrogen peroxide is a versatile and popular bleach in both textile and paper conservation. Research in the paper industry has shown that although hydrogen peroxide is a good bleach, higher levels of cellulose degradation can occur when there are traces of metal contaminants, *e.g.* iron and copper. Acidic hydrogen peroxide decomposes evolving oxygen bubbles; however, this effect is much reduced in alkaline conditions. Hydrogen peroxide in the range 0.5–3% can be made alkaline with a few drops of ammonia solution, although a fully buffered and stabilised solution may also be used, which is preferred by some. Such recommended solutions for bleaching at pH 8–9 may be quite complex containing sodium phosphate, boric acid or sodium carbonate for pH 8, 9 or 10 baths respectively.

Hydrogen peroxide is mostly used in aqueous solution, but it can also dissolve in diethyl ether. If hydrogen peroxide in aqueous solution is placed in contact with diethyl ether, the ether layer floats on top because it is immiscible and less dense. On shaking the mixture some of the hydrogen peroxide goes into the ether. The pigments white lead (basic lead carbonate) and red lead may react with hydrogen sulphide in ambient air and blacken due to the formation of a thin surface layer of lead sulphide. The sulphide can be oxidised by applying the ethereal solution of hydrogen peroxide when the lead sulphide is oxidised to lead sulphate and the original appearance is regained. The desired reaction occurs with both ethereal and aqueous hydrogen peroxide but the ethereal solution has the advantage that no cockling, swelling of paint or localised redistribution of water-soluble materials occurs.

Sodium borohydride is the most widely used of the reducing bleaches. Short treatments can reduce aldehyde and carbonyl groups in the cellulose-type molecules back to hydroxyl groups and cause bleaching and stabilisation against future degradation. Longer treatments reduce carbon–carbon double bonds by adding hydrogen to them. Two other borohydride salts can also be used, tetraethylammonium and the tetramethylammonium borohydrides, but sodium borohydride is the most effective. All three salts are soluble in water and ethanol. Unfortunately, borohydrides decompose in water with the evolution of hydrogen bubbles with associated possible physical damage to the paper and image; the most stable solution is the tetraethylammonium salt. Certain metals present as impurities in the paper or as pigments on the paper can catalyse this decomposition. The borohydrides are used in the concentration range 0.01–2.0% and produce solutions of pH about 9.

Chlorine dioxide bleaching is a process which was developed for industrial paper pulps. Solutions of chlorine dioxide may be produced by mixing 75 ml of 40% formaldehyde solution with 20 g of sodium chlorite in 3 L of water. The yellow/green solution is an effective bleach for paper which has a low oxidation potential and reckoned to cause little damage to cellulose. Chlorine dioxide gas is quite hazardous so the solutions must be used in a fume cupboard and the gas not allowed to concentrate in pockets of air.

Of historic interest is bleaching with 0.5% potassium permanganate solution. In this process the paper is placed in a bath of the purple permanganate solution, the oxidising agent reacts with oxidisable stains and paper constituents and turns the paper brown/black. When the bleaching is deemed to have finished the almost black manganese dioxide is reacted with oxalic acid when the pale pink manganese oxalate is washed from the paper with water. Unfortunately, the oxidising agent is very powerful and capable of degrading the paper, and it is not possible to limit the time of the treatment to the minimum needed as the stains to be removed are difficult to see. If residual manganese salt is left in the paper due to insufficient washing, many years later a re-bleaching in an oxidising solution may turn the manganese salt to black manganese dioxide to the amazement of the conservator!

3.11 Foxing

Old paper may appear uniformly yellowed due to age, but another phenomenon is foxing, where yellow/brown spots appear on the paper (see Figure 3.6). For many years controversy raged as to whether the spots were caused by microbiological agents or were tiny particles of iron in the paper that had corroded to brown iron corrosion products. However, both explanations are valid and under ultraviolet light the two types of foxing can be distinguished. Spots that fluoresce brown, and may have a pale blue edging, are caused by fungi; some spots in this category appear pale blue in UV and are invisible in daylight. Foxing caused by iron particles appears black against the fluorescence of the paper. Iron particles are believed to have been introduced into the paper from the water used for the papermaking process or come from the machinery used.

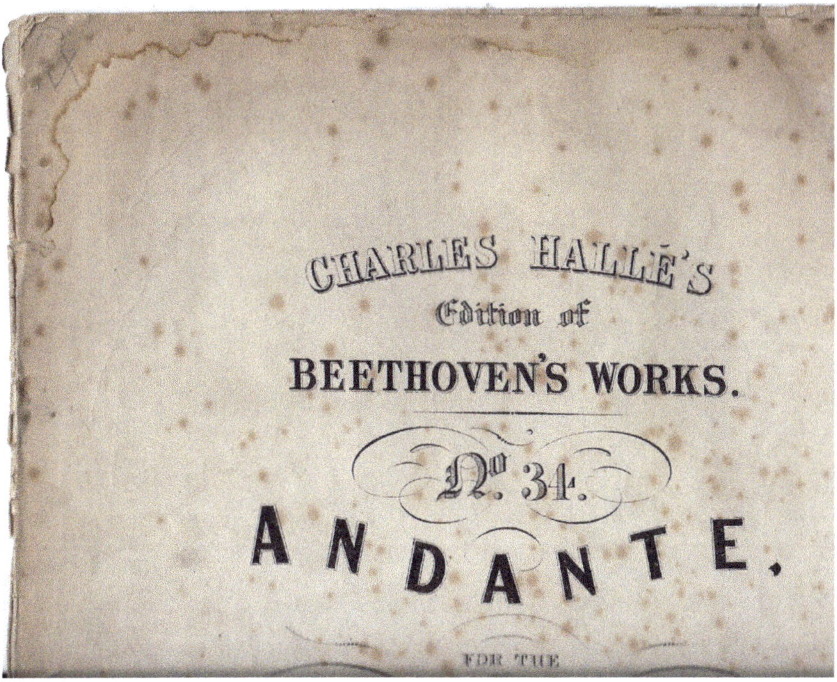

Figure 3.6 Victorian sheet music showing foxing and water staining.

Microbiological foxing is formed very slowly and is thus difficult to reproduce artificially in the laboratory. Besides, identifying the fungi responsible by sampling it from paper is fraught with problems as the samples always include other species in the paper. Whatever micro-organism is causing foxing, the production of spots is slowed down or prevented by storage in low RH conditions.

3.12 Iron Gall Ink

The earliest black inks were made either of carbon pigment in an aqueous binder or of iron gall ink. The latter is made by mixing an iron salt, *e.g.* ferrous sulphate, with a tannin often derived from plant galls. While the former ink is stable on paper, the latter contains not only the black ferric tannate but free acid and ferrous ions which enable both acid-catalysed hydrolysis and oxidation to occur with resulting loss of legibility and stability. In extreme cases there can be loss of the paper due to extensive degradation (see Figure 3.7). The stabilisation of iron gall ink on manuscripts has received a lot of attention in the last few decades and there are several proposed

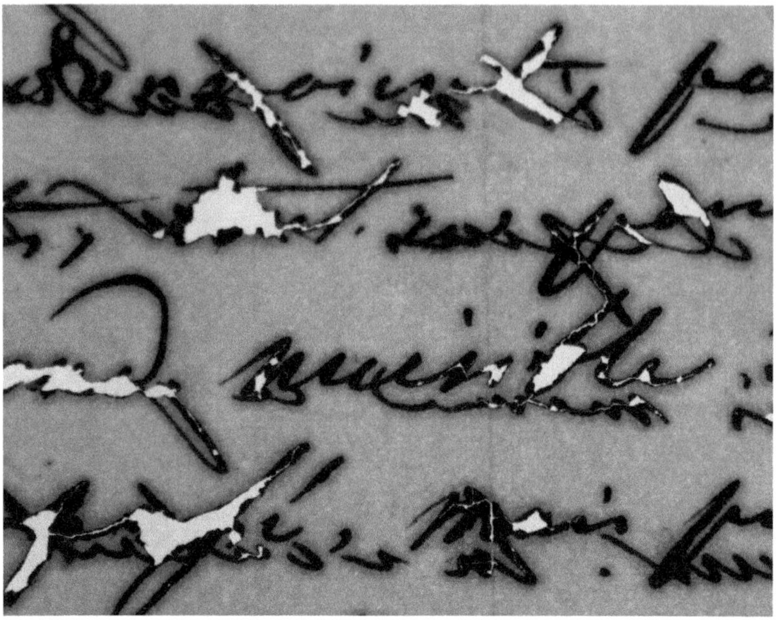

Figure 3.7 An example of iron gall ink on paper, written on both sides, demonstrating areas of loss which correspond to the highest concentrations of the ink.
Reproduced with permission from Paul Garside.

treatments including deacidification alone or in combination with EDTA or phytate to deactivate or remove ferrous ions.

3.13 Safe Environments for Paper

The safest conditions for storage of paper may be easy to define but are difficult to achieve in practice. Oxidation may be eliminated by storage in nitrogen but this is not practical in all but a few instances. Recommended conditions are a compromise between the ideal and the realistic.

At RH greater than about 65% fungal growth may occur, so the RH should be below this. Air circulation is an additional advantage. At low RH, deterioration reactions that need water to proceed will be slowed down but paper and glue become more brittle, so 40% is often the recommended lower limit for general storage. Chemical reactions slow down with decreasing temperature, so ideally, storage should be at as low a temperature as possible. Some institutions have stores maintained well below ambient temperature but objects need lengthy conditioning before they can be transferred to a students' room. International standards generally give the temperature for storage as about 18–20 °C. On exhibition, light levels are recommended to be as low as possible (50–150 lux usually) with no UV or at least less than 75 microwatts per lumen. Paper-based colour photographs are generally stored at lower temperatures on account of the sensitive dyes in the image layer.

3.14 Migration of Materials in and from Paper and Conservation Materials

As the materials which make up a book or a mounted work of art age, they may emit volatile materials which can contribute to the deterioration of adjacent paper. Some poly(vinyl acetate) adhesives and woods emit acetic acid when they age and this can act as a catalyst for the corrosion of lead in seals attached to documents. Some papers emit aldehydes and peroxide that can cause fading of black and white photographic images by attacking the silver micro-particles. To minimise the possibility of works of art becoming contaminated by the migration of undesirable materials, efforts are made to ensure that only high-quality paper and board are used for mounting works of art on paper.

Work on the volatiles from the ageing of paper has shown that substances of low molecular mass such as acetic acid, aldehydes and chemicals like furfural can be emitted, enhancing degradation in adjacent sheets. A pile of paper subjected to accelerated ageing has been shown to deteriorate faster than a single sheet. Paper is an efficient absorber of volatile air pollutants as can be seen at the edges of book pages where air pollution has been absorbed and enhanced degradation. If wood is present, *e.g.* in a picture frame or a book with wooden boards, then resin acids may be emitted. It is also thought that some autoxidising materials evolve hydrogen peroxide which can migrate and oxidise adjacent paper.

The evolution of gases by book and paper materials can be studied by analysing the headspace above the material when stored in a closed space. The volatiles can be concentrated by absorbing them onto a special material and subsequently desorbing them in a gas chromatograph linked to a mass spectrometer. The same technique has been used to analyse the smell of old books.

Non-volatile acids may also migrate in paper assemblages. This is a much slower process than gaseous migration but it is believed that such acids can migrate from one sheet of paper to another with which it is in contact; RH fluctuations may accelerate the process. The migration of liquids in paper is very difficult to study in the laboratory as the processes are very slow and cannot be accelerated successfully.

Old picture frames and inside book covers are good places to see the effects of the slow migration of chemicals.

3.15 Consolidation

Very weak sheets of paper can be made more handleable by laminating them onto a sheet of flexible polymer or paper. The most straightforward method where the weak paper object is pasted to a stronger sheet of paper has severe disadvantages for sheets with information on both sides. If the lamination is performed with a transparent material this problem is eliminated. An early method which used cellulose acetate sheets has shown itself to be unsuitable due to the evolution of acetic acid and mechanical instability of the polymer. Strengthening solutions of polymers include the use of gelatine solution, a commonly used sizing agent in high-quality papers. Other materials used are the cellulose ethers which include aqueous solutions of methylcellulose, the sodium salt of carboxymethylcellulose and hydroxypropylcellulose. There is a controversial but effective method which involves splitting the paper sheet in two and inserting a sheet of paper as a "sandwich filling" between them.

3.16 Impact of Chemistry and Analysis on Conservation, Preservation and Scholarship

A knowledge of chemistry is indispensable for the scientific study of paper conservation and its deterioration, however, a close association with conservators and a collection of paper objects is also valuable. While the effects of ageing on new paper samples can be simulated it must always be remembered that a museum or library collection is a vast experiment itself. Thousands or millions of objects have been made and stored in a variety of different conditions and wait to be examined after having been naturally aged for periods of time, sometimes several centuries. The chemist in close proximity to such a collection and the people working on its conservation will have an advantage over others working remotely.

The first decision in any research project is the selection of a topic to study, and here collaboration with conservators is invaluable as they are the ones with a knowledge of what the problems facing a collection are, either because of the changing condition or problems which occur during conservation. The chemist then has to decide what resources can be brought to bear for studying the proposed problem. Resources include existing knowledge, apparatus and human skills. Frequently the skill set required, the ability to operate apparatus and the ability to read and understand the existing literature is met by a person with a chemistry background.

Chemistry has been responsible for recognising the principal deterioration mechanisms of paper and the development of treatments to slow down that deterioration. Although much good work has been performed by chemists working in universities and museum laboratories much of the science which supports the research has been performed in the paper industry and is available in the chemical literature; similar statements could be made for photographic prints and many other special types of paper-based object. The special needs of library, archive and museum materials has dictated the need for chemists working in those establishments. Much of the work which supports art historical scholarship is derived from analysis of the image media, the chronology for the use of image media can be used to date objects and artists' technique.

3.17 Case Study

The project chosen for the case study demonstrates the need for careful experimental design and shows that complex apparatus is not always needed to make advances in the understanding of conservation practices.

Washing paper in water is one of the most basic processes in paper conservation and one of the most frequently used, however, few fundamental studies had taken place. Washing in water may be used to remove unwanted acidity or other undesirable contamination in the paper, *e.g.* to help reduce water-soluble staining. Several different methods are used by paper conservators ranging from simple immersion in water to the more complex washing on a suction table using ultrasonically generated water mist. It was decided to perform some experiments in which the rate and efficiency of washing would be determined and to vary individual physical parameters to see what the effects would be.

Conservation research projects of this kind are ideally conducted on samples of naturally aged paper, sufficient being available that comparisons can be made of the results of different treatments. An eminent paper historian was contacted to obtain various samples of discoloured paper and it was anticipated that a significant reduction in the yellow/brown discolouration would be made using simple immersion of the paper in water, however, only a small reduction was obtained. An instrumental method of measuring the discolouration used a tristimulus colour meter. Although this was a good method for measuring the discolouration, the measurement error and inhomogeneity of the discolouration in the potentially available samples was such that there was an insufficient difference before and after washing for useful results to be obtained. It was eventually decided that a model artificially discoloured paper would be made by impregnating a piece of pristine paper with a solution of the water-soluble brown extractives from discoloured paperboard. The new model material was made and when washed it was observed that more discoloured material could be extracted than with the naturally aged paper and furthermore, as with naturally discoloured paper, a large amount of the brown discoloured product still remained behind. One interesting observation was that as the samples aged in the laboratory, over several weeks the amount of discolouration which could not be washed out increased. This meant that some groups of experiments had to be done at the same time.

A range of experiments was performed on this paper and of particular interest was the effect of temperature on washing. More discolouration could be removed the higher the temperature of washing. This is might seem obvious to anyone who has washed clothes but it was a hitherto unproven observation for discoloured paper and was important because the vast majority of paper washing is done at room temperature. If the time of washing at room temperature was greatly increased the amount of washing could not exceed that obtained even at a slightly higher temperature. Simple immersion washing was found to be the most efficient but that cannot always be used because of the nature of the paper and image media. Several more conclusions could be made which are described in the published work.

As a result of this work, it was concluded that there was a need for an artificially discoloured paper for use in the training of conservators and, with the aid of dye and paper industry scientists, one was made by dyeing paper with fugitive dyestuffs.

3.18 Conclusions

Paper is a complicated material and objects often contain many other materials, such as image media. It is not practical for the paper conservator to examine all the components and properties of a paper object before it is treated. However, experience of treating a wide variety of paper objects and knowledge of the properties of paper and associated materials enables the conservator to achieve success. Much of this information has been obtained by the scientific study of paper objects. This scientific knowledge enables the conservator to communicate effectively with conservation scientists who can investigate problems outside the conservator's area of specialisation.

Acknowledgements

I would like to thank Nigel Meeks of the British Museum's Department of Scientific Research for obtaining the SEM image of paper.

Recommended Reading

G. Banik and I. Brückle, *Paper and Water, a Guide for Conservators*, Elsevier, Amsterdam, 2011.
B. L. Browning, *Analysis of Paper*, Marcel Dekker Inc., New York and Basel, 1997.
H. Burgess, Practical Considerations for Conservation Bleaching, *J. Int. Inst. Conserv. - Canad. Group*, 1988, **13**, 11.
J. C. Roberts, *Paper Chemistry*, Blackie, Glasgow and London, 1991.
M. Strlič and J. Kolar, *Ageing and Stabilisation of Paper*, National and University Library, Ljubljana, 2005.
S. Zervos and I. Alexopoulou, Paper Conservation Methods: A Literature Review, *Cellulose*, 2015, **22**, 2859.
British Standards Institute, *Guide for the Storage and Exhibition of Archival Material*, PD5454:2012, 2012.
British Standards Institute, *Specification for Managing Environmental Conditions for Archival Collections*, PAS198:2012, 2012.
M. Hey, Paper Bleaching: Its Simple Chemistry and Washing Practices, *The Paper Conservator*, 1977, **2**, 10.

V. Daniels, The Rate of Discolouration Removal from Paper by Washing, in *Paper and Water, A Guide for Conservators*, ed. G. Banik and I. Brückle, Elsevier, Amsterdam, 2011, ch. 10, p. 289.

J. M. Kosek, S. l'Anson, A. Tindel, V. Daniels and M. Sandy, Model Discoloured Paper for the Analysis of Aqueous Washing and Other Conservation Processes, *Restaurator*, **35**, 47.

Reference

1. E. May and M. Jones, *Conservation Science: Heritage Materials*, The Royal Society of Chemistry, Cambridge, 1st edn, 2006.

4 Wood

E. J. Schofield*[a] and A. V. Chadwick[b]

[a] Mary Rose Trust, Main Road, HM Naval Base, Portsmouth PO1 3PY, UK; [b] School of Physical Sciences, Ingram Building, University of Kent, Canterbury CT2 7NH, UK
*Email: e.schofield@maryrose.org

4.1 Introduction

Due to its natural availability and properties, wood has been employed as the material of choice for a variety of applications since records began. From personal items such as combs, bowls and spoons, up to structural masterpieces such as buildings and ships, its versatility has secured its place as an essential material from civilisations past to the current day. Whilst some materials may have replaced wood in certain functions, it is reasonable to assume that it will still be employed significantly in the future due to its adaptability to different uses and its sustainability. The latter issue is becoming more and more relevant as we aim to make wise choices in terms of our impact on the environment. Therefore, it is reasonable to assume that the discovery of wooden archaeological items and the need to maintain ageing historical structures will continue into the foreseeable and long-term future. To do this, it will be crucial to have a thorough understanding of the material properties and degradation mechanisms of wood, alongside possible ways of monitoring and stabilisation.

In terms of cultural heritage, wood can be broadly split into two categories, namely those items that are still in functional use and those that were lost and recovered, *i.e.* historical and archaeological (Figure 4.1). The latter type can be split further into those that are

Conservation Science: Heritage Materials, 2nd Edition
Edited by Paul Garside and Emma Richardson

Published by the Royal Society of Chemistry, www.rsc.org

(a)

(b)

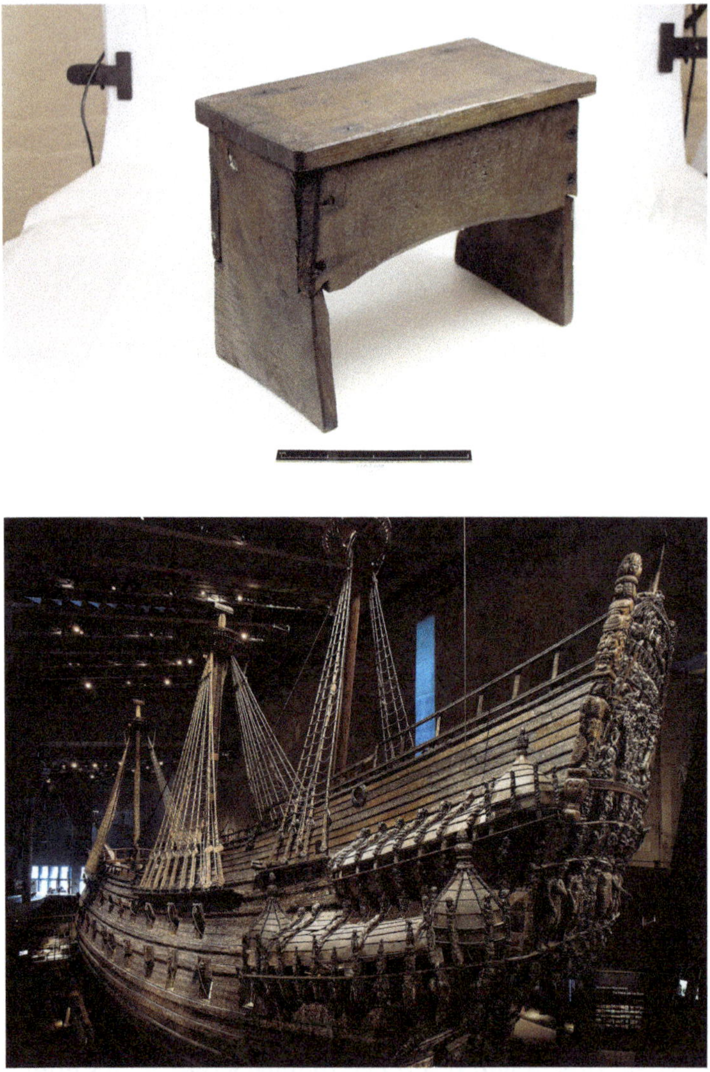

Figure 4.1 (a) Elm nail-board stool from the *Mary Rose*. Photo credit: Mary Rose Trust. (b) The *Vasa*, Stockholm, Sweden. Photo credit: Anneli Karlsson, Vasa Museum/SMTM.

marine or terrestrial, as the chemical environments differ and can impact the resulting material properties. Natural wood is a very complex material and conservation can become even more challenging when mechanical, biological and chemical changes have occurred. These changes can be induced by a variety of factors, such as prolonged exposure to aggressive environments or removal from its stable burial environment. Compared to other materials, such as stone and brick, wood can also degrade due to moisture or biological attack.

This chapter will explore the mechanisms by which wood degradation occurs, discuss the methods by which this damage can be detected and assessed, and give the currently available treatments to mitigate further degradation or compensate for the loss of material and therefore stability. A review will also be given of current research to further our understanding of the material properties of degraded wood, and the development of new treatments. Whilst this review will focus on wooden artefacts from shipwrecks, and include specific examples from the *Mary Rose,* Henry VIII's Tudor warship, many aspects included will be applicable to other areas of wood conservation.

4.2 Chemistry, Structure and Microstructure of Wood

Regardless of whether the wood is from a historical or archaeological context, or the extent of its degradation, it is vital to initially understand the original properties of the material. This section will explore the micro- and macrostructure of wood and how these inform the material properties the wood exhibits.

4.2.1 The Microstructure of Wood

Wood is a complex natural composite material, which contains three major components: cellulose (45–50%), hemicelluloses (20–25%) and lignin (20–30%). Each of these components is composed of carbon, hydrogen and oxygen, and their chemical formulas can be seen in Figure 4.2. The remaining wood mass (0–10%) is made up of extractives. Extractives (terpenoids, fatty acids, phenolics) are usually considered to be soluble in organic solvents and their composition varies with location within the tree. Cellulosic materials, in different contexts, are also discussed in Chapter 3 (paper) and Chapter 10 (textiles).

Cellulose is the most abundant component, which consists of long chains of glucose, with the chemical formula $(C_6H_{10}O_5)_n$. The number of glucose units within a cellulose molecule is referred to as its degree of polymerisation (DP). These chains aggregate into larger units, *via* intra- and inter-molecular hydrogen bonds by the hydroxyl groups, giving cellulose some crystallinity. The mix of crystallinity with other less ordered (amorphous) areas gives rise to complementary properties such as stiffness and flexibility within the same material.

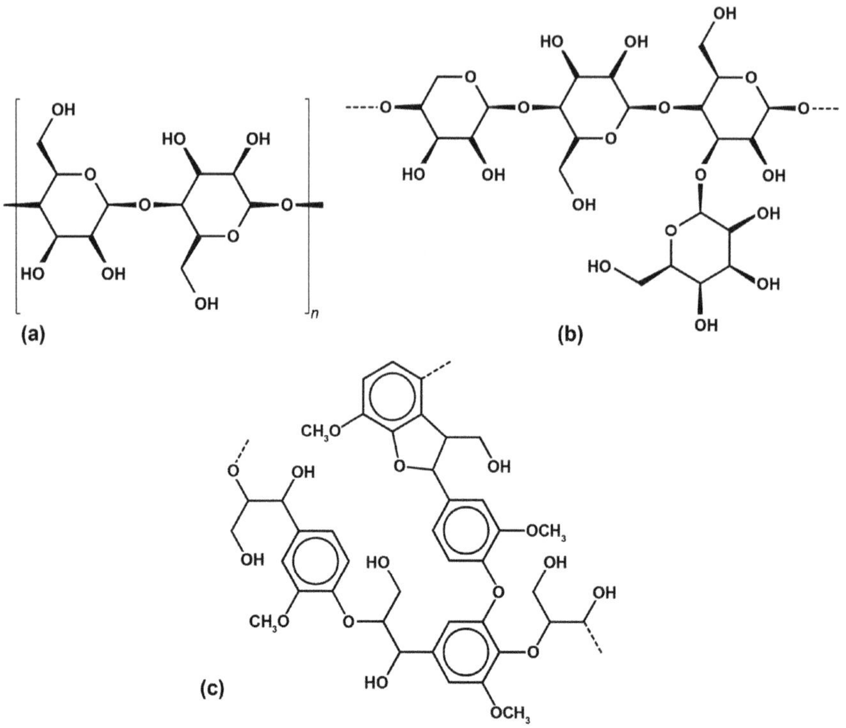

Figure 4.2 The structures of (a) cellulose, (b) hemicellulose and (c) lignin. Photo credit: Paul Garside.

The bulk crystalline cellulose is not accessible to water, chemical or microbial attack, but these mechanisms can occur in the amorphous phase and on the crystalline surface. In some instances, the amorphous phase can be protected by the surrounding hemicellulose and lignin.

Hemicellulose is comprised of a mixture of polysaccharide polymers, such as the sugars D-xylopyranose, D-glucopyranose, D-galactopyranose, L-arabinofuranose, D-mannopyranose, D-glucopyranosyluronic acid, and D-galactopyranosyluronic acid. It has a lower DP than cellulose and consists of an amorphous structure. It has limited strength which means that it is usually the component that is degraded first within the wood. It is soluble in an alkali solution and can be hydrolysed by acids.

Lignin is a highly complex cross-linked macromolecule which consists of different monomeric units. All wood lignins consist mainly of three basic building blocks of guaiacyl, syringyl, and *p*-hydroxyphenyl moieties. Lignin is hydrophobic due to the presence of aromatic rings within its structure. Of the three components, it is the most resistant to degradation.

The extractives present are in the cell wall and consist of fats, fatty acids, fatty alcohols, phenols, terpenes, steroids, resin acids, rosin, waxes, and other minor organic compounds. These exist as monomers, dimers, and polymers. Extractives are responsible for the colour, smell, and durability of the wood. The inorganic content of the wood is usually referred to as its ash content, which gives an indication of the mineral salts that are present after combustion of the wood. These compounds are present after being absorbed through the roots and transported around the tree; therefore, the content can vary drastically depending on the environment where the tree grew.

The proportion of cellulose, hemicellulose and lignin present in wood is relative to the species and this in turn will affect how resistant the wood is to degradation. The structure of wood can be seen in Figure 4.3. Several layers of different components (cellulose, hemicellulose and lignin) with different orientations of microfibrils constitute the cell wall. The secondary cell wall is made up of three layers: S1, S2, and S3. These layers are distinguished by different orientations of the helix of the microfibrils that constitute them. The cells are connected together by the middle lamella, which has the highest concentration of lignin within the wood, and is seen as the backbone of the structure.

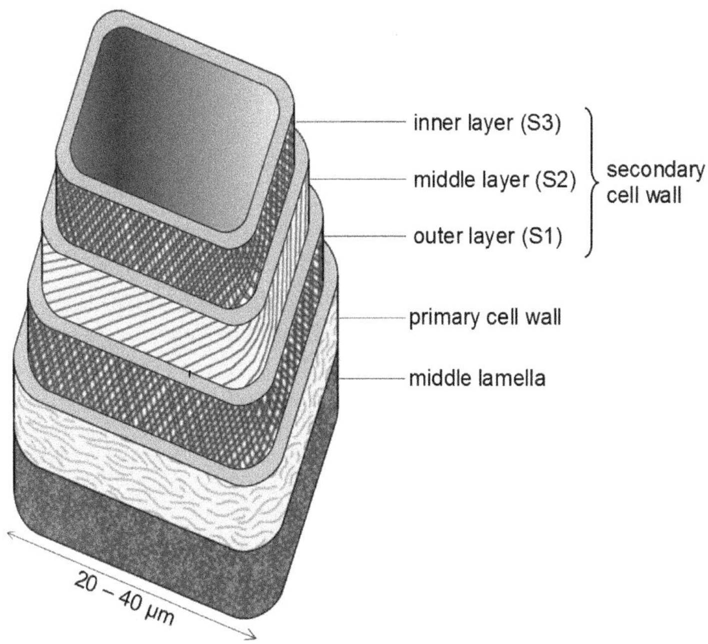

Figure 4.3 The microstructure of wood.
Photo credit: Paul Garside.

4.2.2 The Macrostructure of Wood

Wood can be broadly categorised as either hardwood (angiosperms) or softwood (gymnosperms). Within these two categories lie many species, with over 520 gymnosperms, and over 30 000 angiosperms. Hardwoods are deciduous, usually broadleaf trees such as oak or walnut, and softwoods are usually evergreen conifers such as pine or spruce. The prefix (hard or soft) does not always indicate the relative hardness of the wood, as there are some hardwoods that are relatively soft, *e.g.* balsa wood, and some softwoods that are relatively hard, *e.g.* yew. The two types can be differentiated by the macrostructure they exhibit, with hardwoods possessing the most complex structure.

Hardwoods consist of vessels, fibres (fibre tracheids or libriform fibres), longitudinal parenchyma and ray parenchyma. Thin-walled vessels are used to transport water and nutrients throughout the wood, and can be observed as pores in the structure (Figure 4.4). Larger vessels are formed in the earlywood, with the smaller vessels formed in the latewood, allowing for the identification of the growth period when looking at the structure in detail. Densely packed fibres (Figure 4.4) have thick walls which provide strength to the wood. Whilst the primary function of the fibres is to provide strength, they can also serve to migrate water through pits, which can be present as either bordered pits (fibre tracheids) or simple pits (libriform fibres). Longitudinal parenchyma are used to store food and ray parenchyma can form single or multiple layers and are used to transport food and water radially.

Softwoods consist of tracheids and ray parenchyma. Tracheids are aligned longitudinally and consist of a middle cavity called a lumen (Figure 4.5). They have the dual function of providing mechanical stability and also transporting water and nutrients. The wall thickness of the tracheids changes throughout the growing period, and therefore early- and latewood can be identified. This can be seen with the naked eye as annual growth rings in the cross section of the wood. Softwoods also have a radial system of ray parenchyma which can transport water and nutrients radially.

4.2.3 Tree Growth

Located within a tree trunk are distinct regions which are identified in Figure 4.6. The cambium is a thin layer located beneath the bark and is the front for growth, where new wood cells are produced. In a sapling, when a tree first starts to grow, the entire wood cross section

(a)

(b)

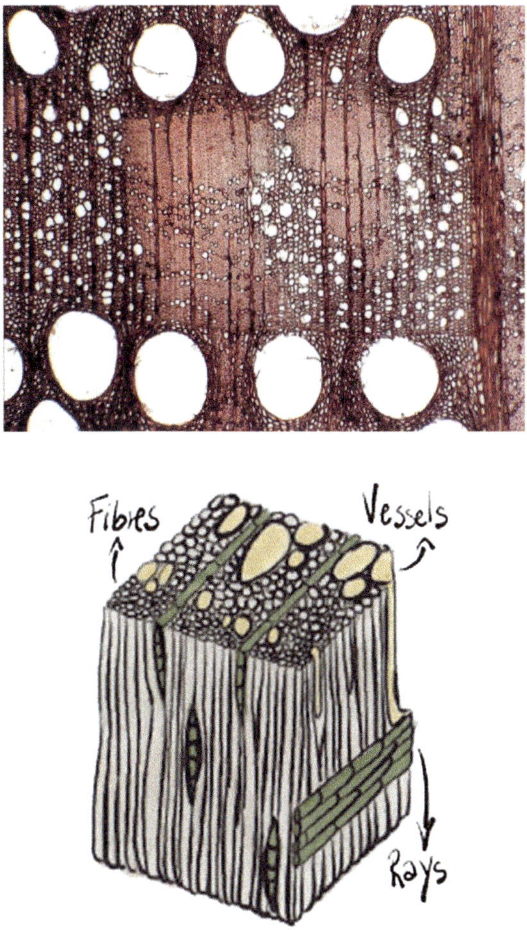

Figure 4.4 Structure of fresh oak (hardwood) depicted by (a) Light microscopy and (b) 3D schematic.
Photo credit: Glenn McConnachie & Eleonora Piva, Mary Rose Trust.

is required for transporting water and nutrients. However, once it has grown sufficiently, the whole cross section is not required, and cells within the centre of the trunk do not act to transport water, but as a mechanical support for the tree. This part of the tree is referred to as the heartwood. The outer layer, which continues to transport water and nutrients, is the sapwood. As the food storing cells die, extractives are produced which remain in the heartwood, making it more durable as the extractives repel wood destroying creatures. Heartwood can also appear visibly darker due to the extractives present. Due to its inferior strength, sapwood is usually removed from the wood before being used in structural/load-bearing applications. Mechanical and

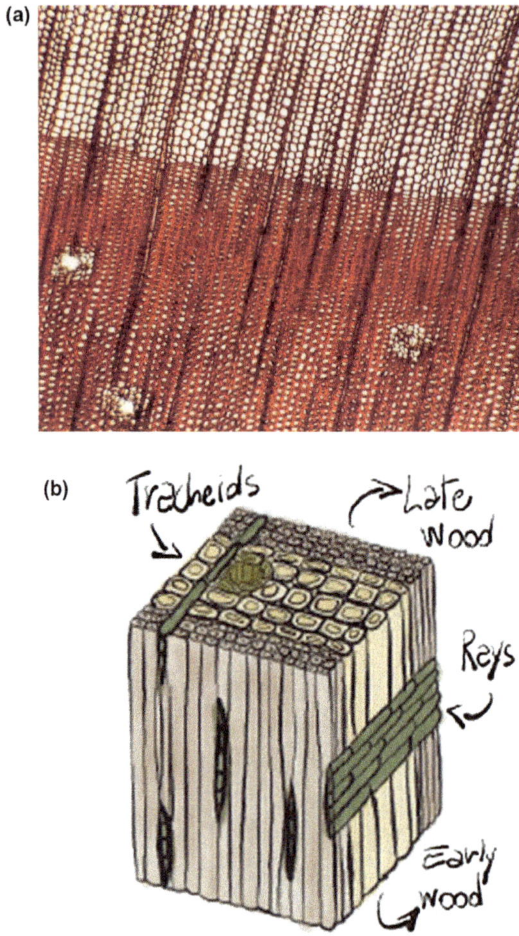

Figure 4.5 Structure of softwoods depicted by (a) SEM micrograph of fresh pine and (b) schematic.
Photo credit: Glenn McConnachie & Eleonora Piva, Mary Rose Trust.

biological modifications can be made to wood to intentionally increase the content of heartwood, as it is the most useable part.

4.2.4 The Properties of Wood

Wood is anisotropic, which means that it will exhibit different properties when tested in different directions. The three axes, longitudinal, tangential and radial, are identified in Figure 4.7. The properties exhibited along these directions are dependent on the cellular composition, structure and arrangement. An obvious example of this is the longitudinal orientation of fibres in hardwoods and tracheids in

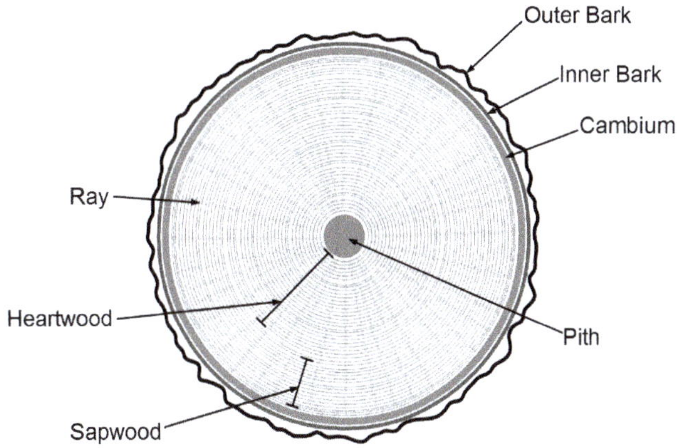

Figure 4.6 Cross section of a tree trunk, identifying heartwood and sapwood. Photo credit: Paul Garside, based on a sketch by Eleonora Piva, Mary Rose Trust.

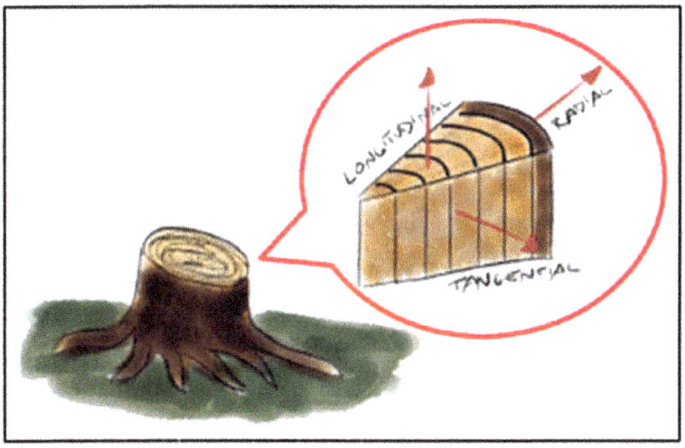

Figure 4.7 The three directional axes in wood. Photo credit: Eleonora Piva, Mary Rose Trust.

softwoods, which means that the greatest load-bearing strength is in this direction. Another key property of wood is its durability, which is dependent on the species and the environment in which it exists. Due to the extractives present, heartwood is more durable than sapwood.

Wood is also hygroscopic, with water being incorporated into the structure as either free or bound water, as shown in Figure 4.8. Water is attracted into the wood by the hydroxyls of its chemical components, *e.g.* cellulose, and is held *via* hydrogen bonds. Once water has saturated the cell walls, it will start to fill any cavities. The point at which the cell walls are saturated but the cavities are empty is referred to as the fibre saturation point. The amount of water contained within wood is dependent on the atmosphere it is in and the extent of any

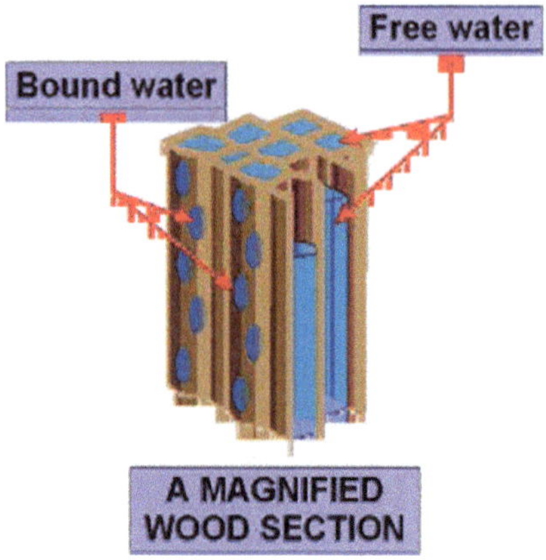

Figure 4.8 Bound *versus* free water in the wood cell structure. Photo credit: Eleonora Piva, Mary Rose Trust.

degradation, and will react to the environment it is in to reach its equilibrium moisture content.

The basic density of wood is measured by the oven-dried weight of the wood divided by the green (*i.e.* freshly cut) volume of the same wood. If the wood in question is not freshly cut it is common to soak the wood to ensure it is fully swelled before measuring the volume. This measurement is important as it gives information on the integrity of the wood which impacts other physical properties.

4.3 Analytical Techniques Appropriate for Wood

Due to the complexity of wood, a great number of techniques have been used historically to yield information on the chemical properties impacting its stability. The best understanding is gained by using a combination of techniques, which may require varying levels of equipment and expertise. Outlined below are some general methods followed by specific analytical methods used in scientific studies to understand wood properties, degradation mechanisms and the resultant impact on the integrity.

For many heritage organisations access to equipment and the necessary expertise to conduct and interpret such experiments can be limited, and other techniques must be relied on that will be discussed here first. Whilst not employing state of the art technology, these techniques can give a good base understanding of the level of degradation. Visual observations can give great insight into the state of a material. Colour changes can indicate damage, as a darkening of the surface can be observed when wood components have been lost, and the interaction of light with the resulting structure is altered. Other straightforward inspections can include seeing how damp the wood appears, and how soft or sponge-like it is to the touch. To determine how soft a set of samples are relative to each other the *pin test* can be carried out, whereby a pin is inserted and it is recorded how far it can penetrate with minimal exertion. Measuring the density and comparing with the known value of a fresh reference sample will also give a good indication of how degraded the wood is. In the case of waterlogged wood, samples can be sectioned wet and completely dried to give a percentage of moisture loss, which can be related to the material lost through degradation. Measuring the environment (temperature and relative humidity) around wooden heritage is often critical and can inform of potentially problematic scenarios, where intervention may be required. Last of all pest infestation can in some instances be visible with the naked eye, therefore prompting immediate action (Figure 4.9). All these observations and tests can give enough information to prescribe a basic conservation treatment, if no other techniques are available.

Determining a species of wood can be done by carefully taking thin sections and examining under a light microscope. In some instances, dye can be employed on the thin section to highlight different regions. Characteristic features of the wood can be identified and compared to reference images (Figure 4.10). For heritage purposes,

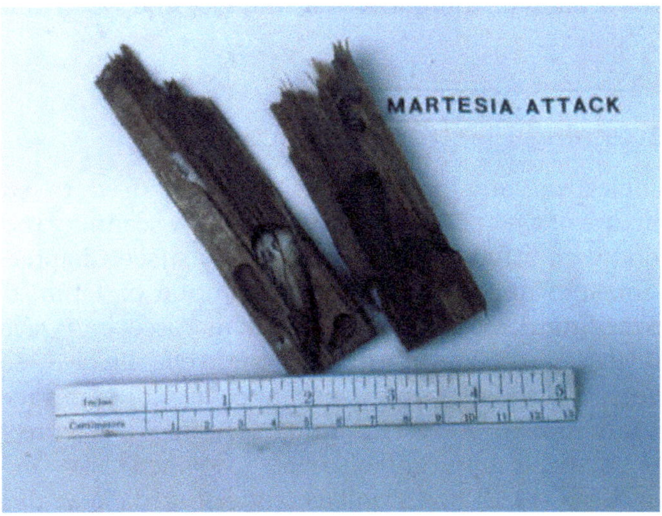

Figure 4.9 Burrows of pholads in wood.
Photo credit: Dr Simon Cragg.

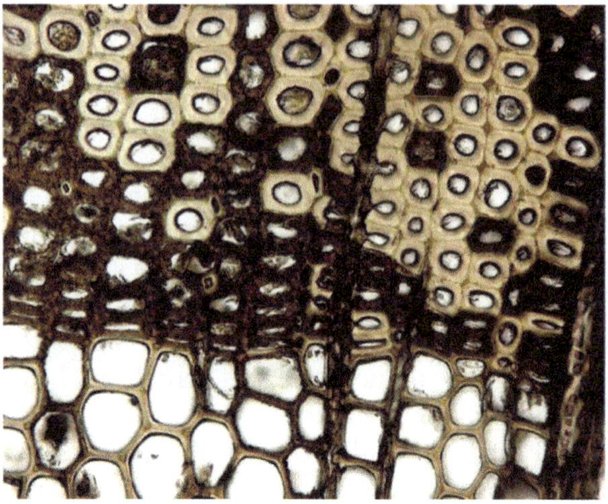

Figure 4.10 Light microscopy image of *Mary Rose* pine. Photo credit: Glenn McConnachie, Mary Rose Trust.

it can also be relevant to determine the age of the wood, which can be done by dendrochronology or carbon dating. The former examines the tree rings and compares them to known standards to determine the age of the tree. The latter measures the radiocarbon (C^{14}) content in the wood. This carbon isotope is incorporated into the tree during photosynthesis, but when the tree is felled and the process no longer continues, the radiocarbon content gradually diminishes as it undergoes radioactive decay. Therefore, the amount in the wood can be correlated back to the point then the tree the wood came from ceased to be alive.

4.3.1 Scanning Electron Microscopy

Scanning electron microscopy (SEM) is often used to study the microstructure of wood. It produces an image by scanning the surface of the wood with a focused beam of electrons (see Chapter 2). It is possible for SEM imaging to reach a resolution of 1 nm, therefore greatly exceeding the capacity of a light microscope. Typical applications related to the conservation of the wood involve looking at cellular degradation, identifying specific biological pathways to degradation and monitoring conservation treatments (Figure 4.11(a)). Elemental distribution can also be obtained *via* an energy dispersive X-ray (EDX) spectrometer collecting data at designated points/areas within the sample. Whilst this technique cannot distinguish the lighter elements that constitute wood from each other

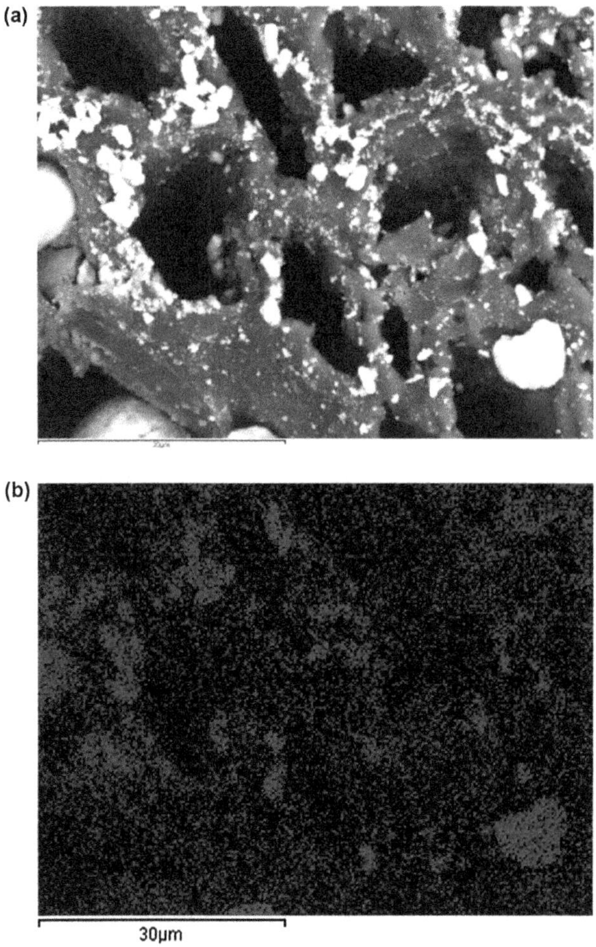

Figure 4.11 (a) SEM image of *Mary Rose* wood treated with $SrCO_3$ (b) EDS analysis highlighting Sr.
Photo credit: Eleanor Schofield, Mary Rose Trust.

(such as carbon, nitrogen and oxygen), it can distinguish heavier elements that may be present, either those natural to the wood, or present as a result of the location in which the item has been buried, or conditions it has been subjected to (Figure 4.11(b)).

4.3.2 X-ray Fluorescence

X-ray fluorescence (XRF) is another way in which elemental analysis can be obtained from wood. In this technique, X-rays are shone onto a material with a high enough energy to excite primary electrons within atoms contained within (Figure 4.12). This results in the emissions of

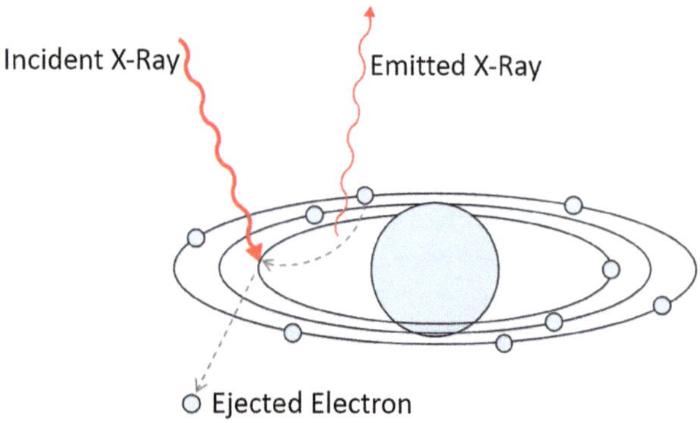

Figure 4.12 Principles of X-ray fluorescence. Photo credit: Paul Garside.

a secondary characteristic X-ray, which can be measured and correlated to a specific atom (see Chapter 2). As with SEM-EDX analysis, care has to be taken with quantitative analysis as it involves the lighter elements of wood (C, O and H). However, with a suitable calibration methodology it is possible to determine concentrations of impurities within the wood. This technique is available in a portable format which lends itself well to analysis of immovable heritage or that which cannot be sampled.

The X-ray based techniques described thus far are completed in typical materials analysis laboratories. A different suite of techniques can be found at a synchrotron source, such as the Diamond Light Source in Oxfordshire (Figure 4.13(a)). This is a circular particle accelerator which uses radiofrequency radiation to accelerate electrons through a sequence of magnets until they reach speeds near that of light. As the electrons are deflected by magnets, energy is lost in the form of X-rays which are channelled down pathways and tailored for specific types of analysis (Figure 4.13(b)).

4.3.3 X-ray Absorption Spectroscopy

X-ray absorption spectroscopy (XAS) is a structural technique that can determine the local atomic structure (the number, types and distances of surrounding atoms) around a target atom to a radius of about 6 Å. This is information similar to X-ray diffraction; however, it does not rely on long-range order and therefore can be applied to both crystalline and amorphous solids, and liquids. The basic experiment involves measuring the X-ray absorption coefficient of a sample as the energy of the photons is scanned across the energy required to eject a

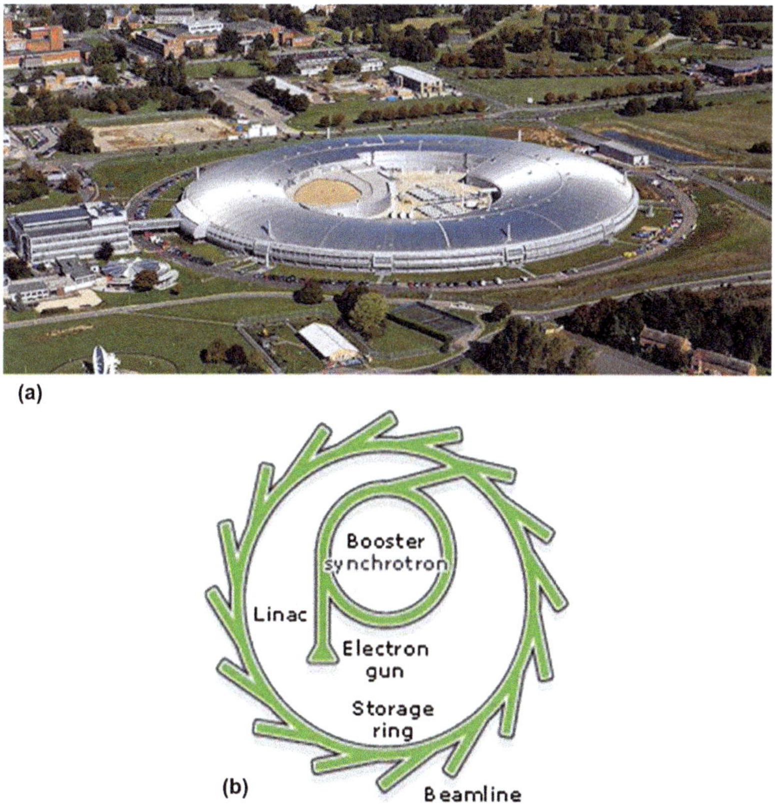

Figure 4.13 (a) Diamond Light Source, (b) how a synchrotron works. Reproduced with permission from © Diamond Light Source.

photoelectron from an inner shell (K or L shell) of an atom in the sample. The experiments are usually conducted at a synchrotron.

A typical XAS beamline is sketched in Figure 4.14. The set-up is relatively simple; the X-ray intensity before (I_o) and after the sample (I_t) are measured with ionisation chambers to yield the absorbance, $\mu = \ln(I_o/I_t)$. The intensity of the ejected photoelectrons and the intensity of the emitted fluorescent X-rays (as the outer electrons drop into the hole left by the photoelectrons) follow the intensity of the absorption offer an alternative means of collecting spectra, termed Total Electron Yield (TEY) and Fluorescence (FXAS) XAS.

A typical XAS spectrum is shown in Figure 4.15, a plot at the K-edge absorbance of a sulfur compound. This shows the rapid increase when the X-ray energy crosses the binding energy of the electron, termed the *absorption edge*, followed by an oscillating absorbance. The spectrum is considered to have two regions: the region extending

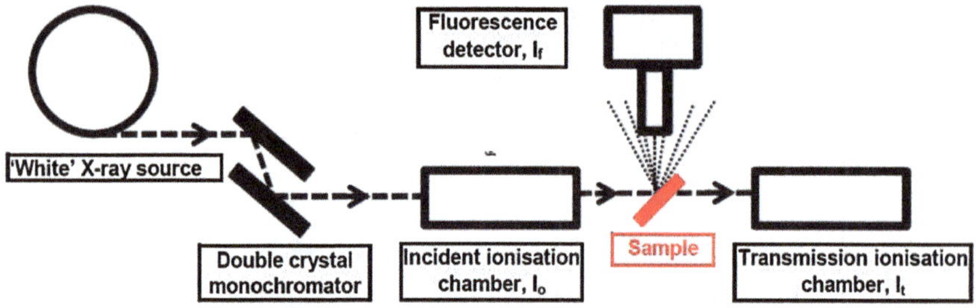

Figure 4.14 A typical XAS beamline at a synchrotron source. Photo credit: Alan Chadwick, University of Kent.

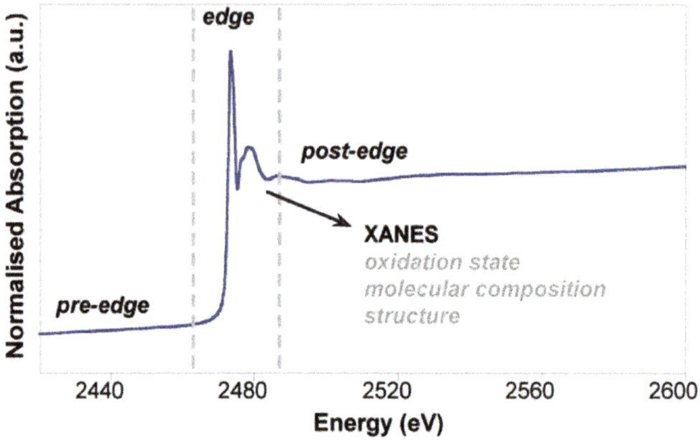

Figure 4.15 The S K-edge spectrum of a sulfur compound. Photo credit: Eleanor Schofield, Mary Rose Trust.

typically 50 eV beyond the edge, termed the *X-ray Absorption Near Edge Structure (XANES)*; and beyond this region the *Extended X-ray Absorption Fine Structure (EXAFS)* which can extend to 1000 eV after the edge. In the XANES region the photoelectron is moving relatively slowly and is interacting with the surrounding electrons. In the EXAFS region the photoelectron is moving fast and in quantum terms is a spherical wave. This wave will be back-scattered by neighbouring atoms and the back-scattered wave will interfere with the out-going wave affecting the final energy state of the absorption process. Since the distance from the target atom to the back-scattering atoms is fixed and the photoelectron wavelength decreases with increasing energy the waves will oscillate between in-phase and out-of-phase interference. This accounts for the EXAFS oscillations and the process is shown schematically in Figure 4.15.

An example of the edge region and XANES spectrum for the S K-edge in various oxidation states is shown in Figure 4.16. Before

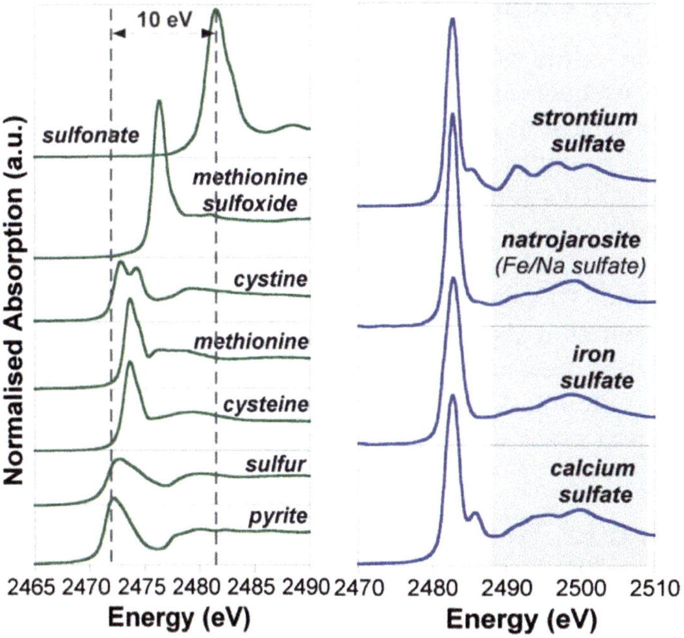

Figure 4.16 The S K-edge XANES in several sulfur compounds. Photo credit: Eleanor Schofield, Mary Rose Trust.

the edge there are small peaks termed *pre-edge features* which are due to excitation of core electrons to higher energy levels without leaving the atom. These features and more specifically the edge position give information on the oxidation state of the target atom. It can be seen that the edge is shifting to higher energy as the oxidation state increases from S(−1) to S(+6) as the charge on the core atom increases the energy to emit the photoelectron increases. XANES can provide a fingerprint for the local geometry and if XANES standards are available for the expected components of an unknown mixture then a linear component analysis of the XANES can be used to determine the composition of the mixture.

EXAFS analysis is more advanced, with interactive computer programmes being used to fit a local model structure to the experimental results. This provides detailed information on nearest neighbours, coordination number and distance from the absorbing element.

At certain beamlines the set-up is configured to produce a beam which can be as low as 1 μm in diameter. This allows for initial mapping of an area using XRF to identify where different elements are located, which can then be used to isolate specific areas for further XAS analysis.

4.3.4 X-ray Computer Tomography

Tomography is the production of a 3D representation *via* the compilation of a sequential series of 2D slices. High intensity X-rays, such as those produced at a synchrotron, enable images to be taken with high spatial resolution and are able to penetrate thick samples non-destructively. This lends itself well to the analysis of heritage materials where non-destructive analysis is always favourable and sometimes essential. An example of tomography on wood can be seen in Figure 4.17. This technique allows not only for the analysis of the wood structure but also of any consolidants that have been used and any impurities present in the wood. In some instances, it is possible to combine the tomography with other changes, *e.g.* mechanical or chemical, and observe the induced response in real time.

4.3.5 Fourier Transform Infra-red Spectroscopy

Fourier Transform Infra-red spectroscopy (FT-IR) is a technique where infra-red radiation is shone through a sample, with some of it being absorbed and some of it being transmitted. What is absorbed or transmitted relates to the chemical bonds within the material and the resulting spectrum can be used to elucidate that information (see Chapter 2). In wood, it enables the distinction between the three main components (lignin, hemicellulose and cellulose) as the different bonds within the three compounds result in different signatures within the spectrum. The technique is usually employed to look at wood degradation, where lignin is the most resistant to degradation, and therefore the increased presence of it relative to the other components signals overall degradation of the wood (Figure 4.18).

4.3.6 Nuclear Magnetic Resonance

Nuclear magnetic resonance (NMR) spectroscopy covers a wide and diverse range of techniques that can be used to (i) provide analytical information on chemical species, (ii) study the dynamics of atomic and molecular motion and (iii) provide 3D images of objects and in some cases their internal structures. The principle of the techniques is that when the nuclei of atoms with a non-integral spin (I) are placed in a magnetic field there will be a Boltzmann distribution of the spins between the energy levels; for example, for nuclei with $I=1/2$ will distribute between levels aligning with or against the field and the sample will have a net magnetic moment. The separation between the

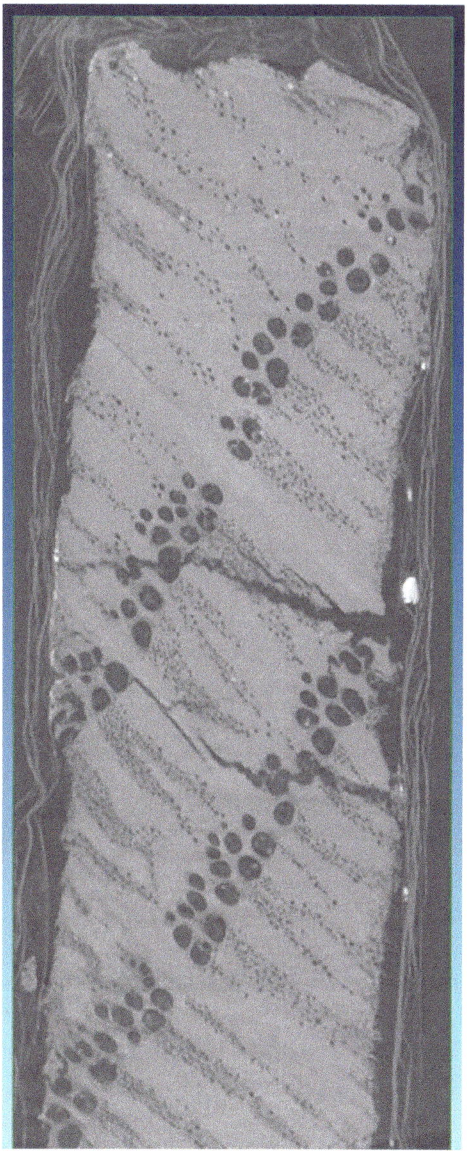

Figure 4.17 Tomography.
Photo credit: Eleanor Schofield, Mary Rose Trust.

levels is small and in the radiofrequency (rf) regime of the electromagnetic spectrum. Hence, irradiating the system with rf power at resonance frequency will disturb the population, change the net magnetic moment, and the system will have to relax back to equilibrium. These changes can be detected by changes in currents in coils around

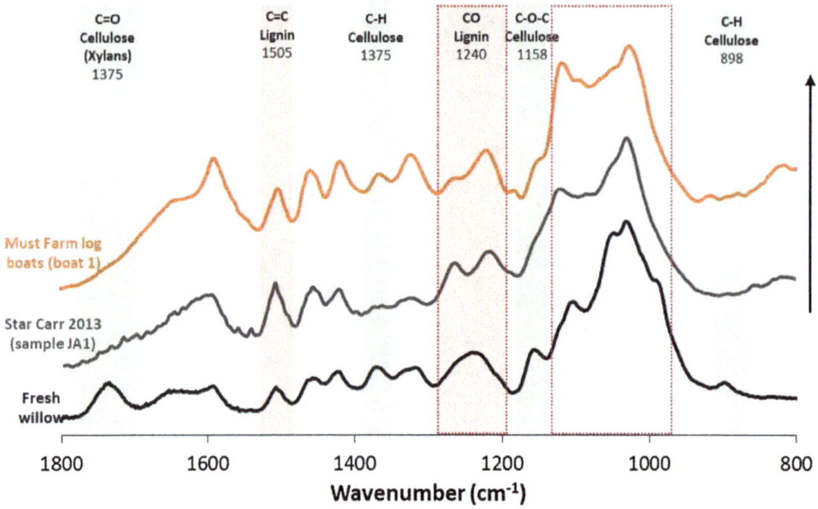

Figure 4.18 FT-IR spectrum of fresh wood *versus* degraded wood. Reproduced from ref. 3, https://doi.org/10.1186/s40494-020-00422-y, under the terms of the CC BY 4.0 licence, http://creativecommons.org/licenses/by/4.0/.

the sample. The relaxation times can give information on dynamic processes as the nuclei need to interact with their surroundings. The energy levels will also depend on local magnetic fields, such as the electrons in chemical bonds, hence the resonance frequency will depend on chemical bonding and structure, providing a means of chemical analysis. Originally NMR was restricted to nuclei with high natural abundance (^{1}H, ^{19}F, ^{31}P, *etc.*) and mainly liquid samples. With the introduction of superconducting magnets and magnetic pulse sequences applied *via* the coils it has been possible to vastly increase the sensitivity and detection of nuclei with low natural abundance is now possible. ^{13}C NMR is a particularly important case in the study of organic materials. The introduction of Magic Angle Spinning (MAS) NMR has meant that the techniques can be applied to solids. Finally, Magnetic Resonance Imaging (MRI) allows large objects to be 'mapped' and the technique is now commonly used in medical diagnostics.

NMR techniques have a long history of materials dating back to the 1950s, however, it is only in the last decade that they have been used to study wood. This is rather surprising given that there have been extensive measurements for the related material, paper. The ^{13}C MAS-NMR spectra of wood samples are usually complex due to the wide variety of components. However, the measurements can be particularly useful as peaks in the spectra from carbohydrates and aromatic groups are strong and easily distinguished. Hence, two areas of use are in estimating relative ages from the ratio of unsaturated (lignin) to saturated (carbohydrate) resonances and assessing the level of decomposition of wood prior to conservation. In addition, the signal from polyethylene glycol (PEG) which is used to conserve waterlogged timber is distinct and NMR can be used to probe the depth and effectiveness of this conservation tool. However, a note of caution has to be sounded when the archaeological wood has been penetrated by paramagnetic species such as ferric ions (from nails, bolts, *etc.*). In this case, the spectra are broadened and information can be lost. A novel rather sophisticated NMR method termed bi-dimensional heteronuclear single quantum coherence (2D-HSQC) is regarded as the best method of characterising lignin in organic matter and hence monitoring decay. Very recently MRI was employed in the study of the rate and of PEG and trehalose penetration into waterlogged wood specimens. In addition, the measurements provided a visualisation of the impregnation. There is little doubt that there will be much more extensive use of NMR methods in future investigations of wood.

4.3.7 GC, GC–MS and py-GC–MS

In the chemical analysis of mixtures of organic materials, the use of chromatographic methods (for example, gas phase chromatography (GC) or high-performance liquid chromatography (HPLC)) to separate the individual constituents followed by their identification in a mass spectrometer (MS) has been a standard procedure termed GC–MS (see Chapter 2). However, this relies on the material being volatile or soluble in a benign solvent and is not suited to materials, such as wood and many other heritage systems like paints and fabrics, which contain cross-linked high molecular weight compounds. In these cases, the technique of pyrolysis GC–MS (py-GC–MS or pyGCMS) has proved to be an invaluable method. Here the sample, typically <1 mg, is introduced into a hot zone where it rapidly volatilised and broken down into fragments. There are several methods of producing the hot zone, including filaments, lasers and micro-furnaces, and in some cases, these can be programmed to control the rate of temperature increase. In general, the spectra produced by the MS can be used as a fingerprint to compare with spectra of known materials or used to 'reconstruct' the molecules in the original polymer. The spectra for archaeological wood are complicated and over 100 compounds have been identified. A very useful feature is that the holocellulose to lignin ratio can be determined and this is the most important factor in determining the degree of degradation of waterlogged wood. A good example are the spectra shown in Figure 4.19 which show the results of archaeological waterlogged wood remains from the roof of a Roman villa buried in AD 79 by the eruption of Vesuvius (Figure 4.19). The area under the peaks is proportional to the concentration and the chemical species of the numbered peaks has been identified. Hence it was possible to relate those originating from holocellulose and lignin and then determine the ratio.

4.4 Impact of Chemistry and Analysis on Conservation, Preservation and Scholarship

Understanding the chemistry of wooden cultural heritage through analytical techniques, such as those outlined in the previous section, plays a critical role in determining the degradation status of these items and thus suitable conservation treatments. This section will explore possible degradation pathways and mechanisms and then also available conservation methodology.

4.4.1 Degradation Mechanisms

The point at which wood decay will start is dependent on the species of the wood, the amount of sapwood present, the use and the environmental conditions it is exposed to. The environment can vary due to oxygen levels, pH, moisture content, temperature, relative humidity (RH), solar radiation, and the introduction of organisms and foreign particles and/or contaminants. These can all impact bonds within the wood, which in turn can reduce the overall strength of the wood. Other damage can include surface roughening, change in colour and warping or twisting. When components of the wood begin to degrade, this can lead to irreparable damage.

All types of wood can be altered and potentially damaged by changes in moisture content, whether this occurs due to heavy degradation allowing elevated moisture levels, or as a result of differences in ambient temperature and RH. As the moisture content varies, wood can expand and contract (or swell or shrink), and the extent to which this happens will vary depending on the axes of the wood which is exposed to the change. A much higher percentage of shrinkage or swelling is observed in the tangential and radial directions, when compared with the longitudinal (see Figure 4.7), simply due to the

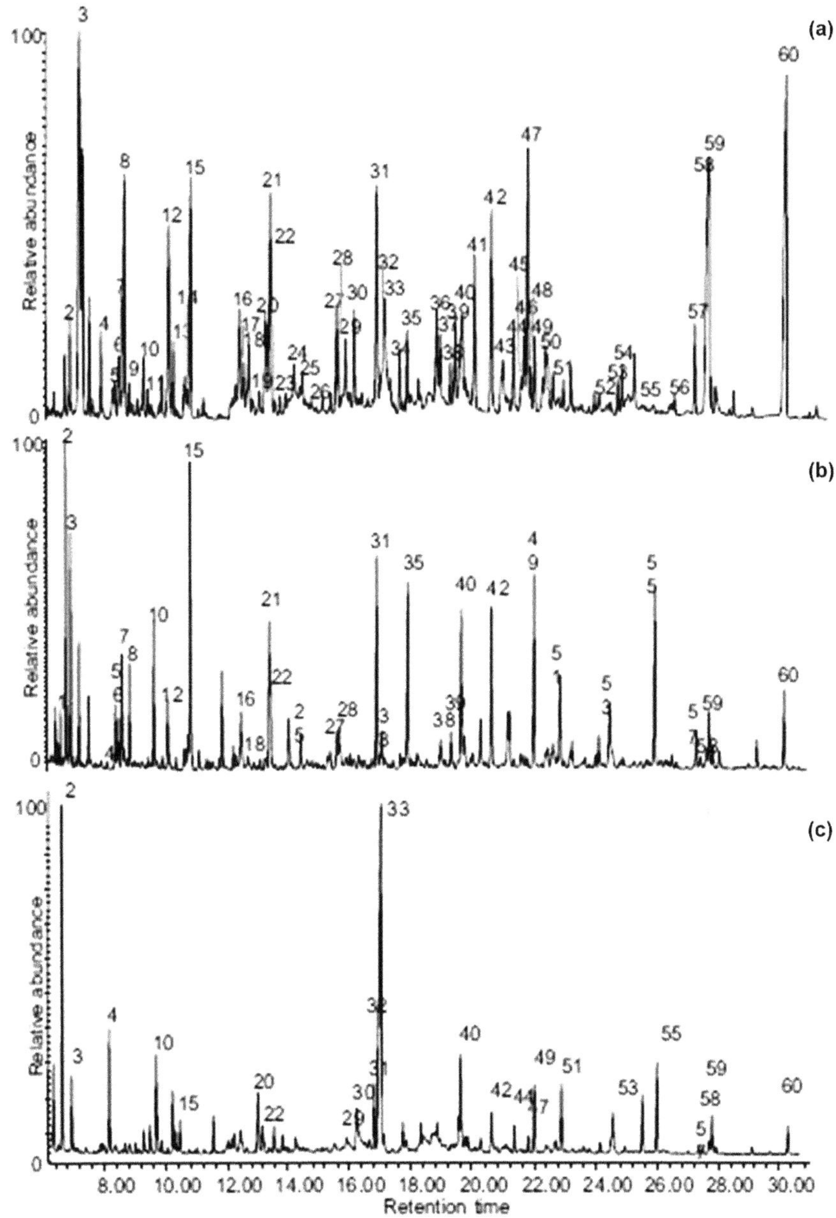

Figure 4.19 Py-GC–MS profiles for waterlogged wood from Herculaneum, (a) reference silver birch, (b) sample of degraded wood showing loss of polysaccharides, (c) sample of degraded wood showing loss of lignin.
Reproduced from ref. 1 with permission from Elsevier, Copyright 2014.

availability of pathways for water in and out of the wood. If there is any loss of wood components, shrinkage can be irreversible. In addition, cracks can occur and warping and distortion can cause failure, as joints and fixtures in the structure are no longer sound. In musical instruments made from wood, change in moisture content can be devastating as it alters the acoustic properties.

In waterlogged wood, the extent of degradation can be determined by the moisture content, as water fills the voids created by the loss of wood components. To do this, wood is sectioned, weighed, completely dried at 100 °C and then weighed. From this the percentage moisture content can be determined.

$$mc = \left(\frac{W_g - W_o}{W_o}\right) \cdot 100\%$$

where:

$mc =$ moisture content
$W_g =$ green weight of the wood
$W_o =$ oven-dried weight of the wood

These can then be assigned levels of degradation, referred to as the De Jong classification system as shown in Figure 4.20. In Class III, well-preserved material dominates, and the moisture content is less than 185% of the dry weight of the wood. In Class II, roughly equal proportions of well-preserved and degraded material are present, and the moisture content is between 185 and 400%. In Class I, degraded material dominates and the moisture content is greater than 400%.

All wood can suffer from biological degradation, in the form of bacterial, fungal or insect attack. It can be difficult to detect as, unlike other materials (*e.g.* stone), the decay can start from the inside and spread out. Also, the rate of decay can be much faster than with other

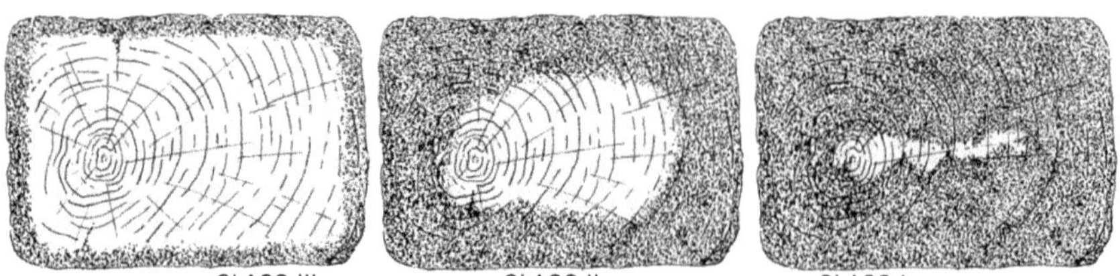

Figure 4.20 De Jong classification of archaeological wood degradation. Adapted from ref. 5 with permission from Elsevier, Copyright 1977.

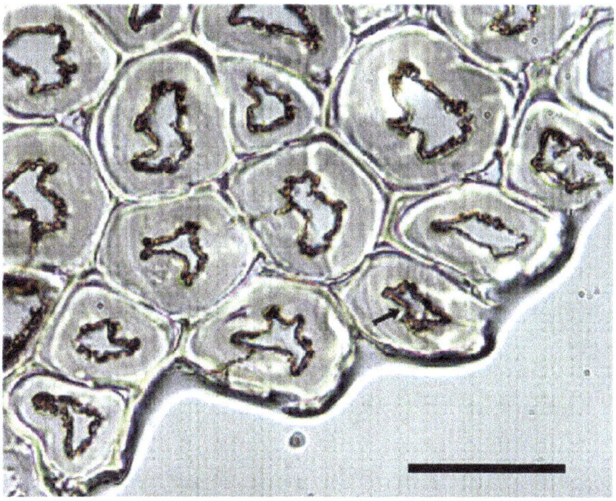

Figure 4.21 Light microscopy of erosion bacteria in *Mary Rose* wood. Photo credit: Glenn McConnachie, Mary Rose Trust.

materials. Bacterial attack can result from the action of erosion, tunnelling or cavitation bacteria. Erosion bacteria decay the wood on the cellular level, attacking the hemicellulose, as can be seen in Figure 4.21. The degradation can be characterised by the formation of erosion troughs or grooves on the lumen surface of the wood cell walls and can occur with minimal oxygen available. Tunnelling bacteria are found in wood exposed to the sea. Like the erosion bacteria, they remove the hemicellulose components of the cell wall; however, they are also known to modify the lignin. The wood appears soft and darkened but this process is thought to be limited to the outer layers. Cavitation bacteria cause decay in the wood cell wall, producing diamond-shaped cavities.

Unlike most plants, fungi lack chlorophyll. They 'feed' on whatever is available to them as they need a steady source of nutrients. The damage from soft rot is largely superficial, and it does not penetrate deep into the wood (Figure 4.22). Brown and white rot can be distinguished by the resulting colour after the degradation. Brown rot rapidly degrades the hemicellulose, but has minimal impact on the lignin, whereas white rot will rapidly degrade all wood components, *e.g.* hemicellulose, cellulose and lignin.

Animals, insects or borers, burrow into the wood and digest the cellulose and hemicellulose (Figure 4.23). Therefore, not only do they consume the wood, they also make pathways for the ingress of other organisms or inorganic contaminants. Some insects show a preference for soft- or hardwoods, whereas others will attack either.

(a)

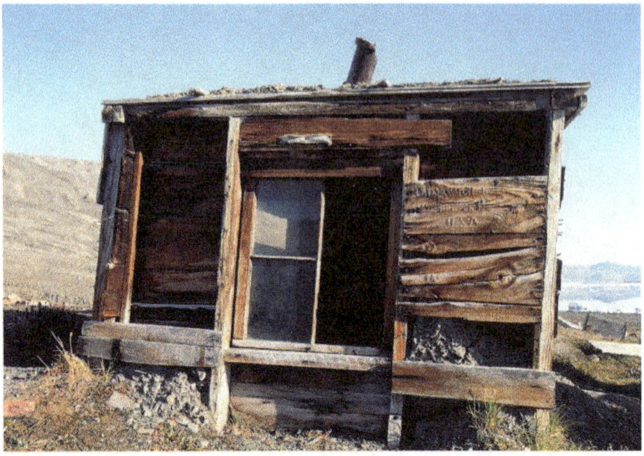

(b)

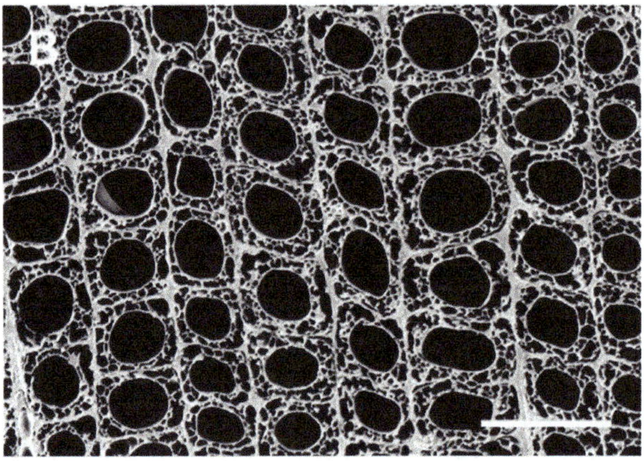

Figure 4.22 (a) One of the Peary expedition shelters, showing signs of wood decay. (b) Advanced stages of soft rot from Fort Conger timbers with extensive soft rot attack and cavity formation within all secondary cell walls (scale bar = 50 μm). Reproduced from ref. 4, https://doi.org/10.1371/journal.pone.0246049, under the terms of the CC BY 4.0 licence, https://creativecommons.org/licenses/by/4.0/.

Termites digest polysaccharides, partially decompose lignin and are usually found in tropical/subtropical climates. Marine wood borers, some of the most destructive wood agents, are able to burrow into the wood, creating a network of pathways through the material, which can be in the order of millimetres in diameter. They can be broken down into different groups: teredinids, pholads, spahaeromatids, and chelurids.

Whether historical or archaeological, there are always ways for acid production to critically damage wood. This can be due to aggressive environments, the evolution of contaminants or the degradation of previous conservation treatments. When exposed to the natural

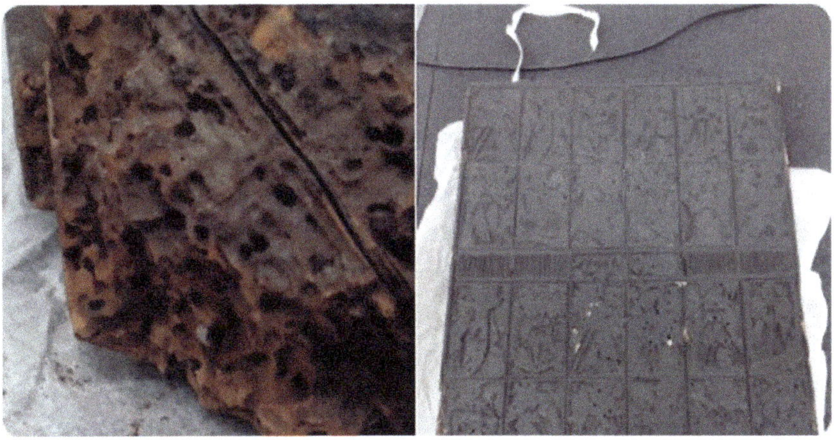

Figure 4.23 Insect attacks of archaeological wood. See also ref. 2. Photo credit: Barry Goodell, Elsevier.

environment, such as buildings, there are many variables that can affect the level of degradation such as moisture, temperature, sunlight, heat, cold, abrasion and biological attack. Weathering is a process that refers to the surface degradation *via* UV radiation, which is able to induce chemical processes that lead to damage. When wood absorbs a UV photon, a free radical can be formed, which in turn reacts with oxygen and water to form a hydroperoxide. The free radical and the hydroperoxide can both result in scission reactions which lead to the degradation of the components of the wood. As the wood weathers, extractives are removed from the surface layers which results in the wood becoming less water repellent. The inclusion of excess water can exacerbate the degradation process by causing cracks and cleavage of the wood at the surface. If sections peel away due to weathering, then in addition to the mechanical damage, pathways are created for the ingress of the culprits of biological attack. Rain can assist in this process as well, as it washes out degraded parts, and exposes more material to be attacked, *via* the hydrolysis of cellulose.

For marine archaeological wood, a major chemical process that threatens the wood is the production of acidic compounds. These can arise from the presence of iron and sulfur, the breakdown of the wood components or the breakdown of consolidants used on them. Sulfur and iron may be present in the wood from seawater and corroded ferrous fixtures, and in the absence of oxygen are found in a reduced form. (It should also be noted that wood does inherently have an organic sulfur component, albeit minimal.) Whilst posing no threat to the wood whilst underwater, once raised and exposed to oxygen the

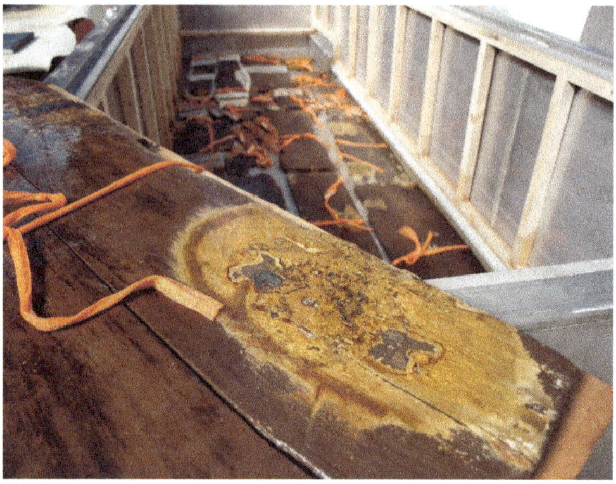

Figure 4.24 Sulfur salts on a *Mary Rose* chest panel.
Photo credit: Glenn McConnachie, Mary Rose Trust.

oxidation of the sulfur compounds is realised and acidic compounds can be formed (Figure 4.24).

$$S_{(s)} + 3/2O_2 + H_2O \rightarrow H_2SO_{4(aq)}$$

$$FeS_{2(s)} + 7/2O_2 + (n+1)H_2O \rightarrow FeSO_4 \bullet nH_2O_{(s)} + H_2SO_{4(aq)}$$

These were first identified in sulfate salts formed on the *Vasa* after a particularly humid summer which aided the oxidation process. This is a difficult problem due to the extent of variation of sulfur distribution and oxidation state that can be found within artefacts.

Iron also plays a role in the breakdown of PEG which can result in the formation of acidic compounds. This is thought to occur *via* a Fenton type reaction where iron(II) reacts with hydrogen peroxide to form a reactive hydroxyl radical which can then react with organic molecules, as illustrated by the following equations:

$$Fe^{2+} + H_2O_2 \rightarrow Fe^{3+} + OH^- + {}^{\bullet}OH$$

$$RH + {}^{\bullet}OH \rightarrow R^{\bullet} + H_2O$$

4.4.2 Conservation

Buildings invariably involve some aspect of wood, whether it is used structurally, for protection, or in a decorative way. It can be employed in a variety of external joinery (eaves, cornices, windows, doors) and internal joinery (wall panels, skirting, doors, floorboards, and staircases and balustrades). If the wood had degraded such that repairs are required, there are a number of options, with pros and cons. A wood reinforcement could be used, but it would need to be of the same species to avoid conflicting material properties between the original wood and repair wood. Epoxies can be used; however, they are generally not used outside due to their potential to interact with moisture, which can result in interfacial damage between the two materials. Last of all steel reinforcements can be used, although again care must be taken to ensure there is adequate ventilation as condensation on the steel could result in an increase in RH experienced by the wood and an acceleration in decay. The moisture on steel could also result in corrosion which could lead to discolouration of the wood.

For outdoor structures, degradation can be mitigated by reducing dampness, which is key to biological attack. Barriers to fungal attack spreading such as metal caps and concrete sections are also possible. Chromic acid (CrO_3) has been used to treat the surface of wood at risk of weathering, as it produces a surface that is both resistant to photochemical degradation and water repellent. Chromium can complex with the wood components to form a highly insoluble product, and therefore reducing the ability of the UV to degrade the wood. Stabilisers and stains also show some ability to slow down the degradation process of weathering.

Light levels will change the appearance of the colour, which can be a problem for furniture. The damage is cumulative and irreversible. It can be removed by sanding but this can only be down a certain number of times and should be approached with caution when dealing with already potentially damaged wood.

For the consolidation of archaeological wood, the most commonly used substance used today is polyethylene glycol ($HO–[CH_2]_n–OH$), often referred to as PEG. This is a water soluble, non-toxic polymer which can be produced at a variety of grades, in other words, polymer chain length (the value of n). This means that the polymer can be tailored to suit the penetration wished to be achieved. Usually, lower grade polymers are used initially as they are able to migrate through the structure more easily and rapidly and can support the cell walls, and then higher-grade polymers are used to seal the surface. Knowing the diffusion rate of the PEG into the wood can be critical to ensure complete filling of the structure. PEG is seen as an advantageous material for this purpose as it is relatively inexpensive and given its water-soluble property, the treatment is always reversible to some degree. An example of *Mary Rose* wood before and after PEG incorporation and drying is given in Figure 4.25. It is clear to see that some of the open vessels are filled with PEG after treatment. In recent

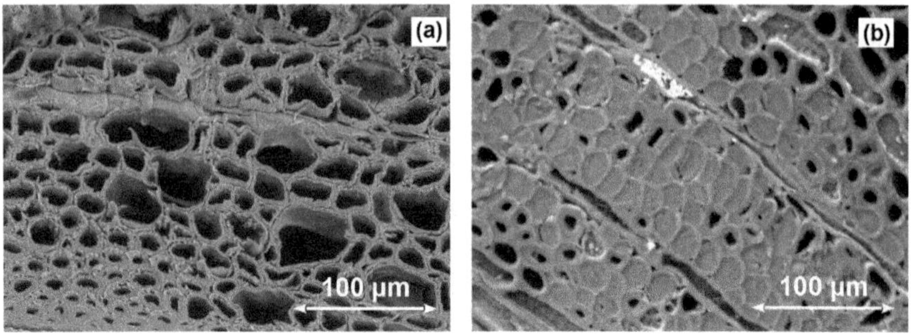

Figure 4.25 Scanning Electron Microscopy image of *Mary Rose* wood (a) without and (b) with PEG.
Photo credit: Eleanor Schofield, Mary Rose Trust.

years some potential problems with the use of PEG have been discovered, which will be discussed in later sections. Alternative treatments to PEG which have been used include silicone oils, sucrose, azaelic acid, silanol, hydrolysed feather keratin, and alum. Following PEG treatment, the residual water needs to be removed which can be done *via* controlled air drying, vacuum freeze drying or *via* chemicals such as mannitol or sorbitol.

A number of methods have been proposed and tested for solving the acid production crisis in these waterlogged archaeological wooden finds. Some of the earlier methods involved applying a borate buffer or an ammonia treatment following the simple chemistry that these would neutralise the acid formed. A more promising and long-term solution has been discovered recently in the fabrication of nanoparticles of suitable neutralisation agents. The advantage of the nanoparticle is that they can easily migrate into the wood and will not threaten the structural stability of the wood. This work was first pioneered for cultural heritage artwork and involved the development of earth-metal hydroxides which combated problematic sulfates *via* the following reaction:

$$Sr(OH)_2 + CaSO_4{\bullet}2H_2O \rightarrow Ca(OH)_2 + SrSO_4 + 2H_2O$$

An alternative method is *via* the use of strontium carbonate nanoparticles. Strontium carbonate is commercially available inexpensively at 'nano' length scales thereby eliminating the need for complex synthesis and reducing costs. An additional advantage is found as the strontium carbonate nanoparticles are able to react with all problematic sulfur, both reduced and already oxidised. This potentially provides a 'one-shot' method which can combat all sulfur and create a product which can reside safely within the wood.

Another avenue explored is to remove the iron from wood, and therefore remove the ability for the oxidation to be catalysed. This has been attempted and to some extent achieved using complexing agents such as ethylenediaminetetraacetic acid (EDTA), diethylenetriaminepentaacetic acid (DTPA) and EDDHMA (ethylenediamine-*N*,*N*′-bis(2-hydroxyphenylacetic acid)).

4.4.3 Scholarship

Continuing scholarship and research will always be required in this field: Wood itself is not fully understood in terms of its properties and behaviour, and as it is such a significant material in our lives that it is still a major topic of investigation and will continue to be explored. The development of new analytical techniques means our understanding is never static, and it is crucial to keep investigating and exploring new ways to study it as a material. Furthermore, there will always be the challenge of aligning the requirements of conservation work with curatorial and ethical considerations.

4.5 Case Study: The *Mary Rose*

The Tudor warship *Mary Rose* sailed the seas for 34 years during the reign of Henry VIII (Figure 4.26). Her construction, commissioned in 1509 shortly after Henry's accession to the throne, signalled the beginning of the Royal Navy. Renovations over the years reinforced the

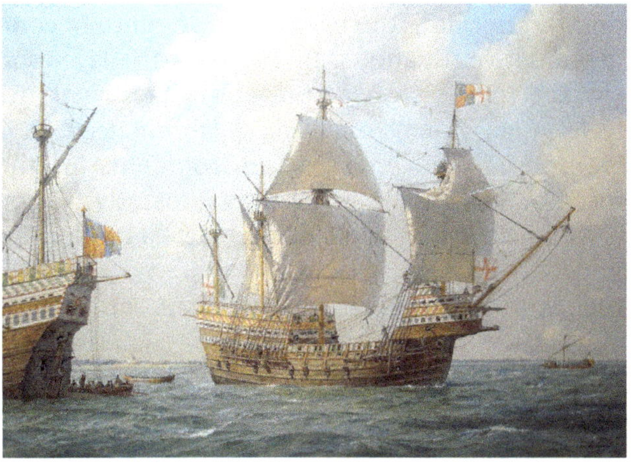

Figure 4.26 Artwork of the *Mary Rose*.
Photo credit: The Mary Rose Trust/Geoff Hunt PPRSMA.

structure in order to accommodate more weapons and increase her capabilities in battle. The years of successful battle were overshadowed by the sudden and dramatic loss during the Battle of the Solent in 1545. Whilst the exact explanation for what happened that day will remain unknown, evidence points to a failure to close the gun ports when turning in bad weather resulting in a mass intake of water, which rapidly sank the ship. This happened so readily as a result of the vessel lying much lower in the water than when initially constructed due to the renovations and larger load on board.

Initial attempts to recover the ship were unsuccessful and the site remained untouched until the 1800s. At this point, the Deane Brothers from Whitstable, Kent, who used pioneering diving equipment, located the site and completed some excavations (starting in 1836). As the site was not protected at this time, it was legal to remove objects and sell them for profit. The site was then untouched for over a century until Alexander McKee, a local historian and diver, and his team of volunteers attempted to locate the site once again, starting in the mid-1960s. This task was not straightforward, as the entire site had gradually been covered in silt, in some areas up to 2 m thick. Acoustic and sonar techniques, state of the art technologies at the time, were used to identify an anomaly under the seabed in the approximate location it was hoped the *Mary Rose* rested. Excavations following this eventually revealed timbers which resulted in the positive identification of the *Mary Rose*.

The thick covering of silt was the key to the survival of the *Mary Rose*, as it acted as a protective layer and a barrier to oxygen permeation. Typical marine organisms could not access the wreck materials and did not have the oxygen they require to complete their biological degradative functions. In addition, the layer anchored the remaining wreck to the seabed, preventing destruction from tidal erosion. However, as soon as excavations commenced, the wreck was once again exposed to aggressive environments in the form of seawater and then air. This meant it was imperative to complete the excavations as quickly as possible and to have a conservation plan in place on land.

Over 19 000 artefacts were excavated from the *Mary Rose*, prior to the raising of the hull in 1982 (Figure 4.27). The artefacts were originally placed in water in order to prevent drying out and the associated shrinkage and cracking that would occur due to the degradation of the wood. Whilst visually in very good condition, which indeed it was considering the age and where it had resided for nearly 500 years, some wood degradation had occurred. This meant that the

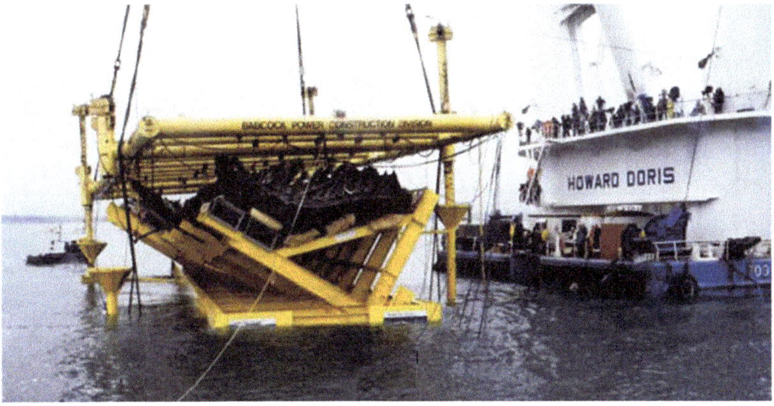

Figure 4.27 The raising of the *Mary Rose*.
Photo credit: Mary Rose Trust.

wood structure was effectively being supported by water occupying regions where there once would have been the wood components cellulose and hemicellulose. The hull was too large to completely submerge in water; therefore, a spray system was constructed around it to douse the hull in water. The water was cold, approximately 5 °C, in order to reduce biological activity.

On reviewing comparable projects, such as the Bremen Cog and the *Vasa*, the treatment methodology of impregnation with PEG was chosen due to the favourable results obtained. Despite many similarities between these projects, there were many subtle differences, such as the type of wood used, the original source of the wood, the length of time the ship was in use and therefore how much 'lifetime damage' it experienced, how long it was in a marine environment and the exact nature of this environment. Therefore, it was essential for research to be completed on *Mary Rose* wood treatment with PEG. The resulting treatment selected, in terms of consolidation and drying, depended on the size of the object and the level of degradation. The latter was in some instances simply determined by calculating the density of an object. By comparing this to fresh wood, it was possible to determine how much of the original material had been lost, and therefore what PEG would be appropriate to introduce. For the hull, samples were taken which again could have the density measured, but also could have thin sections taken to analyse using light and SEM. This enabled detailed cellular analysis of the degradation (Figure 4.28).

The wooden artefacts (*i.e.* all items except the hull) were submerged in tanks of water and low-grade PEG was periodically added into the

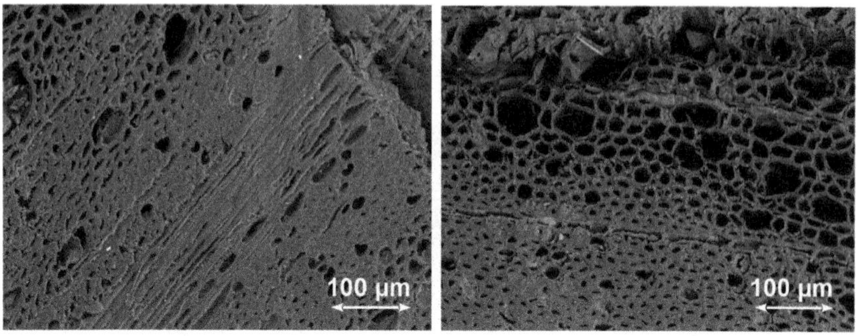

Figure 4.28 SEM image of fresh wood *versus Mary Rose* wood. Photo credit: Eleanor Schofield, Mary Rose Trust.

solution in incremental steps to the required concentration (Figure 4.29(a)). The incremental nature of the PEG inclusion is critical to avoid big changes in the viscosity of the solution, which could result in damage to the cell walls as the PEG-rich solution pushes in to replace the water present in the cell. Once the low-grade treatment was complete, a higher-grade PEG was incorporated in order to seal the more degraded surface of the wood. Heating and pumping had to be employed to ensure that the solid PEG dissolved into the solution and fully circulated the tank. Throughout the process biocide was included in the tanks, to stop bacterial attack from species such as legionella which can be rife in warm water. Following treatment, the artefacts were dried in a freeze drier (Figure 4.29(b)). Freezing the artefacts transforms any remaining water to ice, which is then sublimated out through the application of a vacuum within the drier. The increased speed of drying means that dimensional changes and shrinkage are minimal.

As with the earlier water treatment, the hull could not be placed in a tank, and therefore a spray system had to be constructed around the hull to ensure an even covering of PEG (Figure 4.30(a)). The lower grade PEG first used was a liquid at room temperature, whereas the higher grade, being a solid, required a different spraying system which could cope with the elevated temperature required (28°C). Through the PEG spraying process, the conservation involved routine increases in PEG, checking the spray nozzles to ensure no blockages and cleaning of filters within the system. The content of PEG was monitored daily by taking a sample of the solution surrounding the hull and measuring it using a refractometer. This served as an easy and affordable method that would give an indication of the PEG reaching the hull. To measure inside the ship, core samples were

(a)

(b)

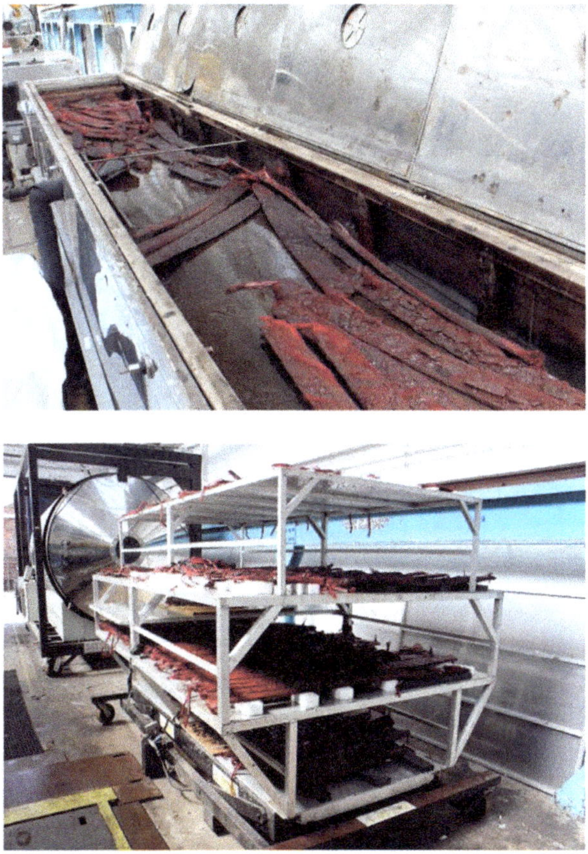

Figure 4.29 (a) Artefacts in tanks (b) Freeze drier. Photo credit: Glenn McConnachie, Mary Rose Trust.

taken and gas permeation chromatography was used to determine the levels of PEG within the wood in terms of concentration, depth and location within the wood structure. The drying technique used for the artefacts could not be employed again here due to the size of the hull. Therefore, a laser scan was taken of the hull structure and computational fluid dynamics (CFD) modelling (Figure 4.30(b)) was used to determine where air drying tubes should be located to ensure that the air velocity, temperature and RH were uniform (Figure 4.30(c)). The temperature and RH levels chosen were deduced from sample studies on *Mary Rose* wood which had experienced the same PEG treatment.

It is a common misconception that once the consolidation and drying of archaeological wood are finished then the conservation is

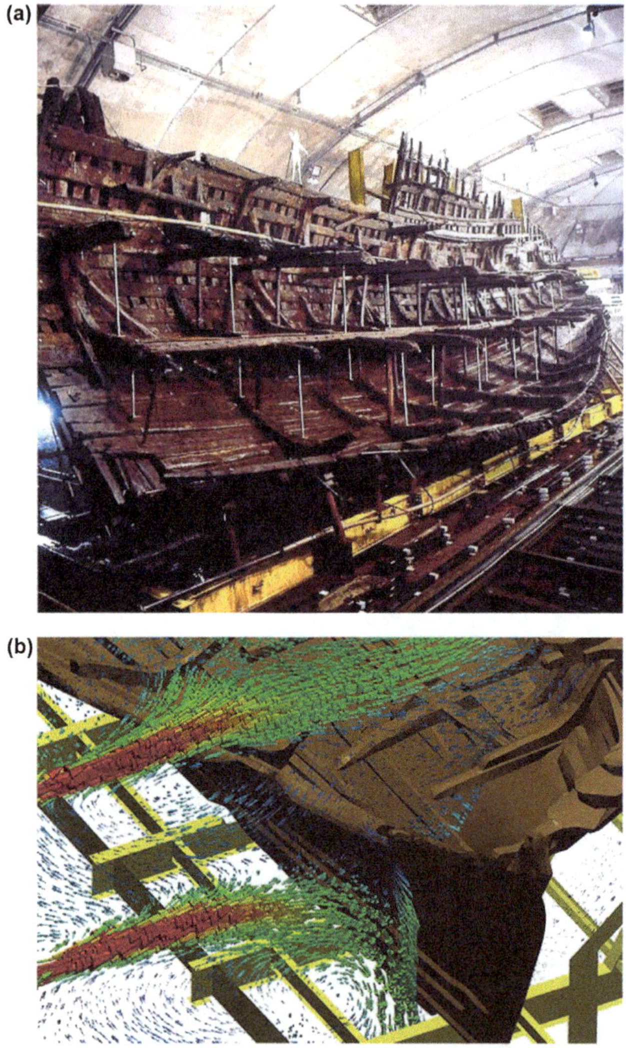

Figure 4.30 (a) Spraying the hull (b) CFD models of drying (c) Drying the hull. Photo credit: Mary Rose Trust.

complete. In truth, what results is a new and complex material that is a mixture of the degraded wood, PEG, and other contaminants that seeped into the wood whilst in the sea and under the seabed. Routine inspection of conserved finds is critical in order to pick up early signs that unwanted reactions are taking place that could be detrimental to the wood. These signs can manifest themselves in colour changes, the appearance of precipitates and visible movement, cracks or distortion. By following the observations made on similar projects, it was possible to set up dedicated monitoring of the *Mary Rose* hull as it dries, and an overview of what techniques have been used, and preliminary insights are given here.

The drying of the *Mary Rose* hull has inevitably led to some physical changes in the structure. These have been monitored in a variety of ways to ensure both local and global movement has been captured. Core samples were taken throughout the drying process, and the percentage moisture content was determined by sectioning, weighing when wet, drying completely and then weighing again. This has allowed us to understand the rate of drying, how it varies around the ship, and how to correlate moisture loss to movement. A series of local markers were installed around the ship on existing cracks and joints to be able to track their evolution in real time as a function of drying time (Figure 4.31(a)). Alongside this, more sophisticated techniques involving laser scanning have given us detailed overall structural information. A total station (a surveying instrument for measuring distances and angles to particular points), installed shortly after drying commenced, locates over 30 targets attached to the hull and records their position relative to reference points on the ship's supporting cradle. Full point-cloud laser scans have also been completed, which then allow not only absolute movement determination but the shrinkage and crack evolution on individual timbers (Figure 4.31(b)).

Alongside mechanical changes, it has been critical to monitor chemical changes within the wood and the potential for destructive acids to form. One particular source of acid comes from sulfur and iron; sulfur is present from the seawater, and iron is present from corroded artefacts. In low oxygen conditions these can combine to form reduced sulfur compounds such as greigite, pyrite and pyrrhotite amongst others. These remain relatively stable whilst in low to no oxygen environments but can rapidly form damaging acids or salts when exposed to oxygen. A programme of work has been underway since the drying commenced, and is ongoing, to monitor the evolution of these potentially harmful compounds, using X-ray absorption spectroscopy at Diamond Light Source.

Further work has focused on trying to find solutions should the sulfur- and iron-related acids become problematic. These treatments focus on trying either to neutralise or to remove the harmful entity *e.g.* iron.

With the *Mary Rose* we also have to be mindful that other acid sources are present, such as the breakdown of the wood itself forming acetic and oxalic acids, and the breakdown of the PEG which can

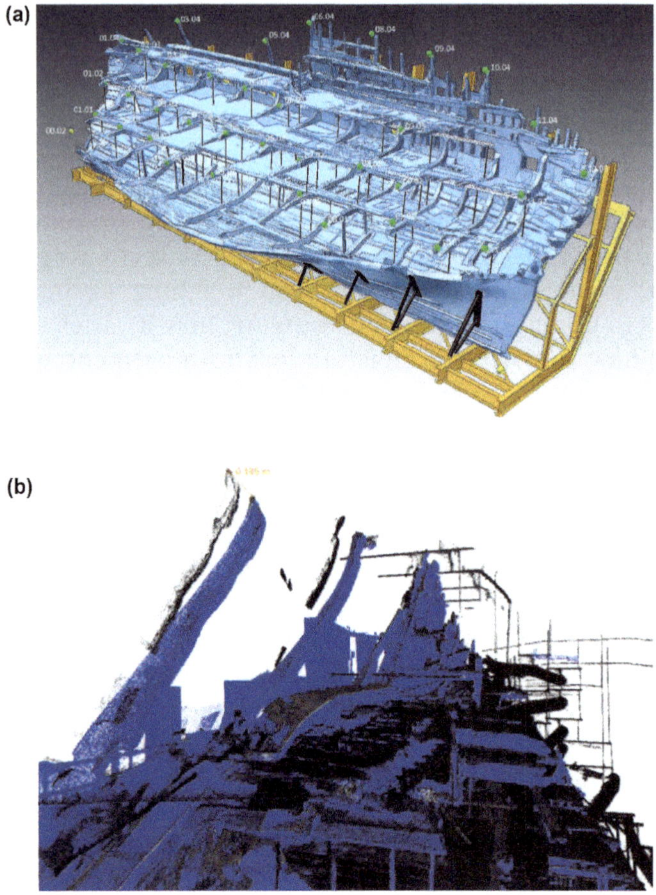

Figure 4.31 (a) Monitoring markers on the *Mary Rose* (b) Laser scan of a beam from the hull.
Photo credit: Mary Rose Trust.

produce formic acid. These will need to be monitored to ensure that limited degradation occurs, and work is underway to determine the most time- and cost-efficient way to do this.

Recommended Reading

Conservation

For Future Generations: Conservation of a Tudor Maritime Collection (Archaeology of the Mary Rose), ed. M. Jones, The Mary Rose Trust, Portsmouth, 2003.

C. Gjelstrup Björdal and D. Gregory, *WreckProtect: Decay and Protection of Archaeological Wooden Shipwrecks*, Archaeopress Archaeology, Oxford, 2012.

NMR

J. B. Lambert, C. E. Shawl and J. A. Stearns, Nuclear Magnetic Resonance in Archaeology, *Chem. Soc. Rev.*, 2000, **29**, 175–182.
M. Bardet, G. Gerbaud, M. Giffard, C. Doan, S. Hediger and L. Le Pape, ^{13}C High-Resolution Solid-State NMR for Structural Elucidation of Archaeological Woods, *Prog. Nucl. Magn. Reson. Spectrosc.*, 2009, **55**, 199–214.
Salanti, L. Zoia, E.-L. Tolppa, G. Giachi and M. Orlandi, Characterization of Waterlogged Wood by NMR and GPC Techniques, *Microchem. J.*, 2010, **95**, 345–352.
M. P. Colombini, J. J. Lucejko, F. Modugno, M. Orlandi, E.-L. Tolppa and L. Zoia, A Multi-Analytical Study of Degradation of Lignin in Archaeological Waterlogged Wood, *Talanta*, 2009, **80**, 61–70.

Py-GC–MS

Degano, F. Modugno, I. Bonaduce, E. Ribechini and M. P. Colombini, Recent Advances in Analytical Pyrolysis to Investigate Organic Materials in Heritage Science, *Angew. Chem., Int. Ed.*, 2018, **57**, 7313–7323.

XAS

S. Calvin, *XAFS for Everyone*, CRC Press, Boca Raton, 2013.
G. Bunker, *Introduction to XAFS: A Practical Guide to X-ray Absorption Fine Structure Spectroscopy*, Cambridge University Press, Cambridge, 2010.
L. Bertrand, M. Cotte, M. Stampanoni, M. Thoury, F. Marone and S. Schoder, Development and Trends in Synchrotron Studies of Ancient and Historical Materials, *Phys. Rep.*, 2012, **519**, 51–96.
A. V. Chadwick, A. Berko, E. J. Schofield, J. F. W. Mosselmans and A. D. Smith, Application of Microfocus X-Ray Beams from Synchrotrons in Heritage Conservation, *Int. J. Archit. Heritage*, 2012, **6**(2), 228–258.

References

1. D. Tamburini, J. J. Lucejko, F. Modugno and M. P. Colombini, Characterisation of archaeological waterlogged wood from Herculaneum by pyrolysis and mass spectrometry, *Int. Biodeterior. Biodegrad.*, 2014, **86**, 142–149.
2. B. Goodell, Wood products: Deterioration by insects and marine organisms, *Encyclopedia of Materials: Science and Technology*, ed. F. Beal, Elsevier Science Ltd, 2001., p. 6.
3. K. E. High and K. E. H. Penkman, *Heritage Sci.*, 2020, **8**, 2479.
4. R. A. Blanchette, B. W. Held, J. Jurgens, A. Steer and C. Dupont, *PLoS One*, 2021, **16**, 1–16.
5. J. de Jong, in Conservation Techniques for Old Archaeological Wood from Shipwrecks Found in the Netherlands, ed. A. H. Walters, *Biodeterioration Investigation Techniques*, Elsevier Science & Technology, London, 1977, pp. 295–338.

5 Leather and Parchment

L. Robinet

Centre de Recherche sur la Conservation (CRC), Muséum national d'Histoire naturelle, Ministère de la Culture, CNRS, 36 rue Geoffroy Saint-Hilaire, 75005, Paris, France
Email: laurianne.robinet@mnhn.fr

5.1 Introduction

Since prehistoric times, man has used animal skin, developing technologies to transform it, to make objects for everyday use, artistic or spiritual expression. Leather has been used for several millennia and appeared in different civilizations.[1] In contrast, the use of parchment as a writing support is said to be more recent, and seems to be dated to the first millennia in Assyria or to the second century BC according to Pliny. Skin is a material that has been largely used due to its unique physical properties, being at the same time flexible and highly resistant to strain, tearing or abrasion. Until the development of woven textiles, it was the only material available in a large format.

Leather and parchment or rawhide are both made from animal skin, also called hide when it comes from larger animals.[2] However their manufacturing process differs, thus leading to two different materials from their appearance, properties, use and reactivity towards the environment. Parchment is an untanned animal skin, dried and scraped under tension. Parchment has been mostly used in Europe as a writing support because of its lightness, light colour and strength properties, however this material has also been employed to make musical instruments, canoes or even windows (see Figure 5.1).

Conservation Science: Heritage Materials, 2nd Edition
Edited by Paul Garside and Emma Richardson

Published by the Royal Society of Chemistry, www.rsc.org

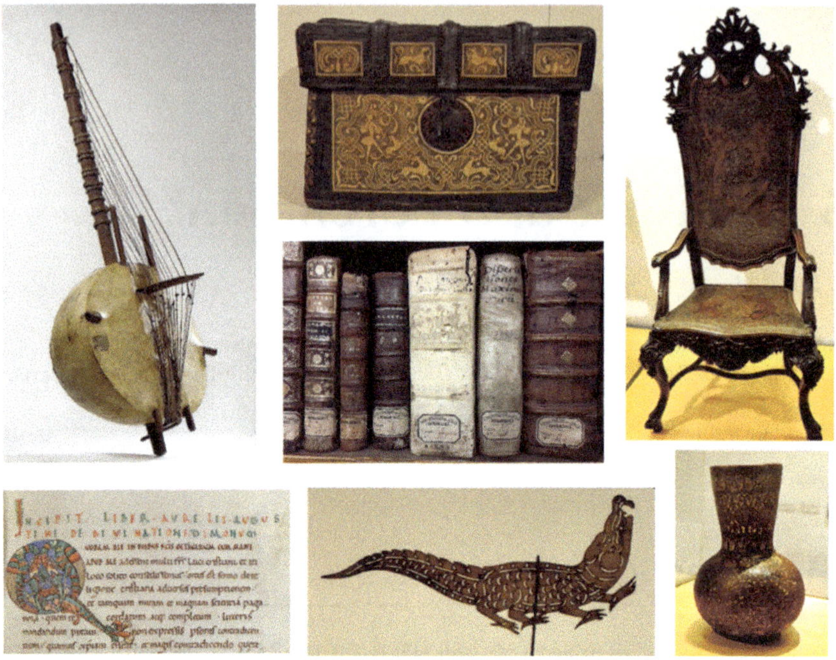

Figure 5.1 Different museum objects made of skin. Top left to right: Harp-lute "kasso" from Senegal and dated from the 20th century, Copyright Claude Germain, Musée de la Musique, Paris; Leather chest, 17th century, Leather museum, Vic; Spanish leather chair, 18th century, Leather museum, Vic. Middle: Books with different skin-based bookbinding, Avranches library. Bottom left to right: Manuscript illumination on parchment from the 11th century, Avranches library; Figure for shadow theatre from Thailand made of untanned skin, Leather museum, Vic; Coptic vase made of cuir bouilli, Leather museum, Vic.

Parchment can only be made from mammal skins, because they produce an unstained and more uniform product. Vellum is a particular type of parchment that is very thin and smooth and has been prepared using the skin of very young animals, often calves. Finally, membranes like stomach tissues have also been employed, in particular to make clothes or musical instruments.

In contrast, leather is an animal skin that has been tanned to protect it from putrefaction and make it resistant to water. Leather has been selected for its strength and flexibility to make clothes, shoes, bookbinding, and elements for mechanical use (*e.g.* belts, harnesses) or decorative purposes (*e.g.* jewellery, gilt leather) (see Figure 5.1). Depending on the intended product, several parameters in the manufacturing of leather can be selected or controlled, such as the skin thickness (dependent on the animal species and age), the

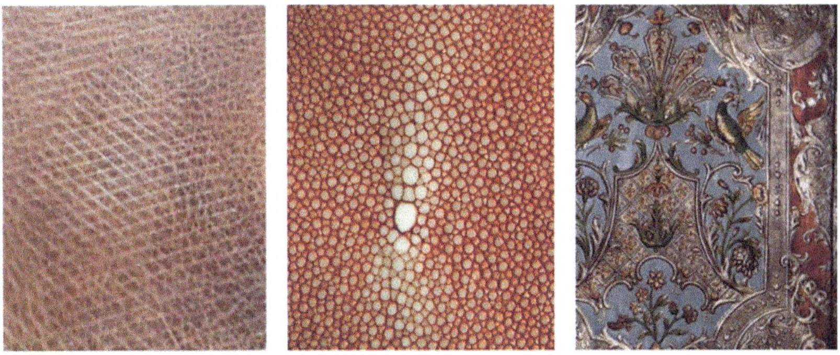

Figure 5.2 Details of the surface of different leathers. From left to right: Russian leather (exported from Russia mostly in the 17th–18th century), shagreen (earliest known example of use dating from the 16th century), and gilt leather (widespread in Europe between 16th and 18th century).

tanning process or the finishing steps.[3] Any animal skin can be used to make leather, mammal or fish, however it relies mostly on materials locally available, and for that reason, generally skins from local domestic animals were used. Thus, European people mainly used goat, sheep or calf skin, whilst Inuit mostly employed seal and whale skin. In the field of leather, specific terms can be encountered that relates to the use of particular animal skins or manufacturing techniques. We could mention *Russia leather* tanned using birch wood and birch oil having a peculiar surface pattern and smell, *shagreen* made from ray or shark skin,[4] *gilt leather* made by adhering a silver leaf to the leather then covered with a yellow varnish, or *cuir bouilli* prepared by heating the vegetable leather in water to mould it and achieve a hard material (see Figure 5.2).

Outside these manufactured objects, other skin-based artefacts are found in heritage collections, such as mummies and natural history specimens. In these cases, different treatments are applied to preserve the skin together with the hair of the complete human or animal body, however they will show similar behaviour towards the environment, and as a result, the same analytical approaches are applied in their study.

5.1.1 Manufacturing Process

Before starting the transformation, the skins or hides are washed in running water to remove the blood, dirt or salt used for preservation. Then, the first step, common to both leather and parchment, is the removal of the hair and the remaining flesh from the skin. Depending on the period and the location, different methods have been applied to make the skin swell and to loosen the hair. Biochemical methods are the most primitive and involve the putrefaction of the skin accelerated by the presence of enzymes and resulting from the fermentation of animal or vegetable substances that are often added. Chemical methods imply the use of alkali, commonly a lime bath (mostly from the 8th century in Europe), but also wood ashes and nowadays sodium sulphide added to lime. Once the hair was loose, the hide was placed on a wooden beam and scraped on both sides with a two handled knife.

At this stage, the manufacturing differs. If the skin is left to dry untanned, it is named rawhide, and such a skin is mainly encountered in ethnographic collections or natural history collections for small animals like birds. To make a parchment, the skin is stretched on a frame, where it is scraped with a circular knife and left to dry sometimes using chalk or lime paste (see Figure 5.3). The drying step is very important for the quality of the parchment; the rate has to be controlled and the skin frequently retensioned. During this step, different pastes made of chalk, gypsum, ashes and/or lime may have been applied to remove the fats, and powdered pumice used to smooth the surface. When the skin is dried, both sides are scraped and shaved again to get a uniform thickness, before the parchment is cut from the frame. Some medieval recipes mention particular treatments, like the sprinkling of alum or vegetable tannin to partially stabilize the parchment or the application at the surface of animal

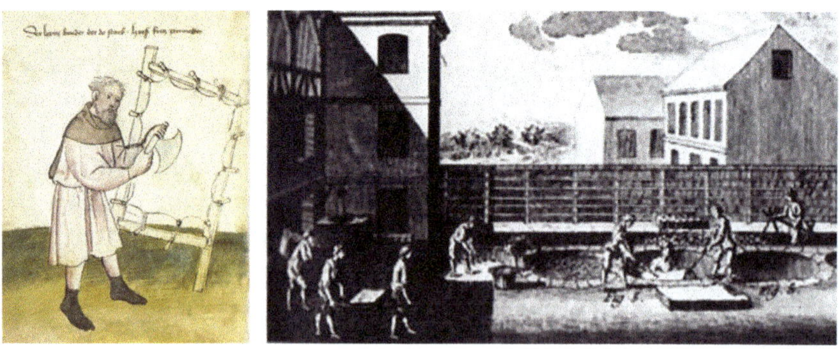

Figure 5.3 Left: Parchment maker - Amb. 317.2° Folio 34 verso (date 1425) Stadtbibliothek Nürnberg. Reproduced with permission from Stadtbibliothek im Bildungscampus Nürnberg, Amb. 317.2°, f. 34v. Right: Tanning pits during the manufacture of leather, Encyclopaedia of Sciences, Arts and Trades, Diderot, 1769. Reproduced with permission from Bibliothèque nationale de France.

glue or egg white to facilitate writing. For special rendering, parchment has also been died using animal or vegetable dyes (see Chapter 8) or made transparent to produce windows by impregnating the parchment with gum, oil or animal glue.

In the making of leather, the important step is tanning that transforms skin into a material resistant to microbial attack, water and heat. Tanning involves a chemical reaction between the collagen in the skin and the chemicals in the tannin that induces the removal of water from the skin, compacts the tissues and increases the fibre strength.[10,11] The choice of the tanning method depends on the characteristics required for the skin regarding colour, softness, rigidity, or stability, but also on the materials available at a time locally. A distinction is made between organic tannins, which include vegetable or oil tanning, and mineral tannins, with alum and chromium tanning.

Vegetable tanning. Until the end of the 19th century, the most common tanning method was vegetable tannin that could be extracted from different parts of a plant: wood, bark, seed, leaves or nuts. Vegetable tanning gives a brown colour to the skin, which hue depends on the plant used. It generally produces a leather with good water resistance and offers the possibility of moulding. With hides a thick and strong leather is obtained. In traditional methods, the skins were put in pits filled with plant fragments, and the process required twelve to eighteen months to turn them into leather (see Figure 5.3). Nowadays, the combination of tannin extract and mechanical action by the use of a rotating tanning drum, greatly reduces the time of the tanning step.

Oil tanning or semi-tanned skin. Also known as wash leather or "chamois", oil-tanned and brain-tanned leathers have recognizable appearance from their yellow colour, softness and high hydrophilicity properties (see Figure 5.4). Oil tanning is one of the most ancient processes and was mostly used by Native American and First Nations communities in North America (buckskin) or in central Asia and the Far East.[7] In this case, the skin is tanned with unsaturated oil such as cod liver oil, brain, or butter, in combination with mechanical work (such as batting or chewing) to make the oil penetrate.

Smoke tanning. This ancient process has mostly been used by First Nations, Indigenous or Aboriginal people. From the smoking, the skin becomes waterproof, rigid, with a smoke smell and gets a colour that can vary from yellow, brown to red depending on the fuel used. This process can be used in combination with oil tanning to accelerate oxidation of the fats.

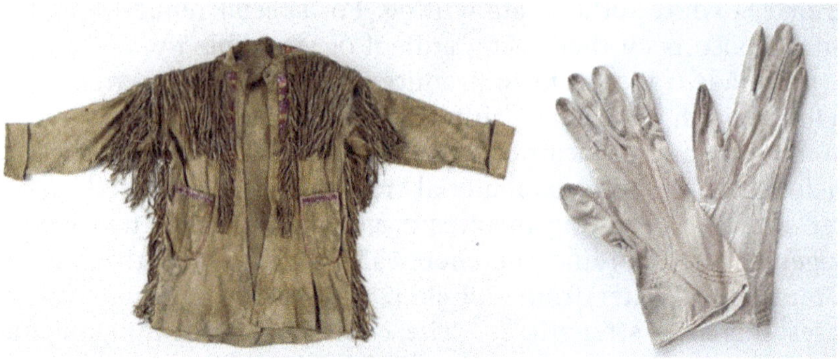

Figure 5.4 Leather objects illustrating different tanning processes. Left: North American leather jacket made of semi-tanned skin, which has been smoked, 1870. Reproduced with permission from CCI, Copyright © Government of Canada, Canadian Conservation Institute. Right: Gloves made of alum tawed skin.

Alum tawing. This technique has also been used for a long time for leather making. It relies on the use of potash alum ($KAl(SO_4)_2{\cdot}12H_2O$), in combination with water, salts, flour and egg yolk, to produce a white and soft material ideal for making gloves (see Figure 5.4). Since the process is reversible in the presence of water, the term tanning cannot be applied here, and tawing is used instead. Such skins were often dyed with vegetable dyes, and in that case, alum also acted as a mordant (see Chapter 8). Alum salts have also largely been used in the preparation of natural history specimens, due to the great flexibility and light colour given to the skin but also because of the antiseptic properties of the salt.

Chrome tanning. This was used for the first time in 1858 and is currently the tanning method primarily employed worldwide, as it represents about 90% of production. It is based on the use of chrome sulfate salt that creates strong links with the collagen molecule responsible for the high heat resistance and water stability. The process is fast and produces a highly flexible leather with a blue-grey colour (and hence called "wet-blue"), but it can be easily dyed in a wide range of colours.

Combination tanning. Depending on the application, tanning methods have sometimes been combined to get the properties of both and improve the characteristics of the leather produced. Thus, the use of vegetable tannins together with alum improve the collagen stability as well as the tensile and shear strength, whilst providing full and soft leathers. Chrome used in combination with vegetable tannin produces leather with increased durability, provided by the chrome, and firmness and water resistance characteristic of vegetable tanning. Moreover, the order in the use of the tannin has its importance, since most of the properties are provided by the first tanning method and the second completes them. Nowadays, the severe restrictions imposed on the use of chromium forces the leather industry to find alternatives, and combination tanning is one of them. Following the tanning process, the leather can undergo different finishing steps, in particular lubrication, with the addition of fats, essential to facilitate the movement of fibres and to soften the skin. Mechanical action known as staking is also used to provide flexibility to the leather.

5.2 Chemistry, Structure and Microstructure of Leather and Parchment

5.2.1 Skin Structure

By nature, a skin has a heterogeneous structure, varying between different areas as well as with depth. Over a whole skin, the thickness, the collagen fibre orientation and the composition will differ depending on the part of the body (neck, belly or spine) as this relates to its functions when the animal was alive. Consequently, it has an impact on the physical and mechanical properties of the material. Notably, the spine area is the strongest and least stretchy part, whilst the belly area is the weakest and the most elastic. In depth, three parts are distinguished in the structure: *the epidermis,* the external layer composed mainly of keratin; *the dermis,* the main part of the skin in the centre mostly made of collagen fibres; and *the hypodermis,* the underlying layer, where fat is stored. Apart from natural history specimens, some ethnographic objects, and fur, where the epidermis with the hair is retained, in the manufacturing of leather or parchment, the outermost and innermost layers are eliminated to retain only the dermis. The dermis layer can be divided into two sub-layers: the grain or papillary region toward the epidermis, composed of fine and loosely arranged collagen fibres, and the corium or flesh or reticular region towards the hypodermis, composed of densely packed and thicker collagen fibres. These two regions are easily recognized in leather, as the grain side has a smoother surface showing hair follicles, whilst the flesh side has a fluffy surface and the fibres can be seen. They are more or less identifiable in parchment.

In leather, the original dermis structure is maintained, fibres are entangled as in a fresh skin, which therefore retains great flexibility.

The animal species and age will affect the skin thickness, size, density and shape of the fibre bundles. As an example, the expected thickness of leather from an adult cow is around 4–6 mm, whilst it is around 1 mm for a one-month-old calf and 1.5 to 2 mm at six months. Due to the manufacturing processes, parchment displays very different morphological characteristics. The stretching step greatly modifies the organization within the dermis structure: the collagen fibres are rearranged in a lamellar plane parallel to the skin surface. The skin becomes thinner, semi-rigid, with a thickness ranging around 200 to 600 μm (0.2–0.6 mm). This distinct structural organization of the fibres is the essential criterion to discriminate between leather and parchment.

5.2.2 Collagen Structure

The dermis layer contains two types of collagen, principally collagen type I (also called fibrilar collagen) with a smaller amount of collagen type III. Collagen has a high hierarchical organization responsible for the mechanical strength of the material.[8,9] As a protein, it is made from a sequence of amino acids characterized by a high glycine content (every third amino acid) and the presence of two rare amino acids hydroxyproline and hydroxylysine. The collagen molecule corresponds to the assembly of three amino acids chains (α chains) into a triple helix that is maintained by hydrogen bonding providing stability and rigidity. The collagen molecules are then aligned parallel, staggered by about one quarter of the molecule length relative to each other and stabilized by cross-links, thus forming fibrils (see Figure 5.5).

This particular assembly produces a characteristic periodic banding with a 67 nm interval in the fibril, which is visible in electron microscopy or atomic force microscopy (see Figure 5.5). Lastly, these fibrils are again assembled into fibre bundles.

Although a large amount of the water is removed during the manufacturing process, leather and parchment still retain some water that provides flexibility to the material. As a simplified approach, two types of water are present, free water, corresponding to multilayer water between the fibres, and molecularly bound water combined to different sites in the protein and believed to stabilize the triple helix. In normal conditions of temperature (25 °C) and relative humidity (50–55%), the water content is around 14% in a new leather and 16% in a new parchment. As the relative humidity fluctuates, the material will absorb or desorb water, particularly in parchment, whilst leather has a lower water absorption capacity because tannins block sites for water fixation.

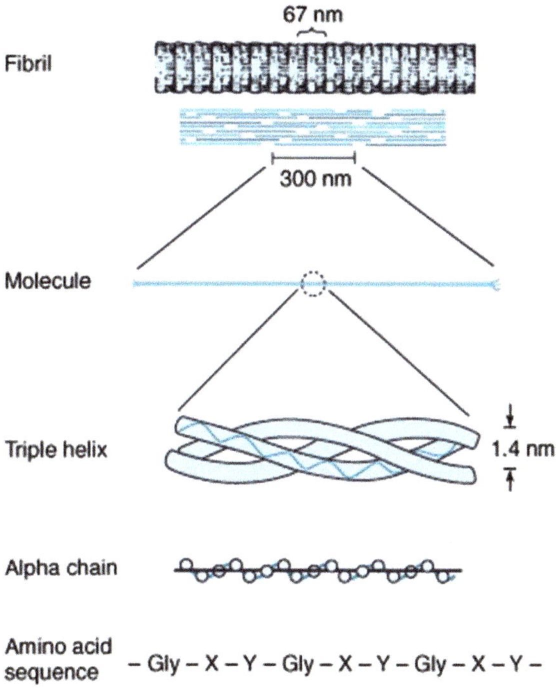

Figure 5.5 Collagen structural organization. The amino acid sequence Gly-X-Y, with Gly for Glycine, X and Y, often proline or hydroxyproline. Image credit: Namrata Chhabra.

5.2.3 Tannin Interaction

In leather, the tannins fill spaces within the structure resulting from the removal of water, proteins and fats, thus the fibres do not collapse when the skin dries. The interaction created between tannin and collagen stabilizes the molecule and increases its thermostability, which is reflected by the measurement of the shrinkage temperature (T_S). The added stability will depend on the tannin and the strength of its interaction with collagen (see Table 5.1).

Vegetable tannins are complex mixtures of polyphenol compounds and two groups are differentiated: hydrolysable and condensed tannins. Plants may contain both in different proportions. Hydrolysable tannins (or pyrogallols) are sugar derivatives that hydrolyse easily into their acids, with gallic acid forming a subgroup called gallotannins (from sources such as sumac, tara, *etc.*), and ellagic acid forming the other subgroup called ellagitannins (from chestnut, oak, *etc.*) (see Figure 5.6). Nevertheless, the identification of ellagic acid does not necessarily indicate the presence of ellagitannins, since the molecule

Table 5.1 Shrinkage temperature (T_S) in different skin-based materials.

Type of tanning	T_s (°C)
Raw skin	65–67
Parchment	55–60
Alum tawed skin	50–63
Fat tanning	65–70
Vegetable tanning – hydrolysable	70–75
Vegetable tanning – condensed	80–85
Chrome tanning	>95

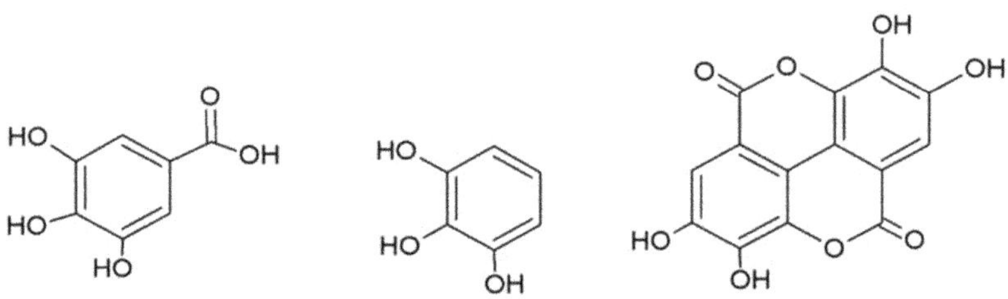

Figure 5.6 Main chemical groups encountered in hydrolysable tannin.

can also be formed from gallotannins by photo-oxidation. Condensed tannins (from mimosa, quebracho, *etc.*) are polymeric flavonoids. They are generally flavonol derivatives by linking two or more molecules through a carbon–carbon bond. Condensed tannins do not undergo hydrolysis but instead polyphenol molecules can aggregate to form phlobaphenes called "reds". For vegetable tannins, hydrogen bonds are the main interaction between phenol groups in the tannins and N–H, C=O and O–H groups in the protein. However, other bonds can also be formed in addition, such as ionic bonds and covalent bonds in the case of condensed tannins. These stronger bonds explain the higher shrinkage temperature measured for the leather tanned with condensed tannins (see Table 5.1).

In the case of oil tanning, the unsaturated fatty acids introduced are oxidized during the process into aldehydes that form strong bonds with collagen. For smoke tanning, it is also aldehydes compounds present in the fumes, as well as phenols, that bond to collagen. With mineral tanning, the metal ion binds to the collagen carboxyl groups with coordination bonds having an ionic and/or covalent character. Aluminium creates mostly ionic bonds producing only weak interaction with collagen that is reversible in the presence of water, for that reason the process cannot be called tanning. In contrast, chromium creates bonds with a covalent character with the collagen molecule that is five times stronger than those formed with aluminium and is responsible for the high stability of these leathers (see Table 5.1).

5.2.4 Degradation

In the 1990s and 2000s, several European projects, focusing on leather,[12,13] and later on, on parchment,[15,16] have increased the knowledge of degradation mechanisms (see also ref. 5, 6 and 14) whilst examining the potential of different non- and micro-destructive analytical techniques, working from the macro- to the nanoscale, to assess the degradation of these materials. These projects are still nowadays the basis for many research studies developing in the conservation of leather and parchment.

The degradation in leather and parchment can be physical, chemical or biological in origin or a combination of each. Because of their distinct manufacturing processes and nature, leather and parchment will not react in the same way to the different environmental factors and agents of deterioration. The degradation will affect, among others, the main constituent, collagen, and generally occur at different length scales in the structure (fibril, triple helix, collagen molecule, amino acids).

The two main chemical degradation mechanisms taking place in collagen are hydrolysis and oxidation. Both reactions cause chain cleavage, leading to the formation of smaller fragments called peptides, thereby reducing the hierarchical stability of the molecule. Oxidative breakdown is mainly induced by heat, light, oxidative pollutants and free radicals. The reaction involves the breakage of the N–C covalent bond within the main chain and the formation of amide groups that can break down into ammonia (see Figure 5.7). The degradation induces a decrease of basic amino acids (arginine, hydroxylysine and lysine) and an increase of acidic amino acids (glutamic, aspartic).

Acid hydrolysis is induced by acids, and the primary agent in the atmosphere is sulfur dioxide (SO_2) from industrial air pollution.

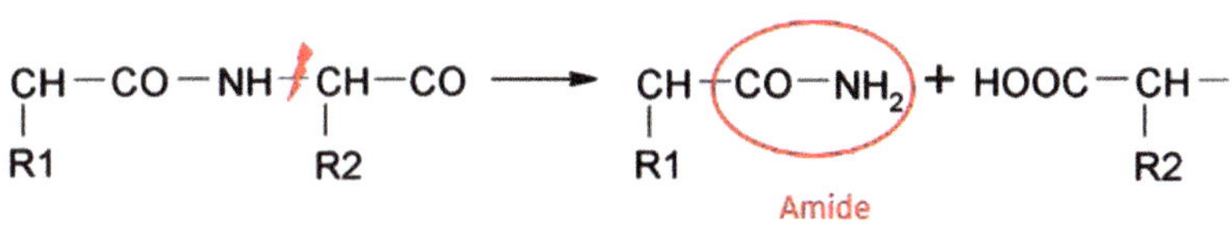

Figure 5.7 Oxidative breakdown of collagen.

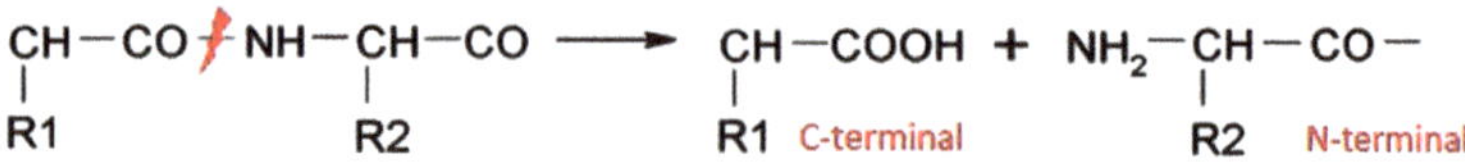

Figure 5.8 Hydrolysis breakdown of collagen.

The reaction involves the breaking of collagen at the C–N linkage resulting in smaller molecules called peptides or the release of free amino acids (N- and C-terminals) (see Figure 5.8). Visually, acid hydrolysis causes in the most extreme cases a complete loss of fibre cohesion, the materials becoming weak and powdery.

Gelatinisation is the most extreme stage of alteration and corresponds to the collagen denaturation to gelatine where the triple helix dissociates to form a random-coiled structure. Gelatinisation takes place in the presence of water and is accelerated by heat. Water provides hydrogen bonds that are competing with existing internal hydrogen bonds holding the collagen molecule. Gelatinisation is favoured if the collagen molecule has already been partially degraded by hydrolysis or oxidation, the breakage facilitating molecule rotation and access to hydrogen bonds. Visually, gelatinisation causes, essentially in parchment, an increase in transparency, rigidity and retraction of the material.

Water is a central element, as an essential constituent of leather and parchment, but it is also the main factor of degradation, as it is involved in chemical, biological and mechanical processes. All alteration processes cause a drop in the water content, explaining why these materials become hard and less flexible over time. In parallel, their absorption capacity is increased and their water sorption/desorption speed is modified, therefore their reactivity to the environment changes. The water sorption/desorption induces dimensional modifications, which may generate physical constrain to the skin or the associated materials such as ink and pigment, possibly leading to cracks, tears, flaking in the object. Microorganisms tend to develop when organic materials are exposed to high humidity (above 65% RH for mould and 90% RH for bacteria) and will be enhanced by heat, the pH at the material surface, and the presence of dust, that may create a favourable environment for them by providing a source of food and water. Microorganisms favour the hydrolysis of collagen from the enzyme they secrete, and the by-products formed during the process generally produce coloured stains on the object. Humidity also encourages insect infection, and insects such as dermestids (*e.g.* hide and carpet beetles) will eat collagen- or keratin-based objects and produce holes that weaken them. In bookbinding, xylophagous insects may also cause damage by attacking the leather to get access to the wooden or cellulosic support underneath. In leather, vegetable tannins and fats will also be affected by oxidation or hydrolysis reactions that will in turn modify the mechanical, chemical and visual properties of leather and affect its stability. In addition, the nature of the vegetable tannin or of the fats (introduced during manufacturing or maintenance treatment), which can deteriorate into fatty acids, can even promote the alteration of collagen as will be shown later on in this chapter. Finally, deterioration may originate from the other materials present on the object, such as inks or pigments in manuscripts that sometimes induce local hydrolysis of the parchment.

Leather can survive burial or waterlogged conditions, however its chemical composition and structure is generally strongly modified. Products such as humic acids, ferrous or copper ions, salts, which derive from the soil, the water or the materials in the surrounding environment, will migrate into the leather structure and interact with the collagen and the tannins. In mineralized leather, the metal ions will precipitate and replace progressively the fibre structure. Thus, the structural features may be preserved even if no collagen has survived.

5.3 Analytical Techniques Appropriate for Leather and Parchment

Leather is a complex material to study due to its multi-scale structure, in particular of the main constituent collagen, the heterogeneity of the raw material, the interaction of other compounds like tannins and fats, and its use in combination with other materials (metal, pigment, ink). Scientific analyses of leathers and parchments can inform us of the manufacturing process, which can subsequently help identify the object's provenance from technological or material characteristics. The analyses would also be essential for the preservation through characterization of the material's state of conservation, sources of degradation and deterioration mechanisms. From the acquired knowledge, optimal conservation conditions can be determined and appropriate conservation treatment or materials can be developed.

The analysis of leather or parchment often requires a multi-technique approach and the need to work at different length scales, from the macro- to the nanoscale, to get a complete characterization of a material or its alteration. As in other fields, the need for less invasive techniques has led to a reduction of sample size, thanks to the increase in sensitivity of the techniques, or the development of non-invasive methods.

5.3.1 Identification of Animal Species

Knowledge of the animal species of the skins used to make leather or parchment is of interest to historians and archaeologists to better understand past practice in their manufacturing in relation to a country, region or the type of object, but this information can also be valuable in the field of conservation to sometimes grasp the origin of an alteration or to assist in restoration.

The simplest and most commonly-used method to identify the animal species is to observe the hair follicle pattern under low magnification (×10 to 20). On the grain side of the skin, each animal displays a specific distribution and frequency of the hair follicles. For example, goat and sheep skins display clusters or rows of large and fine hair holes that are orientated in a row, whilst the hair holes in calf skin are uniformly distributed over the surface and evenly sized (see Figure 5.9). The distinction between sheep and goat skin is often tricky but may be made from the observation of the fine follicles. In goat skin there are alternating rows of large and fine follicles, whilst in sheep skin the fine follicles are grouped. Nevertheless, this identification approach requires access to the grain surface in a good state of preservation, so it can be compromised if the surface is abraded, distorted due to the area of the skin, the stretching during the manufacturing of parchment, or masked as in gilt leather. In these situations, other identification approaches should be sought.

Proteomic analyses rely on the analysis of the protein (*i.e.* collagen) by a method called peptide mass fingerprinting (PMF).[17] The method has the advantage of being simple, rapid, sensitive, specific and the analysis requires only a few small fibres from the skin representing a quantity of about 10–20 µg. The preparation involves putting the sample in solution with an enzyme, which cuts the protein at specific sites. The peptide mixture obtained is then analysed by mass

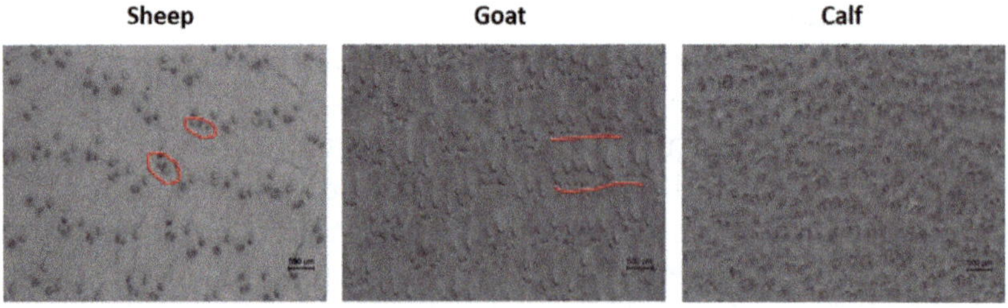

Figure 5.9 Hair follicles patterns for different animal species at the grain surface of leather. The organization of the fine follicles are shown in the goat and sheep skins.

spectrometry, using MALDI (matrix assisted laser desorption/ionization), a soft ionization method to avoid damaging the peptides. Each animal species has characteristic peptide markers, therefore identification of the species is made by comparison of the mass spectrum obtained with reference material spectra, focusing on marker ions characteristic of each species (see Figure 5.10).

Ageing and alteration of the collagen can affect the signal collected from proteomic analysis, with a decrease of its intensities and deamination of the molecule responsible for peak shifts. In this case,

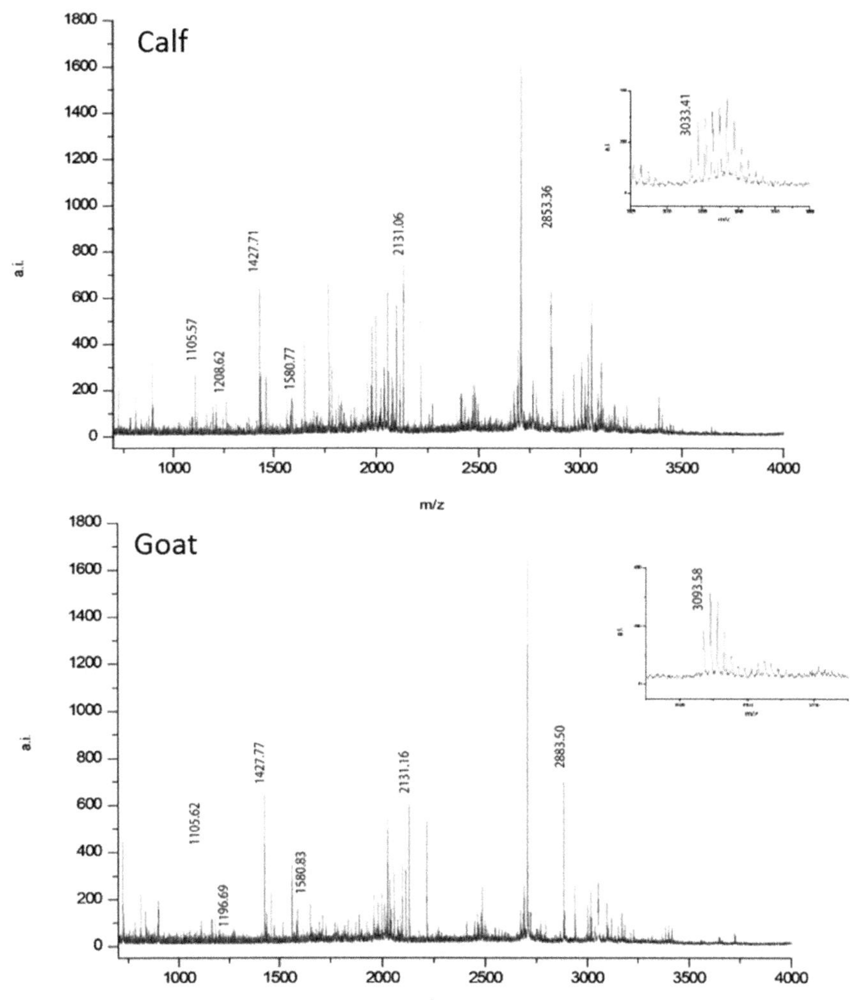

Figure 5.10 Peptide mass fingerprinting spectra obtained on calf and goat parchments.

additional techniques can be applied on the digested samples such as Fourier Transform Ion Cyclotron Resonance (FT-ICR) providing a higher mass resolution, or liquid chromatography coupled to a mass spectrometer (LC–MS) that allows separation and collection of a mass spectrum for each separate peptide. Despite this damage, species identification remains generally possible from proteomic analysis, as proteins are better preserved than DNA.

When genetic data is preserved in parchment and leather, DNA analysis allows more extensive investigations by providing information on sub-species, animal relationships, gender or microbiome which may help better understand past animal husbandry, the varieties and development of breeds, and orientate provenance studies.[25] DNA analyses are more costly and require specialist expertise. The method involves extraction of either mitochondrial or nuclear DNA which is amplified by polymerization chain reaction (PCR).[18] Identification is done by DNA sequencing and comparison with databases of species sequences. Previous studies have demonstrated that nuclear DNA cannot be amplified from leather most likely due to its degradation during the tanning process. From surface sampling techniques, such as the PVC eraser method, the exogenous and/or environmental DNA present at the surface of the document can be picked up and analysed. This DNA might contain microbial traces, which may reveal information on the object's use, history or conservation risk.

5.3.2 Tannin Identification

The identification of tannin is important not only to understand the leather technology or the origin of degradation in ancient leathers but also to assess the suitability of new leather to be used for conservation purposes. Considering the diversity of tanning materials employed in leather, different analytical approaches can be used or combined for their identification. If the leather has not been dyed, its colour can initially provide some clues into the tanning method used, as vegetable tanned leather is generally brown, whilst alum-tawed skin is white. In smoke tanned skin, only the colour, and possibly the smell, can facilitate the interpretation as no chemical analysis can evidence this tanning method. The ash test is a basic method that requires the burning of a piece of leather into ashes (a few grams), thereby tannin type is deduced from the colour of these ashes. Whilst skin tanned with organic compounds (*e.g.* vegetable tannin, fats) and untanned skin always displays black ashes, skins prepared with aluminium or chromium salts give characteristic white and green ashes respectively.

Elemental analytical techniques such as X-ray fluorescence or SEM–EDX are mostly used to identify a mineral tannin like chromium, or alum from the presence of aluminium, potassium and sulfur. Elemental analysis may also be recommended in the case of natural history or ethnographic collections to identify past treatments used to prevent insect and pest damage, such as the use of mercury, arsenic, chlorine or bromine compounds. Fourier transform infrared spectroscopy in attenuated total reflection mode (ATR–FTIR) is a useful screening technique as it is fast and non-destructive; however being a surface technique, one should keep in mind that the measurements will be affected by any surface treatment or deposit.[19] From the 600–1800 cm^{-1} spectral region of the ATR–FTIR spectra it is generally possible to identify the presence of alum, fats, or vegetable tannins but no further specificity can be gained. The technique may also help reveal the presence of other compounds employed during the manufacturing process or formed over time such as gypsum, kaolin, metal soaps, *etc.*

In the presence of vegetable tannins, it may be important to discriminate between the different chemical groups (condensed and hydrolysable tannins) or sub-groups (gallotannins and ellagitannins) and different analytical approaches exist. However, the reactivity of these tests can be affected by the alteration in leather and thus become inefficient. Spot tests are often used by conservators as they are simple, cheap and only require a small sample.[20] The most common are the ferric test and the vanillin test used to characterize the presence of vegetable tannin and condensed tannin respectively. In the case of the ferric test, the detection is based on the reaction of the iron sulfate with polyphenols. These polyphenols present in vegetable tannin are also found in vegetable dyes, therefore this should be kept in mind for interpretation. Other spot tests have recently been adapted from plant tissues analysis to go further: the rhodamine test for gallic acid and gallotannins, nitrous acid test for ellagitannins and acid butanol test for condensed tannins detection. All five spot tests are effective on well-preserved leather and may lead to the identification of the tannin group and subgroup or tannin mixture. The specificity of ATR–FTIR in the identification of the vegetable tannin can be improved by analysing the dried residue of the tannin extracted from the leather. From this approach, it is possible to avoid interference from collagen, and all the vibrational bands associated with the tannin can be seen in the spectra and attribution to the tannin group or subgroup can be made (see Figure 5.11). Again, with this method, any alteration can have a strong impact on the measurements causing a modification of the tannin spectra and/or presence of collagen residues in the extract despite filtration of the solution.

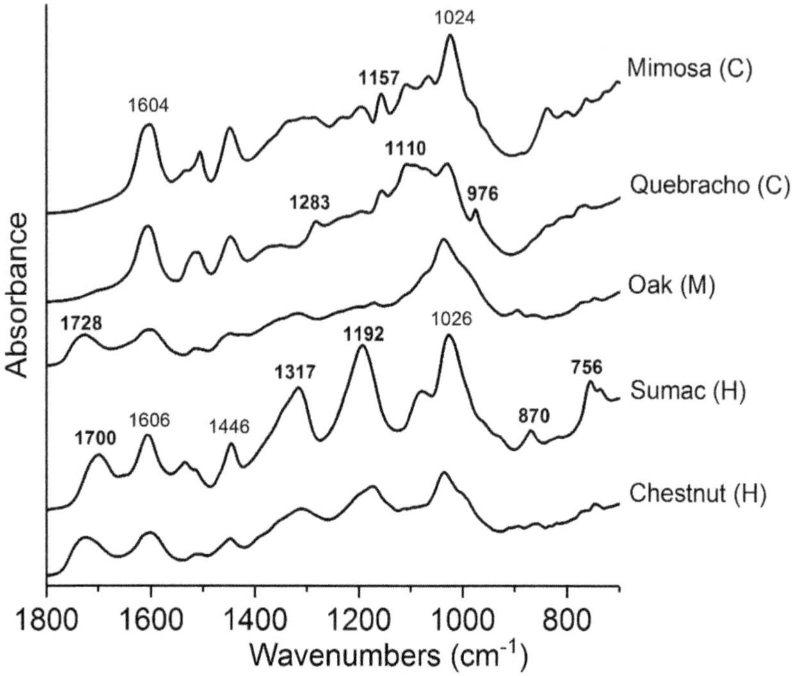

Figure 5.11 Infrared spectra in the region 700–1800 cm^{-1} for different vegetable tannins organized by group hydrolysed (H), condensed (C) or mixed (M). The bands specific to each group are marked in bold.

Although it requires more complex instrumentation, high performance liquid chromatography (HPLC) remains the most sensitive technique for the identification of vegetable tannin even when the leather is degraded. From the analysis of the tannin extracted from the material, the technique allows the identification and quantification of the unit molecule such as gallic acid or ellagic acid, and could possibly lead to the characterization of the source plant. The field of dye analysis has greatly evolved thanks to the rapid evolution in the instrumentation of liquid chromatography techniques and it is believed that tannin analyses could benefit from these developments.

5.3.3 Degradation State Assessment

Knowledge of the leather or parchment degradation state is essential to select suitable conservation treatments or storage conditions. Considering that modification can take place at different structural levels in the skin, from the macroscopic to the molecular scale, different analytical techniques are generally applied. As a first approach, visual observation and soft manipulation to feel the mechanical properties of the material can already provide clues on the condition, especially in the case of leather. For parchment, different characteristics detected by simple visual methods can be used to highlight possible damage, such as the presence of transparent regions, spots or stains, shrinkage, and discolouration. At the microscopic scale, observation of the fibre structure from a sample is also informative on the state of the collagen hierarchical organization and physical properties. For parchment, the fibres sampled are put in water and the damage is assessed based on the observation of different characteristic changes to the fibre structure. In well-preserved parchment, the fibre bundles are compact and display the twisting linked to the helical structure. On ageing the degradation manifests through flattening, splitting or fraying of the fibre with, in the extreme case, the formation of a gel-like material that can dissolve in water. For leather, the assessment relies on the coherence of the dried fibres taken from the corium. Powdery and low coherence in the fibres at the microscopic level is generally correlated with a low mechanical strength of the leather at the macroscopic scale.

The most important indicator of the stability of a skin is obtained by measuring the shrinkage temperature (T_S) of collagen or hydrothermal stability. The shrinkage temperature is measured in the leather industry to assess the manufacturing progress, in particular tanning, whilst in conservation it is mainly used to estimate the degree of deterioration of skin-based objects. The shrinkage temperature corresponds to the temperature at which the material shrinks when heated in water. At the molecular level, this temperature is associated with the energy required to break down the hydrogen bonding holding the coil structure and leading to denaturation. The cross-links created by the tannins increase the stability of the collagen, leading to a rise of the shrinkage temperature dependent on the tannin (see Table 5.1). In contrast, the degradation will weaken the collagen backbone, causing a fall of the shrinkage temperature. The basic method to make this measurement is to hold a piece of leather in a bath of water, warmed up progressively, and note the temperature when the leather shrinks visually. In the case of ancient objects, methods based on the same principle but more reproducible and requiring smaller samples (<1 mg) are used: differential scanning calorimetry (DSC) or the micro hot table (MHT). In the latter case, the measurement is conventionally carried out on samples soaked in water at a controlled heating rate, and the transition is defined as the point where several fibres visually move in MHT or the thermal transition occurs in DSC. In the DSC thermogram, T_S is generally measured

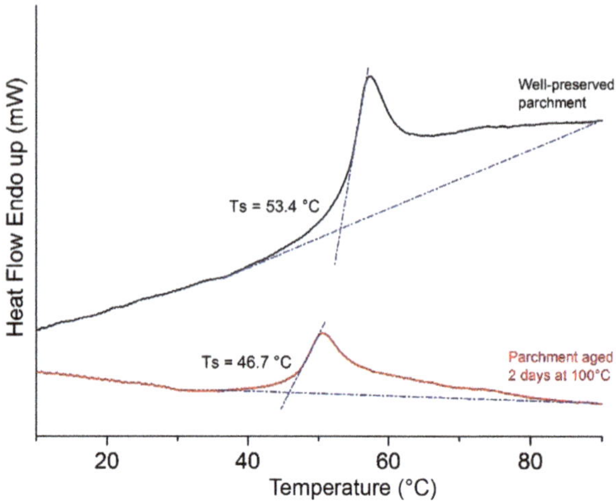

Figure 5.12 DSC Thermogram of well-preserved and altered parchments.

at the intersection of the baseline and the rising portion of the peak, known as T_{onset} (see Figure 5.12). In addition, the measurement of the enthalpy change (ΔH) by DSC informs on the amount of collagen in the measured state and is generally related to the degree of damage.[21] Whilst T_s and ΔH are generally correlated in leather, it is not always the case for parchment, therefore both parameters should be considered. The nature of the skin (animal type, age, location in the skin) and its pH (due to the presence of salts, fatty acids, for example) can affect the value of T_s, however the thermal characteristics are generally stable between pH 3 and 5. It should be noted that archaeological waterlogged leathers show a very different behaviour as their state of deterioration does not seem to be correlated with the hydrothermal stability. Finally, this standard approach may not be adapted to assess the degradation state of heavily damaged leather or parchment (*e.g.* fire-damaged parchment or leather suffering from "red rot"), where the contact with water causes complete denaturation. In such cases, it is necessary to develop a methodology based on dry state measurements.

Nonlinear optical (NLO) microscopy is a recent technique in the field that shows a high potential for the assessment of the conservation state of collagen-based materials in cultural heritage.[22] It is an efficient technique for investigating artworks as it allows non-invasive, three-dimensional imaging with micrometre-scale resolution. This technique is based on the interaction of several photons, generated from a near-infrared laser, with a medium, and producing a unique photon. Different signals can be combined simultaneously such as two-photon excitation fluorescence (2PEF) from compounds containing fluorophores (*e.g.* fat, elastin, tannin) and second harmonic generation (SHG), which enables the visualization of unstained fibrillar collagen. SHG signals are generated only by non-centrosymmetric organizations at the molecular scale (*e.g.* collagen triple helix) and amplified by constructive interferences at the macromolecular scale (alignment of the collagen molecules to form collagen fibrils). SHG microscopy is then a structural probe of collagen hierarchical organization, any disruption of which leads to a loss of SHG signals. In biomedical research, SHG microscopy has emerged as a powerful technique to quantify collagen 3D organization in tissues, in particular in skin dermis. In skin-based historical objects, SHG microscopy allows information to be acquired on the manufacturing as well as the degradation of the skin. The alteration is characterized by the loss of the SHG signal and the increase of a 2PEF signal together with morphological modifications (see Figure 5.13). This imaging approach is particularly efficient on parchment or untanned skin artefacts, because of the absence of interference from fluorescing components such as vegetable tannins or fats. Compared to thermal techniques, which only provides information averaged and localized to the sampled area, NLO microscopy offers the possibility to examine non-invasively the conservation state of collagen in any part of the object and multiply measurements to take into account the heterogeneity. In the future, the development of polarization resolved SHG signals, allowing access to the main orientation and the degree of orientation disorder at the sub-micrometre scale should enable the extraction of quantitative information from the orientation of the fibrils on sub-micrometre to millimetre scales and allow a more precise ranking of the degradation.

To go further in the assessment of the degradation, different analyses can be undertaken to study the alteration mechanisms that may have

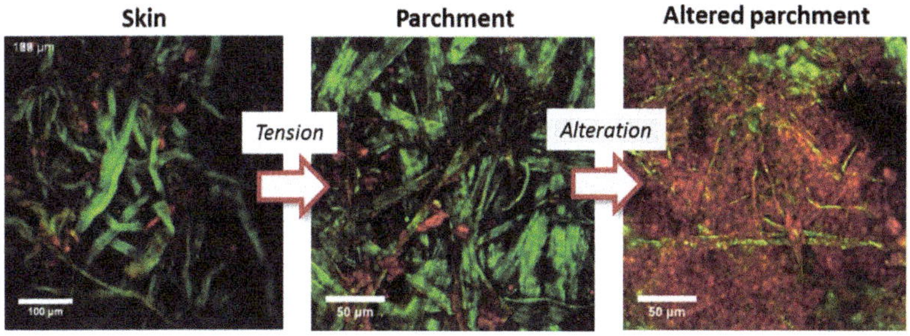

Figure 5.13 Nonlinear optical microscopy image of a dry skin, a well-preserved and an altered parchment.
False colour images: green = SHG signal, red = fluorescence signal.

taken place. Measurement of the pH can indicate if a material suffers from increased acidity that may be harmful and could engender acid hydrolysis of the collagen. Such measurements are mostly done on leather; rarely on parchment having an alkaline reserve, due to its manufacturing process, preventing acidic alteration. Cases exist of strongly deteriorated leathers exhibiting good pH values (above 3.5), as for leather that has suffered from both strong oxidation and acid pollution. In that event, the ammonia released, as a result of the oxidation process, is trapped in the structure as sulfate and "neutralize" the acidity induced by the acid pollution. The most reliable pH measurement approach is based on the cold water extract method that can be used to further assess the presence of harmful strong acids if the leather has been previously exposed to pollution. Using a microelectrode, the pH measurement only requires 3 mg of leather. Raman and FTIR spectroscopies can help characterize the damage at the molecular level, as they provide information on the collagen secondary structure.[23] In infrared spectra, the main features indicator of the degradation process can be seen in the region between 1200 and 1800 cm^{-1}, notably from the position and intensities of the amide I, amide II and amide III bands, and in particular Δv the distance between the amide II and amide I bands (see Figure 5.14). However, the presence of other compounds in the material will generate peaks that may interfere with

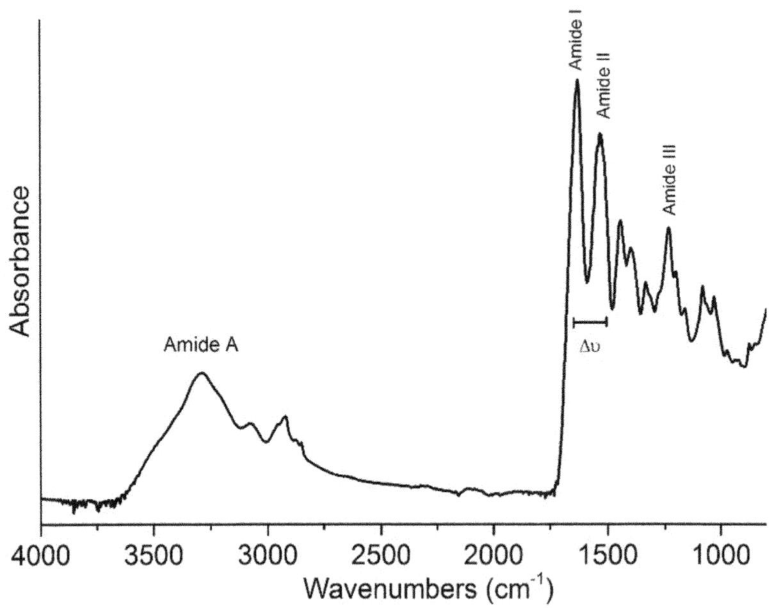

Figure 5.14 FTIR spectrum of a parchment.

the collagen band vibration, as the calcium carbonate peak in parchment or the vegetable tannin and fats in leather, thus rendering the interpretation tentative or impossible. Although it is less commonly applied, Raman spectroscopy appears as a useful technique for describing the hydrolytic breakdown of collagen in parchment.

Amino acid analysis by liquid chromatography is an efficient technique to study the protein and amino acids breakdown due to oxidation or hydrolysis. Oxidative conditions modify the amino acid distribution with a decrease of basic amino acids, and in some cases proline and hydroxyproline, as well as the formation of acidic amino acids. From the calculation of the ratio between the sum of basic and the sum of acid amino acids (B/A), it is possible to tell if an oxidation reaction has taken place in collagen and its degree. In well-preserved parchment or leather, the expected value is around 0.7 and decreases to values below 0.5 are observed in deteriorated skin. Similarly, attempts have been made to assess the degree of hydrolysis from the quantification of C- and N-terminal residues by HPLC, but this proved difficult to achieve.

Many other techniques have been applied to analyse leather or parchment; they will only briefly be mentioned here as they are not routinely used in the conservation field, but may be of interest for more fundamental research on degradation. Thermal and mechanical methods amongst them dynamic mechanical thermal analysis (DMA), allows the measurement of the mechanical properties of the materials under controlled environmental conditions. Atomic force microscopy (AFM), X-Ray diffraction (XRD), small angle X-ray scattering (SAXS) or Transmission Electron Microscopy (TEM) have been applied in particular to examine the periodic banding in the fibrils. Finally, unilateral Nuclear Magnetic Resonance (NMR) has been used to investigate the collagen–water interaction.

5.4 Impact of Chemistry and Analysis on Conservation, Preservation and Scholarship

5.4.1 Impact of Acidity

High acidity in a skin causes, at the molecular level, the hydrolysis of collagen, fats and in some cases tannins that is responsible for the loss of flexibility, delamination, and breaking down into a powder. Although the degradation follows different pathways, the deterioration is known

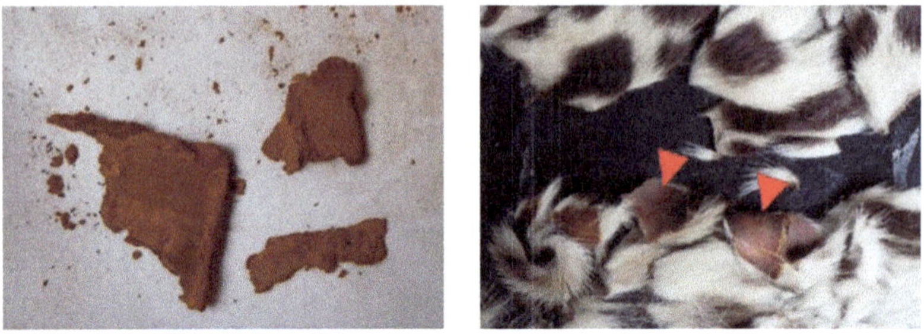

Figure 5.15 Effect of acid hydrolysis on leather. Left: Leather bookbinding suffering from "red rot". Right: "Hydrolysed" cheetah skin that has gelatinized (red arrow).

as "red rot" and that of hydrolysed natural history specimens are good illustrations of acid hydrolysis degradation. Both show high acidity with pH values generally below 3, as well as low shrinkage temperature close to ambient temperature.

"Red rot" is an alteration process affecting vegetable tanned leathers, turning them into a red powder having sometimes a pungent smell (see Figure 5.15). Many leathers now suffering from red rot were produced from the late 19th century or early 20th century, when leather manufacturing processes evolved to respond to an increase in demand. However, this alteration has also been encountered in more ancient leathers. The deterioration is associated with the presence of strong acids, particularly sulfuric acid, either introduced during the process to accelerate the leather manufacturing, or produced by the reaction of sulfur dioxide pollutants in the atmosphere with humidity. Leathers tanned with condensed tannins are more prone to red rot, as they tend to absorb more SO_2 pollutants that react with tannins to form phlobaphenes and anthocyanidin compounds, which are responsible for the red colour.

Hydrolysis also affects alum-tawed skins, but this problem is generally encountered in natural history collections. To preserve these specimens, taxidermists in the past have often used alum salts because the skin retains a light colour and great flexibility which is important to give a feel the specimens are alive. Nowadays, a large number of alum-tawed specimens in such collections suffer from acid hydrolysis, their skin has lost its flexibility, becomes powdery and is at an advanced stage of deterioration, turning into a brown gelatinous material (see Figure 5.15). The loss of mechanical strength can lead to tears of the skin when humidity fluctuations are taking place (see Figure 5.16). The mechanisms associated with this alteration are not yet understood. It relates to the presence of alum and seems

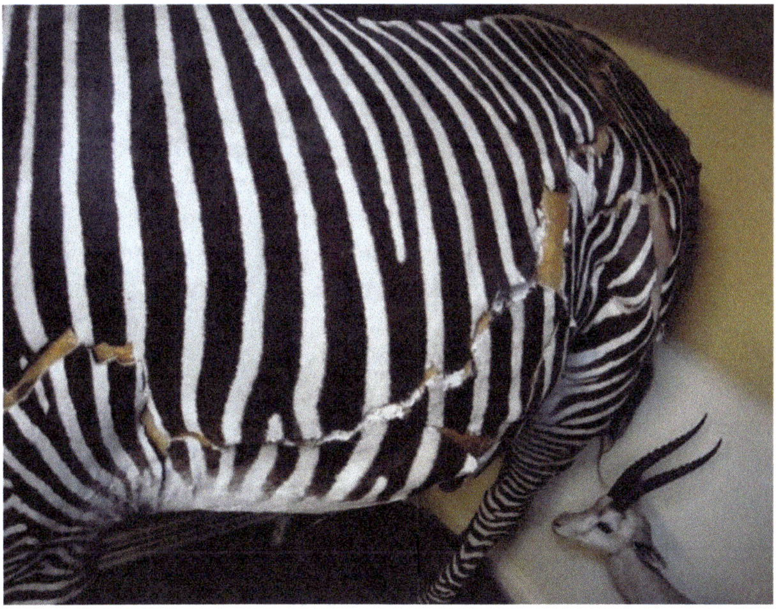

Figure 5.16 Tears to the "hydrolysed" skin of a zebra specimen. Natural history museum, Paris.
Image Credit: Jacques Cuisin, MNHN.

accelerated by the presence of fats, therefore the acidity could possibly come from an incomplete neutralization of alum salts during the skin preparation and/or the production of fatty acids from the fats introduced during the lubrication or remaining from the animal.

In these two acid hydrolysed examples, the high acidity and the low stability of the skin proscribe the use of water-based treatments or the exposure to high humidity as this would favour further hydrolysis and lead to complete gelatinization of the material. The manufacturing process playing an important part in the alteration, it is important to identify the nature of the tannin, the acidity of the skin and determine if sulfates are present. More generally these analyses may help detect objects most at risk to select suitable conservation conditions, but also better understand the degradation mechanisms. In order to estimate how a new material will age or its suitability to be used in conservation for backing or infilling historical objects, similar analyses are recommended.

5.4.2 Impact of High Heat

The exposure of leather and parchment to high heat induces, at the molecular level, loss of water, oxidation of collagen and fats, and the creation of new cross-links in the structure which are correlated with the changes observed at the macroscopic level, shrinkage, flexibility loss, and often darkening of the material. When the oxidation is extensive, subsequent application of water will cause irreversible denaturation of the collagen, which engenders additional shrinkage and rigidity as well as an increase in transparency for parchment.

These different stages are in general clearly seen in skin-based objects exposed to fire and subsequently to water used to extinguish it. Damage is generally inhomogeneous across the object or document due to the variable degree of exposure to the harmful conditions. In order to assess the state of degradation of such objects, it is advisable to examine each of the different areas or at a minimum focus on the most exposed or damaged areas. Nevertheless, it is important to bear in mind that the advanced oxidation in parchment cannot always be appreciated visually, as experiments on artificially aged parchment have demonstrated. Therefore, it is highly recommended before undertaking any humidification treatment to carry out analyses to measure the degradation state. If a shrinkage temperature close to ambient is measured, no water-based treatment should be undertaken.

5.4.3 Impact of Lubricants on Leather

For many years, it was common practice to lubricate leather objects in museums and libraries as though they were still in use. These lubricants, also called leather dressing and supplied as emulsion or wax, contain mainly fats, such as neatsfoot oil and lanolin. The function of a lubricant is to prevent adhesion of the fibrils during drying and to reduce friction between fibrils. Water is the native lubricant in fresh skin and is replaced by tannins and fats in leather. On ageing, leather loses water and fats, therefore the mechanical properties are modified. When a lubricant is reintroduced into the structure, the mechanical properties are improved in most cases, but in the long term these dressings have a detrimental effect. The re-lubrication of ancient leather has a completely different impact compared to original lubrication following the manufacturing process, due to the lack of moisture in the re-lubrication. Over-dressing was recognized to cause darkening and hardening of the leather, to attract dust by creating a sticky surface, favour biological attack, and generate fat spews (see Figure 5.17).

In composite objects, lubricants may interact with the other materials present and induce degradation. This has been observed in gilt leather where the fats migrate through the different layers and interact with the silver leaf, the varnish and the painted layer. In the presence of copper-, zinc- or lead-based metals, the lubricant can react with these metals and

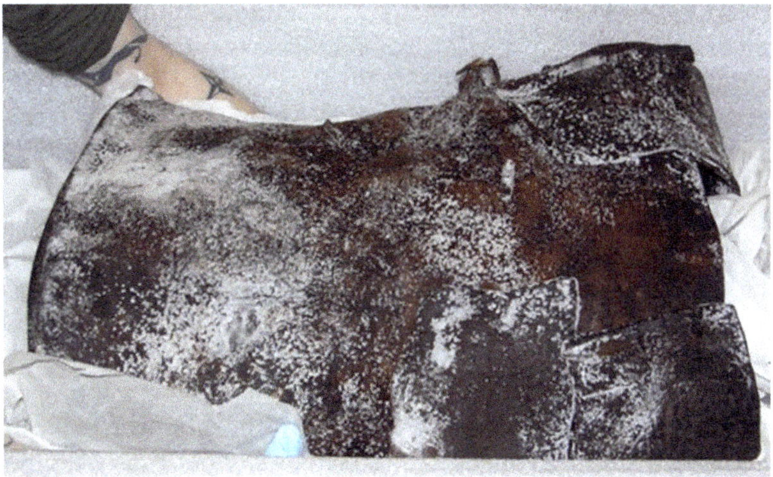

Figure 5.17 Fat spew on a Tlingit hide armour, Sheldon Jackson Museum I-A-449.
Photo Credit: Ellen Carrlee.

form metal soaps. Alterations may also occur in the absence of contact between the leather and the metal due to the sulfur-based vapours emitted by sulfited lubricants that will corrode the metal.

For that reason, current museum practice is to minimize the application of lubricants – the need to lubricate should depend on the mechanical solicitation of the historic leather. Thus, an object in a showcase that is not manipulated would require no lubrication, whilst lubrication may be required for objects heavily manipulated or under tension. In addition, care should be taken when selecting a new leather for restoration to avoid using leather lubricated with sulfited oil.

5.5 Case Study: Chartres Fire-damaged Medieval Manuscripts

Chartres was well known for its famous cathedral school, being among the most important occidental intellectual centres during the 11th and 12th centuries. On the 26th of May 1944, during an allied bombing of the city, a fire irremediably destroyed almost half of the collection. Of the 518 original medieval manuscripts, at least 215 survived the fire but remained in very variable states, from almost intact to entirely calcined. The cumulated actions of fire, water (to extinguish the fire) and the subsequent biological infestation (worms) resulted in different types and levels of alteration. During restoration campaigns, the distorted and agglomerated parchment leaves were separated and flattened.

Considering the fragility of the fragments, a digitization campaign was started in 2006 by the Image Department of the Institut de Recherche et Histoire des Textes (IRHT-CNRS) so the documents could be available again for research. However, in some manuscripts the alteration caused the writings to become partly or totally unreadable (see Figure 5.18), preventing their study by historians.

A research project was started recently at the Centre de Recherche sur la Conservation on these degraded manuscripts conserved at the Médiathèque l'Apostrophe in Chartres. The objectives were multiple: the characterization of the parchment degradation state, the identification of the animal species of the skin and the recovery of illegible writings. Despite the advanced degradation state sometimes encountered, the work demonstrated that current scientific techniques can still gain information from these manuscripts that we thought lost, and open new research perspectives.

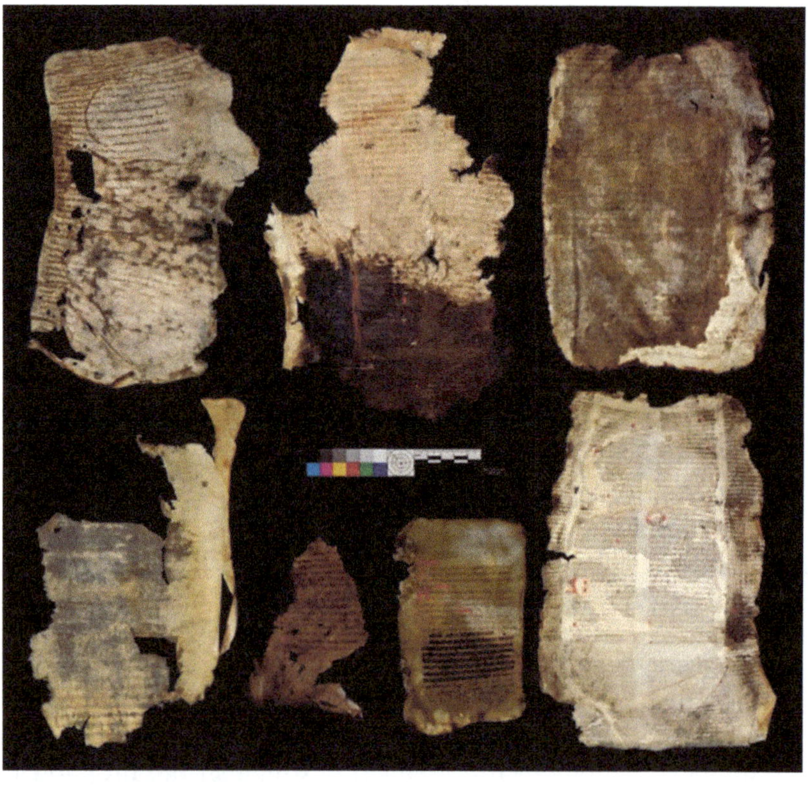

Figure 5.18 Examples of degraded manuscript fragments from Chartres library, France.
Reproduced from ref. 24 with permission from Elsevier, Copyright 2019.

5.5.1 Parchment Conservation State

In the parchments from the Chartres' manuscripts, the collagen suffered oxidation from the exposure to high heat and has, in some cases, subsequently denatured (gelatinization) due to the contact with water used to extinguish the fire. These modifications at the microscopic scale result at the macro scale in a shrinkage of the parchment sheet, and an increase in transparency and rigidity.

Prior to a conservation treatment, particularly a humidity-based treatment, it is essential to assess the degradation state of the collagen in this type of parchment. Although the degradation state may appear appreciated by eye, especially in the Chartres manuscripts, it is not always a good indicator. A parchment visually well-preserved may be revealed to be highly degraded at the molecular scale, as has been shown by work on artificially aged parchments. Moreover, the alteration of a parchment, especially in the case of Chartres manuscripts, is often heterogeneous, as can be seen in the examples shown. Consequently,

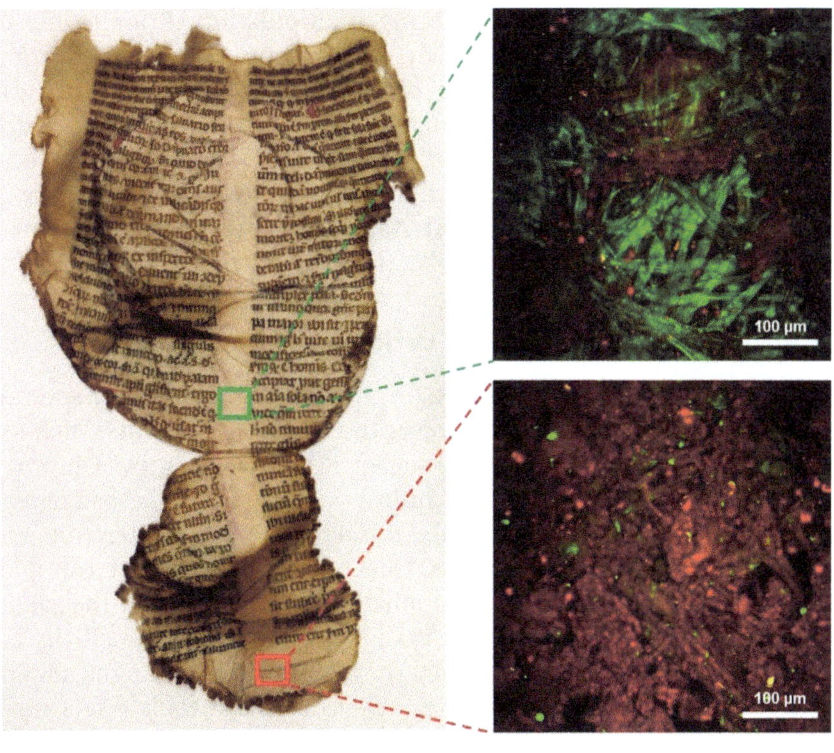

Figure 5.19 Nonlinear optical microscopy images collected from parchment sheet 312 of manuscript 205.
False colour images: green = SHG signal, red = fluorescence signal.

the degradation state should ideally be measured in different areas of the document, therefore the use of on-invasive techniques essential.

Nonlinear optical microscopy was applied to the Chartres manuscripts to examine the collagen fibre morphology and conservation state. The technique confirmed the heterogeneity of the degradation in the parchment sheet 312 of manuscript 205. Measurements collected from the central part show strong second harmonic generation (SHG) signals, attesting to the good preservation of the collagen, whilst signals collected from the edges of the sheet are dominated by fluorescence linked to the presence of gelatin (see Figure 5.19). These measurements were then repeated on a parchment sheet coming from the same manuscript but that underwent a humidification treatment, and it shows similar results, indicating that the conservation treatment did not induce any changes to the collagen.

Proteomic analysis was undertaken to parchments from the Chartres library to assess if the technique was able to recover information on the animal species despite the degradation undergone by the parchment. The analysis was applied to fragments that self-detached from the sheets during manipulation. Prior to these analyses, the degradation state of these fragments was obtained from the measurement of the shrinkage temperature (T_S) by DSC. Amongst the ten fragments analyses, T_s ranged from around 30 to 50 °C with low enthalpy in the most degraded cases. In all the samples analysed, proteomic analyses led to clear identification of the animal species, despite the collagen conservation state, therefore opening up new research perspectives on these collections.

5.5.2 Recovering Illegible Writings

When faced with legibility issues on a historical document, several approaches can be explored to recover the degraded contents, such as hyperspectral imaging, XRF scanning or X-ray tomography. Unfortunately, due to the time, cost and technical skills required, these imaging techniques cannot be applied to the numerous faded or altered documents present in library collections. Hence, there is a need for a simple imaging procedure that could be applied on a large scale, in the same way libraries and archives are being digitized. The great sensitivity of modern digital camera sensors used for digitization campaigns added to the development of simple image post-treatment tools offers new potentialities regarding this issue. In the present research, a fast and low-cost methodology was established to archive the current aspects and reveal hidden contents of documents. It is based on the use of UV fluorescence photography in combination with contrast enhancement treatments (principal component analysis, colour space conversion and decorrelation stretch).

UV fluorescence photography proved to be a powerful tool to (at least partially) reveal initially illegible content in most of the manuscript fragments from Chartres that were studied. In the cases of darkened and/or transparent substrate, the accurate colour restitution of the UV fluorescence was generally sufficient to reveal the text, whilst an extra step of colour contrast optimization was necessary when the inks were washed out (see Figure 5.20). Unfortunately, some fragments are in such an advanced degradation state that the presented methodology did not provide much legibility improvement. Neither would it help when matter is superimposed on top of the writings (*e.g.* fibres from blotting papers used during the emergency conservation treatments). In these cases, it seems that more sophisticated imaging instrumentations, such as X-ray based techniques, will have to be explored.

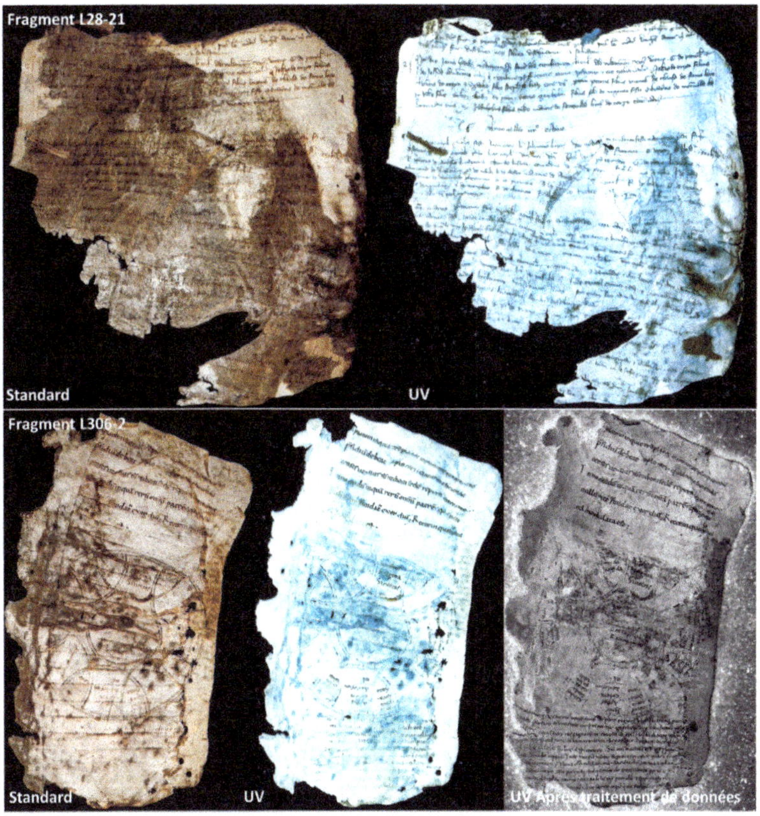

Figure 5.20 Parchment fragment L306-2 under visible light, UV light and after data treatment by PCA.

Acknowledgements

The author would like to thank the different colleagues that contributed to the research projects presented in this chapter: Sylvie Heu-Thao, Anne Michelin, Fabien Pottier from the Centre de Recherche sur la Conservation; Gaël Latour and Marie-Claire Schanne-Klein from the Laboratoire d'Optique et Biosciences, as well as Jacques Cuisin from the Natural History Museum in France. I am also grateful to Céline Bonnot Diconne and Claire Chahine for their attentive reading and their advice. The research on the Chartres fire-damaged medieval manuscripts was supported by the Paris Seine Graduate School Humanities, Creation, Heritage, Investissement d'Avenir ANR-17-EURE-0021–Foundation for Cultural Heritage Science.

Recommended Reading

C. Chahine, Changes in Hydrothermal Stability of Leather and Parchment with Deterioration: a DSC Study, *Thermochim. Acta*, 2000, **365**, 101.

L. Falcão and M. E. M. Araújo, Application of ATR–FTIR Spectroscopy to the Analysis of Tannins in Historic Leathers: The Case Study of the Upholstery from the 19th Century Portuguese Royal Train, *Vib. Spectrosc.*, 2014, **74**, 98.

B. Haines, in *Leather: Its Composition and Changes with Time*, ed. C. Calnan and B. Haines, The Leather Conservation Centre, Northampton, 1991, p. 66.

C. J. Kennedy and T. J. Wess, The Structure of Collagen within Parchment – A Review, *Restaurator*, 2003, **24**, 61.

D. P. Kirby, M. Buckley, E. Promise, S. A. Trauger and T. R. Holdcraft, Identification of Collagen-Based Materials in Cultural Heritage, *Analyst*, 2013, **138**, 4849.

M. Kite and R. Thomson, *Conservation of Leather and Related Materials*, Routledge, London, 2006.

A. D. Covington and W. Wise, *Tanning Chemistry: The Science of Leather*, 2nd edn, Royal Society of Chemistry, Cambridge, 2019.

References

1. C. Chahine, *Cuir & Parchemin ou la métamorphose de la peau*, CNRS Éditions, Paris, 2013.

2. R. Reed, *Ancient Skins, Parchments and Leathers*, Seminar Press, London, New York, 1972.
3. E. Heidemann, *Fundamentals of Leather Manufacturing Eduard Roether KG*, Darmstadt, 1993.
4. J. P. Fournet, *Cuirs dorés, « cuirs de Cordoue », un art européen*, Monelle Hayot, Paris, 2020.
5. C. Dignard and J. Mason, *Caring for Leather, Skin and fur, CCI Notes Series 8*, Canadian Conservation Institute, Ottawa, ON, 2018.
6. M. Kite and R. Thomson, *Conservation of Leather and Related Materials*, Routledge, London, 2006.
7. M. Baillargeon, *North American Aboriginal Hide Tanning The Act of Transformation and Revival*, University of Ottawa Press, 2011.
8. C. J. Kennedy and T. J. Wess, *Restaurator*, 2003, **24**, 61–80.
9. M. D. Shoulders and R. T. Raines, *Annu. Rev. Biochem.*, 2009, **78**, 929–958.
10. A. D. Covington and W. Wise, *Tanning Chemistry: The Science of Leather*, Royal Society of Chemistry Cambridge, 2nd edn, 2019.
11. P. Borgard, J. Brun and M. Picon, *L'alun de Méditerranée*, Publications du Centre Jean Bérard, Naples, 2005.
12. European Commission, STEP Leather Project, Evaluation of the Correlation between Natural and Artificial Ageing of Vegetable-Tanned Leather and Determination of Parameters for Standardization of an Artificial Ageing Method, ed. R. Larsen, *Protection and Conservation of European Culture Heritage*, Research report no. 1, 1994.
13. European Commission, Environment Leather Project, Deterioration and Conservation of Vegetable tanned leather, ed. R. Larsen, *Protection and Conservation of The European Culture Heritage*, Research report no. 6, 1996.
14. B. Haines, Natural Ageing of Leather in Libraries, in *Leather: Its Composition and Changes with Time*, ed. C. Calnan and B. Haines, The Leather Conservation Centre, Northampton, UK, 1991, pp. 66–74.
15. R. Larsen, *Microanalysis of Parchment*, Archetype publications, London, 2002.
16. R. Larsen, *Improved Assessment of Parchment, Assessment, Data Collection and Sharing of Knowledge (IDAP)*, Report 18, European Commission, Luxembourg, 2007.
17. (a) S. Fiddyment, B. Holsinger, C. Ruzzier, A. Devine, A. Binois, U. Albarella, R. Fischer, E. Nichols, A. Curtis, E. Cheese, M. D. Teasdale, C. Checkley-Scott, S. J. Milner, K. M. Rudy, E. J. Johnson, J. Vnouček, M. Garrison, S. McGrory, D. G. Bradley and M. J. Collins, *J. Proc. Natl. Acad. Sci.*, 2015, **112**, 15066–15071; (b) C. P. Kirby, M. Buckley, E. Promise, S. A. Trauger and T. R. Holdcraft, *Analyst*, 2013, **138**, 4849.
18. A. Vuissoz, M. Worobey, N. Odegaard, M. Bunce, C. A. Machado, N. Lynnerup, E. E. Peacock and M. T. P. Gilbert, *J. Archaeol. Sci.*, 2007, **34**, 823–829.
19. L. Falcão and M. E. M. Araújo, *Vib. Spectrosc.*, 2014, **74**, 98–103.
20. L. Falcão and M. E. M. Araújo, *J. Cult. Herit.*, 2011, **12**, 149–156.
21. C. Chahine, *Thermochim. Acta*, 2000, **365**, 101–110.
22. G. Latour, L. Robinet, A. Dazzi, F. Portier, A. Deniset-Besseau and M.-C. Schanne-Klein, *Sci. Rep.*, 2016, **6**, 26344.
23. B. B. Doyle, E. G. Bendit and E. R. Blout, *Biopolym. Orig. Res. Biomol.*, 1975, **14**, 937–957.
24. F. Pottier, A. Michelin and L. Robinet, *J. Cult. Herit.*, 2019, **36**, 183–190.
25. M. D. Teasdale, S. Fiddyment, J. Vnoucek, V. Mattiangeli, C. Speller, A. Binois, M. Carver, C. Dand, T. P. Newfield, C. C. Webb, D. G. Bradley and M. J. Collins, *R. Soc. Open Sci.*, 2017, **4**, 170988.

6 Metals

S. Golfomitsou

University of Gothenburg, Department of Conservation, Box 130, SE 405 30, Gothenburg, Sweden
Email: stavroula.golfomitsou@gu.se

6.1 Introduction

The use of metals started with native metals such as gold, silver and copper, found in metallic form in nature, around 10 000 years ago. Native metals were only available in some regions of the world and were originally simply hammered into shape and used as beads for jewellery, and then later melted and cast. Metallurgy stems from the use of metal-bearing minerals as pigments and decorative stones. Starting in the Palaeolithic period, stones used as beads were heated to improve their properties. Evidence shows that malachite, a common copper ore already in use as a pigment, was collected and heated before use as a decorative stone.[1] The beginning of metallurgy, the extraction of a metal from its ores, lies in the Neolithic period and it could be a discovery with one or multiple independent origins in areas such as Iran and Serbia.[1] The development of metallurgy required a new way of working and living which led to the development of specialisations within a community. Metal specialists produced tools for the farmers who in turn provided their victuals. The extraction of ores for smelting and the further development of furnaces required new labour structures and division of roles which lead to the establishment of professions.[2] Metallurgy required the sourcing of raw materials that lead to the development of trading routes.

Conservation Science: Heritage Materials, 2nd Edition
Edited by Paul Garside and Emma Richardson

Published by the Royal Society of Chemistry, www.rsc.org

Copper was one of the first metals to be extracted from its ores; however, as a soft metal, its use was limited until it was mixed, alloyed, with others such as arsenic or tin that made it suitable for tools and weapons.

As technology evolved, it became possible to extract a wider range of metals from minerals and to mix them to produce alloys. The importance of metals is evident from the archaeological classification of human activities into copper, bronze and iron ages, and continues to the present with the use of advanced metal alloys.[3] As the types of metal available increased, along with the ability to combine them to create alloys with a wide range of properties, so did their potential uses. Alloying can improve working and performance properties such as fluidity during casting or hardness. The discipline of archaeometallurgy investigates historic metal production, distribution and use, in the wider context of the understanding of technological developments and their impacts on civilisation.[3]

Although there are many metals, only a limited number were discovered and used for the fabrication of objects in the past, including gold, silver, lead, copper, tin and iron. The limited number of metals used were partly due to availability, partly due to technical limitations. Metal production was controlled by both availability and technological advancements as well as cultural traditions.[4] In the 18th and 19th centuries, a number of new metals were discovered and extracted. Alloying developed further, with significant alloys such as weathering steel (known as corten steel) being developed in the 20th century. Nowadays, metal alloys are complex and may be engineered to meet a wide range of specifications, amongst which corrosion resistance is a key priority. Metals and alloys share common physicochemical properties. Some of these properties are affected by the way objects are manufactured which in turn affects their long-term corrosion and subsequently their conservation.

6.2 Properties of Metals

The properties of metals are related to the bonds between their atoms. Metal ions become positively charged when losing electrons from zero valence, are closely packed and share electrons which flow freely between them. If an electron flows away, another one will take its place. The strong interaction between the metal ions and this 'cloud' of electrons is called metallic bonding and is the reason behind the characteristic chemical and physical properties of metals. The electrostatic forces between metal ions and electrons are very strong, requiring a lot of energy to break, resulting in high melting and boiling points. The bonding *via* dissociated electrons makes metals easy to form into alloys.

Metals are good heat and electricity conductors. They are malleable and ductile: malleability is the ability of metals to deform under compression which makes them easy to hammer and create objects, and ductility is linked with their ability to be drawn out to form wires. Metals tend to be heavy and dense, although some are relatively light, such as aluminium or zinc. They are solid at room temperature with the exception of mercury and gallium and are generally hard with a few exceptions such as sodium and potassium. Iron, nickel and cobalt have magnetic properties. Metals are lustrous because of their high reflectivity and can be polished. Their colour when polished varies between dark silvery grey and white, with the exceptions of copper and gold, which are reddish and yellow respectively. These properties are why metals have been used from antiquity until the present.

In many textbooks, metals are divided into ferrous (*i.e.* iron-based alloys) and non-ferrous, with the latter being generally considered more resistant to corrosion than the former. In the periodic table, metals are classified as alkali, alkaline earth and transition metals. The majority of the metals extracted in antiquity are transition metals such as copper, lead, iron and silver.

Metal atoms are arranged in a repetitive pattern having a particular crystal structure, most commonly cubic, body-centred cubic structure or face-centred crystal structure (Figure 6.1). The crystal structure of metals is related to the arrangement of the atoms. Metal crystals are also called grains.

The microstructure of metals and metal alloys can be seen microscopically, with the help of light etching to reveal their structure. This microstructure depends on the composition and the way the metal was formed. A number of different techniques have been used over the years to create metal objects, such as casting and hammering. During casting molten metal is poured into moulds and following cooling the surface is cleaned and polished. Techniques such as lost-wax have been used since antiquity and allowed the creation of hollow objects. Large-scale objects can be cast in pieces and welded together (Figure 6.2). Surface finish could involve polishing using abrasives to the level desired. Hammering allows forming objects using tools and it is one of the first methods used to create sheets of metals and then objects. The metal can be hammered cold or it can be heated. During this process, the crystals undergo permanent plastic deformation and the metal can become rigid. Iron,

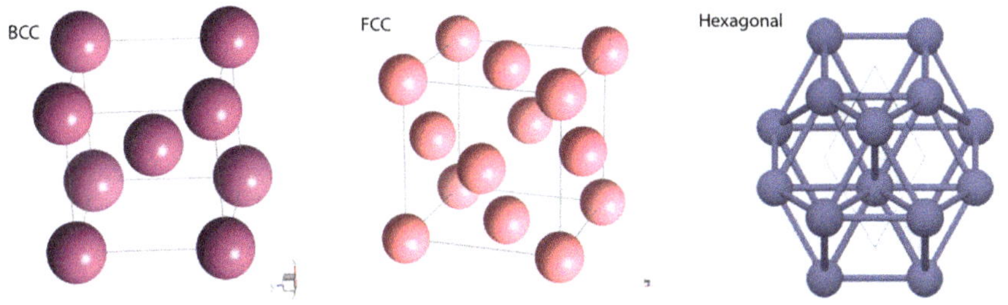

Figure 6.1 Crystal structures of metals where atoms are arranged in a regular pattern. Cubic body centred (BCC) is typical of metals such as Fe, Cr, V and Nb. Adapted from Materials Project, https://doi.org/10.17188/1189317, under the terms of the CC BY 4.0 license, https://creativecommons.org/licenses/by/4.0/. Cubic face centred is typical of Cu, Ag, Au, Al and Ni (FCC). Adapted from Materials Project, https://doi.org/10.17188/1204433, under the terms of the CC BY 4.0 license, https://creativecommons.org/licenses/by/4.0/. Hexagonal is typical for Zn, Ti, Mg and Cd. Reproduced from Materials Project, https://doi.org/10.17188/1307804, under the terms of the CC BY 4.0 license, https://creativecommons.org/licenses/by/4.0/.

because of its high melting temperature, was first reduced out of bloomery iron, which was a mixture of iron with slag. Following hammering to remove the slag and heating, wrought iron was produced.

Metals like other materials exist in different phases (gas, liquid and solid) depending on composition, temperature and pressure. Each phase is homogeneous with clear boundaries and a distinctive crystal structure governing the physical and chemical properties of the material. These can be understood with the use of phase diagrams, which are plots representing the phases of a single metal or an alloy based on the temperature and composition. These show what happens at equilibrium where different compounds or crystalline structures co-exist and are characterised by homogeneous compositions (Figure 6.3). However, in real conditions, phase changes might happen at higher temperatures for melting and lower temperatures for cooling depending on the rate of temperature change.[5] For early metallurgy, homogeneous conditions might have been difficult to achieve, so these diagrams might not be as accurate as they are for modern alloys. The rate of cooling of a molten metal affects its properties and this was used in antiquity to modify properties such as hardness. The composition of the alloys affects the colour of the final material. For example, depending on the composition, bronzes and arsenical bronzes display different colours (see ternary colour diagrams by ref. 6). Historically, a range of objects were made which exploited these properties, such as high-tin bronze mirrors which allowed the reflection of an image.

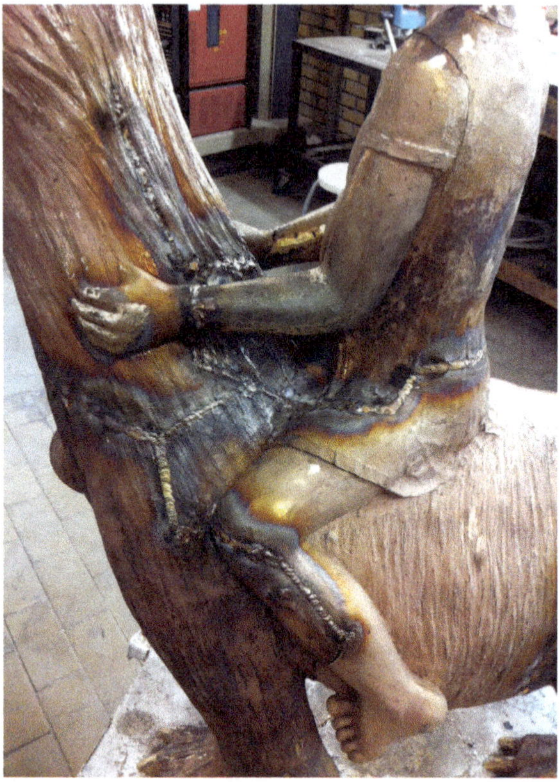

Figure 6.2 Emma Ströde's "Nike's Göteborg" in Kungstorget, Gothenburg, Sweden during manufacturing. The sculpture was cast in pieces and welded together. Seams were polished before being patinated. Reproduced with permission from Emma Ströde.

Metals present different crystal structures depending on the conditions (*e.g.* temperature) under which they solidify. Metal alloys often form different compounds varying in both composition and properties. An example of this is iron–carbon alloys which can contain phases such as ferrite (α-Fe), cementite (FeC_3, iron carbide), graphite (carbon in the form of thin plates) and pearlite (mixture of ferrite and cementite), to name but a few. The grains of a metal can be coarse or fine, may or may not be homogeneous, could have inclusions and pores, and may adopt different orientations; these factors depend on their composition and heating and cooling conditions.[7] In the case of cast metals, as the metal cools and becomes solid, dendrites are formed. Dendrites are tree-like crystals with a main tip and side branches which develop depending on both the temperature and the metal. Impurities concentrate around the dendritic structures causing segregation. The size of dendrites affects the physical, chemical and electrical properties of a metal (*e.g.* hardness,

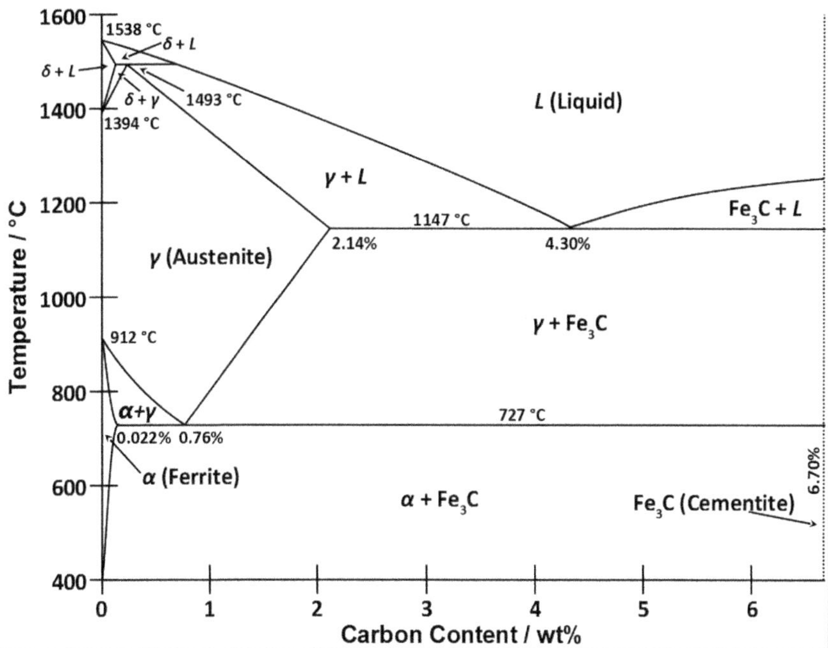

Figure 6.3 Iron–carbon phase diagram showing the iron and carbon phases at different temperatures. L is liquid iron or carbon, α-ferrite, γ-austenite, δ-ferrite has a similar structure as α-ferrite, iron carbide or cementite (Fe_3C) which is a metallic metastable phase hard and brittle. The eutectic point is where the phases coexist. A eutectoid reaction below 763 °C leads to the formation of pearlite, a layered structure of α-ferrite and cementite.

workability). When a metal is worked further or hammered (*e.g.* wrought iron) the grains are deformed, and the metal then needs to be heated in a process called annealing to release some of the energy and relieve the stress that could make the metal fragile. During annealing, recrystallisation occurs which is witnessed by the formation of double lines called annealing twins (Figure 6.4). Some of the stress is evident through strain or slip lines, which are thin parallel lines (slip bands) indicating the shearing of crystals.

6.3 Corrosion of Metals

The majority of metal corrosion takes place through electrochemical reactions that occur between the metal and its environment. Therefore, the major issues to be considered when discussing corrosion are the metal itself (composition, structure, properties), the environment (chemical composition and nature) and their interaction.[8] During

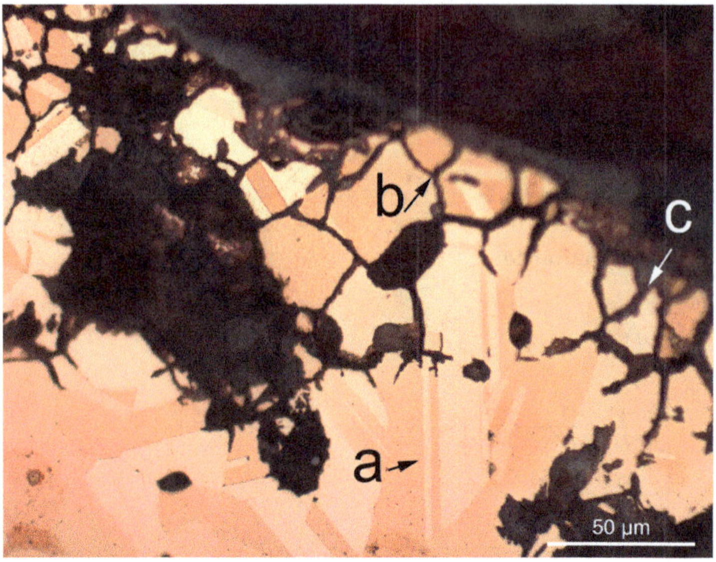

Figure 6.4 Metallographic structure of a copper alloy showing annealing twins (a). Intergranular (b) and transgranular (c) corrosion around and through the metal grains are also evident.

corrosion electrons are transferred, through either an aqueous or a solid electrolyte, causing corrosion of the metal. While the tendency of a particular metal to corrode can be determined thermodynamically, the rate at which this will take place is dependent on the kinetics of the various reactions involved.

6.3.1 Thermodynamics of Corrosion

Corrosion of metals is essentially the result of the transfer of electrons from the metal to the oxidising agent, during which the metal itself becomes oxidised.

This process is traditionally examined in terms of two half reactions, one anodic and the other cathodic, in electrical contact through an electrolyte. The anodic reaction is the oxidation of the metal (resulting in an increase of valence) while the cathodic is the reduction of the oxidising agent (which experiences a decrease in valence). In neutral solutions the oxidising agent is predominantly oxygen. The anodic reaction (6.1) liberates electrons which are then consumed by the cathodic reactions (6.2, 6.3). Consequently, the rates of the anodic and cathodic reactions are the same.

$$\mathrm{M} \rightarrow \mathrm{M}^{n+} + n\mathrm{e}^- \text{ (anode)} \tag{6.1}$$

$$\mathrm{O_2} + 2\mathrm{H_2O} + 4\mathrm{e}^- \rightarrow 4\mathrm{OH}^- \text{ (cathode in neutral solutions)} \tag{6.2}$$

In acidic conditions (pH < 7) the cathodic reaction results in the evolution of hydrogen:

$$2H^+ + 2e^- \rightarrow H_2 \text{ (cathode in acidic solutions)} \tag{6.3}$$

Corrosion will develop slowly from nuclei that are thin and invisible to the eye to thick multi-layered corrosion products of varying compositions and properties (hardness, compactness, *etc.*). If there are no aggressive anions triggering corrosion reactions (such as chlorides in the case of copper and iron alloys) the naturally formed oxides which build up slowly will act as a protective layer for the metal. Corrosion develops through a process which involves dissolution of metal ions in the presence of a liquid, the reaction of metal ions with other ions and, finally, the precipitation of the corrosion products. The repetition of this process leads to changes in composition and properties of the corrosion layers formed, such as that they become more protective and crystalline over time.[9]

In the case of metals with more than one oxidation state such as copper ($^{+,\ 2+}$), iron ($^{2+,\ 3+}$) and several other metals, oxidation happens in two steps. The initial anodic reactions for copper are:

$$Cu \rightarrow Cu^+ + e^- \tag{6.4}$$

$$Cu^+ \rightarrow Cu^{2+} + e^- \tag{6.5}$$

Following initial reactions, there are several different anodic and cathodic reactions which can occur. In addition to promoting an initial oxidation of metal, dissolved oxygen and other elements such as chlorine (Cl) can lead to the formation of other compounds. For example, in the presence of ions such as Cl^-, anodic reactions for copper could include:

$$Cu + Cl^- \rightarrow CuCl + e^- \tag{6.6}$$

$$CuCl + Cl^- \rightarrow CuCl^{2-} \tag{6.7}$$

The tendency of metals to corrode is given by the electromotive force (EMF) potential series (Table 6.1[10]). This series represents the reduction reaction of a metal, relative to hydrogen reduction, in what is called standard hydrogen electrode (SHE). The hydrogen electrode potential is defined as zero. Metals that are oxidised more easily than hydrogen have negative values, and those that are oxidised less than hydrogen have positive values. In general, the more negative the potential value is, the more reactive the metal (active metals), and the more positive the value, the less likely it is to corrode (noble metals).

Table 6.1 Electromotive Force Potential Series (EMF) (modified from ref. 10).

Half reaction	Standard potential, E^0, (V)
$Au^{2+} + 2e^- \rightarrow Au$ *Noble*	1.70
$Au^{3+} + 3e^- \rightarrow Au$	1.50
$Pt^{2+} + 2e^- \rightarrow Pt$	1.2
$Pd^{2+} + 2e^- \rightarrow Pd$	0.987
$Hg^{2+} + 2e^- \rightarrow Hg$	0.854
$Ag^{+} + e^- \rightarrow Ag$	0.800
$Cu^{+} + e^- \rightarrow Cu$	0.521
$Cu^{2+} + 2e^- \rightarrow Cu$	0.337
$2H^{+} + 2e^- \rightarrow H_2$	0.000
$Pb^{2+} + 2e^- \rightarrow Pb$	−0.126
$Sn^{2+} + 2e^- \rightarrow Sn$	−0.136
$Ni^{2+} + 2e^- \rightarrow Ni$	−0.250
$Co^{2+} + 2e^- \rightarrow Co$	−0.277
$Ti^{+} + e^- \rightarrow Ti$	−0.336
$Cd^{2+} + 2e^- \rightarrow Cd$	−0.403
$Fe^{2+} + 2e^- \rightarrow Fe$	−0.440
$Cr^{3+} + 2e^- \rightarrow Cr$	−0.744
$Zn^{2+} + 2e^- \rightarrow Zn$	−0.763
$Cr^{2+} + 2e^- \rightarrow Cr$	−0.91
$Mn^{2+} + 2e^- \rightarrow Mn$	−1.18
$Zr^{4+} + 4e^- \rightarrow Zr$	−1.53
$Ti^{2+} + 2e^- \rightarrow Ti$	−1.63
$Al^{3+} + 3e^- \rightarrow Al$	−1.66
$Mg^{2+} + 2e^- \rightarrow Mg$	−2.37
$Na^{+} + e^- \rightarrow Na$	−2.71
$K^{+} + e^- \rightarrow K$	−2.93
$Li^{+} + e^- \rightarrow Li$ *Active*	−3.05

These reactions are measured in volts (V), and the values represent the equilibrium potential of a metal with its ions, at a 1 M ionic concentration, 25 °C and 1 atm pressure. When two metals are in contact, the metal with the lower potential in the electrochemical series acts as the anode while the more positive is the cathode, therefore the metal with the more negative corrosion potential will undergo preferential corrosion. One drawback to the EMF is that it refers only to pure metals. This behaviour of alloys is reflected by the galvanic series, which represents the reactivity of an alloy in a given environment. Table 6.2 gives the galvanic series for alloys in seawater; the order of reactivity of these alloys in different natural environments does not present any major deviations.[11]

The tendency of a metal to corrode can be evaluated with the use of potential–pH diagrams. These are charts plotting the equilibrium potential against the pH of a solution. Potential–pH diagrams can be used to predict whether or not corrosion, passivation or immunity of a metal in a given environment is thermodynamically possible.

Table 6.2 Galvanic series in seawater (modified from ref. 11).

Platinum	Cathodic (noble)
Gold	
Graphite	
Titanium	
Silver	
Zirconium	
AISI Type 316, 317 stainless steels	
AISI Type 304 stainless steels	
AISI Type 430 stainless steels	
Nickel	
Copper–Nickel (70–30)	
Bronzes	
Copper	
Brasses	
Nickel	
Naval brass	
Tin	
AISI Type 316, 317 stainless steels	
AISI Type 304 stainless steels	
Cast iron	
Steel or Iron	Anodic (active)
Aluminium	
Cadmium	
Zinc	
Magnesium and magnesium alloys	

Immunity refers to the state of the metal where corrosion is energetically impossible to occur. These diagrams provide a useful tool for studying the behaviour of a metal in a corrosive system, as they give an indication of the reactions that could take place and the corrosion products that could be formed. Importantly, these diagrams illustrate the passive or immune areas in a given system, so that the potential of a metal or the pH of the solution can be adjusted in order to prevent corrosion.[11]

6.3.2 Forms of Corrosion

The types of corrosion found on a metal object depend on the properties of the metal or alloy (*i.e.* composition, defects in the crystal lattice, *etc.*), the shape and geometry of the object and the environment to which it has been exposed. To this, one should add conservation treatments, especially cleaning and the application of surface protection, which could also affect corrosion forms.

Uniform corrosion develops over the entire surface of a metal and implies a homogeneity in both the metal and the environment.

Galvanic corrosion happens if two different metals are in contact with each other, and results in the more active one corroding, leaving the more noble metal (the one with the higher EMF potential) intact. Galvanic corrosion can also happen on the same metal alloy, either because of compositional changes between grains or because of metallurgical defects where one area plays the role of the anode and the other the role of the cathode with the subsequent flow of electrons.

Filiform and crevice corrosion are based on differential aeration between different areas of the metal. Filiform corrosion typically happens on surfaces covered with a protective coating which has failed, leading to partial access of oxygen to the metal. It results in the creation of thread-like corrosion which starts on the interface between the metal and the coating, and it can develop further into general corrosion. Crevice corrosion happens when part of a surface is covered, leading to corrosion that starts at the interface and continues under the covered surface.

Pitting corrosion is the most severe form of corrosion. Pits can be narrow or wide, deep or shallow. During pitting corrosion, there is an excess of metal ions in the pit with anions from the atmosphere, such as Cl^-, attracted to maintain electroneutrality. The pH inside the pit is usually very low resulting in the breakdown of stable corrosion products. Pitting corrosion is usually an autocatalytic process and can continue until the total consumption of the metal occurs. The process happens in stages, with initiation leading to development and then to the final stage, which could be either the total consumption of the metal or repassivation where a corrosion layer such as an oxide is formed hindering further corrosion.

6.3.3 Corrosion in Different Environments

As abovementioned, corrosion depends also on the environment to which a metal or metal alloy is exposed. During exposure, chemical, electrochemical and physical processes take place between a metal and its environment. These environments can be largely solid in nature (as will occur with burials), gaseous (such as outdoor conditions) or liquid (if submerged in water), as discussed below.[9]

6.3.3.1 Atmospheric Corrosion

Atmospheric corrosion is complex as it involves reactions at interfaces between the metal and gaseous, liquid and solid environmental phases. Environments are complex and depend on the climate and micro-climate surrounding a metal. The factors that affect the rate and type of corrosion depend on the humidity, temperature, pollutants and airborne particulates, to name but a few. Corrosion is significantly influenced by humidity as a monolayer of water is formed on the surface as soon as a metal is exposed. This layer might be in dissociated form with hydroxyl groups forming on the surface. Subsequent water monolayers are adsorbed on the surface depending on the relative humidity, time of exposure and surface characteristics.[9] The surface of the object, whether it is highly polished or has a texture, or has a protective coating such as varnish or wax, will also affect these reactions. The presence of pollutants and other airborne particulates will affect these reactions as some of them will be hygroscopic thus facilitating water adsorption and oxidation of the metal. Some pollutants, such as nitrogen oxides (NO, NO_2), sulfur dioxide (SO_2), hydrogen sulfide (H_2S), ozone (O_3) and ammonia (NH_3) are in gas form, while others such as sodium chloride (NaCl) or ammonium chloride (NH_4Cl) deposit on the surface as airborne particulates. The reactions of pollutants and particulates with humidity lead to the formation of acids that attack the metal. The acids formed (*e.g.* H_2SO_4) will vary in concentration over the course of wet and dry cycles[9] which themselves can contribute significantly to the formation of corrosion. Depending on the conditions, corrosion will develop very fast but over time the rate will slow down significantly, eventually reaching a steady rate (see Figure 6.14).

In addition to the composition of a metal alloy, and its lattice and manufacturing defects, atmospheric corrosion of metals depends on the type of object, its size, orientation and geometry, its function, whether it is patinated, corroded or covered with a protective coating like paint, wax or varnish.

As a general rule, metals exposed in outdoor conditions are characterised by corrosion representative of the environment they are exposed to. In urban settings, sulfates and sulfides are commonly found, especially on copper-based alloys. In coastal environments, chlorides will prevail forming corrosion products that in the case of certain metals lead to accelerated corrosion rates.

Corrosion rates of metals found inside, regardless of the overarching environment, will be lower than those outdoors. In all cases, there are exceptions, such as those found in very humid environments (as may be the case with damp buildings).

6.3.3.2 Burial Environments

Corrosion in burial environments depends on the composition and physical properties of the soil, its redox potential, resistivity and pH, the amount and the chemistry of water present, and the presence and type of microorganisms.

Soils are complex systems composed of organic materials at different levels of decomposition, inorganic materials of various particle sizes, water, plants, animals, microorganisms and gases such as oxygen, nitrogen, argon, carbon dioxide and methane.[12] Soils vary depending on their composition from those that are largely organic matter to those composed mainly of inorganic material (*e.g.* clay or sandy soils). The composition and the particle sizes of inorganic material affect the physical properties of a soil and in turn the permeability, which influences the way liquids and gases move through the soil. The water content in soil is determined by the water table level (groundwater), which varies seasonally, the ability of the soil to keep water within its pores and water from the environment such as rain or snow.[13] Clay soils will retain water longer than sandy soils. Water will affect the presence of oxygen and other gases and microbiological activity. The corrosivity of a soil depends on the chemistry of the water and the presence of dissolved ions, which can inhibit or accelerate corrosion. Soil pH is a significant factor in corrosion. Soils with high organic matter contents are typically acidic. Other soils can also become acidic over time and are influenced by external factors (such as rain and pollutants) affecting the corrosion of metals such as iron and copper alloys. The redox (reduction–oxidation) potential of the soil indicates whether the conditions in the soil are oxidising or reducing. Another important factor affected by all of the above is the presence of microorganisms, which result in microbiologically induced corrosion (MIC). The most common and severe form is caused by sulfur-reducing bacteria that lead to the formation of sulfides, which in the case of iron alloys can be particularly damaging.

The corrosivity of soils is related to soil resistivity. The lower the resistivity, the more corrosive the condition for a metal becomes.[13] All these factors are interlinked and determine the rate and type of corrosion.

In the case of archaeological objects, the depth of the burial will determine whether there are aerobic or anaerobic conditions. Corrosion products reflect the environment in which they were formed. For example, iron-based alloys in aerobic conditions will primarily form oxides and oxy-hydroxides, while in anaerobic conditions one might expect iron sulfides to be present due to sulfur-reducing bacteria, which could lead to high corrosion rates. In aerobic conditions, the presence of aggressive ions such as chlorides might also lead to high corrosion rates. See Figure 6.5 for an example of heavily corroded iron. It is worth noting that soil changes over time.

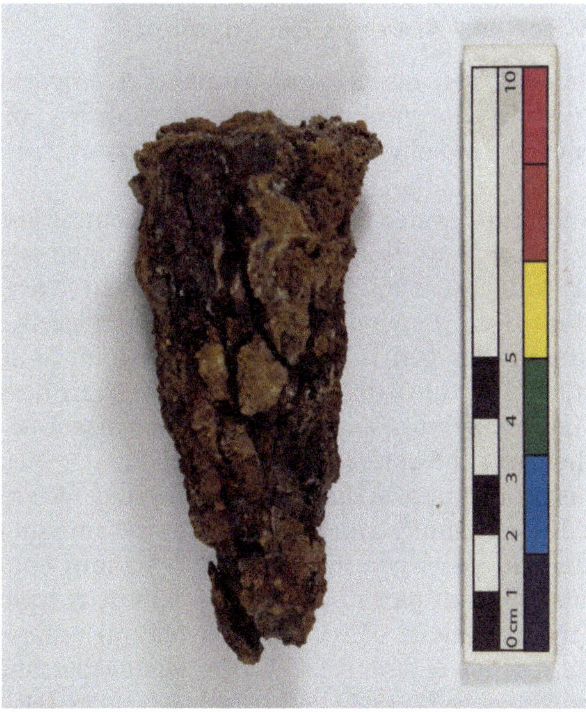

Figure 6.5 Image of an iron object from Sudan. Iron corrosion is voluminous covering the original shape and size of the object. Cracks, soil trapped within corrosion products and losses are also shown. Reproduced with permission from Stephanie Black.

Corrosion is not proportional to the time of the exposure. Corrosion develops relatively quickly (in a few decades) and slows down once the metal reaches equilibrium with its environment.[14,15]

The composition of an alloy will also affect corrosion. Robbiola *et al.* analysed a number of archaeological bronzes and classified them into two types of corrosion depending on whether the original surface is retained or lost.[14] Internal tin oxidation and controlled copper dissolution result in the formation of protective corrosion layers and subsequent preservation of the original surface, while mass migration of aggressive anions from the soil (mainly chlorides and oxygen) lead to the formation of more bulky corrosion.

Silver found in burial environments might become brittle or develop intergranular and transgranular corrosion-induced embrittlement due to stress corrosion cracking (Figure 6.4b and c). This might be due to the manufacturing of the alloy and stresses or insufficient annealing, segregation of copper in silver–copper alloys or extended period of burial.[16]

6.3.3.3 Water and Aqueous Environments

Water is characterised by its composition and properties, such as hardness and the concentrations of minor elements and dissolved gases. Water can be broadly considered as either natural (fresh) water or seawater.

The pH of natural water is neutral or near-neutral and it is characterised by its hardness, from soft to very hard, based on the amount of dissolved $CaCO_3$ present.[15] Dissolved gases include oxygen, carbon dioxide and nitrogen. Depending on the pollution levels, other gases might be present. Oxygen is particularly important as it influences corrosion. Corrosion also depends on the pH, hardness, and temperature. The temperature and the amount of dissolved oxygen are usually higher near the surface.

Seawater is more complex than natural non-salty water. Seawater is characterised by its salinity which is related to the amount of dissolved salts expressed in parts per thousand (‰). Sodium chloride (NaCl) represents a significant part of these salts. Salinity is typically around 35‰, with the exceptions of seas around natural waters (*e.g.* rivers, lakes), where it is lower, and seas like the Mediterranean, where it is higher (approximately 37–38‰). Sulfates, bicarbonate (HCO_3^-), potassium (K^+), calcium (Ca^{2+}), bromide (Br^-), fluoride (F^-), magnesium (Mg^{2+}) and many more may also be found.[15] Chlorinity is related to the amount of chlorides present. Salinity and chlorinity affect the density of the water and its electrical conductivity.

There are seasonal variations in temperature, salinity, dissolved oxygen and pH. Temperature fluctuations are related to the environmental temperature and affect primarily shallow water, not the deep ocean.[15] Dissolved oxygen is significant for corrosion and its concentration varies depending on the depth and currents. The pH of seawater is usually a near-neutral value (around 8); however, ocean acidification due to CO_2 dissolution is resulting in lower values (although still over 7).[17]

Iron alloys corrode faster in relation to other metals and alloys, yet the rate depends on the local conditions in the water and whether the hardness of the water allows for the quick deposition and precipitation of a $CaCO_3$ layer which acts protectively.[7] Organic and inorganic pollutants from industrial waste, food processing to name but a few dissolved in water will also affect corrosion rates.

The corrosion rate is expected to be higher near the surface because of the amount of oxygen and higher temperatures although there might be exceptions due to local conditions.

6.4 Analytical Techniques Appropriate for Metals

Compositional analysis can answer a number of questions regarding metallic heritage. Understanding of metal structure and the different phases present can be done by metallography. There are a number of methods providing elemental chemical analysis (see Chapter 2). X-ray fluorescence (XRF) and scanning electron microscopy coupled with energy dispersive X-ray spectroscopy (SEM–EDS) are the ones most commonly used. Analysis using some of these methods requires a small sample from the object which is then embedded in a resin. These sections can be also used for metallography, or covered with a thin layer of carbon to be analysed using SEM–EDS. Although not covered here, inductively coupled plasma mass spectrometry (ICP-MS) is another method frequently used nowadays, particularly for trace element analysis. When equipped with a laser ablation system (LA-ICP-MS) for sample removal, this method becomes minimally invasive.

6.4.1 X-ray Imaging

Voluminous corrosion products covering the surface of a metal (*e.g.* iron) prohibit visual reading of an object or assessment of its condition. X-ray radiography is an imaging technique based on the use of X-rays to gather information about the form, manufacturing features and condition of an object. The technique is based on passing a beam of X-rays through an object. Metals of different compositions, densities and thicknesses will absorb X-rays to a different degree allowing a black and white image to be captured on film or, more recently, in a digital format. The higher the density and thickness of a metal, the brighter it will appear on the image (see Lang 2005[18]). Corrosion products appear grey in relation to solid metal (Figure 6.6). The image reveals information about the object such as whether it is cast or wrought, the presence of manufacturing defects and joints, compositional variations, decorative features and corrosion patterns.

It is a technique which can be carried out without any preparation of the object being necessary and can be used as a guide for its treatment. For example, it helps conservators define where the original surface was, and the image can be used as a guide for its re-revealing during cleaning. Other issues related to the condition, such as cracks are also revealed. The image obtained will depend on the type of object (form, structure, size, composition and density of the metal, thickness, and additional elements) as well as the source

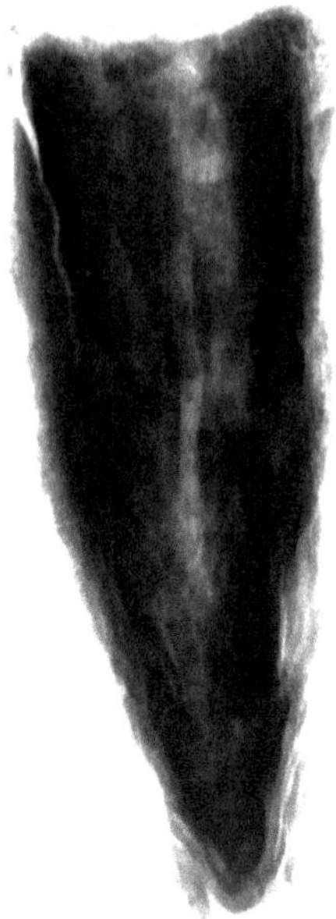

Figure 6.6 X-ray image of the archaeological iron from Sudan (120 kV, 4 mA, 280 ms). More dense areas appear darker and corroded surfaces appear lighter.
Reproduced with permission from Stephanie Black.

of X-rays (intensity controlled by current measured in milliamperes, energy measured in kilovoltage, length of exposure).[19]

X-ray radiography is an easy and relatively quick technique. Nowadays, digital X-ray set-ups are used. Cabinets provide protection from the X-rays; however, they have limitations in relation to the X-ray energy and the size of the object they can accommodate. Other systems can allow the imaging of larger objects; however, they need to be installed in enclosed locations, where lead-lined walls will provide protection for the users. Portable technologies such as those used in hospitals can also be used for *in situ* examination, however, the size and thickness of metals may present safety issues.

There are other types of radiography that can be used for metals. For instance, gamma rays can be used for the imaging of bigger objects. A number of software packages are also available to enhance the quality of the image obtained.

6.4.2 Sampling

Taking a sample from a metal object, especially if corroded, should be done carefully with respect to the object. Heavily corroded metals are fragile and sampling can damage the metal if done inappropriately. Soft metals require a different approach as any sampling could alter their microstructure. Sampling should be from an area that is not visible and employ the least invasive method. A jeweller's handsaw with a diamond blade, a micro-rotating cutting tool or tile cutters can be used. Samples should be representative of the object, but not at the expense of its integrity. Therefore, sampling should be done at a break or an edge if possible. Although sampling is invasive it has the advantage that in most cases, the techniques used do not destroy the sample, which can then be used with other analytical techniques or stored for the future.

For cross sections for elemental analysis such as SEM–EDS or metallography, the sample is embedded in resin (epoxy or polyester), under vacuum if available, and is polished with paper and/or polishing cloths with pastes (*e.g.* diamond, silicon oxide) to as close to a mirror finish as possible. Very corroded samples especially if fragile or very porous might be difficult to impregnate with resin without vacuum. Soft metals can be polished either with waxed papers or by the use of chemical etching.[20]

Other types of sample are required for other techniques, for example ICP-MS, if not equipped with a laser ablation system, requires a liquid sample, so usually a drilled sample is taken and subsequently dissolved. The sample is therefore lost. Neutron activation analysis (NAA), another powerful method that can achieve low detection limits, can in principle analyse whole artefacts without the need for any sample preparation, but the material will remain radioactive for some time.

6.4.3 Metallography

Metallography remains a fundamental method to understand the microstructure of an alloy, the phases formed, the size of the grains to name but a few. It is a destructive or micro-destructive method as it requires a sample to be taken from an object. Following embedding and polishing of the sample, the surface is immersed in a chemical solution, with the aim of etching the surface and revealing the microstructure. Etchants used depend on the alloy. Etching is the light corrosion of the surface. The phases are representative of different properties, therefore are affected at different rates by the chemical solution used revealing the pattern of the structure. Etching can be instant and therefore extra care should be taken in relation to the time (usually seconds) the sample is immersed. Following immersion, the surface is rinsed and dried before observation under a metallographic microscope.

It is possible to carry out metallographic analysis on an object without sampling. For example, metallography can be carried out at an edge of an object that is polished or, in the case of small objects, with the partial or complete immersion of the object in a resin (*e.g.* acrylic) which will be removed after the completion of the study.[20]

The samples are examined using a high magnification reflected light microscope (metallographic microscope). Samples can provide a lot of information about the extent of the corrosion, corrosion stratigraphy and forms of corrosion such as intergranular or transgranular (Figure 6.4b and c).

6.4.4 X-ray Fluorescence

X-ray fluorescence (XRF) is a widely used technique for the analysis of metals as it can identify elements from magnesium (Mg, $Z=12$) to uranium (U, $Z=92$). XRF provides information about the elemental composition of an alloy and it is based on the measurement of the secondary radiation emitted by an atom when irradiated by X-rays (also see Chapter 2). The result is a fingerprint spectrum for each element with intensities representing K, L and M transitions.[21] There are different types of XRF techniques, including energy dispersive (EDXRF) and wavelength dispersive (WDXRF) with the former being used primarily for metals. Micro-XRF allows the analysis of a smaller focus area (down to a few microns). In the last few years, portable or handheld instruments which allow non-contact analysis and scanning of an area have been developed, which were welcomed in the field.

There are several advantages of the technique, with the ease of use, analytical range and short acquisition time being the main ones. It is a relatively affordable solution which has led many museums to acquire instruments. The depth of analysis depends on the composition, density and the energy emitted. Some of these advantages could become disadvantages. For example, a number of end-users might use it without appropriate training. Metal surfaces are rarely homogeneous and after corrosion and conservation, things become even more complex. Understanding the principles of the technique and having appropriate knowledge of the material under analysis and the corrosion, which could affect the outcome, are vital to interpreting the results correctly. Table 6.3 shows the results of the XRF analysis of Damien Hirst's silicon bronze art installation The Miraculous Journey.

Table 6.3 XRF analysis of silicon bronze from Damien Hirst's The Miraculous Journey. Silicon bronze is a high copper alloy with Si around 4% and Mn around 1% with small amounts of Ni, Fe, Zn, Pb, Sn and Al. Analysis was carried out on the surface using an Olympus Innov-X Delta Premium, a handheld XRF Spectrometer with a 4 W, 40 kV Rh anode X-ray tube.

	Cu	Sn	Pb	Zn	Fe	Ni	Al	Si	Mn
Average	92.07	0.09	0.05	0.68	0.66	0.09	1.02	2.76	0.60
St. Dev.	1.51	0.05	0.03	0.64	0.35	0.08	0.54	0.92	0.15
Max	94.26	0.21	0.11	1.56	1.35	0.19	2.53	5.14	1.00
Min	88.85	0.02	0.02	0.17	0.05	0.01	0.45	1.42	0.41

XRF can be used for qualitative analysis to identify the alloy and possible impurities it might have. One of the main drawbacks of the technique is that it cannot identify light elements such as oxygen or carbon, essential components of corrosion products. XRF does not provide any meaningful information for ancient iron alloys where elements such as carbon are an essential component of the alloy (*e.g.* distinguishing between high or low carbon steels) unless there are other elements present such as phosphorus (P) or arsenic (As) that could be detected which are significant indicators of technological innovations, ores used or past conservation treatments. Analysis of light elements (down to Na) can be carried out under a vacuum or a purged atmosphere of helium.

XRF can provide quantitative results with the use of appropriately developed calibrations. In cultural heritage, there have been several attempts to produce standards for alloys such as the copper CHARM set (Cultural Heritage Alloy Reference Material set) which can help heritage professionals and conservators test the performance of their analytical set-up.[22]

6.4.5 Scanning Electron Microscope Energy Dispersive X-ray Spectroscopy

The scanning electron microscope (SEM) is a very useful imaging tool which, when coupled with energy dispersive X-ray spectroscopy (EDS), allows the analysis of metals and alloys. The surface of a sample is scanned by a fine beam of electrons resulting in the emission of X-rays, Auger electrons, cathodoluminescence as well as secondary and backscattered electrons, with these last two providing the two main imaging modes (Figure 6.7). Secondary electrons (SE) are incident electrons irradiated from the sample in energies up to 50 eV; they give information about the topography of the surface analysed.

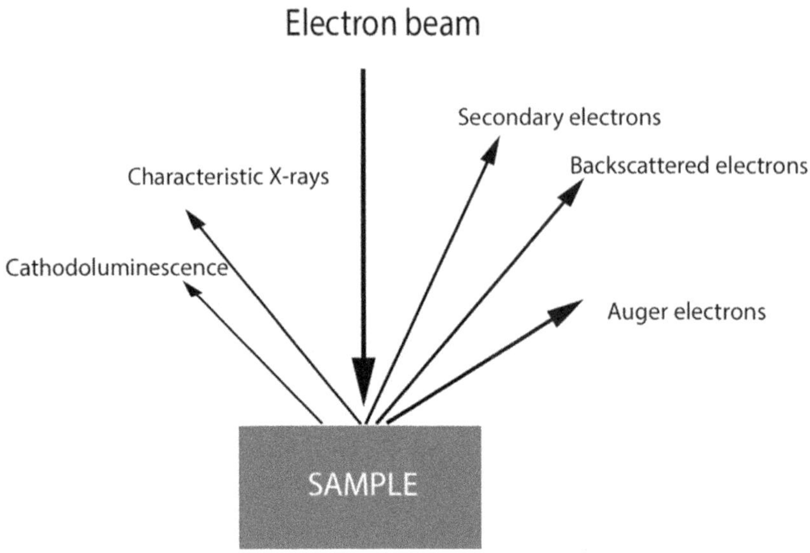

Figure 6.7 Illustration showing the interaction of the electron beam with a sample in SEM–EDX spectroscopy. From the interaction different types of signals emitted such as secondary electrons, backscattered electrons, X-rays, photons, Auger electrons. The secondary and backscattered electrons are used for imaging while the X-rays are used for elemental analysis.

Backscattered electrons (BSE) are the primary electrons scattered within the sample and vary with the atomic number of the analysed phases, providing visual information about the chemical composition of the sample.[23]

EDS analysis provides qualitative and quantitative elemental composition analysis by measuring the X-rays emitted by the specimen. EDS measures the energy and the intensity distribution of the X-rays. An SEM could also be coupled with a wavelength dispersive spectrometer (WDS) providing higher resolution and accuracy for quantitative analysis, albeit usually with longer measuring times.

SEM–EDS provides information about the composition of an alloy and the elemental distribution on a surface as well as distribution of impurities, pores and cracks while in cross section provides information about the stratigraphy of corrosion (Figure 6.8). Bulk, area and spot analyses, line scans and elemental maps can be performed to examine the distribution of the elements (Figure 6.9). The accuracy of the method is in general very good with the exception of light elements (*i.e.* C, S, O, N) which are not as easily detectable. Accuracy depends on the calibration of the instrument. Analysis of corroded surfaces might be misleading in terms of the original composition.

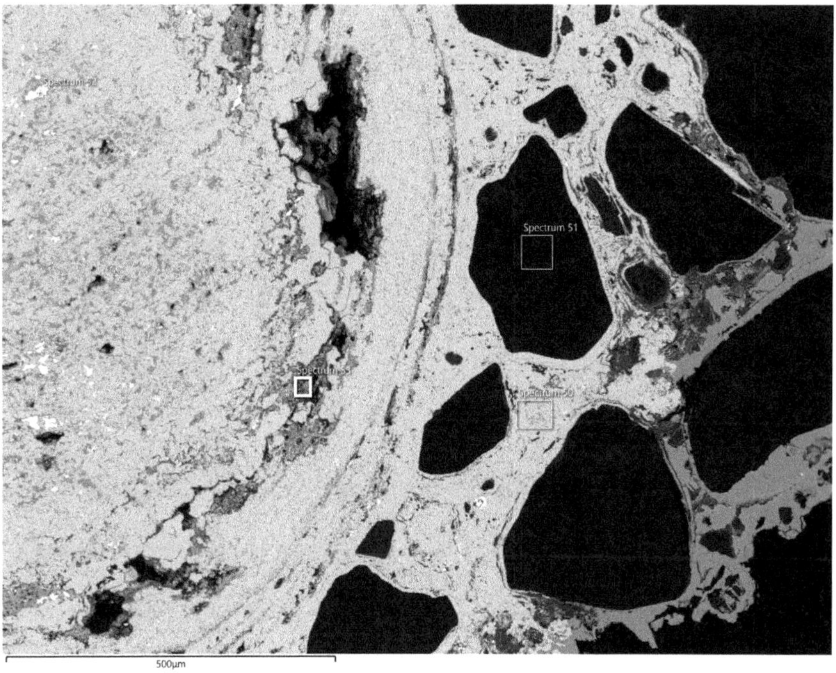

Figure 6.8 SEM backscattered image of a bronze ring from Sudan. Individual spectra show the presence of lead (Pb) and silver (Ag) as well as bromide (Br) and quartz from the burial trapped within corrosion layers. The areas where EDS elemental analysis was carried out are marked on the image.

Detection limits are around 0.1%, depending on the element and matrix in question.

If the technique is used for SEM imaging without also requiring EDS analysis, the samples could be gold-coated, rather than carbon-coated, to obtain better quality images, although this is rarely necessary for metals.

6.4.6 X-ray Diffraction

Analysis of corrosion products allows a better understanding of the environment to which an object has been exposed and, based on the nature of corrosion identified, one can devise the most appropriate conservation plan. There are several methods to be used, with X-ray powder diffraction (XRD) and Raman spectroscopy being two of the most advantageous (see Chapter 2).

XRD is widely used for the identification of the atomic structure of a crystalline compound. It is based on the principle that a crystal diffracts X-rays in a three-dimensional space based on Bragg's Law,

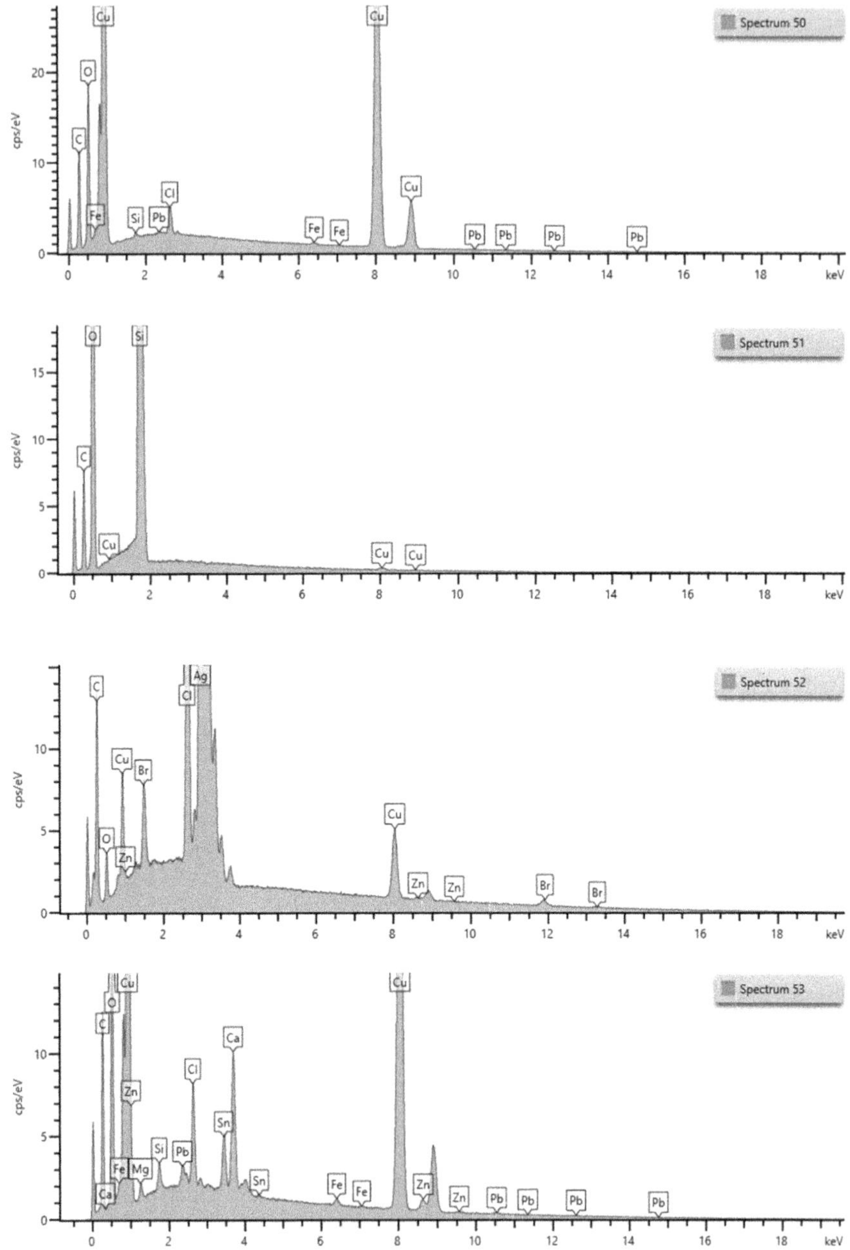

Figure 6.9 Spectra of the areas marked in Figure 6.8. Spectrum 50 shows primarily copper with some chlorides and lead. Spectrum 51 on the black area shows quartz from the soil trapped within corrosion layers. Spectrum 52 is spot analysis and shows the presence of silver. Silver appears brighter in the photo. Spectrum 53 from a corroded area shows the presence of chlorides as well as silica from the soil. Analysis of the corrosion shows the presence of lead (Pb) and silver (Ag) as well as bromide (Br) and quartz from the burial trapped within corrosion layers.

reflecting the atoms it is comprised of. The sample is irradiated with a collimated X-ray beam produced by an X-ray tube and the scattered rays are collected by a detector. The sample is rotated to collect all scattered X-rays. The angle of the diffracted X-rays (θ) and the distance between the crystal lattice planes (d) determine the composition of the compound. The result is presented in a spectrum of intensity *versus* 2θ which allows the spacing between atoms to be determined.[24]

There are different methods, but in most cases, a sample is required to determine the molecular structure. The amount of sample needed is determined by the minimum requirements of the instrument and the available sample holders. If the sample is very small, micro-XRD can be used. The sample is ground and placed in a sample holder.

Single-crystal XRD could also be used, although not widely available thus limiting access to conservation specialists.

The results obtained can be interpreted with the help of established databases such as the ones from the International Centre for Diffraction Data (ICDD, http://www.icdd.com). XRD is a particularly useful qualitative method for determining the nature of corrosion observed, however, there are methods which can be used to quantify different phases within a sample.

6.4.7 Fourier Transform Infrared Spectroscopy

Infrared spectroscopy is based on the vibrations of the atoms of a molecule interacting with infrared light.[25] The different chemical groups in a molecule absorb infrared radiation and the peaks of the spectrum represent chemical bonds expressed in wavenumber (cm^{-1}) on the x-axis. The y-axis can be plotted either in absorbance or in transmittance mode. Absorbance mode corresponds to the amount of infrared radiation absorbed by the molecule and it is used for qualitative analysis. Transmittance mode corresponds to the infrared radiation the molecule transmits, and it is used only for qualitative analysis. The relationship between absorbed and transmitted radiation of a molecule is:

$$A = \log(1/T) \tag{6.8}$$

where A is absorbance and T is transmittance.[26]

A functional chemical group in different molecules absorbs infrared light at the same range with slight shifts depending on the molecule.

Fourier Transform Infrared Spectroscopy (FTIR) is a technique of infrared spectroscopy where the beam coming from the source is divided by a beam splitter before passing through the sample. The beam from the source is partly transmitted and partly reflected by two mirrors placed one across the source and one perpendicular. The mirrors reflect 50% of the beam back to the source and 50% is transmitted back to the beam splitter. The latter beam passes through the sample at 90° and is measured at the detector.[25]

FTIR provides a wide variety of analytical possibilities. Many different preparation methods and techniques can be used to answer different questions and analyse substances in solid, liquid or gases phases, providing information about the chemistry of a molecule. A great advantage of the technique is that only a small sample of a compound is needed for the analysis. It is a fast and comparatively cheap analytical method.

A disadvantage of the technique is that it is not the best method to analyse complex or unknown compounds. If a mixture is analysed, it might be difficult to distinguish and understand which chemical bond belongs to which compound. Certain corrosion products, such as malachite, atacamite or paratacamite in the case of copper, can be identified using FTIR.

The Infrared and Raman Users Group database offers examples of infrared spectra of minerals that can be identified (www.irug.org). These could include copper carbonates such as malachite and azurite, copper chlorides such as atacamite or paratacamite which signify active corrosion of copper-based alloys, lead sulfates and lead carbonates, iron oxides such as hematite, iron carbonates like siderite and hydroxy-oxides like goethite. These have been used extensively as pigments and the majority of the database refers to pigments, although samples collected from objects have also been featured. Unknown compounds and past restoration materials (inhibitors, adhesives, coatings) can be identified using FTIR. Figure 6.10 shows the spectrum of BTA used as an inhibitor for copper alloys.

Potassium bromide (KBr) mixed with the sample can be used in transmittance mode, with one of the key advantages of the technique being that only a small quantity of the sample is needed.[27] Samples of both corrosion products or materials used for the conservation of an object (adhesive, coatings) can be ground using a mortar and pestle, mixed with KBr and pressed into pellets/discs using a hand-operated press. KBr is highly hygroscopic and certain compounds can be affected by the water absorbed. KBr pellets need precision in the way they are made.

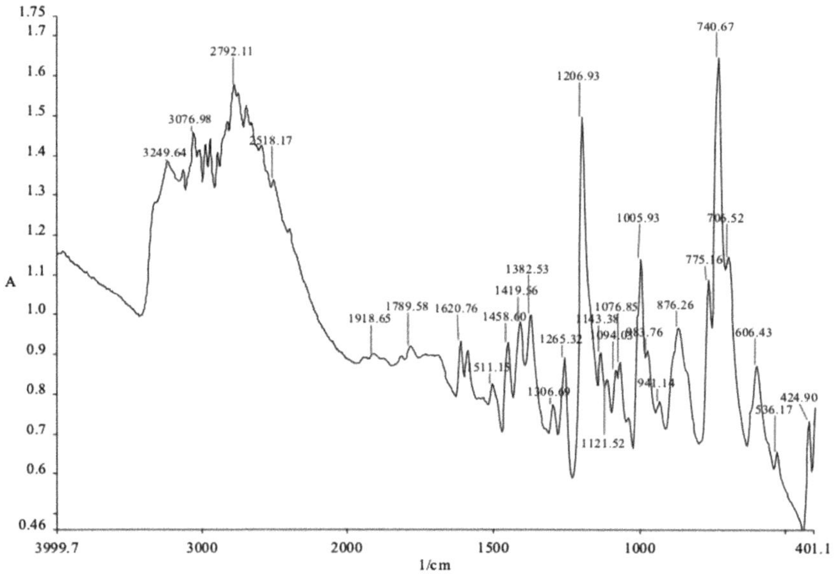

Figure 6.10 FTIR spectrum of BTA with the main peaks of the benzene ring at 740 cm^{-1}, 3100–3000 cm^{-1} (C–H stretching), 2000–1700 cm^{-1}, 1620–1400 cm^{-1} ring modes, 1200–1000 cm^{-1} for C–H in-plane bending and 1000–700 cm^{-1} where the C–H out-of-plane bending takes place. BTA also has the triazole ring that is of great importance as the bond between copper and BTA happens through the nitrogen of the triazole ring. The reaction of BTA with copper is evident by the change in N–H stretching bands at 3200–2800 cm^{-1} and 1600–1000 cm^{-1}.

Attenuated total reflection (ATR) is a common method used based on total internal reflection and it allows the examination of a sample without preparation being necessary. The sample is placed on a crystal with a high refractive index such as diamond or zinc selenide (ZnSe) and pressure is applied to provide good contact. The infrared beam arrives at the crystal at a 45° angle. Only part of the reflected light reaches the sample through an evanescent wave and is recorded by the detector.[25] A good account of the ways FTIR analysis can be carried out can be found in the fourth chapter of Smith's (2011) book on FTIR.[26] Portable set-ups or accessories such as a microscope and a probe can also facilitate analysis directly on an object.

6.4.8 Raman Spectroscopy

Similar to FTIR, Raman spectroscopy is based on the vibrations of molecules. Raman spectroscopy relies on the scattering of photons occurring when a molecule is irradiated by a monochromatic light source. The light interacts with the molecule to polarise electrons

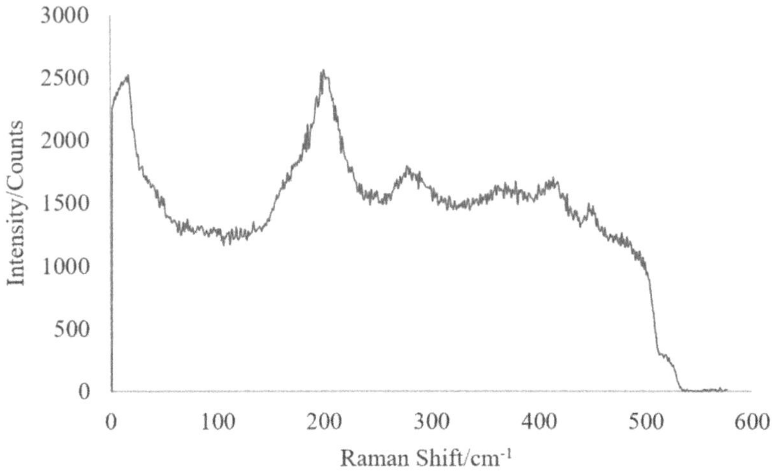

Figure 6.11 Raman spectrum showing the presence of magnetite (668 and 536 cm^{-1}) and lepidocrocite (380, 522 cm^{-1}). The sample was analysed using a Renishaw inVia Raman Microscope, with the excitation source a 532 nm Argon laser. The powdery micro-sample was taken by the Richard Serra installation in Qatar.

which are transformed into a short-lived state called a 'virtual state'. Raman spectroscopy provides information regarding the structure of a molecule which leads to a characteristic 'fingerprint' spectrum.[28]

A number of compounds can be identified by Raman spectroscopy. Analysis can be done *in situ* (on a small object or using a portable set-up) or by taking a small sample.

An example of a sample of powdery corrosion products taken from the Richard Serra weathering steel installations West/East–East/West in Qatar shows the presence of magnetite (Fe_3O_4) and lepidocrocite (γ-FeO(OH)) (Figure 6.11).

6.4.9 X-ray Photoelectron Spectroscopy

X-ray Photoelectron Spectroscopy (XPS) or Electron Spectroscopy for Chemical Analysis (ESCA) is a technique used for surface analysis and can give qualitative and quantitative information about the existing elements. It can provide information about the chemistry, organisation and morphology of all the elements found on the surface with the exception of H and He.[29]

When a surface is irradiated with photons, the energy of the photons is transferred to its core-level electron, followed by the emission of the electron. The electrons are emitted according to the energy, which depends on the atom and molecule present. Each element has a threshold energy level that has to be exceeded in order to have photoemission. Above the threshold energy, the emission of the electrons will be proportional to the rate of the excited photons.[29]

XPS is based on the energy ($h\nu$) of a photon transferred to an electron of an atom resulting in the emission of the electron. The kinetic energy (Ek) of the electron ejected is derived by the equation:

$$\mathrm{Ek} = h\nu - \mathrm{EB} \tag{6.9}$$

where Ek is the kinetic energy of the ejected electron, $h\nu$ is the energy of the X-ray source, and EB the binding energy of the electron in the atom of the surface analysed. The kinetic energy can be easily converted to binding energy ($\mathrm{EB} = h\nu - \mathrm{Ek}$). Binding energy is measured in electron volts (eV) and depends on the type of atom as well as the atoms it is bound to. The binding energy is related to the covalent or ionic bonds of the atoms.[29]

A thin (approximately 10–30 μm) Al or Be window is placed on the sample to eliminate the possible detection of secondary electrons from the X-ray source and contamination of the sample.[30]

The energy levels XPS analysis is based on can be simply classified as core levels, valence levels and Auger series.[31] The core levels peaks are the ones related to the electron state of an atom and do not depend on the fact that the atom is not free but part of a solid surface.[31] Because the core levels are representative in energy terms of the atoms present, the spectrum provides information not only about the composition but also the chemical state of the elements.[31]

The irradiation of a surface using X-rays can result in the emission of Auger electrons. Auger electrons differ from photoelectrons. One of the differences is that the energy of Auger electrons is not dependent on the energy of the source used while photoelectrons are directly proportional to this energy, as can be seen from eqn (6.9).[29] The importance of the Auger series is that they can determine the chemical state of an element. The KLL series are mainly used for this analysis. KLL series are complex and their intensity changes according to the atomic number.[31]

Other peaks found on an XPS spectrum are satellite and ghost peaks. Analysis starts with a wide scan spectrum (survey scan), which covers a large range (more than 1000 eV). Following this, analysis can focus on individual peaks and their features, which cannot be observed from the wide scan. The binding energy is on the x-axis and the counts (intensity) on the y-axis.

Quantitative analysis of the elements found on the surface is also possible. The peak area of each element is measured resulting from the percentage (atomic) of each element. Usually, either the percentage or the ratios between elements is measured. This quantification cannot be taken as absolute, but only as a relative scale. This method assumes that the surface is homogeneous and does not take into account factors such as the contamination layer found on almost every surface, the fact that the different libraries (some of which are theoretical and some empirical) can give different results, and other inconsistencies related to the instrument and/or the operator.

XPS analysis allows the examination of the oxidation state of metals which is useful for the study of corrosion as well as understanding surface changes in case of stabilisation methods such as the use of corrosion inhibitors.

6.4.10 Electrochemical Techniques

A number of electrochemical methods are used to identify and monitor corrosion phenomena in cultural heritage. For example, monitoring the corrosion potential (E_{corr}) over time can be used to understand both corrosion and passivation phenomena of a metal object either *in situ* or during stabilisation treatments. Electrochemical impedance spectroscopy (EIS) has been used to assess the efficiency of corrosion inhibitors and protective coatings[32] with portable options being developed for *in situ* measurements on monuments. Measuring potential requires connecting the metal under investigation (working electrode) against a reference electrode (*e.g.* saturated calomel or silver/silver chloride, Ag/AgCl) and a counter electrode (*e.g.* stainless steel), with electrolyte being an important factor in the process. Potentiodynamic curves have been used to assess the effect of stabilisation treatment on metal coupons.[33]

6.5 Conservation of Metals

Conservation decisions are influenced by various parameters, including the values an object has from archaeological, historical, religious and artistic perspectives; the type of object; the surface properties of the metal or alloy used; the original use and appearance of the object; the way it is to be used or displayed; the context and resources available; and the extent and type of corrosion forms developed which can be destructive for the integrity, interpretation and understanding of an object. Conservation decisions are complicated and become even more so when objects prized for their surface qualities are considered.

Conservation of metals should consider all possible factors affecting their condition. From corrosion, environmental conditions it has been exposed to, to factors such as physical damage, vandalism and wear and tear, to future use and type of exposure. Documentation, including condition assessment and technical analysis, is of particular importance to determine the best way of managing change of an object/artwork through conservation. There are many different ways of carrying this out: however, the main focus here has been on X-ray imaging and analysis to identify composition and manufacturing methods, as well as corrosion products, which are vital in designing the best treatment.

An interesting example where conservation is restricted is situations in which artefacts carry different values and the public interacts with them in ways that result in deterioration (Figure 6.12). In religious objects, wear and tear is often observed because of this kind of interaction. In Figures 6.12a and b the feet of St Peter's statue found in St. Peter's Basilica in Vatican City are worn after years of pilgrims rubbing them. The bronze statue of the Saint blessing the pilgrim is not officially attributed to an artist but is believed to have been cast during the 13th century, although some place it in the 5th century.

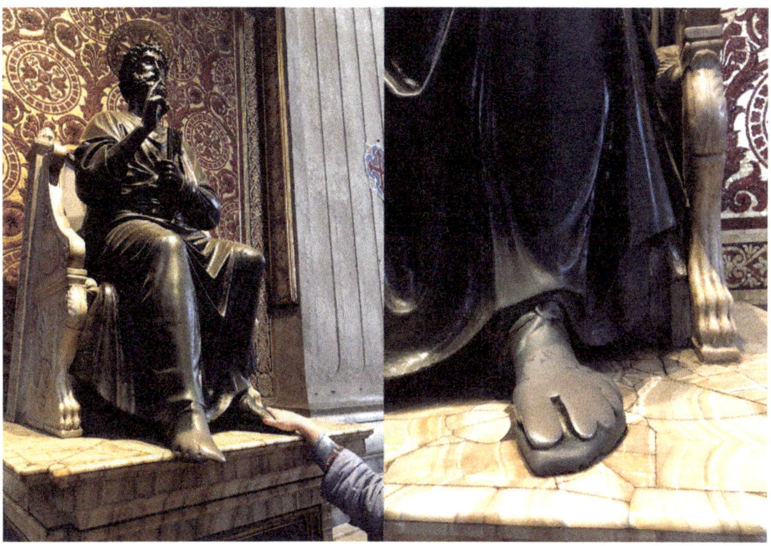

Figure 6.12 Images of St. Peter's statue displaying the wear to the metal caused by centuries of pilgrims touching and kissing the feet. Traditionally people touch the right foot, and its details have now been completely lost; however, it can be seen that the left is in a similar state. Authors own photographs, January 2019.

Conservation of metal objects in museums has the advantage that corrosion could be monitored to allow intervention in the early stages. Conservation of outdoor sculptures or metallic heritage as part of a building is more complicated; decisions often depend on factors out of a conservator's control with a lack of resources for proper study and planning, as well as unrealistic deadlines, challenging best practices. The artist's intent is another important factor to consider as in some cases interaction with the public is desired. An example of this is the artwork by Emma Ströde found in Gothenburg, Sweden (Figure 6.13). The bronze artwork was designed with the intention to allow the interaction of children with the artwork. The original patination has been partially worn out and corrosion oxides have formed.

Objects coming from a burial environment might be covered with corrosion products. Archaeological objects are often considered information carriers as they provide insight into the technology and social practices of a site and as such their treatment might be limited in stabilising active corrosion. The cross section of a ring (see Figure 6.8) shows how trapped soil within corrosion can distort understanding of an object. In this case, X-ray imaging can help the conservator identify an appropriate cleaning method while SEM–EDS analysis shows the presence of chlorides which could lead to

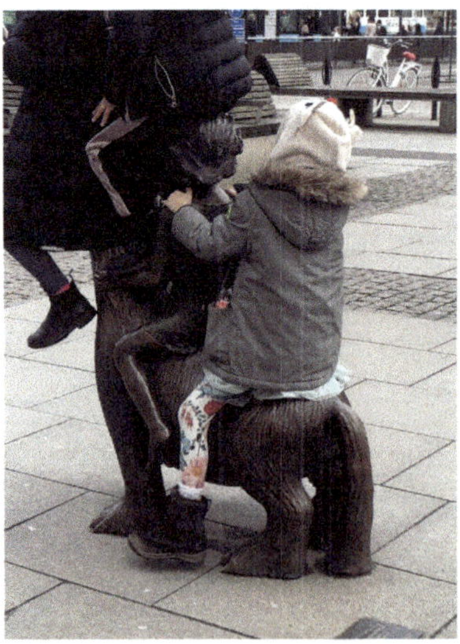

Figure 6.13 Image showing the sculpture "Nike's Göteborg" by Emma Ströde in Kungstorget, Gothenburg, Sweden which is very popular with children. The bronze sculpture was cast in pieces and it was installed in 2015 with the intention of the interaction of children with the sculpture. Author's own photograph. Reproduced with permission from Emma Ströde.

activation of corrosion indicating that the ring should be stored in appropriate environmental conditions (*e.g.* anoxia or RH < 45%) or be chemically stabilised. Such methods are not always available or at the disposal of a conservation laboratory, however, when possible examination and analysis are valuable tools for its treatment.

Large outdoor sculptures such as the installation The Miraculous Journey (2013) by Damien Hirst in Doha, Qatar impose issues related to the environment they have been exposed to. The artwork is comprised of fourteen statues made of silicon bronze depicting the different stages of the development of a foetus inside the womb till birth (Figure 6.14). The artwork is located in front of Sidra Medicine, a hospital dedicated to women and children.

Doha has significant levels of pollutants such as SO_2, NO_2, H_2S, NH_3 and O_3 to name but a few which, in combination with greatly fluctuating temperature and humidity and high levels of dust due to construction and the desert, accelerate corrosion. Cleaning and

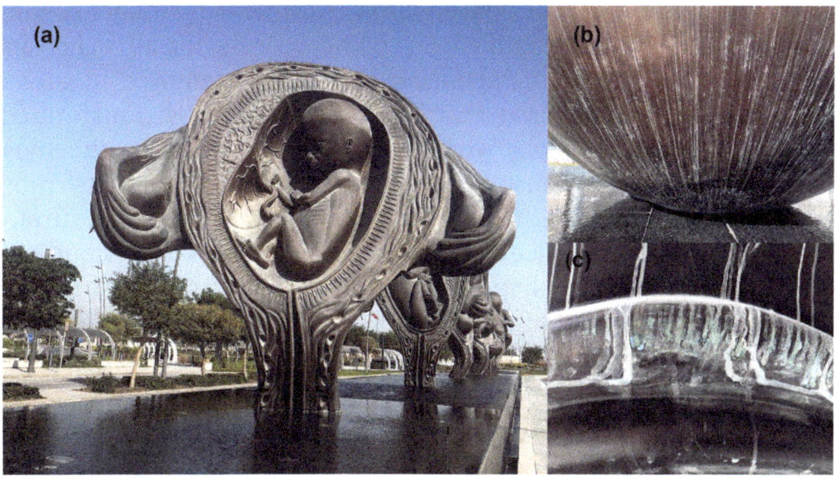

Figure 6.14 (a) shows Damien Hirst's art installation The Miraculous Journey located in Doha, Qatar. The sculptures were patinated in the foundry and were covered with a protective coating. That lead to the formation of a uniform corrosion on the surface. Oblique surfaces are covered with dust. (b) Doha displays high levels of fluctuating RH. Condensation in the early hours of the day leads to the formation of droplets which in combination with chlorides from the sea and air pollutants result in the formation of localised corrosion. (c) Detail of active corrosion with copper chlorides. At the lower part, one can see porosity due to casting. The uniform corrosion is disrupted by copper oxide (red), copper chlorides (light green and greenish blue). Pitting corrosion is also evident.

surface protection carried out at the end of 2018 were not enough to prevent the development of new corrosion (see Figure 6.14b and c). Conservation in such cases should be based on monitoring and maintenance actions such as regular cleaning to remove particulates from the surface to avoid both corrosion and cementation of dust which will require more drastic methods of cleaning. Analysis of corrosion products using XRD and Raman spectroscopy showed the presence of cuprite, copper chlorides (atacamite and/or paratacamite) which as indicative of active corrosion and malachite. XRD analysis of surface dust showed the presence of calcite, dolomite and gypsum while SEM analysis showed also the presence of rubber, possibly because of tyres from the busy road in front of the installation and fibres.

6.5.1 Cleaning

Cleaning of metals involves the removal of dust, dirt, corrosion, and materials used in previous conservation treatments no longer serving their purpose. Cleaning is an irreversible process for all materials, however, in the case of metals, what is primarily removed in such a process is original material that has been chemically transformed. The removal of corrosion, whether necessary or not, is the most laborious step in the conservation of metals, yet it is not the most important for the long-term survival of an object. The original surface is of immense importance when cleaning metals. Voluminous corrosion which hides the shape and size of an object requires removal to reveal the 'original surface' of the object. What this 'original' surface represents remains a variable for each object, whether it is the surface when the object was manufactured, abandoned or in some cases the surface of the object since its last clean. Bertholon (2013) defined it as the surface at the time it was abandoned and suggests the use of 'limits of the original surface' or 'limitos' possibly preserved within other corrosion layers and could be located based on the physicochemical characteristics of the different corrosion layers.[34]

Methods to remove corrosion include mechanical, chemical and electrochemical methods, each of which presents different challenges and affects the perception of the object and its material differently. Mechanical methods are favoured as they are considered more controllable than the other methods. In the case of soft metals such as lead or silver, chemical and electrochemical methods are used. The type of object (archaeological, historical, contemporary artwork) plays a significant role in these decisions.

Electrolytic reduction has been used on silver and lead artefacts as a method of 'cleaning' to avoid losing surface details or original material (*e.g.* the use of electrolytic pen to reduce silver sulfide back to silver) or as a stabilisation method where reduction is carried out using cathodic polarisation at constant potential.[35] Electrolytic reduction has been used since the beginning of the 20th century. The surface of the reduced metal might not be compact as the corrosion products reduced are usually porous. The length of the cleaning/consolidation treatment will depend on the thickness of the corrosion layer and the surface will need protection as the surface could get tarnished or corroded fast.

In artworks where the surface finish plays an important role (*e.g.* the mirror image finish of Anish Kapoor's Cloud Gate at the Millennium Park in Chicago) cleaning becomes very important for the perception of the object, therefore the idea of preserving a patina does not apply. In addition, there are a number of surface finishes which impose new challenges in terms of conservation and corrosion removal. Mirror finish needs to maintain the shine and cleaning to a mirror polish is required on a regular basis. Another example is the polishing of the surface to create a specific effect, such as that of fine vertical lines found on Robert Adams' sculpture 'Two' at the Getty in Los Angeles. The stainless steel surface was cleaned using acidic solutions assisted with the use of mild mechanical means to maintain its finish.[36] Spray paints and markers often witnessed on public art can cause aesthetic disfigurement and may be difficult to remove without damaging the patina or surface of the object.

6.5.2 Stabilisation and Surface Protection

Stabilisation of active corrosion is the most crucial step in ensuring the preservation of a metal object. Stabilisation of corrosion involves the use of chemicals which transform 'harmful' corrosion products into more stable forms, often with the simultaneous release of the ions which initially stimulated the corrosion (*e.g.* chlorides in the case of copper and iron alloys). A number of methods including the use of sodium sesquicarbonate ($NaHCO_3 \cdot Na_2CO_3$) for the stabilisation of copper or the use of sodium hydroxide (NaOH) and alkaline sulfite (NaOH and Na_2SO_3) for the stabilisation of iron are used. Deoxygenated alkaline solutions have also been tested for iron with positive results.[37] Electrochemical methods have been used to both monitor and assist in the removal of chlorides. Electrochemical methods might cause chemical and physical changes, which may not be acceptable or desirable in some cases. They also entail expertise not always available amongst conservators. Stabilisation can also be achieved with the use of corrosion inhibitors, which create complexes with the metal, thus halting corrosion. Corrosion inhibitors have been in use for a long time as they provide a cheaper, easier and faster way to retard corrosion without drastic alterations of existing patinas. Inhibitors such as benzotriazole (BTA) have been used since the 1960s for the stabilisation of copper alloys.[38] Research has been directed towards environmentally friendly compounds such as carboxylates which have been tested for iron, copper and other metals (see ref. 39). Stabilisation treatments are important especially when preventive measures are not possible.

Customarily metals are covered with protective coatings such as acrylic or nitrocellulose resins and microcrystalline waxes to protect them from the environment. In the past, coatings used on metals included oils, epoxy resins, polyurethane varnishes, natural and synthetic waxes, and a number of other coatings including paints, which created a new set of problems. Some of these were not effective (*e.g.* oils), others were not reversible (epoxy resins, polyurethane varnishes, waxes) while others create issues such as filiform corrosion (acrylic, epoxy and other polymers).

Modern metal alloys are a challenge for conservators as not much is known about their long-term corrosion behaviour. For instance, there are alloys which were developed to be corrosion resistant and have been widely used in art and modern architecture, which however have proven not to be resistant if exposed to unsuitable environments. An example of this is weathering steel, which does not behave well when exposed to fluctuating relative humidity with high levels of chlorides, typically found at coastal sites.[40]

Painted metal surfaces present problems related to both the base metal and the paint layer. The application of the paint and the primer are of extreme importance in this case. A compact, well-adhered paint layer acts protectively to the metal. If the paint layer is no longer adhered to the substrate and there are cracks or losses, then oxygen can access the metal and corrosion can result in further loss of the paint. Painted surfaces are often difficult to preserve, especially in the case of outdoor sculptures where environmental parameters such as rain, sunlight and pollutants affect their long-term preservation. In such cases, the paint might be replaced with more resistant paint (see ref. 36).

Archaeological objects are often treated as information carriers and artworks are treated with the aesthetic outcome in mind, while there are others where none of the traditional methods can be applied. The example of the statue of St. Peter in St. Peter's Basilica shows how difficult it is to prevent the wear of the metal. Solutions like the application of a sacrificial layer of a surface coating might not be effective over the long term without maintenance and replacement of the coating at regular intervals.

Study of corrosion and conservation methods for metallic heritage continue to develop as new methods of analysis are developed, new alloys are used, and surface finishes continue to challenge conservation practices. Understanding the material and how it interacts with its environment and the aim of its conservation will determine the best methods to investigate it and conserve it.

Recommended Reading

R. Bertholon, Characterisation and Location of Original Surface of Corroded Metallic Archaeological Objects, *Surf. Eng.*, 2013, **17**(3), 241.

R. A. C. Cottis and L. L Shreir, *Shreir's Corrosion*, Elsevier Science, Amsterdam, 4th edn, 2010.

P. Letardi, Electrochemical Measurements in the Conservation of Metallic Heritage Artefacts: An Overview, in *Corrosion and Conservation of Cultural Heritage Metallic Artefacts*, ed. P. Dillman, D. Watkinson, E. Angelini, and A. Adriaens, Woodhead Publishing Limited, Cambridge, 2013, p. 126.

C. Leygraf, T. Graedel and E. Society, *Atmospheric Corrosion*, Wiley, Hoboken, 2000.

L. Robbiola, J. M. Blengino and C. Fiaud, Morphology and Mechanisms of Formation of Natural Patinas on Archaeological Cu–Sn Alloys, *Corros. Sci.*, 1998, **40**(12), 2083.

D. A. Scott, *Copper and Bronze in Art: Corrosion, Colorants, Conservation*, Getty Conservation Institute, Los Angeles, 2002.

D. A. Scott and R. Schwab, *Metallography in Archaeology and Art*, Springer International Publishing, Basel, 2019

Th. Rehren, Metals, in *Archaeological Science – An Introduction*, ed. M. P. Ruchards and. K. Britton, University Press, Cambridge, 2020, p. 365.

Th. Rehren and E. Pernicka, Coins, Artefacts and Isotopes – Archaeometallurgy and Archaeometry, *Archaeometry*, 2008, **50**(2), 232.

R. Wanhill, Stress corrosion cracking in ancient silver, *Stud. Conserv.*, 2013, **58**(1), 41.

References

1. Th. Rehren, in *Archaeological Science - An Introduction*, ed. M. P. R. K. Britton, University Press, Cambridge, 2020, pp. 365–386.
2. *An Encyclopaedia of the History of Technology*, ed. I. McNeil, Routledge, London, 1990.
3. Th. Rehren and E. Pernicka, *Archaeometry*, 2008, **50**, 232–248.
4. Th. Rehren, Chemical Analysis, in *Encyclopedia of Archaeology*, ed. D. M. Pearsall, Academic Press, New York, 2008.
5. F. Campbell, *Phase Diagrams - Understanding the Basic*, ASM International, Ohio, 2012.
6. M. Radivojević, J. Pendić, A. Srejić, M. Korać, C. Davey, A. Benzonelli, M. Martinón-Torres, N. Jovanović and Ž. Kamberović, *J. Archaeol. Sci.*, 2018, **90**, 106–119.
7. D. A. Scott, *Copper and Bronze in Art: Corrosion, Colorants, Conservation*, Getty Conservation Institute, Los Angeles, 2002.
8. *Shreir's Corrosion*, ed. T. J. A. Richardson, Elsevier Science, 2010.
9. C. Leygraf, T. Graedel and E. Society, *Atmospheric Corrosion*, Wiley, 2000.
10. B. D. Craig, *Fundamental Aspects of Corrosion Films in Corrosion Science*, Plenum Press, New York, 1991.
11. D. A. Jones, *Principles and Prevention of Corrosion*, Prentice Hall Inc., Upper Sandle River, New Jersey, 2nd edn, 1996.
12. A. R. Conklin, *Chemical Analysis, Introduction to Soil Chemistry*, John Wiley & Sons, Inc., Hoboken, NJ, USA, 2005.
13. T. R. Jack and M. J. Wilmott, in *Uhlig's Corrosion Handbook*, ed. R. W. Revie, Wiley, US, 3rd edn, 2011.
14. L. Robbiola, J. M. Blengino and C. Fiaud, *Corros. Sci.*, 1998, **40**, 2083–2111.
15. R. A. Cottis and L. L. Shreir, *Shreir's Corrosion, NL*, Elsevier Science, NL, 4th edn, 2010.
16. R. Wanhill, *Stud. Conserv.*, 2013, **58**, 41–49.
17. S. Flecha, F. F. Pérez, J. García-Lafuente, S. Sammartino, A. F. Ríos and I. E. Huertas, *Sci. Rep.*, 2015, 5.
18. J. M. Lang and A. Middleton, *Radiography of Cultural Material*, Elsevier Butterworth-Heinemann, Oxford, 2nd edn, 2005.
19. V. M. Fell, Q. E. Heritage and R. White, *Guidelines on the X-radiography of Archaeological Metalwork*, English Heritage, 2006.
20. D. A. Scott and R. Schwab, *Metallography in Archaeology and Art*, Springer, Cham, Switzerland, 2019.
21. J. P. Potts, in *Portable X-ray Fluorescence Spectrometry: Capabilities for In Situ Analysis*, ed. P. J. Potts and M. West, Royal Society of Chemistry, Cambridge, 2008.
22. A. Heginbotham, J. Bassett, D. Bourgarit, C. Eveleigh, L. Glinsman, D. Hook, D. Smith, R. J. Speakman, A. Shugar and R. Van Langh, *Archaeometry*, 2015, **57**, 856–868.
23. S. V. D. Ammelinckx, J. Van Landuyt and G. Van Tendeloo, *Electron Microscopy: Principles and Fundamentals*, 1st edn, 1997.
24. R. L. Snyder, in *X-ray Characterization of Materials*, ed. E. Lifshin, Wiley-VCH, Weinheim; New York, 1999.
25. B. Stuart, *Modern Infrared Spectroscopy*, John Wiley and Sons, Chichester, 1996.
26. B. C. Smith, *Fundamentals of Fourier Transform Infrared Spectroscopy*, CRC Press, Boca Raton, Fla, 2011.
27. L. D. Norman Colthup and S. Wiberley, *Introduction to Infrared and Raman Spectroscopy*, Academic Press, 3rd edn, 1990.

28. E. Smith and G. Dent, *Modern Raman Spectroscopy: A Practical Approach*, John Wiley & Sons, Incorporated, Newark, 2019.
29. B. D. Ratner and D. G. Castner, *Surface Analysis–The Principal Techniques*, Wiley, Chichester, 2009.
30. D. P. Woodruff and T. A. Delchar, *Modern Techniques of Surface Science*, Cambridge University Press, Cambridge, 2010.
31. D. Briggs and J. C. Riviere, in *Practical Surface Analysis: Auger and X-Ray Photoelectron Spectroscopy*, ed. D. Briggs and M. P. Seah, John Wiley & Sons, Chichester, 2nd edn, 1994.
32. D. Christian, *J. Solid State Electrochem.*, 2009, **14**, 353–361.
33. P. Letardi, in *Corrosion and Conservation of Cultural Heritage Metallic Artefacts*, ed. P. Dillmann, D. Watkinson, E. Angelini and A. Adriaens, Woodhead Publishing Limited on behalf of the European Federation of Corrosion, Cambridge, 2013.
34. R. Bertholon, *Surf. Eng.*, 2013, **17**, 241–245.
35. C. G. Degrigny and R. Le, *Stud. Conserv.*, 2013, **44**, 157–169.
36. B. Considine, J. Wolfe, K. Posner and M. Bouchard, *Conserving Outdoor Sculpture: The Stark Collection at the Getty Center*, Getty Conservation Institute, Los Angeles, 2010.
37. M. Rimmer, D. Watkinson and Q. Wang, *Stud. Conserv.*, 2012, **57**, 29–41.
38. S. M. Golfomitsou and J. F. Merkel, Metal 07: Interim Meeting of the ICOM-CC Metal WG: Amsterdam, 17–21 September 2007, Amsterdam, 2007.
39. S. M. Hollner, F. Rocca, E. Reguer and S. Corrosion, *Eng., Sci. Technol.*, 2013, **45**, 362–366.
40. S. Golfomitsou, *Stud. Conserv.*, 2016, **61**, 55–60.

7 Glass, Ceramics and Enamelled Objects

P. Colomban

Sorbonne Université, Faculté des Sciences et de l'Ingénierie, CNRS, MONARIS umr8233 4 Place Jussieu, F-75005 Paris, France
Email: philippe.colomban@sorbonne-universite.fr

7.1 Introduction

The mastering of fire is associated with the early development of humankind. The first potteries appear to have been produced as early as 12 000 BC in different continents (Japan, Siberia, Africa), before or just at the Neolithic Transition from nomadic to sedentary life.[1–6] However, artefacts made of fired clay are much older (about 25–30 000 BC for the Dolni Vestonice Venus sculpture).[7] The first potters could have been itinerant, as pointed out by archaeological and ethnographic studies,[8] and their main activity may have been the production of containers of differing shapes. The word *pottery* comes from the Latin *potum*, the name of a drinking utensil. *Ceramic* is from the Greek *keramos* (*κεραμοσ*) which means *horn*, indicative of the matter then used for drinking utensils. *Glass*, *glaze* and *enamel* (and smalt) are thought to come from the Proto-Germanic '*glasa*' (to shine) and 'smaltjan' (to smelt), respectively. In the Chinese language the word for pottery, *tao ci* (陶瓷), is related to a more philosophic meaning: *the way*, the principle of life, and this fact could explain the importance of Ceramic Art in Asian Cultures. A good knowledge of pottery and enamelling technologies is needed to analyse ancient

Conservation Science: Heritage Materials, 2nd Edition
Edited by Paul Garside and Emma Richardson

Published by the Royal Society of Chemistry, www.rsc.org

artefacts. Here we will try to give an overview of the background required, as well as useful references.

Ceramic tools were required for the development of metallurgy and both techniques require the building of kilns and the careful selection of raw materials. Figure 7.1 schematizes the technological evolution from the production of pottery consisting of a porous body, free of waterproofing coating, up to an advanced high dense porcelain

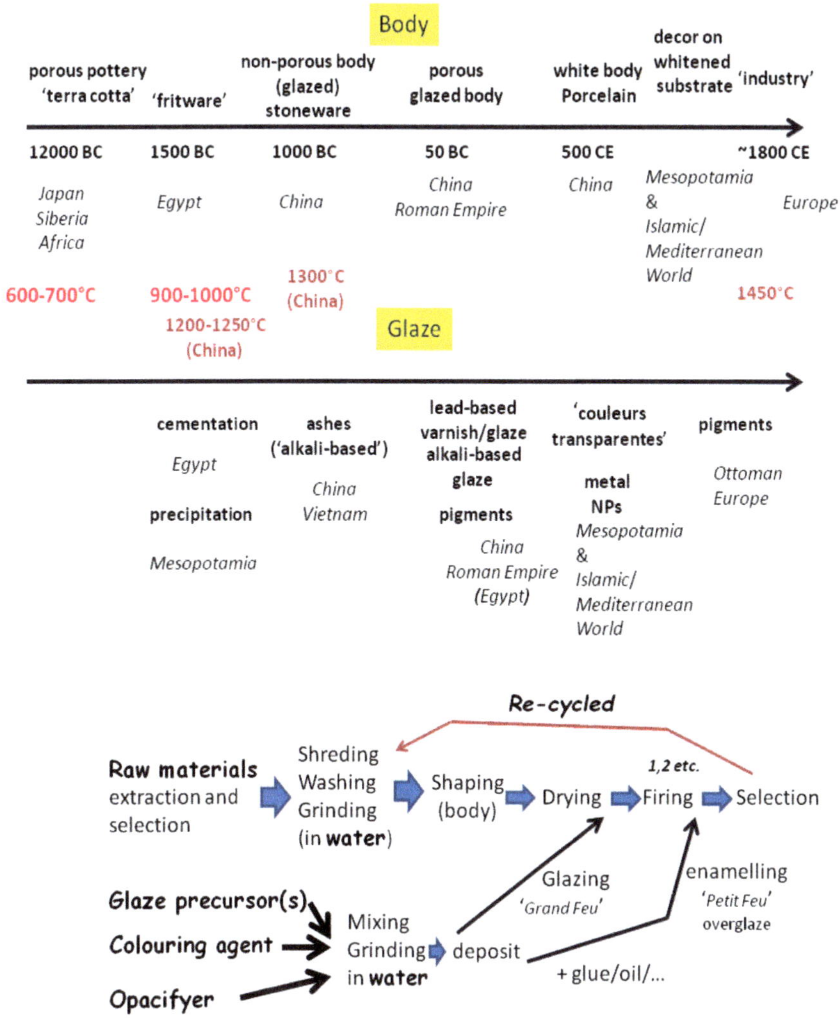

Figure 7.1 Timeline of the development of pottery technology (top) and flowchart of ceramic manufacturing operations (bottom). Required firing temperatures are indicated in red. The development of technologies associated with kiln production determines the evolution of ceramic production.

enamelled with a complex decor, and then to special ceramics and glass having specific chemical, thermomechanical or electrical/magnetic properties.[5,6]

The deposit of a waterproof coating, a glassy silicate called 'glaze', only started with the development of an urban or elite culture in Egypt and Mesopotamia, *ca.* 1500 BC. The preparation of glassy paste is much older.[3–6]

The first classification of the different types of pottery (Table 7.1) was established in 1844 by Alexandre Brongniart, head of the French Imperial (established as a Royal Factory by Louis XV in 1756, then as a National Factory during the French Revolution and presently) Sèvres Factory and in 1805 founder of the first ceramics museum coupled to a factory.[9,10] Brongniart's classification is based primarily on the mechanical properties of the fired body and on the composition of the glaze. This slightly modified classification (Table 7.2) remains the reference to this day.[11,12]

Consequently, studies conducted on ancient pottery and enamelled artefacts (on glass or on metal) are aimed at documenting the technical history and production methods, identifying characteristics related to:

(i) the raw materials [nature, provenance, selection, processing (washing/grinding/heating)],
(ii) the body and its processing/firing (microstructure, phases, compositions),

Table 7.1 Types of pottery (after Brongniart' classification).[9,10]

Class	Type	Glaze	Remarks
Soft-paste body	Terracotta	No	—
	Lustre pottery	Silica-alkaline composition	Thin 'coloured' glaze
	Varnished pottery	Lead-rich composition	
	Glazed pottery (faience)	Tin oxide opacification	—
Hard-paste (non-fuse) body	Fine faience	Lead-based composition	Colour-less body
	Grès-cérame (stoneware)	No or silica-alkaline composition	Coloured body
Hard-paste (translucent) body	Hard-paste porcelain (China)	Feldspar-based composition	Use of kaolin
	Soft-paste 'natural' porcelain (fritware, English bone China)	Lead-(borax-) based composition	Use of clays or kaolin or bones
	Soft-paste 'synthetic' porcelain (fritware)	Lead-based composition	Use of marl and frit

Table 7.2 Professional classification of ceramics.[11,12,34]

Class	Type	Body open porosity%	Glaze	Remarks
Terracotta		20–35	No coating	
	Varnished/glazed		Lead-based	
Faience		15–20	Tin oxide opacification	
	Fine faience	10–15		Colourless body
Fritware		<15	Lead-based	A slip can be added in-between body and glaze
Stoneware	Salt varnish	<8	Alkaline (Salt)	
	vitreous	<3	Feldspar based	White body
Porcelain	Hard-paste (China)	<1		Feldspar and kaolin-based body
	Bone China	<1		Phosphate-based body
	Soft-paste	<2	Lead-based	Vivid colours
Technical	Oxide	<0.5	In general, no glaze	Physical/chemical properties controlled
	Carbide			
	Nitride			
	Carbon			

(iii) the decor/coating (nanostructure, composition, colouring agents, pigments),
(iv) the firing cycle and kiln technology (fuel, casing . . .),
(v) the relationship with other/previous centres of production.[2–6]

Such information informs our understanding of trade routes and technology transfers.

The interest in the documentation of technologies is long-standing. A first foray in the matter can be found in Pliny the Elder's History.[13] A few rare medieval manuscripts dealing with the subject have survived, such as those of Theophilus,[14] del Rey Alphonso[15] or Jean d'Outremeuse[16] in Western countries and those of Persian potters in the Islamic World.[17,18] Subsequently, the development of the printing press led to an abundance of books concerning the Fire Arts during the 16th century,[13] in particular those of Agricola, Piccolpasso, Neri, *etc.*[13,19,20] The technical literature was then broadened during the 18th century.[21–23] The first archaeometric studies on porcelain were published by François de Réaumur,[23] in light of the information collected in China by Jesuit d'Entrecolles during his visit to the Jingdezhen Factory, one of the leading Chinese centres in porcelain production.[24] The interest at that time for all Fine Arts techniques was large as proved by the publication of the Encyclopédie.[25] However, it is with the works of Brongniart,[10] Bastenaire-Daudenart (especially in faience),[26] Brongniart and Bontemps (glass),[27,28] and then of Salvetat,[29] Deck[30] and others[31–34] that very useful technical and scientific reference books became available. More details can be found in ref. 13.

7.2 Raw Materials and Processing (Grinding, Shaping, Firing, Enamelling)

We will first define some terms. Raw materials are the rocks selected for the preparation of the artefact (unfired or 'green' ceramic body) and of the glaze/glass/enamel precursor (Figure 7.1). A rock is generally the association of different minerals: for instance, a pegmatite (*petungtse*) is a rock made of big crystals of feldspars, quartz and micas. The composition of minerals is very variable and the composition of a centimetre-long feldspar crystal can vary from potassium-rich (orthoclase or microcline) at one tip to sodium- (albite) or calcium-rich (anorthite) solid solution at the other one. Clays are hydroxylated (earth)alkali aluminosilicate hydrates in the form of lamellar nanoparticles, the interlayer cations being more or less solvated by water molecules. A clay (rock) can be made of different clayed minerals (illite, smectite...). Kaolin (rock) and the associated mineral (kaolinite) belong as clay to the phyllosilicate group but do not have interlayer hydrated (earth)alkali cations and hence a much higher melting temperature than clays, and a much lower plasticity. To eliminate impurities that can colour the matter, raw materials are carefully washed (and mixed and stored) before use.

Artefacts are prepared by hand(wheel)- or mould-shaping from clay-based paste (water content ~20–30 wt%).[10–12,34] The addition of water leads to an aqueous suspension (sol) that can be used to make a vessel using a porous (*e.g.* plaster) mould; part of the water is trapped, and a millimetre-thick layer forms at the mould surface by sol–gel transition; pouring the remaining liquid slip and drying allows recovery of the green body. Complete drying and firing are then performed.

Enamel is a glassy coating on a substrate, metal, glass or ceramic; a glaze is an enamel on a ceramic body; the French terms '*glaçure*' and '*couverte*' correspond to a glaze on a porous body (and hence deposited on an already fired body) and on porcelain/stoneware/vitreous (deposited on a 'green' body and fired at the same time). A frit is a mixture melted and quickly cooled in water in order to make easier powdering of the resulting glassy mixture. Frit can be used as an ingredient of the body (fritware/stonepaste/soft-paste) and glaze (Tables 7.1 and 7.2).

The number of raw materials required to produce pottery, glass/glaze and enamelled artefacts is variable. For *terracotta* and certain faience and stoneware only one 'clay' was used. However, in the majority of cases, the addition of sand is required: the large shrinkage of the clay-water mixture (typically 30 wt% water is needed to obtain a paste ready for shaping[11,12]) requires the mixing of the clay with an inert phase to avoid the formation of 'elongated' cracks during the drying. Broken and powdered ceramic waste (also called grog) can partially replace sand. In a sand- or fireclay-containing paste, small shrinkage cracks form around inert grains that prevent the formation of long cracks. These small cracks disappear during the firing/sintering stage. More complex potteries require the use of different clays, each of them giving specific properties: good plasticity for shaping, good mechanical strength in the green state (*i.e.* before firing), good sintering, *etc.*[11,34] For sophisticated ceramics such as fine faience and porcelain many clays/kaolin, feldspar and sands are used.

Consolidation during firing (called 'sintering') results from the reaction between adjacent grains. A liquid phase forms at the grain surface and accelerates reaction kinetics (atom diffusion rates are orders of magnitude faster in liquid states and dissolution–precipitation phenomena at the corrugated surface of grains help to form cement in between adjacent grains, Figure 7.2). Good compaction of clay, sand, and feldspar grains maximizing the number of inter-grain contacts is mandatory to obtain good sintering and low porosity artefact with good mechanical properties. Contrarily to glass preparation that requires the melting of the entire batch, ceramics are achieved with a limited volume of molten phase: fluxing materials such as chalk, feldspars, intrinsically, or by contact, melt at low temperature (<1100 °C). The addition of frit, lead oxide, borax, iron oxide, *etc.* lowers the melting temperature up to 500–700 °C.

In order to decrease the amount of undesirable phases (colourants, sources of gas) and/or to avoid coarse grains, grinding, washing and flotation can be undertaken. Kaolin and clays are rocks formed by the corrosion and sedimentation of mother rocks such as gneiss, granites, and pegmatites (a mixture of feldspars, quartz, and micas). This leads to naturally fine grain materials (the largest dimension ranges from a few nanometres to a few microns) but composition varies from place to place. Deterioration of the quality of the production of a factory, and then its disappearance, results often from the impossibility of the potter to adapt the processing to the shift in composition. Specific trace elements (Sr, Rb, rare earths, U, *etc.*) or minor components (Fe, Ti, Zr, *etc.*) can be used to identify particular raw materials with the same provenance.[35–41] The more sophisticated the technology of production, the easier the identification of place or date of production will be. Indeed, innovations can be used as *post quem* milestones.[13] As an example Figure 7.3 shows the plot of Rb *vs.* Sr and Ti *vs.* Fe content measured with a portable X-ray fluorescence (pXRF) instrument (see Chapter 2) in the body of '*Boccaro*' wares (early 18th-century red stoneware assigned to J. F. Böttger, Meissen, Saxony).[36,37] Data have been normalized to the Si signal in order to take into account the variation of geometrical conditions of the measurements, such as distance and angle of analysis, owing to the different shapes of the objects. These '*Boccaro*' wares try to imitate *Yixing* Chinese contemporary productions, highly prized by the European elite of the 17th and 18th centuries.[36,37] For instance some teapots have been prepared by moulding the original *Yixing* wares and confusion is possible when employing only visual criteria to distinguish them.[36]

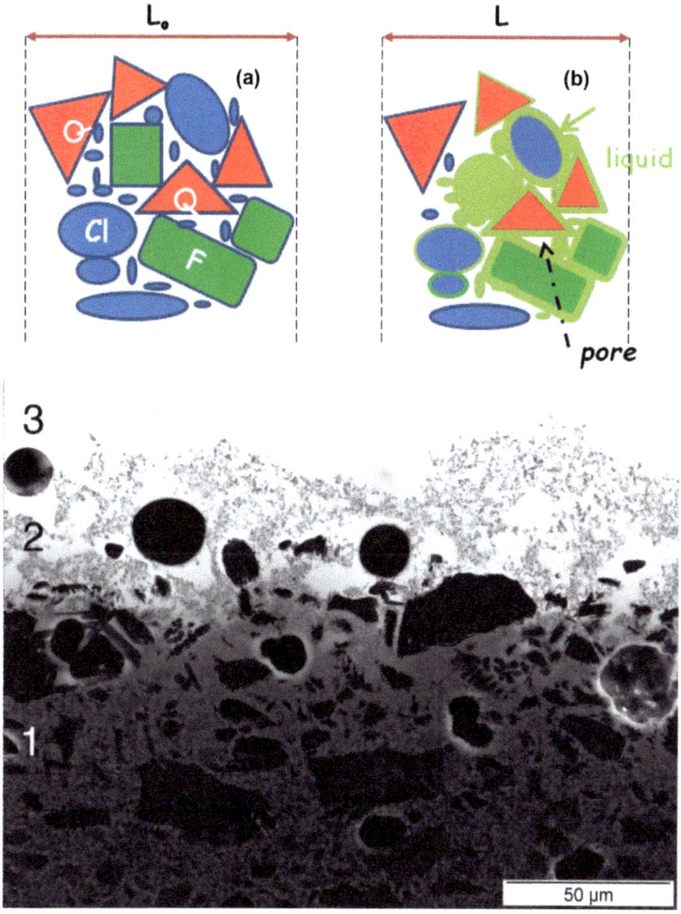

Figure 7.2 Sketch of the transformation taking place during sintering: (a) compacted matter; (b) a liquid phase forms, wet the grains and cement them on cooling (Q: quartz, F: feldspar, Cl: clay aggregates); the densification (L < Lo) and the formation of new phases, amorphous and crystalline on cooling cements the grains. SEM-BSE photomicrograph of a cross-section from the body (1) to the Pb-rich glaze (3) with a reaction zone (2) in between. The body shows some large roundish pores (black), relictic quartzes surrounded by circular cracks (black), subidiomorphic tridymite laths (black), many small subidiomorphic crystals of wollastonite (white) and a glassy matrix (grey). The outermost parts of the body lack wollastonite and is also part of the reaction zone. The 1–2.5 μm long, prismatic shaped CaO-rich crystals (grey) of the reaction zone are too small for a quantitative chemical analysis. Soft (frit) porcelain plate from the Royal manufacture of Sèvres, France, 1781.
Reproduced from ref. 93 with permission from Prof. M. Maggetti.

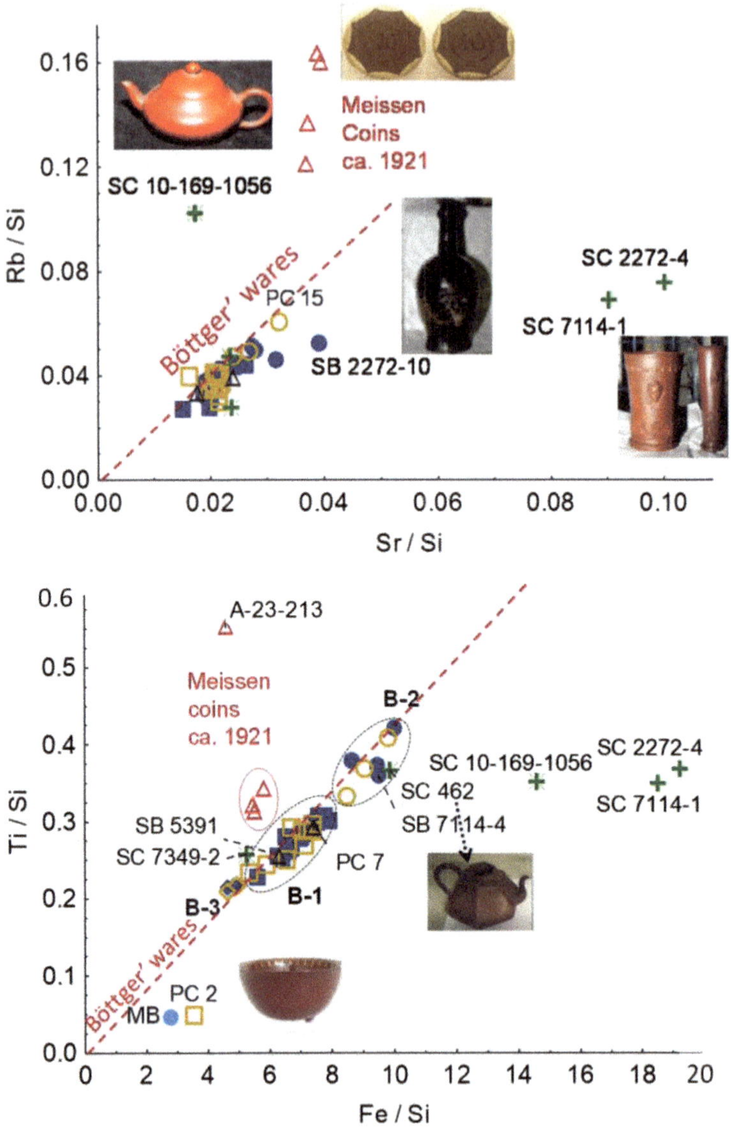

Figure 7.3 Examples of categorization obtained from the elemental ratio plots, here measured with a portable XRF instrument (after ref. 36); Böttger productions made with the same raw materials providing Rb and Sr, or Ti and Fe elemental ratios are located along the straight line starting from the origin. Other artefacts are Chinese (+), modern Meissen (triangle) and fake items (MB,PC2).

Matter made with variable proportions of the same raw materials will give data aligned on a straight line starting from the origin. For instance, Rb and Sr elements are associated with calcium (from chalk or alabaster or gypsum) and Ti and Fe elements are typical impurities of clays and kaolin. Differentiation between Böttger, aligned with the calibration line, and Chinese *Yixing*, modern Meissen and other artefacts is thus obvious.[36,37] For instance the SC462 teapot from Sèvres Collection was a true Böttger '*Boccaro*', whereas the SC 10-169-1056 was actually the Chinese model. These elements have been selected by considering the impurities *a priori* characteristic of the raw materials reported to have been used at Meissen (red kaolin, alabaster, *etc.*).[42,43] Another approach is the use of algorithmic techniques such as principal component analysis (PCA), Euclidean cluster variation method and supervised PCA[44] to identify the most pertinent parameters, statistically and independently of the considerations regarding the ceramic or glass technology.

The preparation of enamels requires the availability of fine (micronic or submicronic) powder(s).[45] Grinding needs a lot of energy and before the mastering of hydraulic energy (~12th century in Western Europe, 9th century in the Islamic World), potters and glassmakers took advantage of the thermal decomposition of 'hydrated' compounds to obtain fine powders; for example $Ca(OH)_2$ from calcareous rocks or shells, calcium phosphate from bones and complex compositions ready for glazing from wood and plant ashes.[45–47]

The main mechanism of the sintering of pottery is so-called liquid sintering:[11] at a given temperature one of the raw materials melts, or at the contact between two different raw materials, a liquid phase appears, although the other component grains are not modified.[11] Sintering takes place (Figure 7.2): first, the liquid phase dissolves corrugations at the grain surface; the liquid phase forms a neck between grains in contact, forming cement on cooling with the precipitation of new phases, and promotes densification due to the rearrangement of the grains by capillarity or wetting effect.[11,12] Figure 7.2a and b show a sketch of the body microstructure before and after the formation of a liquid phase by reaction between feldspar grains and clay aggregates. The liquid phase wets the grains, forming liquid necks in between that bring the grains closer by capillarity action. Dissolution of small grains and corrugations modifies the liquid composition, and leads to new phases that precipitate; the composition of the liquid phase also evolves towards a less fusible composition and the liquid phase becomes more viscous and solidifies. A bridge is formed in between adjacent grains. The micrograph in Figure 7.2 shows a polished section of glazed porcelain observed with a scanning electron microscope

(SEM); the backscattered electron (BSE) mode is sensitive to the Z number (matter made of heavier elements appears whiter). At the contact between molten glaze and solid substrate, part of the body is dissolved, saturates the glaze and small crystals appear. In some cases, the liquid glaze can separate into two liquid phases that lead to an opalescent glaze.

Anybody can experiment with the efficiency of a liquid phase to facilitate densification: a hard and dense ball of snow is easily obtained with snow at a temperature close to 0 °C (water-assisted densification); on the contrary, it is almost impossible to make a ball with cold snow (−15 °C or more).

Hydroxylated compounds (such as clays) undergo dehydration/decomposition around 600 °C, giving a very reactive and fine material (called metakaolin if kaolinite is present).[11,45] The first liquid phase forms at about 750 °C between silicates and FeO–Na_2O,[48] *i.e.* in a reducing atmosphere, and reacts with the high reactive product arising from clay/kaolin dehydroxylation. *Terracotta* with rather good mechanical strength[49] can be thus obtained a millennium BCE ago with very simple kilns made from pits covered with branches and straw.

Lead oxide melts at ~900 °C, but with the addition of alkali or alkaline earth-rich materials or borax it melts at a lower temperature, as low as about 500 °C.[48] Most of the liquid phases of SiO_2, Al_2O_3, CaO, Na_2O, K_2O *etc.* mixtures form between 900 and 1200 °C,[48] the typical range of temperature needed to produce pottery. Wood and plant ashes and a mixture of earth-alkali and alkali hydroxide with small amounts of silica and phosphorus oxide were the first glaze precursors.

Table 7.3 Crystalline phase identification methods. Representative references are given.

Methods	Resolution (usual)	Sample preparation	References
(Polarised) Optical microscopy	0.5–10 μm	Fracture/cut/polishing	35, 38, 49, 93, 106, 107, 113
SEM(EDS)	1–10 nm	Fracture/cut/polishing	69, 93, 132, 145
TEM	0.1–1 nm	Cut/polished/FIB	86–90
(μ)XRD	>(10 μm) mm	Cut/powder	52
μXRD (Synchrotron)	1–10 μm	Cut/polishing	81–84
FTIR–ATR	0.1–2 mm	Cut/polishing/powder	94, 101
μFTIR–ATR	10 μm	Cut/polishing/powder	102
μRaman	0.5 μm	No/cut	51, 54 (see Table 7.6)

For instance, typical oxide compositions of faience are given in Table 7.4. Potters do not use elemental compositions because the raw materials are natural rocks or minerals and precise composition measurements only date from the 19th century. They use mass (wt%) or volume (vol%). A typical composition of faience is clay (26 vol%), marl (a mixture of clays and limestone, 50 vol%) and clayish sand (25 vol%); another example: clay 60 vol%, sand 20 vol% and limestone 20% vol.[34] For a (hard-paste) porcelain the proportions are kaolin 50 wt%, feldspar 25 wt% and sand 25 wt%.[10] This almost corresponds to 70 wt% SiO_2, 26 wt% Al_2O_3 and 4 wt% K_2O. Regarding phase content, the proportions are (undissolved) quartz ~22 vol%, mullite 16 vol% and glass 62 vol%.

During the sintering of pottery, only a limited part is molten, from a few vol% or less for a *terracotta* up to ~60 vol% for porcelain.[11,12,50] Porcelain requires a temperature higher than 1250 °C–1300 °C (up to 1450 °C) for the growth of long acicular mullite crystals.[50–52] These crystals build a 'birds nest' that retains by capillarity the molten (earth)alkali-rich silicate phase that gives a glassy and translucent phase during cooling. Consequently, a pottery body is always heterogeneous and made of many phases: coarse grains (*i.e.* relics of raw materials) of up to a few millimetres can subsist and new phases resulting from the reaction between the grains form around them.

On the other hand, the glassy matrix of enamels and glazes should be homogeneous or its heterogeneity controlled at the micro- or nanoscale.[52–54] The resolution of the human eye is close to 1–10 μm; if coloured grains are larger, the colour will appear punctuated, granulated, not united. Three types of colouring agents are used to colour a glass:[53–55]

(i) 3d (transition metal) or 4f (rare earth) ions having an electronic transition in the visible range dissolved in the glassy enamel; as a function of the element, 0.05 to 2 wt% are sufficient to colour a glassy silicate.
(ii) Pigments dispersed in the glassy phase or formed by precipitation on cooling. Pigments are crystalline phases coloured by the previously mentioned coloured ions not degraded by the molten glass during the process; the colouration power of pigment is higher than that of dissolved ions.
(iii) Metal or semi-conductor nanoparticles with a specific organization or optic properties. The colouration efficiency is huge, 0.2 wt% is sufficient to blacken a millimetre-thick glass plate.

We will discuss further the most suitable techniques to identify the colouring agents.

7.3 Micro/nanostructure and Elemental Analysis

An exhaustive list is beyond the scope of this publication, but here we try to provide references representative of the different objects and analytical techniques. Table 7.3 lists the different methods of analysis and points out some of their advantages and drawbacks. Invasive or destructive methods are commonly used in materials science studies, but we will address those methods more especially devoted to rare pottery and enamelled objects, which tend to be non-invasive, mobile methods. Glass and ceramics are easily broken and, therefore, artefacts of exceptional quality are very rare so sampling is prohibited! The use of mobile, non-invasive instruments is needed when shards are not available.[55–64] Optical techniques are optimal for the study of the coloured material. When fragments for sampling are available, scanning electron microscopy with energy dispersive spectroscopy (SEM–EDS) facilities (see Chapter 2) and inductively coupled plasma mass spectroscopy (ICP-MS) are the most used methods to determine elemental composition. ICP-MS, a (micro)destructive technique has for a long time been to analyse transparent or translucent archaeological glass fragments, available in large numbers and made of a homogeneous matter.[65–68] Some authors neglect the fact that many coloured glazes and even glasses are actually composite materials, and hence the representativeness of the composition probed by ICP-MS can be questioned.[69] The great advantage of advanced ICP-MS is the identification of isotopes.[70–72] For instance, Pb isotopes can be used to associate glaze production and mining places.[70] Laser induced breakdown spectroscopy (LIBS) is a micro-destructive technique, damaging for enamelled artefacts due to the generation of thermal shock, but offers accurate measurement of composition profiles from the surface to the body.[73,74] Neutron activation analysis (NAA)[75,76] and ion beam analytical techniques (PIGE: particle-induced gamma-ray emission, PIXE: particle-induced X-ray emission, RBS: Rutherford backscattering spectroscopy, NRA: nuclear reaction analyses) are non-destructive techniques; they can be used to study large objects using an extracted refocused beam line, but instruments are rare and expensive (as well as the issue of transportation of masterpieces from secure room to laboratory).[77–79] Furthermore, the irradiated sample can become radioactive. The microbeam of synchrotron or of rotating anode X-ray sources can be used for local analysis (μXRF or μXRF).[80–85] Transmission electron microscopy (TEM), in particular coupled with focused ion beam (FIB) for the final preparation of the sample, is a unique technique to study microstructure at nanometre and sub-nanometre scale. However, it is expensive and time-consuming to study many samples and therefore the representativeness of the measurements is not obvious.[86–91] XRF, in particular mobile XRF, is a very good method for semi-quantitative elemental analysis.[36–38] Finally, a combination of optical and electron microscopies as exampled in ref. 39, 92 and 93 remains the most common approach when fragments are available. The use of vibrational spectroscopies, in particular FTIR[94–99] and Raman,[58–64] is increasing, not only to identify crystalline and amorphous inorganic phases but also to identify organic residues in a non-invasive manner,[100] historically made possible by mass spectroscopy and chromatography[101–103] on matter extracted from the pottery pores (see Chapter 2). These residues inform of the use of the utensils.

7.3.1 Porosity

The first parameter to be considered in the study of pottery is the porosity, the main parameter to classify and name the pottery. Open porosity is commonly measured by the Archimedes method:[104] the comparison of the mass measured in the dry and wet state (pores filled by water under vacuum) with that measured for the wet sample immersed in water, thus allowing calculation of the open porosity.[11,104] Water can be replaced by other fluids for low porosity materials.[105] Closed porosity can be obtained by comparing the density of the sample with that measured with a pycnometer on ultrafine powder (the perfect wetting of the powder with the liquid makes the measurement is not distorted by residual porosity), or by comparison with the theoretical density calculated from the unit-cell volume measured by XRD.[11] Typically, open porosity ranges between ~30% for coarse *terracotta*[49] to less than 0.5% for porcelain[50] and technical ceramics (Table 7.1). The porosity level directly controls the mechanical properties and the reactivity with the environment. Porous bodies should be enamelled to become tight.

7.3.2 Determining the Firing and Glazing Temperatures

Contrary to crystal and glass that reach a homogeneous state, ceramics, with the exception of single-phase technical ceramics sintered after a long high temperature annealing, do not approach the equilibrium state and their microstructures remain very heterogeneous.[108] Much information can be extracted from the microstructure; thus, it is very important to have access to a fracture or the polished cross-section of enamelled artefacts. Pottery is prepared by reactive sintering and the progression of the reactions will be controlled by the degree of grinding, which determines the grain size distribution, and the firing conditions.[11,12,45,108,109] The final quality of a fired artefact depends on the many steps of the process (Figure 7.1). Obviously, everybody knows that the cooking of an object in hot water, for instance an egg, depends on both time and temperature, especially at the core; the surface temperature is limited by that of boiling water, whatever the temperature of the heat source. After all the water has evaporated, the temperature will depend on the heating. Similar events take place during the firing of a ceramic: for instance, if hydroxides lose water at *e.g.* 400 °C, the temperature of an untransformed hydroxide grain will be kept at 400 °C up to complete dehydration, even if adjacent grains are at a higher temperature. The same feature takes place for the reaction between a solid phase and a liquid one. Essentially, the composition of the liquid is slightly modified and the viscosity shifts with the dissolution of grains in the liquid phase.[11,48] Consequently, only the combination "grain size x temperature x time/heating rate" makes sense: for a given value of atom diffusion rate, the matter transportation will be the same if we increase the time or the temperature or if we change their combination (slower heating rate); the area in contact between grain being increased when the grain size decreases, advancement of the reaction will be higher for a compact assembly of fine grains than for one of made coarse grains. This also explains why the core of many grains is preserved in its raw state, with only the periphery and the matter around grains being modified,[108] with the appearance of new phases (Figure 7.2). It is thus difficult to reduce the heterogeneity of a ceramic made of coarse and fine grains. Typically, for *terracotta* and many faience, majolica, *etc.*, sand grains are larger than a few millimetres although the size of clay particles are several orders of magnitude smaller (a few nanometres): the sizes of newly formed phases are thus very variable and a multiscale analysis is required.

As indicated before, grinding, firing and enamelling are critical technologies. The design and construction of kilns were the most critical issues[3–6,110] and the evolution of the techniques shows a progressive increase of the temperature and a decrease of the firing cycle duration. In medieval China firing in a Dragoon Kiln lasted many weeks, whereas present-day firing needs only a few hours or even a few minutes for small artefacts like tiles! The measurement of the thermal expansion or shrinkage of the 'green' body is made by professionals to establish and optimize the heating–cooling cycle.[11,34] Figure 7.4 shows typical curves measured on already fired pieces, a good 'tool' to analyse phase content and to determine the firing temperature of pottery.[34]

For a clay-rich paste, a linear expansion is observed up to the formation of the first liquid phase and the associated drastic contraction known as liquid-assisted sintering (Figure 7.3). The microstructure of a type of ceramic will not depend on the mean composition, but on that of raw materials, their grain size, and of the firing cycle (heating rate, atmosphere, duration): the fraction of reaction and the nature of

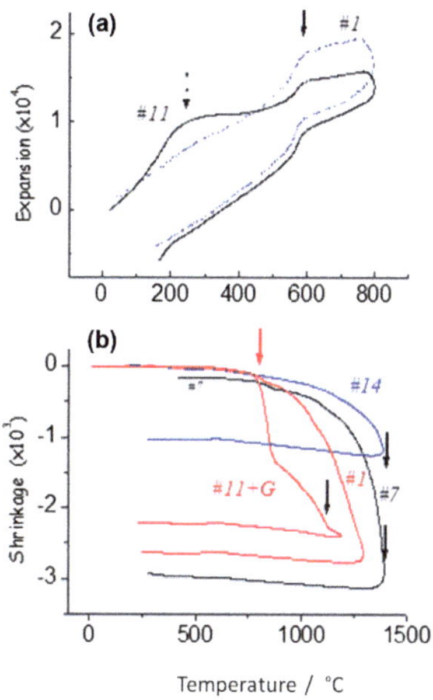

Figure 7.4 Examples of thermal expansion measured on stoneware and porcelain:[112,113] (a) A comparison between the thermal expansion of two body of Vietnamese celadons (#1, 13th c. and #11 16th c.) heated below their firing temperature: the bumps observed arise from reversible phase transitions of quartz (573 °C alpha to beta quartz transition). (b) The shards are heated at higher temperatures; the Y scale is 10 times smaller, so the expansion up to ~800 °C appears almost flat; the glazed sample #11 + G show a shrinkage jump at the melting temperature of the glaze (arrow), then the slow evolution arises from the reaction of the molten glaze with the body substrate and the second arrow indicates the sintering of the body; the sintering of #7 and #14th –16th c. Vietnamese porcelains shards takes place at higher temperatures, ~1450 °C.
Adapted from ref. 112 with permission from John Wiley & Sons, Copyright © 2000 John Wiley & Sons, Ltd.

formed phases can be thus very variable.[108] The microstructure of the body and the micro- and nanostructure of the glaze constitute the fingerprint of the used technologies.

Briefly, we can consider that firing 'reveals' the reactivity of the material. Due to the anharmonicity of chemical bonds, the bond length increases with temperature; consequently, the matter expands quasi-linearly with temperature, the thermal expansion coefficient ranging typically between 10×10^{-7} K^{-1} (amorphous silica) and 120×10^{-7} K^{-1} (metals),[11,34,53,108] values being between 60 and 100×10^{-7} K^{-1} for most crystalline oxides. Phase transitions modify the structure of solids and lead to instant contraction or expansion. Therefore, the measurement of the thermal expansion provides information on the nature of the crystalline phases.[11,34] For this reason, a fragment of the body in its green state, or already fired in the case of archaeometric studies, can be cut and placed in between a fixed support and a sensor of a dilatometer measuring the displacement of the sample. A quartz-rich mixture shows a huge expansion at 573 °C (Figure 7.4a), due to the α–β phase transition of quartz, and this event is used by ceramists to measure quartz content in clays, paste and a fired body.[11,34] When all the liquid phases have reacted with infusible grains *i.e.* the material is kinetically stable, the thermal expansion starts again. Reheating a previously fired ceramic specimen will give a linear curve up to the previous firing temperature. As sketched in Figure 7.3 the reactions between grains are limited, especially in *terracotta*, and faience. The reaction is more complete in stoneware and especially in porcelain body, which gives rise to its translucency. Consequently, materials being heterogeneous, the reactions will start again when the firing temperature will be over the previous maximal temperature achieved during the firing cycle and the shrinkage develops again. The sintering can be limited or stopped when reactions that lead to expansion take place at the same moment, for instance, the formation of a new crystalline phase having a less-compact structure or some elements are volatilized.[111] Decarbonation starts around 900 °C, but temperatures up to 1300 °C could be necessary to obtain the complete transformation of coarse grains of a few millimetres because of the phenomenon explained previously. It is thus very important to limit the heating rate when reactions take place, *e.g.* the departure of carbonates in the 900–1100 °C temperature range. Figure 7.4b compares the thermal expansion measured on different pieces sampled from stoneware[112] and porcelain shards.[113] If a glaze coats the body, the measured thermal expansion is the addition of that of the glaze (a shrinkage jump is observed at ~700 °C in

Figure 7.4b when the ash-based glaze melt) and of the body (shrinkage above 1200 °C).

The jump at ~200 °C observed during the first heating cycle, but not on cooling, arises from soaked water present in ancient shards excavated from the soil. Cristobalite (a polymorph of silica) phase transition also has an event in this temperature range. The main sintering mechanism of pottery is the liquid sintering that induces a drastic shrinkage as exampled in Figure 7.4b: the vertical tangent gives the firing temperature, a vertical tangent being observed when the volume of the liquid phase is important, as for porcelain and stoneware. For instance, the sintering temperature of Vietnamese Celadon ranged between 1200 °C and 1300 °C and that of porcelain was ~1350 °C.[38,113,114] *Terracotta* sinters below 800 °C.[49] This clearly points out that the degree of sintering and associated properties depend on many factors.

Figure 7.5 (left) shows the thermal expansion of a glass: just before the melting temperature, after the glass transition (T_g) a strong expansion takes place (annihilation of the stress) which can be decreased or even removed by the applied load. The shrinkage is very drastic due to the complete fusion and low viscosity of the glaze. Note, this behaviour depends both on composition and on grinding. If the thermal expansion is measured on a piece where one or two sides are glazed, the measurement will show the behaviour of both materials (proportionally to their thickness). For instance, in curve #11 + G in Figure 7.4 the shrinkage jump at ~700 °C indicates the melting (and firing) temperature of the Celadon glaze.

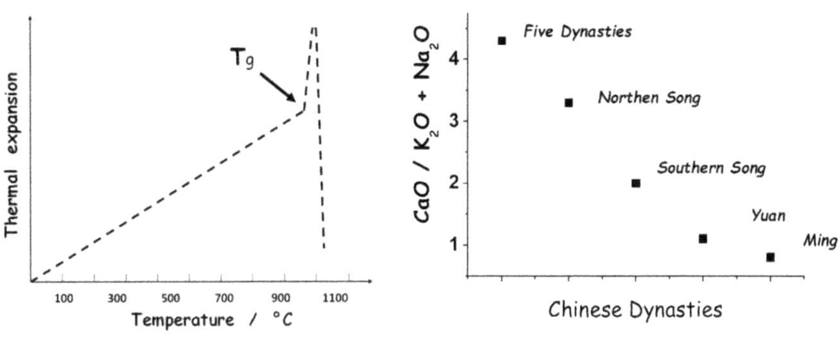

Figure 7.5 Thermal expansion of a glaze (schematic) up to the melting temperature. Evolution of the relative CaO content in Chinese porcelain and Celadon glazes from the Five Dynasties (10th century) to the Ming Dynasty (16th century).[114]
Adapted from ref. 114 with permission from John Wiley & Sons, Copyright © 2005 John Wiley & Sons, Ltd.

$PbO–SiO_2$ glaze composition, the common one of many *terracotta* glazes, is fired at 900 °C (010a Segers cone). The composition can be balanced in order to adapt the thermal expansion and/or the viscosity, *etc.*: *e.g.* composition SiO_2–0.1Al_2O_3–0.90PbO–0.05CaO and 0.05 K_2O needs to be fired at 900 °C. If the potter limits the use of PbO to have a more chemically stable glaze (Pb-rich glazes are dissolved by acetic acid, and hence hazardous), a composition such as SiO_2–0.05B_2O_3–0.1Al_2O_3–0.1K_2O–0.1Na_2O–0.1CaO–0.5PbO is possible.[34] A composition to be fired at 1300 °C (10 Segers cone) will be 6SiO_2–0.6Al_2O_3–0.7CaO–0.3K_2O. These examples illustrate the complexity of mastering enamelling technology. Figure 7.5 shows the evolution of the CaO content *versus* other flux content with time for Chinese Celadon and porcelain glaze.[114] The decrease of CaO content arises from the replacement of ashes by a mixture of different minerals and is correlated to the increases of the firing temperature and the mastering of kiln technology (Table 7.4).

The mismatch between the thermal expansion of the enamel and the substrate should be controlled,[34,53] with the former being a little lower than that of the substrate to avoid the enamel being put under tension and cracking. However, the stress resistance of glass and ceramic is much higher under compression than under tension. The thermal expansion of the ceramic body being very variable, in particular as a function of quartz content, a huge variety of compositions of glaze have been explored by potters.

Some authors attempt to determine the firing temperature from the presence of specific phases requiring certain temperatures to be formed, without taking into account the problem of "grain size×time×temperature" conditions required to have a significant (*trans*)formation of the phases. For ceramics made with coarse grains (2–10 μm or more), this approach is not reliable. As a function of the raw material, the phase formation can be delayed and the conclusions should be considered with caution: for instance carbonates (decarbonization temperature 900 °C) and carbon (combustion temperature 700 °C) may persist in ceramics fired at 1300 °C.[115,116] Likewise, the upper temperature is that required for the reaction between oxygen and carbon atoms in carbide.[117]

7.3.3 Determining the Origin of Raw Materials

Contrarily to gems or even rare metals whose mining places are limited and have been known for centuries due to their formation under very specific geological conditions, most of the minerals used

Table 7.4 Examples of faience compositions.[34]

Oxide	Darius Palace[a] (4th BCE)	Persia[a] (19th)	Della Robia[a] (16th)	Lunéville (18th)	Modern (20th)	Fine (20th)	Vitreous (20th, comparison)
'H_2O,CO_2'	4.36	2.48	8.58				
SiO_2	90.53	85.17	49.65	67.39	68.0	67.85	70.80
Fe_2O_3	0.54	1.26	3.70	2.01	1.0	1.25	0.52
Al_2O_3	0.30	1.44	15.50	16.0	10.58	29.40	23.34
TiO_2	—				1.04		
CaO	3.98	7.29	22.40	13.16	12.72		0.52
MgO	0.04	0.04	0.17	1.02	6.08	0.06	0.15
Na_2O	0.21	0.2			0.38		4.56
K_2O					0.25	1.43	

[a]Ancient or excavated shards retain water.

to make pottery bodies are rather common. Furthermore, clays are sedimentary detritic rocks, made by mixing different sources. The identification of minor phases and traces is possible but it is often difficult to associate them with a precise quarry when the local geology is not known precisely. Nevertheless, analysis of traces and minor phases is useful to categorize groups.[36–38] XRF and ICP-MS easily identify traces such as Rb, Sr, rare earths, uranium, *etc.*[36–38,65] It has been shown that Raman mapping[41] is more efficient and often faster than XRD to identify minute amounts of specific minerals in ceramic bodies. Due to the large heterogeneity of *terracotta* and faience (grain size >a few mm) the sampling should be sufficiently large to be representative. Powdering the material guarantees better homogeneity and representativeness.

Raw materials required to make coloured and opacified enamels should be selected with caution and hence the number of sources is more limited and some minor/major/traces elements associated can be very characteristic.[118–123] Glazes and enamels belong to glassy silicates as glass but, due to the high number of characteristics required, much more than those required for glass objects, the range of compositions experimented by enamel-makers is huge.[124–128] Determining their elemental composition, namely major (~15% or more), minor (1–10%) and trace element (<0.5%), is very useful, but it is very important to consider how representative the composition is in relation with the heterogeneity of the coloured/opacified glass-ceramic.

7.3.4 Analysing the Enamels and Glazes

If the definition of enamel is "a glassy (silicate) film on a substrate", the microstructure of the film can be complex: many glazes are a mixture of different amorphous phases (*e.g.* the *Jun* and '*pigeon blood*' glaze[52,129]) and glazes coloured and opacified with pigments are composite materials, called glass–ceramics.[69] During the glazing, the molten coating reacts with the substrate and forms an interphase (Figure 7.2). Furthermore, some elements are volatile (B, Pb, Li, Na, *etc.*) and the surface can have a different composition with respect to the film bulk.

Figure 7.6 shows representative sections of potteries. The porosity of the body can be quantitatively evaluated by measuring the pore size and density by surface unit and comparing and calibrating with the water absorption measurement (Archimedes method). The combination of optical microscopy and EDS–SEM is efficient for phase identification. Raman microspectrometry identifies both crystalline

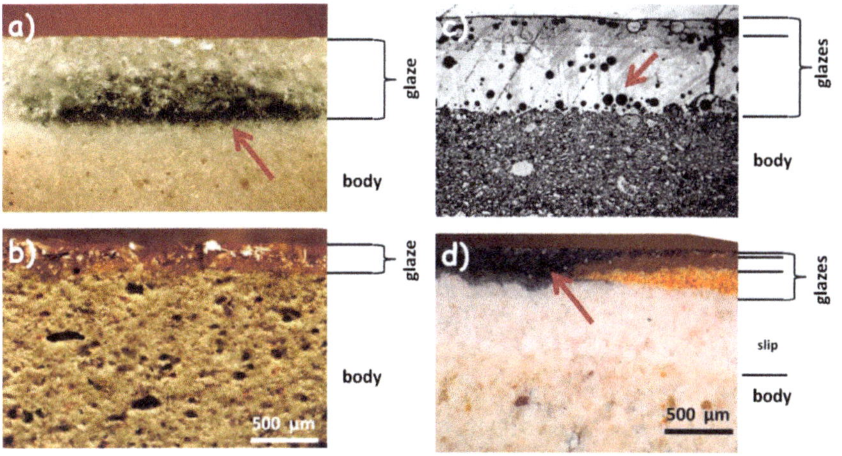

Figure 7.6 Examples of pottery sections: (a), (b) and (d) are optical micrographs of sliced shards; (c) is an SEM image of a polished shard; (a) Chinese porcelain with a blue underglaze decor (15th century); (b) *terracotta* covered with a lead-rich glaze (14th century, Oman Sultanate); (c) Ottoman lead-alkaline multilayer glaze on porous clay-rich body tile tempered with quartz grains (15th century); (d) Ottoman Iznik glazed tile: note the thick white quartz grain-rich slip over the body (fritware) and the multilayer glazed blue and red overglaze (16th century).

and amorphous phases.[51,64,126–130] One or more layers coat the body, and in addition, an intermediate slip (aqueous suspension of fine white grains, of clays or quartz, typically) can be deposited in between the body and the glaze(s) in order to mask the colour of the body (as made for Iznik ceramic). When sampling is not possible, especially for the study of rare objects, only Raman microspectrometry allows the non-invasive, on-site profile analysis from the surface to a few millimetres, *i.e.* up to the glaze/body[130] or enamel(patina)/metal interface.[131] Raman profiling is a useful technique to measure the thickness of the coating up to a few hundreds of microns. Note that a carbon film or a crack often forms at the metal/coating interface and modifies the Raman intensity.[131] Attaching strips of a conducting carbon-rich tape on the sample section to isolate the spot to be analysed helps to compare data collected by optical techniques [Fibre Optics Reflectance Spectroscopy (FORS), UV–Visible eflectance, FTIR–ATR and Raman][92] with SEM–EDS measurement (carbon-rich tape is required if a conducting film has not been deposited).[69,132]

Conservation treatments have been applied to many objects, for example archaeological conservators often cover shards with Paraloid®[133] that is detrimental to vibrational analysis. Waxes and agglomerated dust also cover many objects. The latter mixture is also detrimental to the analysis but a good local cleaning can be obtained with a few minutes of illumination with a blue or green laser beam. In any case, a prior cleaning conducted by a conservator is recommended.

Crystalline Phases

Table 7.5 lists the techniques to be used for the identification of phases and their advantages and drawbacks. Sampling points should be distributed on a sufficiently large area, *i.e.* at least 10 times larger than that of bigger grains, to give representative data. Consequently, measurements on powdered samples by XRD or Raman techniques, or mapping on a large area, are recommended but not always possible. Due to the molten state of glaze at the firing temperature, the substrate is dissolved at the glaze–body interface and new phases are formed on cooling (Figure 7.2).[93] The formed phases are very characteristic of the raw materials and of the firing cycle and this region should be analysed specifically, at different scales. The diffusion coefficient of ions is very variable[108] and depends on the matrix; small size, high charge cations diffuse faster and play an important role in the redox equilibrium in the solid and liquid states.[45,55] For instance, as shown in Figure 7.6a, the diffusion of the blue colour due to Co^{2+} ions of the drawing painted on the body before deposition of the glaze powder (underglaze decor) is much larger in the glaze (molten at the top firing temperature) than in the porcelain substrate. In Figure 7.6d the colour (*i.e.* the Co cation) diffuses significantly in the fritware substrate, through the glassy phase added to cement the quartz grains.

Only transmission electron microscopy (TEM) provides topological information at the submicron scale.[52,82–89] Figure 7.7 shows some examples. Lustre decor arises from the controlled dispersion of metal nanoparticles in the glaze close to its surface.[45,55,87,88] The alternation of metal nanoparticle-rich and -free layers leads to visible light diffraction phenomenon, dependent on the relative orientations of the observer, the object and the light source. This type of colouration technique is common in nature (rainbow, opals, beetle carapace, butterfly wing, cuttlefish skin, *etc.*) but rare for manmade objects (gratings, advanced optical filter). In the other directions of observation, the colour arises from the light absorption (here by a plasmon, cloud of electrons at the surface of metal nanoparticles).[55] The non-absorbed colour is received by the eyes of the observer. A homogeneous dispersion of copper nanoparticles leads to different red glazes (*ox* or *pigeon blood*, *sacrificial red*, *etc.* as a function of the size, shape, concentration

Table 7.5 Analytical techniques useful for pottery, glass and enamelled artefact studies. References are given as an example.[a]

Methods	Advantages	Drawbacks/preparation	Best resolution	Information collected	Remarks	References
Optical microscopy	Colour identification	Smooth surface required (thin slice for polarization measurement)	0.5 μm	Phases, grain and pore size	Mobile	35, 38, 93
SEM-EDS	Large depth of field Large magnification Local composition	Cannot see the colour (Very) light elements not measured	>nm	Composition, grain, crystal and pore size		69, 93
TEM-EDS	High resolution	Sample preparation, destructive, time consuming	0.1 nm	Structure, phase, composition		52, 83, 86–90
XRD	Phase identification	Smooth surface required	$>$mm^2	Phase identification		39–41, 52
μXRD	Phase identification	Expensive (Synchrotron, Rotating anode)	μm	Phase, stress		52, 80–91
ICP-MS	High precision	Destructive, representativeness	>μm	Elemental composition (major, minor and traces)		65–69, 72
XRF	Mobile and fixed instruments	Light elements not measured	$\gg$mm^2		Mobile	36, 37
μXRF	Local analysis	Expensive (Synchrotron, Rotating anode)	μm	Local composition		91
LIBS	Erosion – sampling	Micro-destructive Thermal shock	A few μm	Composition profile	Mobile	73, 74
Raman	Mobile, non-invasive No preparation, mapping	Laser selection Not quantitative	0.5 μm	Phase, structure, stress, *etc.* Organic residues	Mobile	44, 54–64
UV–Vis–NIR	Low cost	Smooth/glossy surface required or polishing	<mm	Colouring agent, site	Mobile (FORS)	40, 90, 92

Table 7.5 (*Continued*)

Methods	Advantages	Drawbacks/preparation	Best resolution	Information collected	Remarks	References
IR	Low cost, quantitative	Sample preparation	>10 μm	Phases, organic residues	Mobile	94–102
Thermal expansion	Firing temperature	Sample preparation, destructive	mm	Phase, firing temperature, softening temperature		38, 49
DTA/DSC	Melting temperature	Sample preparation, destructive	mm	Phase, melting temperature		106
NAA	Composition Non-destructive	radioactivity	Large volume	Elemental composition (major, minor and traces)		75, 76
Ion beam Analysis (RBS,PIXE,PIGE, NRA)	Composition (profile) Non-destructive	Rare facilities, expensive	<μm	Elemental composition (major, minor and traces)		77–79, 107
NMR	organic residues	Solubilization required		Composition		99, 102
Chromatography	organic residues					109
Scale	porosity					49, 104, 105

[a]DTA: Differential Thermal Analysis; DSC: Differential Scanning Calorimetry; EDS: Energy Dispersive Spectroscopy; IR: Infrared transmission/absorption/reflexion; LIBS: Laser Induced Breakdown Spectroscopy; NAA: Neutron Activation Analysis; NIR: Near IR; NRA: Nuclear Reaction Analysis; PIGE: Proton Induced Gamma Emission; PIXE: Proton Induced X-ray Emission; Raman: Raman scattering, SEM: Scanning Electron Microscopy; Transmission Electron Microscopy; UV: Ultra-violet; XRD: X-ray diffraction; XRF: X-ray Fluorescence; Vis: visible; μ: micro.

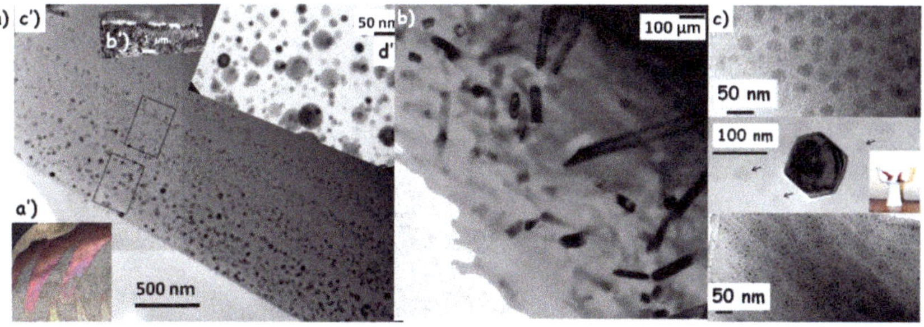

Figure 7.7 TEM images (reproduced courtesy of Ph. Sciau, CEMES-CNRS, Toulouse) of lustre pottery (a), porcelain body (b) and red enamel on porcelain (c) sections; lustre pottery shows iridescent colours (a–a′) due to layer alternation with or without metal NPs (c′); only shining is observed by optical microscopy (b′); mullite needles are visible in (b); note in (c) top the binary phase separation in the red glaze surface and the formation of hexagonal copper crystals; closer to the porcelain substrate a higher density of smaller NPs is observed.
Reproduced from ref. 52 with permission from Elsevier, Copyright 2016.

of copper nanoparticles and of the degree of phase separation of the glassy phase).[52] A velvety glaze is obtained by controlling the phase separation in the glaze [two glasses with slightly different compositions, and different optical indices, form in the liquid state and heterogeneity is conserved on cooling (Figure 7.7c)].[52,129] In the porcelain body, temperatures above 1250 °C lead to the formation of an interconnected array of mullite needles (Figure 7.7b, composition: $3Al_2O_3$ 2 SiO_2) that retain the silica-richer molten phase and is the origin of porcelain translucency.[50,52] XRD and Raman microspectrometry easily identify crystalline phases but only the latter technique can characterize the structure of glassy silicates.[134–139] Databases and reference works can be used for band and phase assignment.[64] However, the band intensity of a Raman spectrum depends on many parameters such as laser wavelength, amount of chromophore, crystal orientation, local structure *etc.*; consequently databases should be used with caution for the identification of coloured matter. Raman scattering provides a lot of information, but an understanding of the Raman phenomenon is required (see Chapter 2).

Amorphous Phases

A large volume of glaze and enamel consists of glassy silicate. Amorphous phases are also present in the body such as fritware, vitreous, stoneware and (soft-paste and hard-paste) porcelain. If in most cases the micron scale is the pertinent resolution to study and identify crystalline phases, a much higher resolution is needed to study the glass and nanocrystalline precipitates. As shown in the TEM image (Figure 7.7c), phase separation takes place at the submicron scale[52,129] and hence glaze that looks homogeneous at micron scale is actually heterogeneous. Glaze composition also varies from the top surface (fluxes like Na_2O, Li_2O, PbO, B_2O_3 are highly volatile and easily contaminate the environment) to the centre and the bottom of the glaze (reaction and diffusion in the substrate). Surface composition can be obtained by XRF, SEM–EDS and ICP-MS. In many cases XRF measurements remain qualitative or semi-quantitative because of the difficulty of measuring light elements and controlling the analysed volume; notwithstanding, the qualitative information obtained is very useful to assign complex Raman spectra. Composition profiles can be obtained by cross-section analysis (fracture or polished section) using SEM–EDS and ion beam analyses or from the top by LIBS abrasion (in Laser Induced Breakdown Spectroscopy a laser pulse volatilizes the matter[73,74] and drills inside from surface to in-depth; caution should be taken for enamelled artefacts because of the huge thermal shock). Vibrational spectroscopy (infrared and Raman scattering) is very effective at identifying glassy phases and in many case can differentiate the contribution of amorphous and crystalline phases.[124–134] The FTIR–ATR technique explores the surface layer: the probed in-depth thickness typically ranges between about 2 and 10 μm as a function of the wavenumber range. The technique is thus a very efficient way to study the corrosion layer if the surface is sufficiently flat. Preparation of the sample is thus often required (powdering, polishing *etc.*). Note that the low hardness of Ge μATR crystal (20×20 μm^2, typically) facilitates the contact with a corrugated surface, but promotes its rapid degradation. The crystal should be replaced periodically. Only Raman microspectrometry provides a suitable non-invasive tool to study glassy phases and composites made of both glassy and crystalline phases. Furthermore, mobile instruments are available.[44,54,135]

The high efficiency of Raman scattering for the study of amorphous materials arises from the very local probe of Raman scattering, the polarizability, in other words, the variation of the electron distribution across the chemical bond in between atoms. Consequently, there is no need to have an ordering of atoms, but it is necessary to have an electron cloud in between atoms, *i.e.* to have a covalent bond. The Raman intensity is directly related to the number of electrons involved and almost no spectrum is observed for ionic bonded compounds; the spectrum of just a small amount of carbon (<1–5%) can obscure those of major phases.[116,117] For metal, the penetration of light is very limited and special features take place, observed at a very low wavenumber range.[35]

Since strong covalently-bonded structures have Raman spectroscopic signatures orders of magnitude larger than those of ionic ones, the Raman spectrum of a silicate consists, as a first approximation, solely of the Si–O network signature (Si–O stretching, bending, and librational/collective modes, *i.e.* the so-called Boson peak).[5,16,39,41] The basic topological and vibrational unit of crystalline and glassy silicate is the SiO_4 tetrahedron. Its tetrahedral (T_d) geometry is almost unperturbed and the symmetric stretching mode (observed between ~750 and 1200 cm^{-1}) is much stronger than the asymmetric ones (same wavenumber range) that could be neglected.[126] Its position varies with the degree of polymerization, *i.e.* if the Si–O bond is connected to an adjacent silicon atom as in pure silica. The second important band arises from the numerous bending modes (observed between 350 and 650 cm^{-1}, with E and F characters from the symmetry point of view, that means 2 and 5 bands after symmetry distortion) and from the distribution of librational (hindered rotational) and translational modes (observed below 300 cm^{-1}). Thus, below 300 cm^{-1}, a very large and asymmetric peak is seen and called the Boson peak which reflects the short-range structure/disorder. See refs. 124–126 and 134 for a more complete discussion. Consequently, the Raman spectrum of a glassy silicate is the fingerprint of the SiO_4 network. Raman parameters are directly related to the nature and arrangement of the chemical bonds: the stretching mode wavenumber depends on the mechanics (atom mass and bond strength); the bending modes wavenumbers are very sensitive to the bond angle modifications and hence to the environment (presence of fluxing atoms: Pb^{2+}, Ca^{2+}, Na^{+}, K^{+}, Li^{+} ...). Thus the Raman spectroscopic signature (peak position and intensity) of crystalline and amorphous silicate depends on the silicate nanostructure. It has been demonstrated empirically[124–127] and by the Density Functional Theory (DFT) calculation[134] that the area ratio of the bending *versus* stretching mode provides a good approximation of the degree of polymerization, being very low (<0.5) for glass compositions having a high amount of flux (*e.g.* $PbO–SiO_2$ where SiO_4 tetrahedron are mostly isolated), and very high for silica-rich glass and pure silica (>6). The degree of polymerization is directly related to the melting and processing temperature:[126] fewer Si–O bonds, lower the melting temperature of the composition. Figure 7.8a illustrates the strong variation of the relative intensity of bending and stretching SiO_4 bands with the modification of the polymerization degree of the Si–O–Si network, from highly polymerized porcelain glaze (#7, where the bending band is very strong) to scarcely polymerized lead-rich *terracotta* glaze (#1). Figure 7.8b shows the evolution of the bending to stretching mode area ratio for a large number of glasses and glazes with various compositions. The highest value is measured for amorphous silica (melting temperature ~1730 °C) and amorphous alkali and alkali earth aluminosilicates (melting temperature >1400 °C). The lowest values are measured for lead-rich glass and glaze melting at a temperature of less than 650 °C. A similar feature is obtained for crystalline silicates but the curve shows a step-by-step increment with the type of silicate (nesosilicate, inosilicate, cyclosilicates and tectosilicates) whereas a continuous line is observed for glassy silicates in Figure 7.8b.[124]

Since the symmetric stretching mode is much stronger than the asymmetric ones, the components of the band can be assigned to the different types of SiO_4 tetrahedra: isolated (Q_0), connected with one (Q_1), two (Q_2), three (Q_3) or full connected (Q_4).[44,51,126,128] Mobile/portable instruments do not exhibit the same characteristics that bigger and more expansive fixed instruments available at the laboratory do. The replacement of gratings by filter(s) increases the background that should be removed under the control of the operator

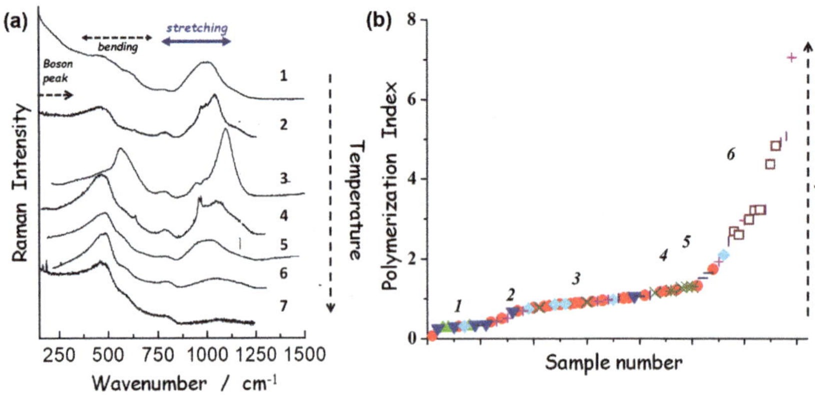

Figure 7.8 Illustration of the variation of the Raman signatures of a selection of glazes (a) and evolution of the ratio of the bending *versus* stretching band area for series of glazes and glasses (b), after ref. 124, 126 and 128. Typically, the melting temperature ranges from ~600 °C to 1450 °C.[126,128]
Adapted from ref. 124 with permission from Elsevier, Copyright 2003.

or automatically by the software. If the recording of narrow bands is easy with portable instruments, having a reliable and robust procedure to subtract the background is not trivial, especially for recovering broad bands. When comparing data obtained with different spectrometers, the same procedures should be used, as exampled in Figure 7.9.[124–128,135] The segmented line shows how the "background" subtraction was done to preserve the main contributions (the Boson peak and molecular signature of all SiO_4 tetraedron forming the aluminosilicate glassy network). Note, the narrow components of nanocrystals dispersed within the glaze has to be considered.

A very simple abacus built with the maximal (or band centre of gravity) wavenumber of bending and stretching SiO_4 bands allows classifying the different types of glass as shown in Figure 7.10. Chemometry (PCA, Euclidean cluster classification, *etc.*) are also useful.[44,126,136,137]

The association of different phases on a substrate leads to interdiffusion at the different interfaces but also contamination by surface or gas transfer. Atomic diffusion is enhanced at the surface and some fluxes such as PbO, Bi_2O_3, B_2O_3 and Na_2O, and ions such as Cr and Li, are highly volatile, leading to contamination of the surface of the artefact or the neighbour artefact. To avoid contamination each porcelain artefact is put in a ceramic box called saggar. The gaseous atmosphere should be controlled (CO/CO_2, H_2/O_2 ratio); this is obtained by controlling the wood water content (green wood), pouring water on coal and opening/closing of air entrance and fume exhaust shutters.

7.3.5 Coloured Matter

Glass coatings on potteries, glass and metals are applied for technical reasons, namely the tightening of porous bodies and hardening of their surfaces, but also serve a decorative purpose. Although the colouration of a glass object body is easy undertaken by dissolving as little as a few 0.1 wt% of a 3d transition metal or 4f rare earth ion, the colouration and/or opacification of enamels is more complex: an enamel/glaze layer is much thinner (10 to 500 µm, typically) and hence specific colouring agents are needed and their concentrations must be higher. To display a high gloss and vivid colour, the dispersion of a 'white' phase in the glaze is useful. XRF technique is a good method to identify colouring agents in a thick homogeneous glass (Table 7.6). On the contrary, because of the variable and uncontrollable penetration of the beam, the differentiation of colouring agents in objects covered by many layers of enamel (*e.g.* Figure 7.6d)

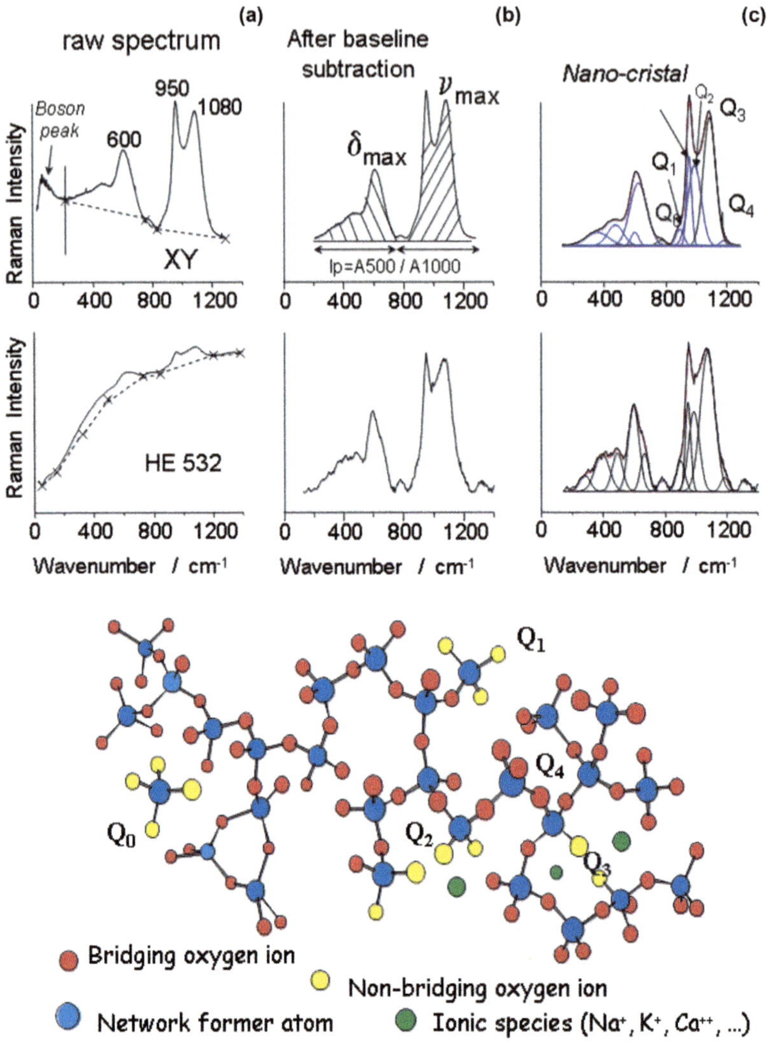

Figure 7.9 Schematic of the 'fitting' procedure for silicate spectra recorded with a high resolution fixed XY Dilor spectrometer measuring up to low wavenumber range (top spectra, Rayleigh elastic scattering at ultra-low wavenumber is rejected using holographic gratings) and with a mobile HE532 Horiba instrument with degraded performances below 200 cm^{-1} (bottom spectra, Rayleigh scattering is suppressed by absorption with an Edge filter); (a) as-recorded spectra; the dashed line show the separation between the inelastic Raman molecular scattering and the background; (b) spectra obtained after background subtraction; the bending and stretching modes of SiO_4 tetrahedra are observed below and above ~780 cm^{-1}; (c) the different components arising from isolated (Q_0), connected by 1(Q_1), 2(Q_2), 3(Q_3) and 4(Q_4) oxygen bridging atoms (see the sketch of a incompletely polymerized glass) are shown for the stretching band. Reproduced from ref. 126 with permission from the Royal Society of Chemistry.

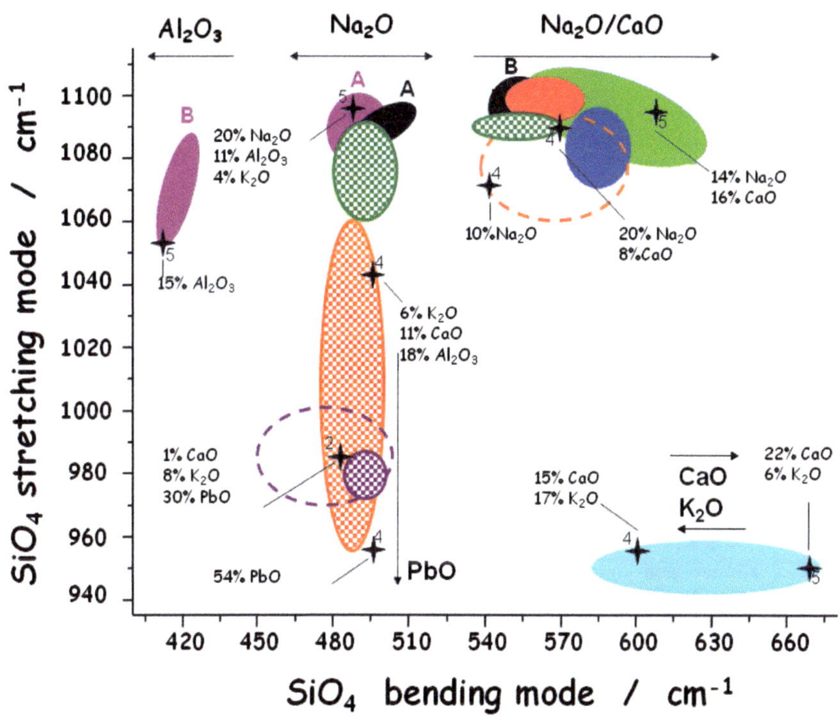

Figure 7.10 Plot of SiO_4 stretching *vs.* bending wavenumber giving the glass type.[136,137]
Adapted from ref. 136 with permission from John Wiley & Sons, Copyright Copyright © 2014 John Wiley & Sons, Ltd. Adapted from ref. 137 with permission from the Royal Society of Chemistry.

requires the use of microbeam (µXRF produced by rotating anode X-ray tube or by Synchrotron Source) and the preparation of the sample through cutting and polishing. The method is however very convenient for the study of surface gilding, even if traces are not visible by eye.[37] It should be kept in mind that in many cases shards of exceptional artefacts such as those outlined in Table 7.6[92,121,136–139] are rarer than some intact objects and hence analytical data very scarce.[136–141] UV-Visible Reflectance and its mobile equivalent technique FORS identify very well colouring ions and the symmetry of the occupied site.[39,40,92] Hyperspectral imaging cameras may provide similar information (see Chapter 2), but the technique is in its infancy for such applications.[142]

Obtaining a large palette of colours, especially for a material heated at a high temperature, requires the use of inorganic pigments.[10,30,53,54] Such pigments are in the crystalline phase, coloured

Table 7.6 Examples of on-site studies with mobile Raman/XRF set-ups.

Type	Objects	Date (century)	Place of measurement	Methods	References
Glass	Ptomelaic enamelled glass beakers (Begram Hoard)	~1st CE	Paris (Mnaa-Guimet)	Raman	137
	Enamelled/ coloured glass	5th BC - 19th CE	Sèvres (City of Ceramics)		146, 148
	Stained glass window	13th–19th CE	Paris (Sainte-Chapelle) France		44, 145, 147
	Mamluk mosque lamps and bottles	13th–14th CE	Paris (Louvre museum)		137, 144
	Brocard and Gallé objects	19th CE	Paris (Fine Art museum)		136
Enamel on metal	Limoges enamels	15th–19th CE			151, 152
	Chinese *cloisonnés*	15th–19th CE			152, 153
Patina	Chinese and Japanese bronzes	10th BC–19th CE	Paris (Cernuschi museum)		131
Pottery	Iznik fritware	15th–16th CE	Sèvres (City of		138
	Kütahya fritware	17th–18th CE	Ceramics) +		150
	Medici porcelain	16th CE	Chicago		149
	French soft-paste porcelains (Rouen, St-Cloud, Mennecy, Paris, Sceaux, Chantilly)	17th–18th CE	(Art Institute)		140, 154
	Meissen	18th CE		XRF, Raman	36, 37, 43
	Chinese porcelain	16th–19th CE	Paris (mnaa-Guimet)	Raman	132, 154

due to the presence of 3d or 4f ions, and stable in molten glass or precipitating on cooling.[53,54] Metal (Au, Cu, Ag) and semi-conductor (CdS–CdSe–X solid solution) nanoparticles are also good pigments.[35,55,88,92] The best analytical technique to identify these pigments is Raman scattering. If the laser colour corresponds to the wavelengths of absorption then resonance Raman spectroscopy takes place: the sensitivity is huge with traces up to micro-mole (μM) commonly detected, but the Raman intensity of scattering modes is drastically modified as demonstrated in Figure 7.11. Figure 7.11a compares the spectra recorded on a colourless crystal of sphene (malayite) and a pink one due to Cr doping (Cr-doped $CaSnSiO_5$).[143] The spectrum of the pink pigment looks very different because the modes involving Sn atom coupled with chromium one (forming the

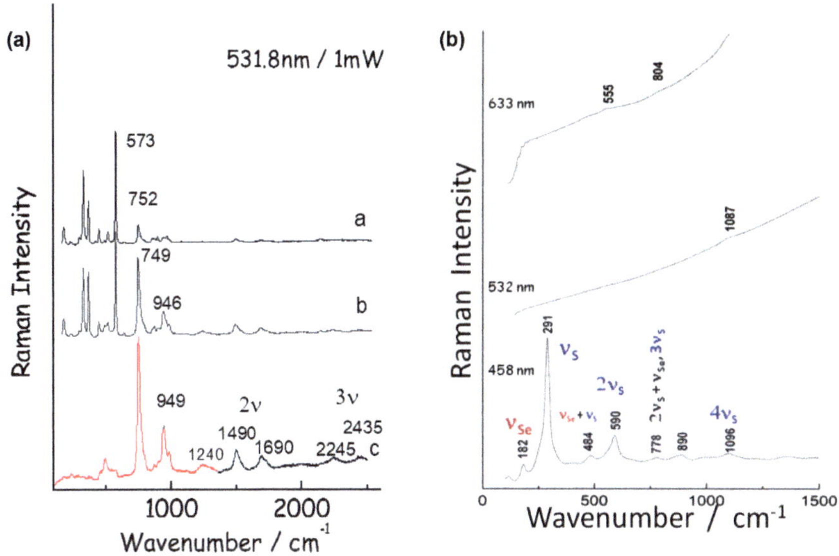

Figure 7.11 (a) Comparison between the Raman spectra of tin-based sphene, a and b colourless sphene along two polarizations and c, after 3 wt% chromium doping to obtain the pink pigment; note the peak wavenumbers are almost at the same value but mode involving Sn atoms in interaction with Cr atoms and their combination/overtone bands are enhanced. Adapted from ref. 143 with permission from John Wiley & Sons, Copyright © 2003 John Wiley & Sons, Ltd. (b) Comparison of Raman spectra recorded with red, green and blue laser on a red 'Liberty' stained glass coloured with nanoparticles of CdSe-CdS solid solution.[95] Adapted from ref. 92 with permission from John Wiley & Sons, Copyright © 2015 John Wiley & Sons, Ltd.; note the combination/overtone bands.

chromophore) are enhanced, but their wavenumber is slightly shifted. Obviously, databases[64] are not efficient for many coloured compounds. The observation of overtones (2ν, 3ν, *etc.*) and combination bands demonstrates that the Raman spectrum is resonant. A good example is the unexpected identification of lapis lazuli in many blue enamels,[144] from the Ptomelaic period (1st century CE)[137] to Böttger porcelain (18th century),[43] including Frederician,[137] Lajvardina[144] and Mamluk[137,145] glass and/or pottery (13th–15th century). The biggest advantage of resonance Raman scattering is the huge sensitivity to detect coloured phases that are not detected by other techniques. The ultramarine colour of lapis lazuli is understood to arise from S_n ions trapped in some sites of the feldspar structure. Under blue to red laser illumination, the S_n ions resonance Raman signature covers the spectrum of the feldspar host matrix, visible only under UV excitation.[119]

Colouring a glass red is also difficult and the method used by medieval glassmakers[146] was to disperse copper nanoparticles, leading to a huge absorption of the light. Therefore, thin layers of (Cu°) metallic copper-containing and copper-free glass should be deposited alternately on a colourless glass plate[139] to combine vivid colour and transparency. The contribution of the glass interacting with metal nanoparticles is enhanced (resonance Raman scattering),[145] which gives a different spectrum from that of the metal nanoparticle-free matrix. Gold (Au°) was also used, first by Bernard Perrot, a French-Italian glassmaker[147,148] and then by 18th century French and Chinese potters to produce *Camaïeu* decor (*Cassius' purple*)[140] and *Famille Rose*[132] porcelains, respectively. At the end of the 19th century, the discovery of CdS–CdSe solid solution with a colour ranging from yellow to red allowed the production of red to yellow glass combining vivid colour and good light transparency.[92] The colour depends on the composition, the size and the concentration of NPs.[90] For red (633 nm) and green (532 nm) laser excitations (*i.e.* when the laser light does not interact with chromophores), the Raman spectrum is that of the glass matrix (more than 98% volume) (Figure 7.11b). Using a blue (458 nm) laser, the spectrum shows two modes at *ca.* 182 and 291 cm^{-1} corresponding to symmetric stretching of Cd–Se and Cd–S bonds, respectively, plus combination and overtone modes. The exact wavenumber of each mode depends on the composition.[92]

Cations belonging to glass networks can be exchanged with protons in the presence of moisture, which leads to a degradation of the very surface of glazes and enamels, inducing cracks and loss of optical quality,[144,147] affecting the Raman intensity.[44,64,148] Proton Raman signature and as-recorded Raman intensity can be used for object dating.[44,64,149,150,152]

7.3.6 On-site Study with Mobile Instruments

Since the beginning of civilization, emperors, kings and princes have been on the receiving end of exceptionally crafted objects. Furthermore, the fragility and scarcity of many masterpieces have limited the number of surviving objects. Consequently, analytical information on the materials and technologies used to make these precious artefacts is also rare, if not inexistent. For instance, in 1815, H. Davy – founder of archaeometric studies with R. A. Ferchault de Réaumur[13,23] – needed large fragments of murals found in Pompeii and pigments of Egyptian blue for their analysis. More recently, in 1922, H. A. Eccles and B. Rackham asked again for several parts of a porcelain service to study the composition of their enamels[13] (an analysis that employed solubilization, specific element precipitation, drying and weighing of the final product). However, present-day ethics require non-destructive, non-invasive techniques. Moreover, the need to keep precious items in secure areas leads to the costly displacement of artefacts; the use of mobile and contactless instruments is required. The ongoing miniaturization of laser, X-ray sources, spectrometers, detectors and computers has led to the popularization of mobile set-ups and important increases in their performances.[13,56–64] For instance, the identification of lapis lazuli as a blue pigment in many glass and glaze is a good example of the potential of the method. Other examples are the identification of Han glass for ear-stoppers expected to be in jade and of Arsenic-rich European cobalt in some painted enamelled porcelains made at a Qing Court workshop with the guidance of Jesuits,[121] with ores similar to those used in France for soft-paste 17th/18th century porcelain[121] and additionally for Limoges enamels.[153,154] The availability of a mobile set-up in a museum allows the study of many artefacts and the collection of unexpected data, as demonstrated in Figure 7.12. For instance the comparison of the nanostructure of enamels covering Chinese *cloisonné* objects[155] with those of Limoges enamels first detected the technology transfer made by Jesuits to the Qing workshop.[121,126,156] Since 2000, a series of measurement campaigns have been made with XRF and Raman mobile set-ups. Table 7.6 lists a selection of studies of enamelled objects, pottery, glass and metal performed with mobile set-ups. Details about the procedure can be found in the literature[36–38] (XRF) and[64] (Raman microspectrometry). The main advantage of a mobile set-up is the possibility of studying all objects without any risks: no contact with the last generation of mobile instruments. The specific characteristics of Raman microspectroscopy are: (i) the high spatial resolution of up to the limit of diffraction of the light (about 0.3 μm using advanced microscope objective and blue laser) on-site; (ii) penetration depth up to 0.1–10 mm as a function of the transparency; (iii) the availability of long working distance objectives with distance to the object surface ranges between ~13 mm for ×200 objective to tens of centimetres or more for low magnification optics; (iv) it is possible to perform analysis through a glass cover;[150] (v) the good Raman signal for covalently-bonded compounds, which constitute most of the phases found in glass and pottery, including restoration ingredients; and (vi) the unique possibility to get information on the structure and composition of amorphous phases. XRF and Raman measurements provide

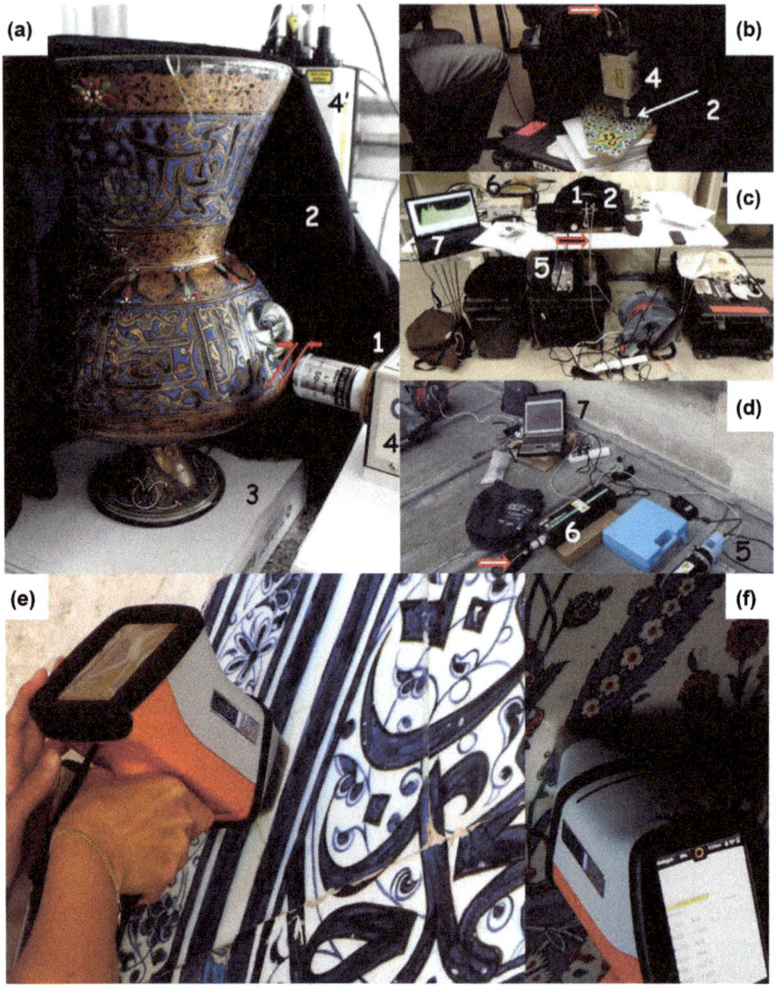

Figure 7.12 On-site analysis with a mobile Raman set-up. Measurements are made in museum storage rooms: (a) horizontal analysis, (b) analysis from the top with a mobile support, (c) view of the whole set-up (1, long working distance microscope objective (the distance between the object surface and the front lens is indicated); 2, black textile; 3, paper rams to adjust the height of the object and to suppress vibrations; 4, remote optical head; 5, Raman spectrometer and CCD detector (in blue); 6, laser; 7, computer) and (d) on the roof of Sainte-Chapelle Palatine Church, Paris. Reproduced from ref. 44 with permission from Elsevier Masson SAS, Copyright 2007. (e) and (f) Examples of measurements with an handheld XRF pistol: in e) the screen allows controlling the analysed spot; the instrument gives after collecting data (a few minutes) a first look of the elemental composition.

complementary information.[36–38,141] The main drawbacks of mobile XRF set-ups are: (i) the combination of the impossibility to measure light elements with (ii) the difficulty to control the in-depth measured volume and the fact that only flat or convex surface can be measured because of the (quasi) contact between the instrument and the analysed artefact. These weaknesses are almost suppressed by using appropriate procedures: the consideration of the elemental ratios is more informative than 'absolute' elemental compositions measured by destructive methods such as ICP-MS or LIBS. FORS is a good mobile technique for the identification of colouring agents.[92]

7.4 Conclusions

The number of publications on the subject is increasing rapidly. Up until 2000, most of the studies were composition analyses of pastes of archaeological samples of Roman, Greek and Islamic origin. Except for the pioneering work conducted by Kingery's MIT Group, by Zhang's Shanghai Institute of Ceramics,[3–6] and by Wood[46] and Tite[157,158] (British Museum/Oxford University), the studies of sophisticated pottery were limited until the development of non-invasive, mobile XRF and Raman instruments in the 2000s. This has led to a rapid increase in the number of studies dedicated to sophisticated enamelled objects, which offers an increasingly better understanding of the connected history of ceramic and pottery technologies and exchanges.

Acknowledgements

The author thanks all colleagues and students involved in the studies as well as Institutions and Curators offering the opportunity of studying rare objects and sharing information about them.

Recommended Reading

P. Colomban, The Destructive/Non-Destructive Identification of Enamelled Pottery and Glass Artifacts and Associated Pigments – A Brief Overview, *Arts*, 2013, **2**(3), 111.

P. Colomban, Natural Nanosized Raw Materials and Sol–Gel Technology: The Base of Pottery Since Millenniums in *Nanosciences and Cultural Heritage*, ed P. Dillmann, L. Bellot-Gurlet and I. Nenner, Atlantis Press, Paris, 2016, p. 59.

C. Fornacelli, P. Colomban and I. Turbanti Memmi, Toward a Raman/FORS Discrimination between Art Nouveau and Contemporary Stained Glasses from CdSxSe1-x Nanoparticles Signatures, *J. Raman Spectrosc.*, 2015, **46**(11), 1129.

Modern Methods for Analysing Archaeological and Historical Glass, ed. K. Janssens, J. Wiley & Sons, Chichester, 1st edn, 2013.

Ceramics and Civilization, Vol. I: Ancient Technology to Modern Science, ed. W. D. Kingery, The American Ceramic Society, Columbus, 1984.

Ceramics and Civilization, Vol. II: Technology and Style, ed. W. D. Kingery, The American Ceramic Society, Columbus, 1986.

Ceramics and Civilization, Vol. III: High-Technology Ceramics – Past, Present and Future, ed. W. D. Kingery, The American Ceramic Society, Columbus, 1987.

M. Maggetti and A, d'Albis, Phase and Compositional Analysis of a Sèvres Soft Paste Porcelain Plate from 1781, with a Review of Early Porcelain Techniques, *Eur. J. Mineral*, 2017, **29**(3), 347.

P. Sciau, Nanoparticles in Ancient Materials: The Metallic Lustre Decorations of Medieval Ceramics, in *Nanotechnology and Nanomaterials – The Delivery of Nanoparticles*, ed. A. A. Hashim, INTECH, ch. 25, http://www.intechopen.com/books/the-delivery-of-nanoparticles/nanoparticles-in-ancient-materials-the-metallic-lustre-decorations-of-medieval-ceramics, (accessed January 2018).

M. S. Tite, T. Pradell and A. Shortland, Discovery, Production and Use of Tin-Based Opacifiers in Glasses, Enamels and Glazes from the Late-Iron Age Onwards: A Reassessment, *Archaeometry*, 2008, **50**(1), 68.

N. Wood, *Chinese Glazes*, University of Pennsylvania Press, Philadelphia, 1999.

References

1. E. Cooper, *Ten Thousand Years of Pottery*, University of Pennsylvania Press, Philadelphia, 4th edn, 2000.
2. P. M. Rice, Pottery Analysis, *A Sourcebook*, The University of Chicago Press, Chicago, 1987.
3. *Ancient Technology to Modern Science, Ceramic and Civilization*, ed. W. D. Kingery, The American Ceramic Society, Columbus, 1984, vol. I.
4. *Technology and Style, Ceramic and Civilization*, ed. W. D. Kingery, The American Ceramic Society, Columbus, 1986, vol. II.
5. *High Technology Ceramics – Past, Present, and Future: The Nature of Innovation and Change in Ceramic Technology, Ceramic and Civilization*, ed. W. D. Kingery, The American Ceramic Society, Westerville, 1986, vol. III.
6. *Cross-craft and Cross-Cultural Interactions in Ceramics, Ceramic and Civilization*, ed. P. E. McGovern, M. D. Notis and W. D. Kingery, The American Ceramic Society, Westerville, 1989, vol. IV.

7. P. B. Vandiver, O. Soffer, B. Klina and J. Svoboda, The origin of ceramic technology at Dolni Vestonice Czechoslovakia, *Science*, 1989, **246**, 1002–1008.
8. J. R. Alden and L. Minc, Itinerant potters and the transmission of ceramic technologies and styles during proto-Elamites period in Iran, *J. Archaeol. Sci.: Rep.*, 2016, 7, 863–876.
9. A. Brongniart and D. Riocreux, Description Méthodique du Musée de Céramique de la Manufacture Royale de porcelaine de Sèvres, A. Leleux Libraire-Editeur, Paris, 1845, http://gallica.bnf.fr/ark:/12148/bpt6k6430171d/f9.image (accessed 21th December 2017).
10. A. Brongniart, Traité des arts céramiques ou Des poteries considérées dans leur histoire, leur pratique et leur théorie, Paris, Béchet jeune, A. Mathias, 1844, 3 vols, http://catalogue.bnf.fr/ark:/12148/cb36023945s.public (accessed 21th December 2017). Third edition is more complete: 3rd Edition, avec Notes et Additions par A. Salvetat, P. Asselin–Libraire de la Faculté de Médecine, 2 vol., Paris, 1877.
11. C.-A. Jouenne, Traité de Céramique et Matériaux Minéraux, Editions Septima, Paris, 2001.
12. M. Haussone, *Technologie Générale. Faïences, Grès, Porcelaines, Bibliothèque Professionnelle*, J-B Baillère & Fils, Paris, 1969.
13. P. Colomban, The destructive/non-destructive identification of enamelled pottery and glass artifacts and associated pigments – A brief overview, *Arts*, 2013, **2**(3), 111–123.
14. Theophilius (*ca.* 11th–12th centuries) Schedula diversarum artium. Traduction by Dodwell C.R., Theophilus: the various arts, London, 1961.
15. Anonymous, El Lapidario del Rey Alphonso X, translation in Spanish by King Alfonso in the year 1279, see in N. Heaton: J. British Society of Master Glass-Painters, vol. 48, 1947, p. 9.
16. A.-F. Cannela, *Gemmes, verre coloré, fausses pierres précieuses au Moyen âge, Le quatrième livre du "Trésorier de Philosophie naturelle des pierres précieuses" de Jean d'Outremeuse*, Bibliothèque de la Faculté de Philosophie et Lettres de l'Université de Liège, Fascicule CCLXXXVIII, Librairie Droz SA, Genève, 2006.
17. Abû al-Qâsem Kâshâni, Arâyes al-javâher, ed. I. Afshâr, (Teheran, 1966). Translation by J.W. Allan, Abû'l-Qâsim's Treatise on Ceramics: Iran, vol. 11, 1973, p111.
18. Y. Porter, *Les techniques du lustre métallique d'après le Jowhar-Nâme-ye-Nezâm* (1196), Actes du VIIe Congrès International sur la Céramique Médiévale en Méditerranée, Thessaloniki, 11–16 October 1999 (Caisse des Recettes Archéologiques, Athènes 2003), p427.
19. C. Picolpasso, *Li Tre Libri dell'Arte del Vasaio*, 1557, first translated and edited by C. Popelin, *Les trois Livres de l'Art du Potier*, (Paris, 1881). See also *The Three Book of the Potter's Art*, La Revue de La Céramique et du Verre, Fac-Similé Edition, 2007.
20. A. Neri r.p., L'Arte Vetraria, Stamperia de' Giunti, Florence, 1612 (https://play.google.com/books accessed 12th January 2018).
21. M. Macquer, *Elemens de Chymie-Pratique*, HERISSANT (J.-T.), vol. 2, Paris (1751). Dictionnaire de Chymie, vol. 2, Didot, Paris (1766, 1777).
22. W. Lewis, *Glass and Enamel by Preparations of Gold, Commercium Philosophico-technicum; or the Philosophical Commerce of Arts: Designed as an Attempt to Improve Arts, Trades, and Manufactures*, London, 1763, p. 170.
23. R. A. Ferchault de Réaumur, Observations sur la matière qui colore des perles fausses et sur quelques autres matières animales d'une semblable couleur, à l'occasion de quoi on essaie d'expliquer la formation des écailles de poissons. Mémoires Académie des Sciences, Paris (1716). Idem, Idée générale des différentes manières dont on peut faire la porcelaine et quelles sont les véritables

matières de celle de la Chine, ibidem (1727). *Idem, Second mémoire sur la porcelaine ou suite des principes qui doivent conduire dans la composition des porcelaines de différents genres et qui établissent les caractères des matières fondantes qu'on ne peut choisir pour tenir lieu de celle qu'on emploie à la Chine, ibidem (1729). Idem, Mémoire sur l'art de faire une nouvelle espèce de porcelaine par des moyens extrêmement simples et faciles ou de transformer le verre en porcelaine,* ibidem (1739).
24. Père d'Entrecolle's letters from Ching-te-chen in 1712 and 1722, translated in R. Tichane, Ching-te-Chen, New York State Institute for Glaze Research, Painted Post, New York, 1983.
25. D. Diderot and J. d'Alembert, Encyclopédie ou Dictionnaire raisonné des Sciences, des Arts et des Métiers, 1751–1772 (28 vol.) (encre.academie-sciences.fr/encyclopedie/accessed 12th January 2018).
26. F. Bastenaire-Daudenart, *L'Art de fabriquer la faïence*, La Librairie Scientifique et Industrielle, De Mahler et Cie, Paris, 1827.
27. A. Brongniart, *Mémoire sur la Peinture sur Verre*, Imprimerie Sellingue, Paris, 1829.
28. G. Bontemps, *Guide du Verrier-Traité historique et pratique de la fabrication des verres, cristaux, vitraux*, Librairie du Dictionnaire des Arts Manufacturés, Paris, 1868.
29. L. A. Salvetat, *Leçons de Céramiques professées à l'Ecole Centrale des Arts et Manufactures*, Malet-Bachelier, Paris, 1857.
30. T. Deck, *La Faïence*, Maison Quantin, Paris, 1887.
31. A. Jacquemart, *Histoire de la céramique*, Librairie Hachette et Cie, Paris, 1875.
32. H. Bertran, Nouveau manuel complet du Porcelainier, Faiencier, Potier de terre, Encyclopédie-Roret, l. Mulo Libraire-Editeur, Paris, 1898.
33. H. Bertran, Nouveau manuel complet de la Peinture sur verre, sur Porcelaine et sur Email, Encyclopédie-Roret, l. Mulo Libraire-Editeur, Paris, 1913.
34. P. Munier, *Technologie des Faiences*, Gauthier-Villars, Paris, 1957.
35. P. Colomban and A. Truong, Non-destructive Raman Study of the Glazing Technique in Lustre Potteries and Faiences (9th-14th centuries): Silver ions, Nanoclusters, Microstructure and Processing, *J. Raman Spectrosc.*, 2004, **35**(3), 195–207.
36. G. Simsek, F. Casadio, P. Colomban, K. Faber, L. Bellot-Gurlet, G. Zelleke, V. Milande and E. Moinet, On-site identification of earlier Meissen Böttger red stonewares using portable XRF: 1, body analysis, *J. Am. Ceram. Soc.*, 2014, **97**(9), 2745–2754.
37. G. Simsek, P. Colomban, F. Casadio, L. Bellot-Gurlet, K. Faber, G. Zelleke, V. Milande and L. Tilliard, On-site identification of early Böttger red stonewares using portable XRF/Raman instruments: 2, glaze and gilding analysis, *J. Am. Ceram. Soc.*, 2015, **98**(10), 3006–3013.
38. G. Simsek, P. Colomban, S. Wong, B. Zhao, A. Rougeulle and N. Q. Liem, Toward a fast non-destructive identification of pottery: the sourcing of 14th-16th century Vietnamese and Chinese ceramic shards, *J. Cult. Heritage*, 2015, **16**(2), 159–172.
39. L. F. V. Ferreira, A. Gonzalez, M. F. C. Pereira, L. F. Santos, T. M. Casimiro, D. P. Ferreira, D. S. Conceicao and I. F. Machado, Spectroscopy of 16th century Portuguese tin-glazed earthenware produced in the region of Lisbon, *Ceram. Int.*, 2015, **41**, 13433–13446.
40. L. V. Ferreira, M. V. Gomes, M. F. C. Pereira, L. F. Santos and I. F. Machado, A multi-technique study for the spectroscopic characterization of the ceramics from Santa Maria do Castelo church (Torres Novas, Portugal), *J. Archaeol. Sci. Rep.*, 2016, **6**, 182–189.
41. B. Minceva-Sukarova, A. Issi, A. Raskovska, O. Grupce, V. Tanevska, M. Yaygingöl, A. Kara and P. Colomban, Characterization of pottery from Republic of Macedonia. III, A study of the comparative mineralogical efficiency

using micro-Raman mapping and X-ray diffraction, *J. Raman Spectrosc.*, 2012, **43**(6), 792–798.
42. W. Goder, W. Schulle, O. Wagenbreth, H. Walter, R. Sonnemann and E. Wächtler, *La découverte de la porcelaine européenne en Saxe*, Meissen, Böttger J. F. (1709–1736), Pygmalion-Gérard Watelet Presse, Paris, 1984.
43. P. Colomban and V. Milande, On Site Analysis of the earliest known Meissen Porcelain and Stoneware, *J. Raman Spectrosc.*, 2006, **37**(5), 606–613.
44. P. Colomban and A. Tournié, On-site Raman Identification and Dating of Ancient/Modern Stained Glasses at the Sainte-Chapelle, Paris, *J. Cult. Heritage*, 2007, **8**(3), 242–256.
45. P. Colomban, *Natural Nanosized Raw Materials and Sol-Gel Technology: the Base of Pottery Since Millenniums in Nanosciences and Cultural Heritage*, ed. P. Dillmann, L. Bellot-Gurlet and I. Nenner, Atlantis Press, 2016, pp. 59–74, DOI: 10.2991/978-94-6239-198-7.
46. N. Wood, *Chinese Glazes*, University of Pennsylvania Press, Philadelphia, 1999.
47. P. Colomban, A. Tournié, D. de Montmollin and L. Krainhoefner, Vegetable ash as raw material in the production of glasses and enamels, for example the contemporary vegetable ashes from Burgundy, France, https://arxiv.org/ftp/arxiv/papers/1012/1012.1504.pdf.
48. The American Ceramic Society, *Phase Diagrams for Ceramists*, Westerville, 1971, 15 vol.
49. P. Colomban, G. Sagon, C. Roche, D. N. Khoi and N. Q. Liem, Microstructure and Processing of the Sa Hyunh and Cham Ceramics, *J. Cult. Heritage*, 2004, 5(2), 149–155.
50. W. M. Carty and U. Senapatti, Porcelain-raw materials, processing, phase evolution, and mechanical behavior, *J. Am. Ceram. Soc.*, 1998, **81**, 3–20.
51. P. Colomban and F. Treppoz, Identification and Differentiation of Ancient and Modern European Porcelains by Raman Macro- and Microspectroscopy, *J. Raman Spectrosc.*, 2001, **32**, 93–102.
52. P. Sciau, L. Noé and P. Colomban, Metal nanoparticles in contemporary potters' master pieces: Lustre and red "*pigeon blood*" pottery: Models to understand the ancient technology, *Ceram. Int.*, 2016, **42**(14), 15349–15357.
53. R. A. Epler and D. R. Epler, *Glazes and Glass Coatings*, The American Ceramic Society, Westerville, 2000.
54. P. Colomban, G. Sagon and X. Faurel, Differentiation of Antique Ceramics from the Raman Spectra of their Colored Glazes and Paintings, *J. Raman Spectrosc.*, 2001, **32**(5), 351–360.
55. P. Colomban, The use of metal nanoparticles to produce yellow, red and iridescent colour, from Bronze Age to Present Times in Lustre pottery and glass: Solid state chemistry, spectroscopy and nanostructure, *J. Nano Res.*, 2009, **8**, 109–132.
56. P. Vandenabeele, H. G. M. Edwards and L. Moens, A Decade of Raman Spectroscopy in Art and Archaeology, *Chem. Rev.*, 2007, **107**(3), 675–686.
57. C. Miliani, F. Rosi, B. G. Brunetti and A. Sgamellotti, In situ noninvasive study of artworks: the MOLAB multitechnique approach, *Acc. Chem. Res.*, 2010, **43**(6), 728–738.
58. P. Colomban, The on-site/remote Raman analysis with portable instruments - A review of drawbacks and success in Cultural Heritage studies and other associated fields, *J. Raman Spectrosc.*, 2012, **43**, 1529–1535.
59. S. M. Shackley, Portable X-ray Fluorescence Spectrometry (PXRF): the Good, the Bad, and the Ugly, Archaeol, Southwest Magazine, 26 [2] (2012). Retrieved from www.Archaeologysouthwest.org (accessed May 19, 2014).
60. P. Vandenabeele, H. G. M. Edwards and J. Jehlicka, The role of mobile instrumentation in novel applications of Raman spectroscopy: archaeometry, geosciences, and forensics, *Chem. Soc. Rev.*, 2014, **43**(8), 2628–2649.

61. J. M. Madariaga, Analytical Methods in the field of cultural heritage, *Anal. Methods*, 2015, **7**, 4848.
62. E. Stanzani, D. Bersani, P. P. Lottici and P. Colomban, Analysis of artist's palette on a 16th century wood panel painting by portable and laboratory Raman instruments, *Vib. Spectrosc.*, 2016, **85**, 62–70.
63. P. Vandenabeele and M. K. Donais, Mobile Spectroscopic Instrumentation in Archaeometry Research, *Appl. Spectrosc.*, 2016, **70**(1), 27–41.
64. P. Colomban, On-site Raman study of artwork: Procedure and illustrative examples, *J. Raman Spectrosc.*, 2018, **49**(6), 921–934.
65. P. Robertshaw, M. Wood, E. Melchiorre, R. S. Popelka-Filcoff and M. D. Glascock, Southern African glass beads: chemistry, glass sources and patterns of trade, *J. Archaeol. Sci.*, 2010, **37**, 1898–1912.
66. J. Koch and D. Günther, Review of the State-of-the-Art of Laser Ablation Inductively Coupled Plasmarometry, *Appl. Spectrosc.*, 2011, **65**(5), 155A–162A.
67. *Modern Methods for Analysing Archaeological and Historical Glass*, ed. K. J. Janssens, Wiley & Sons, Chichester, 1st edn, 2013, 2 vol.
68. L. Dussubieux, M. Golitko and B. Gratuze, *Recent Advances in Laser Ablation ICP-MS for Archaeology*, Springer, Berlin, 2016.
69. E. Neri, C. Morvan, P. Colomban, M. P. Guerra and V. Prigent, Late Roman and Byzantine Mosaic opaque "Glass-ceramics" *Tesserae* (5th-9th century), *Ceram. Int.*, 2016, **42**(16), 18859–18869.
70. S. Wolf, S. Stos, R. Mason and M. S. Tite, Lead Isotope Analyses of Islamic Pottery Glazes from Fustat, Egypt, *Archaeometry*, 2003, **45**(3), 405–420.
71. P. Degryse and J. Schneider, Pliny the Elder and Sr-Nd isotopes: tracing the provenance of raw materials for Roman glass production, *J. Archaeol. Sci.*, 2008, **35**(7), 1993–2000.
72. P. Degryse, J. Henderson and G. Hodgins, *Isotopes in Vitreous Materials*, Leuven University Press, Leuven, 2009.
73. A. W. Miziolek, V. Palleschi and I. Schechte, *Laser Induced Breakdown Spectroscopy*, Cambridge, 2006, Cambridge University Press.
74. R. Gaudiuso, M. Dell'Aglio, O. De Pascale, G. S. Senesi and A. De Giacomo, Laser Induced Breakdown Spectroscopy for Elemental Analysis in Environmental, Cultural Heritage and Space Applications: A Review of Methods and Result, *Sensors*, 2010, **10**(8), 7434–7468.
75. A. Bonneau, J.-F. Moreau, R. Auger, R. G. V. Hancock and B. Émard, Analyses physico-chimiques des perles de traite en verre de facture européenne: quelles instrumentations pour quels résultats ? *Archéologiques*, 2014, **26**, 109–132.
76. M. L. Sempowski, A. W. Nohe, J.-F. Moreau, I. Kenyon, K. Karklins, S. Aufreiter and R. G. V. Hancock, On the transition from tin-rich to antimony-rich European white soda-glass trade beads for the Senecas of Northeastern North America, *J. Radioanal. Nuclear Chem.*, 2000, **244**(3), 559–566.
77. G. Poupeau, F.-X. Le Bourdonnec, T. Carter, S. Delerue, M. S. Shackley, J. A. Barrat, S. Dubernet, P. Moretto, T. Calligaro, M. Milic and K. Kobayashi, The use of SEM-EDS, PIXE and EDXRF for obsidian provenance studies in the Near East: a case study from Neolithic Catalhayuk (Centralk Anatolia), *J. Archaeol. Sci.*, 2010, **37**(11), 2705–2720.
78. G. Padeletti, G. M. Ingo, A. Bouquillon, S. Pages-Camagna, M. Aucouturier, S. Roehrs and P. Fermo, First-time observation of Mastro Giorgio masterpieces by means of non-destructive techniques, *Appl. Phys. A: Mater. Sci. Process.*, 2006, **83**(4), 475–483.
79. G. Padeletti, P. Fermo, A. Bouquillon, M. Aucouturier and F. Barbe, A new light on a first example of lustred majolica in Italy, *Appl. Phys. A: Mater. Sci. Process.*, 2010, **100**(3), 747–776.

80. S. Coentro, L. C. Alves, C. Relvas, T. Ferreira, J. Mirao, J. Molera, T. Pradell, R. A. A. Trindade, R. C. Da Silva and V. S. F. Muralha, The Glaze Technology of Hispano-Moresque Ceramic Tiles: A Comparison Between Portuguese and Spanish Collections, *Archaeometry*, 2017, **59**(4), 667–684.
81. T. Pradell, G. Molina, J. Molera, J. Pla and A. Labrador, The use of micro-XRD for the study of glaze color decorations, *Appl. Phys. A: Mater. Sci. Process.*, 2013, **111**(1), 121–127.
82. T. Pradell, J. Molera, N. Salvado and A. Labrador, Synchrotron radiation micro-XRD in the study of glaze technology, *Appl. Phys. A: Mater. Sci. Process.*, 2010, **99**(2), 407–417.
83. Y. Leon, P. Sciau, M. Passelac, C. Sanchez, R. Sablayrolles, P. Goudeau and N. Tamura, Evolution of terra sigillata technology from Italy to Gaul through a multi-technique approach, *J. Anal. Atom. Spectrom.*, 2015, **30**(3), 658–665.
84. T. Wang, T. Q. Zhu, Z. Y. Feng, B. Fayard, E. Pouyet, M. Cotte, W. De Nolf, M. Salome and P. Sciau, Synchrotron radiation-based multi-analytical approach for studying underglaze color: The microstructure of Chinese Qinghua blue decors (Ming dynasty), *Anal. Chim. Acta*, 2016, **928**, 20–31.
85. G. Padeletti and P. Fermo, Significant findings concerning the production of Italian Renaissance lustred majolica, *Appl. Phys. A: Mater. Sci. Process.*, 2013, **113**(4), 825–833.
86. J. Pérez-Arantegui, J. Molera, A. Larrea, T. Pradell, M. Vendrell-Saz, I. Borgia, B. G. Brunetti, F. Cariati, P. Fermo, M. Mellini, A. Sgamellotti and C. Viti, Luster pottery from the thirteenth century to the sixteenth century: A nanostructured thin metallic film, *J. Am. Ceram. Soc.*, 2001, **84**, 442–446.
87. C. Mirguet, P. Fredericks, P. Sciau and P. Colomban, Origin of the Self-Organization of Cu°/Ag° Nanoparticles in Ancient Lustre Pottery. A TEM study, *Phase Transitions*, 2008, **81**(2–3), 253–266.
88. P. Sciau, Nanoparticles in ancient materials: The Metallic Lustre Decorations of Medieval Ceramics, in *Nanotechnology and Nanomaterials "The Delivery of Nanoparticles"*, ed. A. A. Hashim, INTECH, ch. 25, http://www.intechopen.com/books/the-delivery-of-nanoparticles/nanoparticles-in-ancient-materials-the-metallic-lustre-decorations-of-medieval-ceramics (accessed 12th January 2018).
89. P. Sciau, Transmission Electron Microscopy: Emerging Investigations for Cultural Heritage Materials, *Adv. Imaging Electron Phys.*, 2016, **198**, 43–67.
90. C. Fornacelli, P. Sciau and P. Colomban, CdS_xSe_{1-x} quantum dots as colouring agents of Art Nouveau and contemporary stained glass: a combined transmission electron microscopy and Raman study, *Philos. Trans. R. Soc., A*, 2016, **374**, 20160045.
91. P. Sciau, P. Salles, C. Roucau, A. Mehta and G. Benassayag, Applications of focused ion beam for preparation of specimens of ancient ceramic for electron microscopy and synchrotron X-ray studies, *Micron*, 2009, **40**(5–6), 597–604.
92. C. Fornacelli, P. Colomban and I. Turbanti Memmi, Toward a Raman/FORS discrimination between Art Nouveau and contemporary stained glasses from CdSxSe1-x nanoparticles signatures, *J. Raman Spectrosc.*, 2015, **46**(11), 1129–1139.
93. M. Maggetti and A. d'Albis, Phase and compositional analysis of a Sèvres soft paste porcelain plate from 1781, with a review of early porcelain techniques, *Eur. J. Mineral.*, 2017, **29**(3), 347–367.
94. L. Damjanovic, V. Bikic, K. Saric, S. Eric and I. Holclajtner-Antunovic, Characterization of the early Byzantine pottery from Caricin Grad (South Serbia) in terms of composition and firing temperature, *J. Archaeol. Sci.*, 2014, **46**, 156–172.
95. I. Memmi Turbanti, Pottery production and distribution: the contribution of mineralogical and petrographical methodologies in Italy. State of the art and future developments, *Periodico di Mineral.*, 2004, **73**(3), 239–257.

96. S. Bruni, V. Guglielmi, E. Della Foglia, M. Castoldi and G. B. Gianni, A non-destructive spectroscopic study of the decoration of archaeological pottery: from matt-painted bichrome ceramic sherds (southern Italy, VIII-VII BC) to an intact Etruscan cinerary urn, *Spectrochim. Acta, Part A*, 2018, **191**, 88–97.
97. M. Ostrooumov and A. Gogichaishvili, Raman and Infrared reflection spectroscopic study of pre-Columbian Mesoamerican pottery, *Eur. J. Mineral.*, 2013, **25**(5), 895–905.
98. S. Akyuz, T. Akyuz, S. Basaran, C. Bolcal and A. Gulec, Analysis of ancient potteries using FT-IR, micro-Raman and EDXRF spectrometry, *Vibr. Spectrosc.*, 2008, **48**(2), 276–280.
99. T. F. M. Oudemans, J. J. Boon and R. E. Botto, FTIR and solid-state C-13 CP/MAS NMR spectroscopy of charred and non-charred solid organic residues preserved in roman iron age vessels from the Netherlands, *Archaeometry*, 2007, **49**(3), 571–594.
100. A. Raskovska, B. Minceva-Sukarova, O. Grupce and P. Colomban, Characterization of Pottery from Republic of Macedonia, II: Raman and IR Analyses of Glazed Pottery Finds from Skopje Kale, *J. Raman Spectrosc.*, 2010, **41**(4), 431–434.
101. M. P. Colombini, G. Giachi, F. Modugno and E. Ribechini, Characterisation of organic residues in pottery vessels of the Roman age from Antinoe (Egypt), *Microchemical J.*, 2005, **79**(1–2), 83–90.
102. D. Tanasi, E. Greco, V. Dui Tullio, D. Capitani, D. Gulli and F. Ciliberto, H-1-H-1 NMR 2D-TOCSY, ATR FT-IR and SEM-EDX for the identification of organic residues on Sicilian prehistoric pottery, *Microchem. J.*, 2017, **135**, 140–147.
103. H. R. Mottram, S. N. Dudd, G. J. Lawrence, A. W. Stott and R. P. Evershed, New chromatographic, mass spectrometric and stable isotope approaches to the classification of degraded animal fats preserved in archaeological pottery, *J. Chromatogr. A*, 1999, **833**(2), 209–221.
104. Y. B. P. Kwan and J. R. Alcock, The impact of water impregnation method on the accuracy of open porosity measurements, *J. Mater. Sci.*, 2002, **37**(12), 2757–2761.
105. C. Waters, M. Salih and S. Ajinola, Porosity comparative analysis of porous copper and OOF modelling, *J. Porous Mater.*, 2015, **22**(4), 989–995.
106. P. Holakooei, A technological study of Elamite polychrome glazed brick at Susa, South Western Iran, *Archaeometry*, 2014, **56**(5), 764–783.
107. P. Colomban, T. Calligaro, C. Vibert-Guigue, N. Q. Liem and H. G. M. Edwards, Accrochage des dorures sur les céramiques et tesselles anciennes, *Rev. Archéométrie-Archeosciences*, 2006, **29**, 7–20.
108. W. D. Kingery, H. K. Bowen and D. R. Uhlmann, *Introduction to Ceramics*, J. Wiley & Sons, New-York, 2nd edn, 1976.
109. P. Colomban, Gel Technology in Ceramics, Glass-Ceramics and Ceramic-Ceramic Composites, *Ceram. Int.*, 1989, **15**, 23–50.
110. J. Coudamy, *Le feu, les fours, la porcelaine*, Gaz de France, Paris, 1987.
111. E. Bruneton, J. Bigarré, D. Michel and P. Colomban, Heterogeneity, nucleation, shrinkage and bloating in sol-gel glass ceramics. The case of LAS compositions, *J. Mater. Sci.*, 1997, **32**, 3541–3548.
112. N. Q. Liem, G. Sagon, V. X. Quang, H. V. Tan and P. Colomban, Raman Study of the Microstructure, Composition and Processing of Ancient Vietnamese (Proto)porcelains and Celadons (13-16th centuries), *J. Raman Spectrosc.*, 2000, **31**, 933–942.
113. N. Q. Liem, P. Colomban, G. Sagon, H. X. Tinh and T. B. Hoanh, Microstructure, Composition and Processing of the 15th century Vietnamese Porcelains and Celadons, *J. Cult. Heritage*, 2003, **4**(3), 187–197.
114. L. C. Prinsloo, N. Wood, M. Loubser, S. M. C. Verryn and S. Tiley, Re-dating of Chinese celadon shards excavated on Mapungubwe Hill, a 13th century Iron Age site in South Africa, using Raman spectroscopy, XRF and XRD, *J. Raman Spectrosc.*, 2005, **36**(8), 806–816.

115. N. Osman, I. A. Talib and H. A. Hamid, Properties of sol-gel prepared BaCeO(3) solid electrolyte using acetate precursors, *Ionics*, 2009, **15**(2), 203–208.
116. M. Havel, D. Baron, L. Mazerolles and P. Colomban, Phonon confinement in SiC nanocrystals: Comparison of the size determination using transmission electron microscopy and Raman spectroscopy, *Appl. Spectrosc.*, 2007, **61**(8), 855–859.
117. S. Karlin and P. Colomban, Micro-Raman study of SiC fibre-oxide matrix reaction, *Composites, Part B*, 1998, **29B**, 41–50.
118. B. Gratuze, I. Soulier, M. Blet and L. Vallauri, De l'origine du cobalt: du verre à la céramique, *Rev. d'Archéomètrie*, 1996, **20**, 77–94.
119. P. Colomban, Rocks as blue (green and black) pigments/dyes of glazed pottery and enamelled glass artefacts – The potential of Raman microscopy, *Eur. Mineral. J.*, 2013, **25**(5), 863–879.
120. R. G. V. Hancock, J. McKechnie, S. Aufreiter, K. Karklins, M. Kapches, M. Sempowski, J. F. Moreau and I. Kenyon, Non-destructive analysis of European cobalt blue glass trade beads, *J. Radioanal. Nucl. Chem.*, 2000, **244**(3), 567–573.
121. P. Colomban, Y. Zhang and B. Zhao, Non-invasive Raman analyses of *huafalang* and related porcelain wares. Searching for evidence of innovative pigment technologies, *Ceram. Int.*, 2017, **43**, 12079–12088.
122. R. Giannini, I. C. Freestone and A. J. Shortland, European cobalt sources identification in the production of Chinese Famille Rose porcelain, *J. Archaeol. Sci.*, 2016, **80**, 27–36.
123. J. Van Pevenage, D. Lauwers, D. Herremans, E. Verhaeven, B. Vekemans, W. De Clercq, L. Vincze, L. Moens and P. Vandenabeele, A combined spectroscopic study on Chinese porcelain containing ruan-cai colours, *Anal. Methods*, 2014, **6**, 387–394.
124. P. Colomban, Polymerisation Degree and Raman Identification of Ancient Glasses used for Jewellery, Ceramics Enamels and Mosaics, *J. Non-Cryst. Solids*, 2003, **323**(1–3), 180–187.
125. P. Colomban, A. Tournié and L. Bellot-Gurlet, Raman Identification of glassy silicates used in ceramic, glass and jewellry: a tentative differentiation guide, *J. Raman Spectrosc.*, 2006, **37**(8), 841–852.
126. P. Colomban and L. C. Prinsloo, Optical Spectroscopy of Silicates and Glasses, in *Spectroscopic Properties of Inorganic and Organometallic Compounds*, ed. J. Yarwood, R. Douthwaite and S. B. Duckett, RSC Publishing, The Royal Society of Chemistry, Cambridge, 2009, pp. 128–149.
127. P. Colomban and A. Slodczyck, Raman intensity: an important tool to study the structure and phase transitions of amorphous/crystalline materials, *Opt. Mater.*, 2009, **31**(12), 1759–1763.
128. P. Colomban, Pottery, Glass and Enamelled Artefacts: How to Extract Information on their Manufacture Technology, Origin and Age? in *Analytical Archaeometry*, ed. H. G. M. Edwards and P. Vandenabeele, The Royal Society of Chemistry, Cambridge, 2012, ch. 8, pp. 245–267.
129. C. X. Zhu, Y. B. Wang, Q. Lu, H. X. Zhao, X. L. Zhu, W. J. Fa and Z. Zheng, Reproduction of Jun-red glazes with nanosized copper oxides, *J. Am. Ceram. Soc.*, 2017, **100**(10), 4562–4569.
130. L. D. Kock and D. De Waal, Raman studies of the underglaze blue pigment on ceramic artefacts of the Ming Dynasty and of unknown origins, *J. Raman Spectrosc.*, 2007, **38**(11), 1480–1487.
131. P. Colomban, A. Tournié, P. Meynard and M. Maucuer, On-site Raman and XRF analysis of Japanese/Chinese Bronze/Brass Patina – The search of specific Raman signatures, *J. Raman Spectrosc.*, 2012, **43**(6), 799–808.
132. P. Colomban, F. Ambrosi, A.-T. Ngo, T.-A. Lu, X.-L. Feng, S. Chen and C.-L. Choi, Comparative analysis of wucai Chinese porcelains using mobile and fixed Raman microspectrometers, *Ceram. Int.*, 2017, **43**(16), 14244–14256.

133. M. Asquier, P. Colomban and V. Milande, Raman and Infrared analysis of glues used for pottery conservation treatment, *J. Raman Spectrosc.*, 2009, **40**(11), 1641–1644.
134. V. Labet and P. Colomban, Vibrational properties of silicates: A cluster model able to reproduce the effect of "SiO_4" polymerization on Raman intensities, *J. Non-Cryst. Solids*, 2013, **370**, 10–17.
135. P. Colomban, On-site Raman identification and dating of ancient glasses: procedures and tools, *J. Cult. Heritage*, 2008, **9**(Suppl.), e55–e60.
136. M. C. Caggiani, C. Valloteau and P. Colomban, Inside the glassmaker technology: Search of Raman criteria to discriminate between Emile Gallé and Philippe-Joseph Brocard Enamels and Pigment Signatures, *J. Raman Spectrosc.*, 2014, **45**(6), 456–464.
137. M. C. Caggiani, P. Colomban, A. Mangone, C. Valloteau and P. Cambon, Mobile Raman spectroscopy analysis of ancient enamelled glass masterpieces, *Anal. Methods*, 2013, **5**, 4345–4354.
138. P. Colomban, V. Milande and L. Le Bihan, On-site Raman Analysis of Iznik pottery glazes and pigments, *J. Raman Spectrosc.*, 2004, **35**, 527–535.
139. P. Colomban, A. Tournié and P. Riccardi, Raman spectroscopy of copper nanoparticles-containing glass matrix: the ancient red stained-glass windows, *J. Raman Spectrosc.*, 2009, **40**(12), 1949–1955.
140. D. Mancini, C. Dupont-Logié and P. Colomban, On-site identification of Sceaux porcelain and faience using a portable Raman instrument, *Ceram. Int.*, 2016, **42**(13), 14918–14927.
141. B. Kirmizi, O. Unsalan, K. Bayraktar and P. Colomban, On-site pXRF analysis of glaze composition and colouring agents of "Iznik" tiles at Edirne mosques (15th and 16th-centuries), *Ceram. Int.*, 2019, **45**, 595–605.
142. B. Sober, S. Faigenbaum, I. Beit-Arieh, I. Finkelstein, M. Moinester, E. Piasetzky and A. Shaus, Multispectral imaging as a tool for enhancing the reading of ostraca, *Palestine Explor. Quart.*, 2014, **146**(3), 185–197.
143. X. Faurel, A. Vanderperre and P. Colomban, Pink Pigment optimisation by resonance Raman Spectroscopy, *J. Raman Spectrosc.*, 2003, **34**(4), 290–294.
144. P. Colomban, Lapis Lajuli as unexpected blue pigment in Lâjvardina ceramics, *J. Raman Spectrosc.*, 2003, **34**(6), 420–423.
145. P. Colomban, A. Tournié, M.-C. Caggiani and C. Paris, Pigments and enamelling/gilding technology of Mamluk mosque lamps and bottle, *J. Raman Spectrosc.*, 2012, **43**(12), 1975–1984.
146. P. Colomban and H. Schreiber, Raman Signature Modification Induced by Copper Nanoparticles in Silicate Glass, *J. Raman Spectrosc.*, 2005, **36**(9), 884–890.
147. P. Ricciardi, P. Colomban, A. Tournié and V. Milande, Non-destructive on-site identification of ancient glasses: genuine artefacts, embellished pieces or forgeries ? *J. Raman Spectrosc.*, 2009, **40**, 604–617.
148. Exhibition Catalogue, 'Bernard Perrot, 1604-1709, Screts et Chefs-d'oeuvres des verreries royales d'Orléans, Musée des Beaux-Arts, 2010, Orléans.
149. A. Tournié, P. Ricciardi and P. Colomban, Glass corrosion mechanisms: a multiscale analysis, *Solid State Ionics*, 2008, **179**, 2142–2154.
150. D. Mancini, A. Tournié, M.-C. Caggiani and P. Colomban, Testing of Raman spectroscopy as a non-invasive tool for the investigation of glass-protected miniature portraits, *J. Raman Spectrosc.*, 2012, **43**(2), 294–302.
151. P. Colomban, V. Milande and H. Lucas, On-site Raman Analysis of Medici Porcelain, *J. Raman Spectrosc.*, 2004, **35**(1), 68–72.
152. P. Colomban, R. de Laveaucoupet and V. Milande, On Site Raman Analysis of Kütahya fritwares, *J. Raman Spectrosc.*, 2005, **36**(9), 857–863.
153. B. Kirmizi, P. Colomban and M. Blanc, On-site Analysis of Limoges enamels from 16th to 19th century, *J. Raman Spectrosc.*, 2010, **41**(10), 1240–1247.

154. P. Colomban, L. Arberet and B. Kirmizi, On-Site Raman Analysis of 17th and 18th Century Limoges Enamels: Implications on the European Cobalt Sources and the Technological Relationship between Limoges and Chinese Enamels, *Ceram. Int.*, 2017, **43**(13), 10158–10165.
155. B. Kirmizi, P. Colomban and B. Quette, On-site Analysis of Chinese Cloisonné enamels from 15th to 19th century, *J. Raman Spectrosc.*, 2010, **41**(7), 780–790.
156. P. Colomban, T.-A. Lu and V. Milande, Non-Invasive on-site Raman study of blue-decorated early soft-paste porcelain: the use of Arsenic-rich (European) cobalt ores – Comparison with huafalang Chinese porcelains, *Ceram. Int.*, 2018, **44**, 9018–9826.
157. M. S. Tite, T. Pradell and A. Shortland, Discovery, production and use of tin-based opacifiers in glasses, enamels and glazes from the late-iron age onwards: A reassessment, *Archaeometry*, 2008, **50**(1), 68–84.
158. M. S. Tite, J. Freestone, R. Mason, J. Molera, M. Vendrell-Saz and N. Wood, Lead glazes in Antiquity, Methods of production and reasons for use, *Archaeometry*, 1998, **40**(2), 241–260.

8 Dyes and Pigments

M. Bouchard,*[a] M.-O. Hubert[b] and B. Gandini[c]

[a] Centre d'Analyses et de Recherche en Art et Archéologie -CARAA-, 41 ter rue A. France, 94270 le Kremlin Bicetre, France; [b] 11 rue Désiré Charton, 93100 Montreuil, France; [c] 8/10, Square du Docteur Blanche, 75016 Paris, France
*Email: mbouchard@caraa.fr

8.1 Introduction

8.1.1 State of Art

This chapter has the ambition to reset the record regarding the definition, and often confused, use of the terms pigments and dyes. This will be discussed in relation to a specific conservation-based case study associated with organic and inorganic pigments. Specific pigments or dyes and particular chemical classes are not discussed in detail; the reader is directed to literature dedicated to this purpose.[1–5]

Significant confusion exists in specialised literature regarding the naming of pigments and dyes. This is sometimes due to misinterpretation of the physical or chemical properties of these materials, but mostly arises from differing uses of language based on historical, linguistic or commercial conventions. A quotation from Michel Pastoureau perfectly illustrates this matter:

> *"[. . .] The colour of the painter is not that of the physicist and that of the physicist is not that of the poet".*[6]

Conservation Science: Heritage Materials, 2nd Edition
Edited by Paul Garside and Emma Richardson

Published by the Royal Society of Chemistry, www.rsc.org

Our knowledge of existing pigments or dyes is based not only on the interpretation of old texts and classical authors such as *Pliny the Elder*, *Dioscorides*, *Theophrastus* or *Vitruvius*, but also on the scientific examination and analysis of these pigments.[7–10] Figure 8.1 shows a historiated initial 'C'(olor) of an artist mixing colours.

The contradictions in these various historic and contemporary sources have therefore created some confusion in the naming of pigments and dyes. As an example, the name "minium" nowadays applied to red lead (lead tetroxide, Pb_3O_4, C.I. 77578) has been formerly used during antiquity to designate cinnabar (mercury sulphide, HgS, C.I. n/a) whereas lead tetroxide was called *minium secondarium.*[11] Moreover, during antiquity and the Middle Ages, minium was also confused with realgar (arsenic sulphide, As_4S_4, C.I. n/a).[12]

One can also mention the case of Emerald green which in English refers to a copper arsenate pigment ($Cu(C_2H_3O_2)_2{\cdot}3Cu(AsO_2)_2$, PG21, C.I. 77410) but which in French refers to a chromium oxide pigment (Vert emeraude $Cr_2O_3{\cdot}2H_2O$, PG18, C.I. 77289), with Chromium oxide in English described as Chrome green, Viridian or Vert Guignet. Anecdotally, this confusion in the naming of Emerald green is reflected in Philip Ball's book translation from English into French (p. 228).[12] Many other examples exist and result from undetermined compositions, imprecise translations or cultural naming conventions.[13,119]

This is not to speak about other languages…

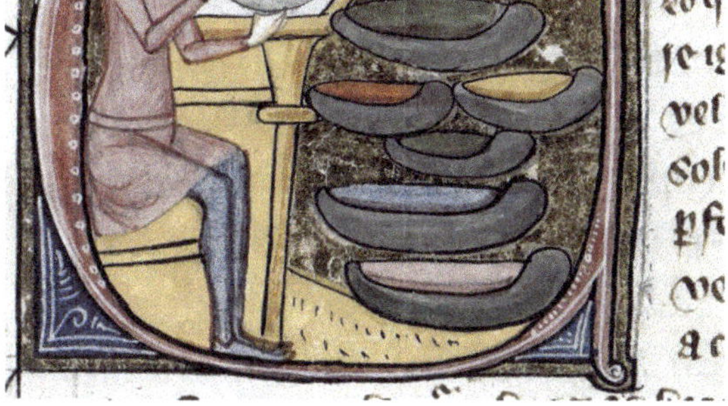

Figure 8.1 Detail of a historiated initial 'C'(olor) of an artist mixing colours, from James le Palmer's Omne Bonum, England (London), *ca.*1360–*ca.*1375.

Furthermore, the complexity can go much further. Chrome green in English has been attributed to Chromium oxide as discussed above, but also to the co-precipitation of two phases: Prussian blue ($C_{18}Fe_7N_{18}$, PB27, C.I. 77510) and Chrome yellow ($PbCrO_4$, PY34, C.I. 77600/77603). This confusion has been referred to as "mixéolytique", introduced by V. Kletzinsky.[14]

A more recent example is Hooker's green, referring to a mixture of Gamboge (NY24, C.I. n/a) and Prussian blue;[3] however, Winston and Newton manufactured two other Hooker's green pigments: one from a mixture of Acid green 1 (C.I. 10020) and Acid yellow 23 (C.I. 19140) and another, made of Quinacridone Deep (PO49, C.I. 73900) and Phthalocyanine Green (PG36, C.I. 74265).[15]

Therefore, it is clear that the naming of pigments and dyes is far from precise and consulting the list of synonyms alone is enough to show the complexity of this area. Only the classification of the Colour Index (published by the Society of Dyers and Colourists[16]) offers a certain guarantee.

Finally, often the term colourant is used in a random manner. It is also a sliding of language. According to Herbst and Hunger[1] a colourant can be either a pigment or a dye.

In this chapter pigments and dyes will be named according to their generic appellation, their chemical or mineralogical name followed in brackets, by their abbreviated pigment designations (*e.g.* PG, PB respectively for Pigment Green and Pigment Blue) and their Colour Index constitution number (*e.g.* C.I. 74265).[16] As an example: chlorinated copper phthalocyanine is designated by Phthalocyanine green (PG7, C.I. 74260).

8.1.2 Definitions

Let us first recall the definitions as well as some of the main characteristics of pigments and dyes.

Solubility: A pigment (organic or inorganic) is a coloured, fluorescent, or pearlescent insoluble substance in a medium; the medium allows the coloured material to dry so it adheres to the underlying support and gives a cohesion to the pigment, so the paint can be applied (note that this definition can vary from those adopted in biology, ceramics or for food additives). Dyes, on the other hand, are soluble. This is the most important difference between the two classes of materials.

Colour: The colour properties of pigments depend on the physical characteristics of the particles and they therefore impart colour by selective absorption, interference and/or scattering of light. Dyes on the other hand are almost exclusively defined by their chemical structure and therefore mainly impart colour by selective absorption of light.[2] The performance of most pigments is therefore dependent on their interaction with their vehicle system or medium. Properties, such as tinctorial strength, photochemical stability, hiding power, transparency, heat and solvent resistance, brightness, particle size, dispersibility and other relevant properties are discussed in dedicated literature.[1,17]

Origin: Pigments and dyes can be of natural origin (mineral, such as ochre, cinnabar or lapis; vegetable, such as indigo, saffron or madder; and animal, such as cochineal or murex) or synthetic origin. Some synthetic pigments and dyes are industrial copies of natural compounds, whereas others constitute a real panel of artificial molecules and minerals.

8.2 Dyes

Dyes are intensely coloured organic substances used in solution to confer a more or less permanent colour on fabrics, leathers, wood, bones, ivories and other kind of materials (see Chapters 3, 4, 5 and 10). It should be noted that not all coloured organic substances are dyes; only certain complex organic molecules are useful as colourants.

The colour of a dye is due to the selective absorption of radiation of a specific wavelength in the spectrum of visible light. The absorbed radiation induces electronic transitions within the molecule of the dye: the differences of energy involved in these transitions determine the colour of the dye. The light reflected or transmitted by the dye produces the colour effect observed by the eye, and the perceived colour is complementary to that absorbed by the dye.

At the molecular level, the colour of a dye is due to groups of atoms that function as chromophores,[17,18] and selectively absorb one or more wavelengths in the visible spectrum. One can principally cite the chromophores shown in Figure 8.2.

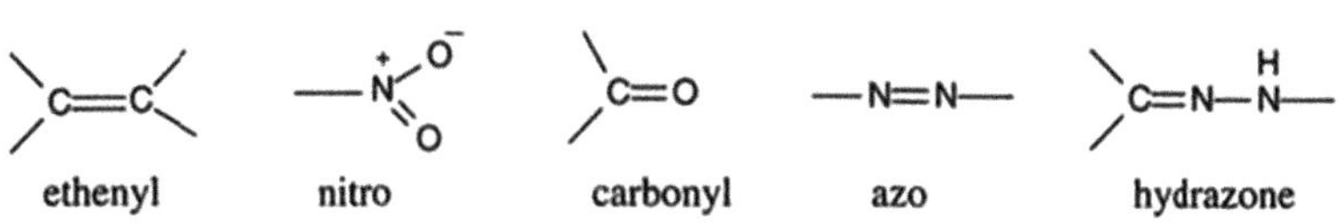

Figure 8.2 Example of chromophore chemical groups.

pH, temperature and the presence of other groups called auxochromes can also influence the wavelength of absorption by distorting electrons. Some of the auxochrome groups are shown in Figure 8.3.

The greater the number of conjugated double bonds the higher the wavelength of absorption of the molecule *e.g.* 1,4-diphenylbutadiene is colourless, 1,6-diphenylhexatriene is coloured Figure 8.4). Generally, the longer the conjugated system, the longer the wavelength of the absorbed visible light.

The spectrum of a polyene below illustrates how each additional double bond in the conjugated pi-electron system shifts the absorption maximum toward the longer wavelengths. The molar absorptivity (ε) also increases with each new conjugated double bond.

These shifts towards longer wavelengths are defined as "bathochromic" in absorption (the plural absorption peaks for a given chromophore is common for highly conjugated systems and is often solvent dependent).

Numerous examples and details regarding the electronic state of the molecules in dyes or organic pigments are given in Herbst and Hunger.[1]

8.2.1 Classification(s) of Dyes

There are several ways of classifying dyes. Some rely on the chemical 'family' of dyes, both natural and synthetic. Among the 25 structural classes listed,[16] one can cite:

Anthocyanins (*e.g.* cyanidins), anthraquinones (*e.g.* Alizarin), arylmethanes (*e.g.* Ethyl blue), azo (*e.g.* Congo red), carotenoids

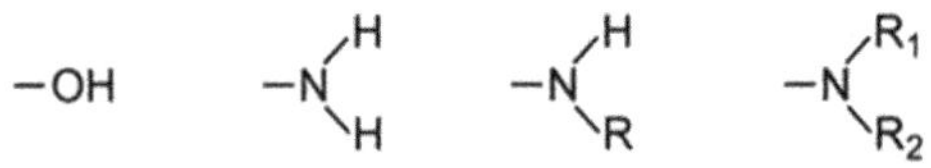

Figure 8.3 Example of auxochrome chemical groups.

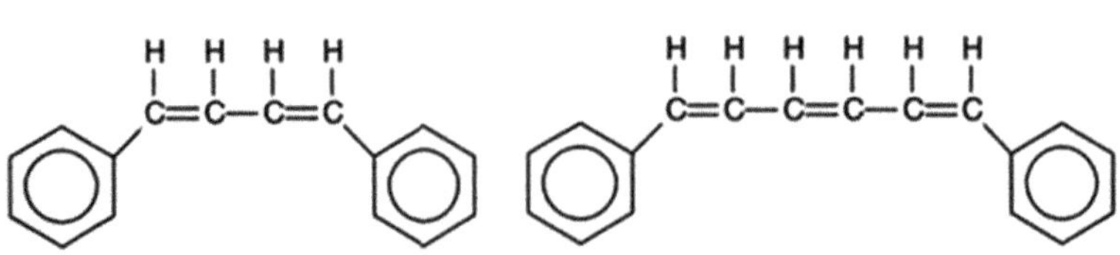

Figure 8.4 Left, colourless 1,4,4-diphenylbutadiene; Right coloured 1,6 diphenylhexatriene.[17]

(*e.g.* Saffron), cyanine (*e.g.* Indocyanine green), flavonoids (*e.g.* Buckthorn), neoflavonoids (*e.g.* Logwood), naphthoquinones (*e.g.* Henna), tannins (*e.g.* tea), and phthalocyanine (*e.g.* Copper *phthalocyanine*).

Some classifications are based on the "natural" or "synthetic" origin of the dye, and finally they may also be separated according to their method of application or industrial uses: direct dyes, mordant dyes, vat dyes, acid dyes, azoic dyes, basic dyes, *etc.*

The classification of dyes according to their chemical class is thoroughly discussed in the literature, for example, K. Hunger.[4] We will discuss the two last classifications in this chapter.

8.2.1.1 According to Origin

8.2.1.1.1 Natural Dyes

The first known colourants by humankind were limited to those from natural sources: plants (from barks, roots, flowers, berries or fruits), insects or molluscs.[19,20] However many of these first colourants were generally fugitive, meaning that their colour is diminished when exposed to daylight.

The earliest known example of dyed fibres was reported in a prehistoric cave (34 000 B.C.) in the Republic of Georgia.[21] Other traces of dyed textile were reported from the Neolithic period in southern Anatolia.[22] Dyed textiles from the 3rd millennium B.C. have been found in China and Pakistan.[23,24] Many dyed textiles dating from the Iron Age have also been found. Saffron, one of the oldest known dyes, was found on linen from Egypt dating *ca.* 1050 B.C.[2] Tyrian purple was known in minor Asia as far back as the 15th c. B.C; Indigo is suspected on a Tibetan dress from 3000 B.C. and has been found on the linen bands from Egyptian mummies from 2400 B.C.[12]

The availability of ancient dyes was strongly dependant on climate and geography. They were rather impure and it was almost impossible to obtain reproducible results. Other factors which influence the final colour of the dye included the part of the plant or the animal that was used, the method of applying the dye to the substrate (*e.g.* temperature), the material of the container (*e.g.* wood or metal), the kind of mordant (for mordant dyes), *etc.*

Nevertheless, these primitive dyes remained predominant until the first synthesis of natural dyes in the early 19th century. At that time, many scientists tried to extract and isolate the dye molecules from different plants such as Alizarin from madder root, isolated in 1826 by Colin and Robiquet.[25,117] This was an essential first step in the complete synthesis of these dyes.

8.2.1.1.2 Synthetic Dyes

Alizarin was later synthesised and commercialized by BASF (1868 and 1869 respectively).[18] At about the same period, the first new synthetic dye, the coal-tar based Mauveine was also manufactured (1856);[26] initially it was incorrectly named Tyrian purple. This dye became the favourite colour of the French Empress Eugenie and Queen Victoria.[12,118]

Hence, synthetic dyes include the chemical synthesis of natural colourants as well as the synthesis of new colouring molecules. Following the discovery of Mauveine by Perkins in 1856, another aniline based dye, Fuscine (also called Magenta or Roseine) was discovered in 1859.[12] Aniline pigments tend to be very fugitive and the first paints based on such pigments (such as those used by Edgar Degas or Vincent Van Gogh) tended to deteriorate significantly as a result.

In 1861, Lauth discovered the basic dye Methyl violet, in 1862 Martius and Lightfoot developed Bismarck Brown and, in 1863, Lightfoot Aniline black. (Picric acid and murexide were discovery during the 18th c. respectively by P. Woulfe and C.W. Scheele but did not meet a real success until the mid-19th c.)

The 1870s saw the discovery of Methyl green (1872, Lauth and Baubigny), Methyl blue (1876, Caro), Malachite green (1877, Dobner and Fisher).[28]

Hence, the period up to 1900 was characterized by the discovery and development of other dyes derived from coal-tar intermediates, *i.e.*, aniline based dyes (the azo dye family).[17]

The first half of the 20th century continued to be an intense period for research and industrialisation of new dyes and, following significant developments, the discovery of reactive dyes in 1954 heralded a major breakthrough in the dyeing of cotton.[4,120]

Due to the great similarity in their chemical structures, details about 20th c. synthetic dyes will be discussed in the section dedicated to synthetic organic pigments.

8.2.1.2 According to the Manufacturing Process

The quality of fixation of the dye on a material or substrate will depend on the nature of the bond formed between them. Certain dyes are held by chemical bonds, while others by physical interactions, and the process of preservation varies according to the chemical nature of the substrate.

Classification of dyes according to their manufacturing process includes: direct dyes, acid dyes, basic dyes, azoic dyes, mordant dyes,

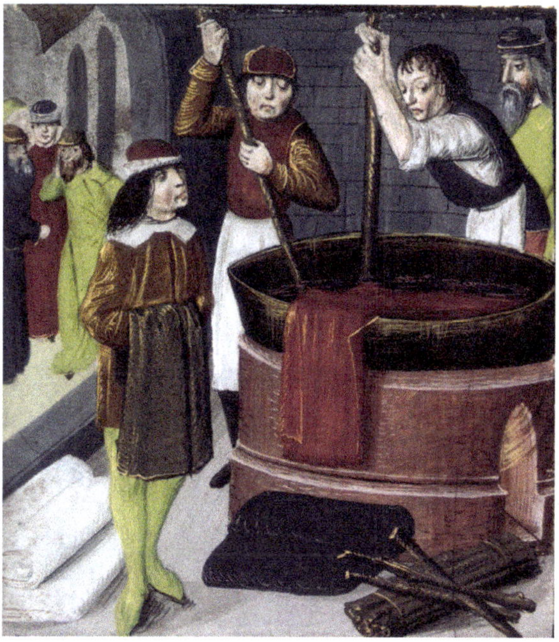

Figure 8.5 Dyeing wool cloth, from "Des Proprietez des Choses" by Bartholomaeus Anglicus, British Library Royal MS 15.E.iii, folio 269.

reactive dyes, vat dyes (Figure 8.5), disperse dyes, solvent dyes and sulphur dyes.[29]

8.2.1.2.1 Water-soluble Dyes (Direct, Acid, Basic and Azoic)

Certain organic materials are easily "wettable", which denotes their capacity to absorb water. If such a material is dipped into water in the presence of a dye, a proportion of the dye will remain attached to the material, imparting a specific colour.

- Direct dyes (also called substantive dyes) are water-soluble anionic dyes which can be applied to cotton, regenerated cellulose, paper, viscose and leather (*e.g.* Figure 8.6). Most direct dyes are azo compounds (azo compounds comprise the functional group diazenyl $R-N{=}N-R'$, where R and R' can be either aryl or alkyl) *e.g.* Direct Orange 25 (C.I. 22135) or Direct Blue 71 (C.I. 34140).
- Acid dyes are water-soluble anionic dyes applied in pH neutral-to-acid baths to nylon, silk, wool and modified acrylics. Azos, anthraquinone and phthalocyanine are the most important chemical types of acid dyes.
- Basic (cationic) dyes were amongst the earliest synthetic dyes. They are water-soluble cationic dyes which can be applied to

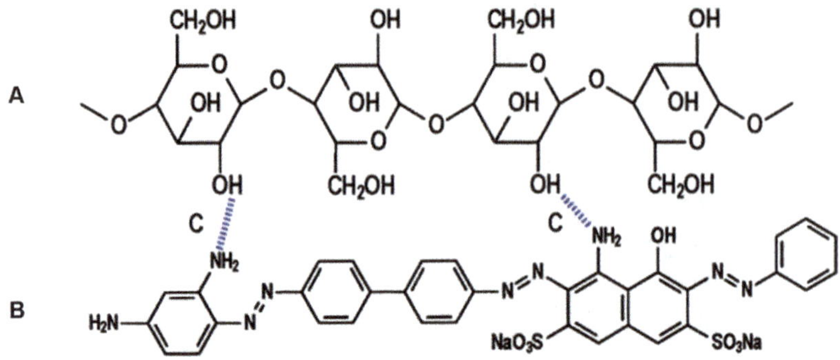

Figure 8.6 Representation of H-bonding (C) between a cellulose (A) and a direct dye (B).
Reproduced from ref. 30 with permission from the International Agency for Research on Cancer, Copyright 2010.

paper or acrylic fibres, and also sometimes to wool or silk. Examples include Mauveine and Basic Yellow 28 (C.I. 48054).

- Finally, azoic dyes or naphthol dyes (because naphthol compounds are used in their synthesis) do not exist *per se*. They are directly applied to fibres during a reaction between diazoic and a coupling components and the combination of the two molecules forms a water-insoluble dye. They are mainly used on cotton fibres.

When the strength of the bond between the dye molecule and the fibre is weak, the dye cannot be applied directly and hence mordant or vat dyes are used instead.

8.2.1.2.2 Mordant Dyes

Certain compounds are used to improve the fastness of the dye against water. In these cases, dyes are fixed to fibres by the interaction of the dye or lake with specific metallic salts to form insoluble metallic complexes. This process is known as mordanting and the metallic salts used are called mordants.

One can draw a parallel between mordant dyes and pigment lakes, which rely on the same principle of association between a colourant and an inorganic salt.

Mordants were used early in the history of dyes. Alum ($KAl(SO_4)_2{\cdot}12H_2O$), natron ($NaCO_3$), soluble iron salts (*e.g.* sulphates) or tin salts were among the most common and oldest mordants. Other kinds of mordants reported are barium sulphate, lead sulphate, bismuth salts, *etc.* In numerous cases, the type of mordant used will not only be crucial for the binding process but will also strongly influence the final colour development of the dye.[31,32]

Pliny describes marvellously the mordanting technique used in Egypt:

> "*In Egypt, too,* they employ a very remarkable process for the colouring of tissues. After pressing the material, which is white at first, they saturate it, not with colours, but with mordants that are calculated to absorb colour. This done, the tissues, still unchanged in appearance, are plunged into a cauldron of boiling dye, and are removed the next moment fully coloured. It is a singular fact, too, that although the dye in the pan is of one uniform colour, the material when taken out of it is of various colours, according to the nature of the mordants that have been respectively applied to it: these colours, too, will never wash out."[33]

Logwood is an example of a dye extensively used by the indigenous people of Mexico, and later by the Spanish in Europe, and which can produce a variety of colours, according to the metallic salt that is used as a mordant. Lac lake (NR25, C.I. 75450), Saffron (NR25, C.I. 75450), Madder (Alizarin Crimson, NR8, C.I. 58205), Kermes (NR3, C.I. 75460) and Carmine dye (cochineal, NR4, C.I. 75470) are also strongly influenced by the mordant used in association (Figure 8.7).

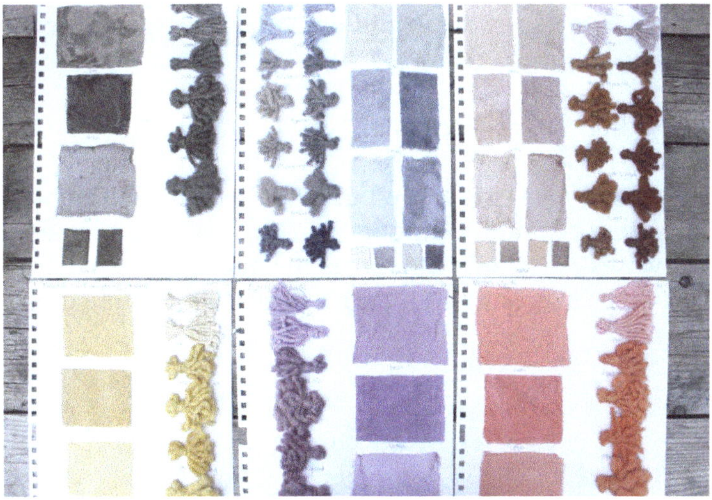

Figure 8.7 In this image, the top row was mordanted with iron, while the bottom row was mordanted with alum. Each column is dyed with the same plant; from left to right, marigold, cochineal and madder root.
Reproduced from ref. 34 with permission from Westlake Knits & Fibres, www.westlakeknits.ca.

C. Cooksey *et al.*[28] detail the complex lo-kao (Chinese green dye, NG1, C.I n/a) which can have the apparent behaviour of a vat dye or a mordant dye.

8.2.1.2.3 Vat Dyes

Other water-insoluble colouring molecules cannot be used unless they are subjected to a process of long fermentation. The term vat dyes is related to the secular way of preparation of these types of dyes in vats or tanks.

During this process, an insoluble complex polycyclic molecule is reduced by decomposition in presence of a mild alkali (acid or basic solutions of soda, potassium hydroxide *etc.*), and forms a pale or colourless 'leuco' derivate compound. This leuco compound is soluble in alkali and can penetrate into fibrous materials in order to colour them.

After this first step, and in the presence of air, the leuco compound re-oxidizes, and returns to its original form of an insoluble and strongly coloured molecule (Figure 8.8).

The most commonly used vat dyes belong to the anthraquinone and indigoid chemical classes. Indigo (NB1, C.I. 75780) and Tyrian purple (NV1, C.I. 75800) are typical examples of vat dyes.[35]

C. J. Cooksey and R. S. Sinclair[36] report the colour variation in Tyrian purple dyeing and its correlation (among other factors) to the intermolecular forces between the dye molecules and different substrate fibres.

8.2.1.2.4 Other Water-insoluble Dyes (Solvent, Sulphur, Disperse and Reactive)

Solvent dyes are water-insoluble but soluble in alcohols, chlorinated hydrocarbons, or liquid ammonia and are used to colour organic solvents, waxes, plastics, and other hydrocarbon-based non-polar materials. Some azo, anthraquinone and phthalocyanine dyes belong to this family.

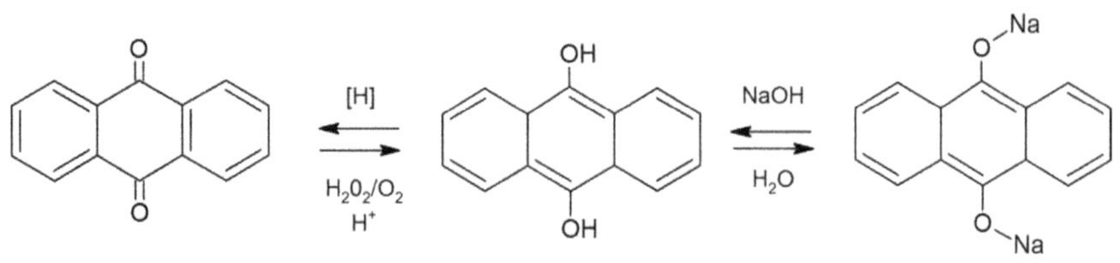

Figure 8.8 Reaction of a vat dye. Reduction and oxidation reactions of an anthraquinonic dye.

Sulphur dyes are water-insoluble, low cost and have good colour fastness. Sulphur Black 1 (C.I. 53185) and Sulphur Blue 7 (C.I. 53440) are among the most widely used sulphur dyes.

Disperse dyes are water-insoluble non-ionic dyes mainly used on polyester, polyamide and acetate fibres.

Reactive dyes contain chromophores that react with the substrate. They are mainly applied to materials such as cotton, flax, wool or nylon. Azos, anthraquinones and phthalocyanines are of this type of dye.

8.3 Pigments

8.3.1 Classification of Pigments

Different classifications also exist for pigments (according to their use or application, or according to their chemical composition) but they can essentially be separated into two main categories:

- Inorganic pigments (natural and synthetic)
  Natural inorganic pigments, including metallic and salt compounds, are produced from ores and used with or without modification[†] whereas synthetic inorganic pigments are manufactured.
- Organic pigments (natural and synthetic)
  Organic pigments are molecules based on a carbon skeleton structure; they may also contain metal ions or atoms (referred to as organo-metallic compounds), and these materials are strongly related to dye production. Organic pigments can also have a natural or a synthetic origin.

The relative merits of organic and inorganic pigments in different industrial applications (*e.g.* artist's paint, architecture, ink, automotive, glass colouration, anti-corrosion coatings or plastic colouration) are based on their physical properties (*e.g.* hiding power, rheology, particle size, strength, thermosensitivity or acidity) and are discussed in detail by various authors such as Willy Herbst and Klaus Hunger[1] or Teofil Jesionowski and Filip Ciesielczyk.[37]

[†]Carbon black pigment is in the elemental state and is often grouped along with the inorganic pigments.

8.3.1.1 Inorganic Pigments (Natural and Synthetic)

Natural inorganic or mineral pigments were the most widely used pigments until the beginning of the industrial era. Iron or manganese ores are some of the most well-known and are extracted from surface sedimentary rocks deposits (Figure 8.9). They exhibit various shades of yellow, red, orange, brown and black.

The vivid colours associated with heavy metals, such as copper oxides, sulphurs, carbonates, sulphates, *etc.* were not readily available in prehistoric times. From the 3rd millennium B.C. vivid colour minerals were traded and used. The earliest of these pigments are cinnabar, azurite, malachite and lapis lazuli.[1]

Although nature has provided a large set of inorganic pigments in the form of ores, humans have also succeeded in synthesising some of these pigments in order to obtain the considerable amount needed.

Synthesis includes simple crystalline modification under specific conditions *e.g.* the heating of yellow goethite to convert it to red haematite. This transformation has probably been carried out since prehistoric times[38] and was common during antiquity. Vitruvius indicates the method of heating yellow ochre to obtain a violet-garnet pigment:[39]

> "... *it is therefore prepared, good ore (sil) is calcined to the point that it is incandescent then, it is extinguished in vinegar and it becomes purple* ..."

Other examples of ancient transformations of pigments include lead white to red lead, metallic lead to lead white, metallic copper to verdigris, *etc.*

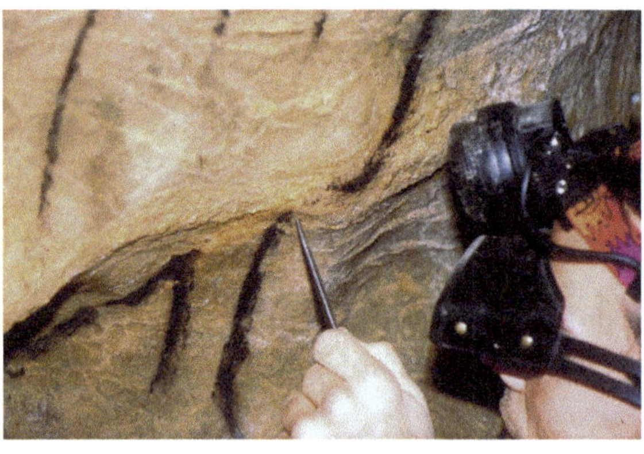

Figure 8.9 Prehistoric pigments – manganese based oxide – during sampling for analyses. Mayenne Science Cave, Mayenne, France. Reproduced with permission from Dr. R. Pigeaud.

More complex syntheses were also achieved during Antiquity. The famous Egyptian blue pigment (calcium copper silicate, based on cuprorivaite) was synthesised around 2000 B.C. and perhaps earlier.[3]

In fact, each period of history had its new synthetic pigments. As an example, Prussian blue, the first modern synthetic pigment, was synthesised in 1704 and became available to artists in 1724.

However, these discoveries can be considered sporadic in comparison with the larger number of synthetic compounds produced during the industrial era (c. 1760–1840) and afterwards.

Emerald green (PG21, C.I. 77410) was synthesised in 1778, Cobalt green (PG26, C.I. 77344) in 1780, Zinc white (PW4, C.I. 77947) in 1782, Cobalt blue (PB28, C.I. 77346) in 1802, Cerulean blue (PB35, C.I. 77368) in 1860; other inorganic pigments such as chrome derivatives (Chrome yellow (PY34, C.I. 77600, 77603), Chrome green (*e.g.* PG17, C.I. 77288)), Cobalt yellow (PY40, C.I. 77357), cadmium based pigments (Cadmium yellow (PY35, C.I. 77205) and later on Cadmium red (PR108, C.I. 77202, 77196)) or synthetic iron oxide followed during the 19th and early 20th century.

More recent synthetic pigments are titanium based (anatase and rutile forms – PW6, C.I. 77891, respectively 1916 and 1936), molybdate based pigments [Molybdate red (PR104, C.I. 77605; 1936), Manganese blue (PB33, C.I. 77112; 1935),[3] Titanium yellow (PY53, C.I. 77788; 1960)] or cerium sulphide pigments (*e.g.* PO75, C.I. 77283:1).[1,40]

The most frequent white (natural or synthetic) inorganic pigments are [zinc, tin, antimony or titanium] oxides, [barium, calcium or lead] sulphates, [lead, barium or calcium] carbonates, [zinc or strontium] sulphides, *etc.* whereas, black (natural or synthetic) inorganic pigments are very often carbon based (bone black, vine black, lamp black, graphite, *etc.*) or iron, manganese, cobalt or molybdenum oxide-based material.

Other coloured pigments include a wide variety of metal salts and oxides, *e.g.* copper carbonate, copper chlorides, cobalt aluminates, chromium oxides, cadmium sulphurs, *etc.*

Special groups of inorganic pigments also exist: "lustred" pigments (mirror effect), metallic pigments (*e.g.* copper, gold, aluminium), pearlescent pigments (*e.g.* transparent lamellar particles), interference pigments (*e.g.* thin oxide layers deposited on mica), anticorrosion pigments or others with specific absorption properties such as fluorescence (*e.g.* radioactive luminescent pigments) or phosphorescence (*e.g.* zinc sulphide).[37,40]

If some inorganic pigments are presumably more inert and more resistant to temperature and extreme pH than their organic counterparts,[37] some of them are also unstable and will be discussed in Section 8.5.

8.3.1.2 Organic Pigments (Natural and Synthetic)

The term organic pigment refers to colourants that have a carboxylic ring skeleton (hydrocarbon-based) as part of their structure.[5]

Confusion may arise from the fact that organic pigments and dyes are often molecularly similar, and a single structural skeleton can act either as a dye or as an organic pigment (*e.g.* indigo, saffron or phthalocyanine).[1] As mentioned in the section dedicated to dyes, the mobile electrons in this structure favoured by aromatic ring functional groups or conjugation will impart a colour to the molecule.

As for inorganic pigments, one can distinguish **natural organic pigments** from **artificial or synthetic organic pigments.**

Natural organic pigments are mainly extracted from plants (roots, flowers, fruits, *etc.*) or animals, and have been known since antiquity.

Natural organic pigments, like almost all other pigments, are insoluble. Some are naturally insoluble after extraction, *e.g.* gamboge, obtained from the resin of *Garcinia* trees. Others, named lakes, are made insoluble by precipitation of a water-soluble dye with an inorganic cation or substrate such as aluminium hydroxide (this process can be related to the mordanting technique for dyes).[3,17]

One can cite Tyrian purple lake (NV1, C.I. 75800), Saffron (NY6, C.I. 75140/75100), Madder lake (Alizarin)[‡] (natural *e.g.* NR9/10/11, C.I. 75330/75420 or synthetic PR83, C.I. 58000), Persian Berries lakes or Brasilwood. Conversely, some pigments can be converted into water-soluble dyes by sulphonation or chloromethylation for dyeing textiles, for example (*e.g.* copper phthalocyanines).[1,41]

Some pigments are also produced from the vat process (see Section 8.2.1.2). As an example, Indigo has been detected as a pigment in Roman parade shields from 200 A.D[3] but was available much earlier as a dye (CI 75780, Nat Blue 1).[§]

Finally, some recipes also report secondary processes for pigment making from original dyes. These processes, called *cimatura* (or *vloken* or *bourre*),[2] consist of extraction from dyed wool and break-down by quenching in a strong alkaline solution and neutralized. Indigo, Tyrian purple, madder or kermes are examples of pigments that were obtained through this process.[2]

Synthetic organic pigments can be either molecules that have a natural counterpart and are synthesized in a laboratory (such as

[‡]Madder root extracts also contain purpurin. Therefore the presence of this additional molecule in a pigment differentiates natural madder lake from synthetic.

[§]The synthetic form of indigo (BAYER-1880-) is listed as Pig blue 66 or Vat blue 1 (CI 73000).

Indigo or Alizarin) or novel synthetic molecules that have no counterpart in nature (such as phthalocyanines).

Synthetic organic pigments were originally derived from coal-tar based dyes during the dramatic development of organic chemistry in the 19th century. Examination of the catalogues of the large sale houses of pigments such as Winsor & Newton, Schmincke or Soc. Lefranc and Bourgeois clearly shows that half or even two-thirds of the listed pigments (organic and inorganic) were discovered during the 20th century.[42]

Whereas only a dozen natural organic colourants were used in antiquity and mediaeval times, more than 4000 synthetic organic pigments and dyes are used nowadays.[12]

Historically, synthetic organic pigments were first manufactured after the discovery of diazonium salts in 1858 and as byproducts of the first azo dyes in the 1880s. They were produced by precipitation on metallic salts (such as calcium, barium, strontium, or manganese salts). However, these first synthetic organic pigments had poor lightfastness, had very bad reputations and caused a great deal of damage.[1,5,43]

Circa 1880–1912 was the period in which the class of Naphtols AS, b-Naphtols and Arylides were discovered. Para red (PR1, C.I. 12070) was introduced in 1880, Lithol red R (PR49, C.I. 15630) was developed in 1899, Toluidine red (PR3, C.I. 12120) in 1905, dinitroaniline (PO5, C.I. 11725) in 1907 and Hansa yellow (PY1, C.I. 11680) in 1909.[2,5,27,42] Diarylides were patented in 1911 but only commercialized in 1935.

Circa 1913–1920: Superior fastness pigments were developed during this period.[17]

Brilliant lakes of complex heteropoly acids of phosphorus, molybdenum and tungsten with basic dyes like Rhodamine B (C.I. 45170), Victoria blue (C.I. 42595), or Methyl violet (C.I. 42535/42555) and Pigment green B (PG8, C.I.: 10006), an iron complex of nitroso-B naphthol, were patented shortly after in 1921.

Circa 1935: The largest single advance in pigment technology after World War I was the discovery of the non-azo (polycyclic) pigments such as copper phthalocyanines in the 1930s,[¶] which were characterized by excellent brightness, strength, bleed resistance and lightfastness.

Circa 1950: After World War II the most important discovery was the family of red violet quinacridone pigments (in the 1950s),

[¶]Some synthetic pigments contain metal or metal ions. However because of the carboxylic skeleton of the molecule these are considered organic pigments.[5]

followed by the mostly yellow-orange benzimidazolones (1960) and isoindolinone pigments (1964, although patented in the 1940s), and the most recent development of the pyrrolopyroles (*e.g.* PR264, C.I. 561300) discovered in 1983. These pigments belong to a group referred to as high performance pigments (HPPs).[1,5] Azo condensation pigments were discovered in 1951.[17]

Other non-lake synthetic pigments are vat pigments. Only a few of the very large number of vat dyes have found applications in the pigment field. One can cite Perylenes, Perinones, Thioindigo (1913 for dyes, 1950 for pigments), Indanthrone blue (*e.g.* PB60, C.I. 69800) or Flavanthrone yellow (*e.g.* PY112, C.I. 70600).[1,17]

A more comprehensive description (taxonomy, physical, chemical, artistic or industrial properties) of these synthetic organic pigments is available in specialized literature.[1,5,42,44,45]

8.4 Inks

Ink is defined as a liquid solution used for writing, printing, stamping or lithography.[3] Inks usually contain colouring agents (a pigment or a dye) and a binder, both uniformly scattered in a liquid carrier. Modern inks can also include other complex additives such as resins, solubilizers, surfactants, lubricants, fluorescents, *etc.*

8.4.1 Ancient Inks

During antiquity, the medium was usually vegetable gum or animal glue, while the vehicle was almost always water. Among the most common are two types of black inks: carbon based ink and ferro-gallic ('iron gall') ink. Both were used in antiquity.

Carbon based ink consists of lampblack, soot or carbonized organic material (wood, bone, ivory, *etc.*), mixed with water, glue or gum (*e.g.* gum Arabic), boiled and filtered under heat.

Chinese and Egyptian carbon based inks can be traced back to the 23rd century B.C. Dioscorides, Vitruvius and Pliny also indicate that ink in antiquity was carbon based.

Metallo-gallic ink (of which ferro-gallic ink is an example) is made from an infusion of crushed oak galls (a source of gallic acid), green vitriol (an iron salt, $FeSO_4$) or blue vitriol (a copper salt), and gum Arabic diluted with water. The soluble metallic salts form dark coloured solutions on reaction with the tannin (*i.e.*, gallic acid) extracts.

It was widely supposed that the use of metallo-gallic ink was introduced after the 7th century A.D.[46] However, the antique world knew very well that the combination of oak gall extract and copper produced black. References to this mineral ink are found, for example, in Philon of Byzantium in the 2nd century B.C. Some iron based inks are also found well before the 7th century A.D. on manuscripts on vellum. One of the oldest examples is a fragment of the manuscript of *Demosthenes* dating to the 5th century B.C. Another example is the Sillaiticus Codex (one of the first Bibles, 4th century B.C.), which offers a good example of early iron based ink use.[47]

8.4.2 Modern Inks

Modern inks used in printing or pens are more complex than ancient traditional inks. Depending on the final use, inorganic and organic pigments can be used as well as dyes. The pigments or dyes employed can be as varied as red Eosin dye (C.I. 45380), Crystal violet (C.I. 42555), titanium dioxide (PW6, C.I. 77891), carbon black (PBk6, C.I. 77266), metallic powders as well as luminescent, fluorescent or pearlescent pigments.

In addition to the traditional components (solvent and pigment or dye), modern inks may contain many additional ingredients such as resins, lubricants, solubilizers, surfactants, biocides, thickeners, pH modifiers, drying retardants, defoamers, surfactants and other materials.[48]

Comprehensive literature concerning the different types of inks and their different applications such as writing, offset printing, gravure printing, screen printing, security printing, *etc.*, can be found in the section on further readings (Section 8.9).

8.5 Dye and Pigment Degradation

Dyed fibres are very often sensitive to light, pH, washing and other causes of colour deterioration. If the colour resists these constraints, it may be referred to as a fast colour. If it changes rapidly or does not settle well, on the other hand, it is termed a fugitive colour. Flavanthrone yellow (PY24, C.I. 70600) for example tends to darken upon light exposure at masstone.[‖] [5]

[‖]The full colour pigment or coating, when applied thickly.

George Field explored and warned against the degradation of many synthetic and organic pigments in the mid-19th century:

> *"Without a strong constitution there is no hope for [painting]; no chemistry can strengthen the sickly frame, restore the faded* colour, stop the ravages of consumption: science stands helpless before dying Art".[12,51]

Padfield and Landi[49] also explored in the 1960s the light fastness of natural dyes (*e.g.* Brazilwood, Madder or Cochineal) on different fabrics such as wool, silk and cotton. Many other studies have been made since on this topic. Kohara *et al.*[50] report the degradation and colour fading of dyed fabrics according to the dye and mordant used. The effects of iron *vs.* alum mordants on the fading and the fabric tensile properties are detailed in this study.

A great number of discoveries have recently been made about the relationship between the chemical composition of organic pigments or dyes and their lightfastness or weatherfastness. The number of electron acceptors and their environment in the molecular structure of pigments and dyes is a key point in this kind of degradation. The number and nature of substitutions in the carbon skeleton, or the kind of metal involved in forming complexes, is also an important factor.[1]

Lomax and Learner[5] note that many early synthetic organic pigments or dyes are not particularly lightfast and that their identification in a work of art can provide information about the original appearance of the work, or raise a concern about its display.

Inorganic pigments for their part have long been considered extremely stable regarding lightfastness or weatherfastness. However, this rule is no longer absolute and certain pigments appear to be unstable in particular with light or in reaction with other materials. Indeed, Pliny[33] and later on Cennini[52] had already noted the poor stability of red lead when used on wet frescos:

> *"A colour such as red lead (...) is only good to work on panel, if you use it on a wall, it will rapidly darken when exposed to air and losses its colour".*[52]

The darkening of cinnabar was also known since Antiquity and Vitruvius advises coating frescos with wax in order to avoid this phenomenon.[39]

Many recent studies are now dedicated to the degradation of inorganic pigments in works of art.

8.6 Analytical Techniques Appropriate for Dyes and Pigments

The analytical techniques suitable for the study of pigments and dyes are to a certain extent dependant on the chemical nature of these materials. Some techniques are more efficient with organic materials (*e.g.* PyGC–MS, FTIR, HPLC), some with inorganic materials (*e.g.* PLM, XRF, XRD), and a few are applicable to both (*e.g.* FTIR, Raman) (see Chapter 2).

A specific chapter in this book (Chapter 2) is dedicated to the analytical techniques used in the conservation and cultural heritage field. We will just focus in this section on the techniques that are more appropriate to pigments and dyes.

A. Verhecken[53] reports the evolution of dyes analyses until the 20th century. The author differentiates predominantly utilitarian analyses (mainly employed by industry and trade dye professionals) from academic analyses. According to the author, in utilitarian analyses, only the class of the dyes is researched (*e.g.* substantive, acid, basic, mordant or vat) while in academic analyses eventually the exact molecular constitution of the dyes chromophores is investigated. The author also lists a comprehensive list of publications of technical and analytical studies made on dyes alone, on various supports, or in complex mixtures.

Table 8.1 reports some of the most common analytical techniques used for the study of pigments and dyes. The information collected, the advantages and drawbacks of each of these techniques along with the reference studies are listed in Table 8.1.

8.7 Case Study: Pigments and Dyes in Le Corbusier Decorative Art and Architectural Works

We have chosen to illustrate this chapter with a case study that includes the two categories of pigments discussed earlier, namely organic and inorganic. The case study concerns the material used by Le Corbusier (Charles-Édouard Jeanneret; 1887–1965) in his furniture and architectural works.

8.7.1 The Context

The full project concerns buildings built by Le Corbusier (*e.g.* Immeuble Molitor in Paris, Unités de Camping, Baraque-atelier in Roquebrune-Cap-Martin), wall paintings, furniture, wooden shutters, *etc.*

Table 8.1 Some of the most common analytical techniques used for pigment and dye analyses. References are given as an example.

Methods	Advantages	Drawbacks	Best spatial resolution	Information collected	References
Microchemical tests	Versatility. It can be applied to inorganic, organic and complex substances Very sensitive	Very few specific existing tests Most of them are for inorganic ions rather than organic functional groups or compounds	—	Chemical composition (organic & inorganic)	64–66
Optical microscopy	Colour identification	Smooth surface required	0.5 μm	Crystallinity and physical properties of the pigments Distribution in various vehicles	1, 2, 67
SEM-EDS	Large depth of field Large magnification Local composition	Absence of colour (Very) light elements not measured	>nm	Crystallinity and Chemico-physical properties of the pigments Distribution in various vehicles	68–70
TEM	High resolution	Sample preparation Destructive Time-consuming	0.1 nm	Structure, phase, composition	68
TLC	Very little equipment is used Very little time for separation Sensitive	Plate length is limited Takes place in an open system sample and may be affected by humidity and temperature	—	Structure, molecular identification	71–74
HPLC	Versatile and extremely precise Small internal volume Accuracy	Cost Complexity	—	Structure, molecular identification	75–78

Fluorescence	Spectral and Imaging technique	Not quantitative Mixtures are complex to interpret	—	Composition and characterisation of organic and inorganic compounds	78–81
Py-GC	Reduced time analysis No sample pre-treatment Small sample amount required Analysis of complex matrices	Destructive technique	—	Qualitative and quantitative analysis Structure and composition	82–86
DTMS	Sensitive to Wider range than Py-GC-MS Particularly sensitive to polar or material with high boiling point (*e.g.* many organic pigments)	destructive technique No Chromatography stage, Mass spectra can become overlaid	—	Structure and composition	86, 87
Laser desorption mass spectrometry (MALDI)	Soft Ionization technique (no molecule fragmentation), High molecular weight analyte can be ionized Molecule need not be volatile Sub-picomole sensitivity easy to obtain Wide mass range, relatively unaffected by small amounts of salts and contaminants	MALDI matrix cluster ions Obscure low m/z species (<600) Signal reproducibility is low Coupling MALDI with chromatography can be difficult	—	Complementary technique to GC-MS for analysis of paint samples Identification of organic structures and compositions	88, 89
TOF Mass spectrometry	Fast acquisition rates, spectral continuity, and exceptional dynamic range Detection limit Good mass accuracy, High mass resolution, High mass range	Fast digitizers used in TOF can have limited dynamic range	—	Qualitative and quantitative analyses	90

Table 8.1 *(Continued)*

Methods	Advantages	Drawbacks	Best spatial resolution	Information collected	References
Mass spectroscopy	Small sample size Fast, Can differentiate isotopes Can be combined with GC-LC etc	Needs pure compounds Difficult with no volatile compounds Not be able to distinguish between isomers of a compound having the same charge-to-mass (*m*/*z*) ratio	—	Molecular formula	91–93
XRD	Simplicity of sample preparation Rapidity of measurement The ability to analyse mixed phases	Size limitations, do not interact strongly with light elements	>mm^2	Crystalline composition and structure	1,94–96
FTIR	Low cost Quantitative Versatile	Single beam while dispersive instruments generally have a double beam	>10 μm	Molecular composition	44, 97–100
Raman spectroscopy	Mobile Non-invasive No preparation Distinction between polymorphs	Not quantitative	0.5 μm	Molecular and crystalline composition	101–109
SERS effect	Sensitivity for organic compounds	Sample preparation Database Not quantitative	0.5 μm	Molecular and crystalline composition	110–116

This chapter focuses more specifically on some of the materials used in his Paris studio-apartment at 24 rue Nungesser and Coli, 75016, Paris, France (Figure 8.10).

In 1934 Le Corbusier and his wife Yvonne moved into the two top floors of the Molitor building, which was designed by himself and Pierre Jeanneret (cousin and partner of Le Corbusier). Le Corbusier lived in this apartment until his death in 1965.[54–57] Its interior design and polychromy presents a particularly interesting case study of the evolution of the architect's style,[58,59] a kind of permanent "work in progress", which echoes to other projects (*e.g.* La Chapelle Notre-Dame-du-Haut, Ronchamp, France, or Chandigarh, India).

Since the 1930s, Le Corbusier had been developing colour swatches with the Salubra company (a Swiss wallpaper manufacturer)[60,61] under the name of "peinture à l'huile en rouleaux". He attached particular importance to colour in architecture and worked with different brands and types of paintings at the time, which have been identified in different projects (see below).

In the 1950s, he regularly bought Berger's paints (a British *paint* manufacturer) in the Matroil and Matone ranges (used, for example, in the Cabanon, Roquebrune Cap-Martin, the "Unités de Camping", Roquebrune-Cap-Martin, and Unité d'habitations, Marseille); these are matt exterior paints, the former based on emulsion oil and the latter on polyvinyl paints.

The apartment nowadays presents four "historic" states and five redecorations since the death of Le Corbusier; post-1965 repairs are clearly visible in cross-section but will not be detailed within this chapter. The main phases are as follows:

- 1934: Use of neutral and light tones, some touches of colour in the room.

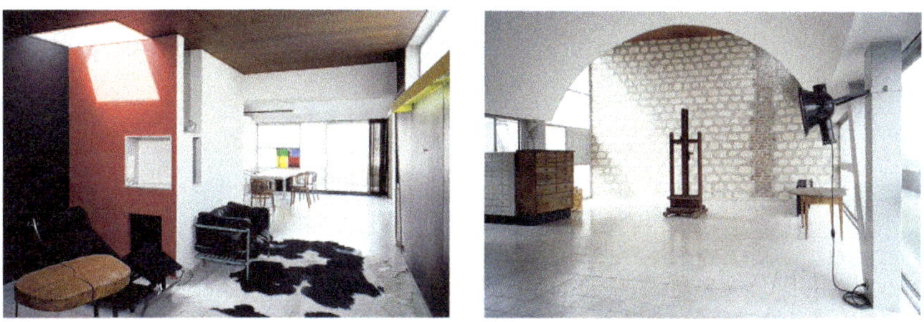

(a) **(b)**

Figure 8.10 (a and b) Photographs of Le Corbusier apartment 24 rue Nungesser and Coli, 75016, Paris, France. Ph. OMG/FLC/ADAGP.

- c. 1939: Contribution of wood siding, first repainting.
- 1946–1950s: Colourful profusion of paint.
- 1965: Last historical state. Colours change to more acidic colours, reduction of the palette and return of clear and neutral colours for some rooms and objects.

The repairs, completed by the Foundation Le Corbusier, have simplified the colours and moved away from the painter's original palette.

The overall purpose of the actual study was therefore to reveal the different historical states of the interior painting (walls and furniture) and restore the interior paintings at the last state wanted by Le Corbusier in 1965.

8.7.2 Analytic Means

The analytical techniques used in this project were X-ray fluorescence (XRF), micro-FTIR, Raman microscopy and SEM-EDX.

The portable XRF spectrometer used was a NITON GOLDD (SDD detector-silicon drift detector-Ag tube, 15-50 KV, 40 μA). This instrument has an integrated camera to visualize the area of analysis.

The FTIR apparatus used consisting of a Nicolet Continuum microscope coupled with a Nexus FTIR spectrophotometer (Thermo-Nicolet); measurements were recorded in transmission.

The Raman microscopy analyses were carried out using a JY Labram HR800 spectrometer, with an RL 785 nm HPNIR laser. The acquisition times were 60 sec with 2 to 3 accumulations over a spectral range of 100 to 4000 cm^{-1}. The spectra were calibrated with a sample of silicon crystal (main band at 520.5 cm^{-1}).

The SEM-EDX apparatus used in this study was a JSM-840A (Oxford) Analytical LZ-5 coupled to a Tetra detector. The analysis conditions used were 10 to 15 kV and 3×10^{-10} A. The spectra were processed with ISIS software.

8.7.3 Results

A first exhaustive stratigraphic survey was undertaken in 2014,** with the aim of documenting the successive layers of paints and proposing

**Campaign supported by the Getty Foundation, the French Ministry of Culture and founded by the "Keeping It Modern" grant.

a faithful restitution of the polychromies to the last historical state of the apartment-studio of Le Corbusier in 1965.

However, after comparison of the surveys with historical data (archives of the Foundation Le Corbusier, Paris)[62] some doubts or uncertainties remained.

In order to address these uncertainties, several types of examination and analysis were recently undertaken:

- A comprehensive survey with conventional stratigraphy in the apartment, in order to confirm or deny the number of layers (especially for layers of identical colours but of different compositions).
- Chemical analysis in order to establish the exact nature of each layer (pigment, fillers and binder) for dating or to correlate similar layers found in different locations.
- Understanding the ageing of certain "historical" layers in order to refine the interpretation of the original colour (we observed for instance certain historical oil-based layers that change colour after re-exposure to daylight) (Figure 8.11).

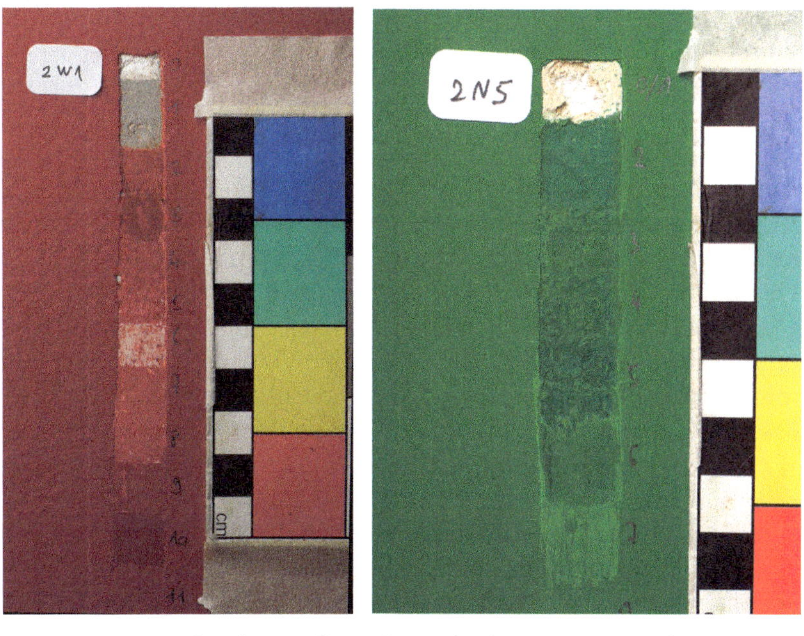

Figure 8.11 Example of soundings from the living room wall in Le Corbusier's apartment (left 2K/2W1; right C10/2N5).

Table 8.2 lists some of the selected samples analysed in this project. The results report the organic and inorganic pigments characterized in the different layers. The pigments are listed according to their elemental or molecular characterization during the study (*e.g.* Titanium white or calcium salts for EDS characterization, and rutile or calcite for Raman microscopy analysis).

The listed layers for each sample in Table 8.2 are correlated to the different historical or repair painting campaigns.

8.7.3.1 White and Yellow Pigments

The white of the ceilings and walls (7F) corresponds to the original 1934 painting scheme. It is a mixture of barite and zinc white with linseed oil in the first layer and a glue/oil mixture in the second (see Figure 8.12). Despite the slight ochre appearance, no iron could be detected, even after thorough analysis by SEM–EDS.

The analysis of the paint on the outside wall (EVN2) revealed a yellow ochre composition (see Table 8.2) and confirmed the original off-white colour of the wall, rather than pure white. The analysis also confirmed the presence of a silicate binder, as had been documented at the time.

The yellowish colour of the door (C13) is mainly related to a minor amount of yellow ochre observed in certain layers, but moreover to phenomena of yellowing of the oil medium.

8.7.3.2 Black, Grey and Brown Pigments

The umber earth (brown) of the doors (7E) also corresponds to the original paints (1934) and is mainly composed of iron oxides, specifically haematite and magnetite (see Figure 8.13).

The brown of the north wall (3A) of the living room is made of haematite and barite. This brown layer dates from 1934 and remained in place until the early 1960s. The upper layers correspond to repainting campaigns after the death of Le Corbusier in 1965.

The first greys (mainly on an original wooden stick) on the south wall of the living room (2H) date from 1934 and are based on barite and carbon black.

The black pigment on the living room wall (2L) is attributed to the ~1965 repainting campaign and is primarily carbon black. The upper layers are also carbon black based pigments (with minor phthalocyanines or fillers) applied after the death of Le Corbusier in 1965.

Table 8.2 Composition of some of the architectural and furniture pigments characterized in Le Corbusier's apartment in Paris.[a]

Upper layer Colour	Location	Samples #	Material	Painting campaign	Pigments identified
Whitish	Outdoor walls	EVN2	Wall painting	c. 1934	Silicates, Aluminosilicates, Calcium salts, Iron oxides
Whitish	Indoor walls	7F	Wall painting	c. 1934	Barite, Zinc white
Yellowish	Door	C13	Wood painting	1969	Titanium white (rutile), Barite, Calcite
				1965	Titanium white (rutile), Barite, Calcite, Iron oxides
				1950	Titanium white (anatase), Barite, Calcite, Zinc white
				1950	Barite, Calcite
				1934	Barite, Zinc white, Aluminosilicates, Iron oxides, Lead oxides, Silicates
Brown	Doors	7E	Wood painting	Post 1965	Phthalocyanine Blue (PB15:4), Calcite, Aluminosilicates
				Post 1965	Iron oxides (haematite), Phthalocyanine Blue (PB15:4)
				Post 1965	Phthalocyanine Blue (PB15:4), Carbon black, Aluminosilicates, Titanium white
				1934	Iron oxides (haematite, magnetite)
				1934	Aluminosilicates, Iron oxides, Barite
Brown	Living room	3A	Wall painting	Post 1965	Titanium white (anatase), Calcite, Gypsum, Barite
				Post 1965	Calcite, Gypsum, Barite
				Post 1965	Calcite, Barite
				1934–1960	Haematite, Calcite, Barite, Gypsum
				1934	Calcite, Barite, Gypsum
				1934	Titanium white, Talc, Calcite, Barite
				1934	Titanium white (rutile), Calcite, Barite, Talc
Black	Living room	2L	Wall painting	Post 1965	Carbon black, Calcite, Barite
				Post 1965	Carbon black, Calcite, Phthalocyanine Blue (PB15:4)
				Post 1965	Carbon black, Calcite

Table 8.2 *(Continued)*

Upper layer Colour	Location	Samples #	Material	Painting campaign	Pigments identified
				Post 1965	Carbon black, Calcite
				Post 1965	Carbon black, Barite
				Post 1965	Carbon black, Calcite
				Post 1965	Phthalocyanine Blue (PB15:4)
				1965	Carbon black, Calcite, Barite
				1958	Carbon black
				1934	Barite
Grey	South wall living room	2H	Wood painting	1934	Barite
					Carbon black
Grey	Door	7A	Wood painting	Post 1965	Iron oxides, Calcite
				Post 1965	Titanium white (rutile), Calcite, Talc
				Post 1965	Carbon black, Calcite, Iron oxides, Phthalocyanine Blue (PB15:4)
				1965	Barite, Calcite, Ultramarine blue, Haematite, Carbon black
				1962	Carbon black, Calcite, Magnetite
Red	Living-room wall	2K	Wall painting	Post 1965	Pigment Red 3
				Post 1965	Pigment Red 3, Pigment Red 4
				Post 1965	Titanium white (rutile), Calcite
				1965	Pigment Red 112, Barite
				1950	Pigment Red 4
				1948	Pigment Red 104, Barite
					Pigment Red 104, Barite
				Und.	Gypsum
Red	Outdoor walls (Balcony)	C1	Wall painting	1965	Titanium white (rutile), Pigment Red 112, Pigment Red 3, Calcite, Talc

Green	Office wall	C7	Wall painting	1965	Phthalocyanine Blue (PB15:3), Phthalocyanine Green (PG7)
				1939	Phthalocyanine Green (PG7), Phthalocyanine Blue (PB15:3)
				1939	Barite, Calcite
Green	Living room wall	C10	Wall painting	1965	Phthalocyanine Green (PG7), Phthalocyanine Blue (PB15:3)
				1965	Phthalocyanine Blue (PB15:3)
				Poss. 1948	Phthalocyanine Green (PG7)
				Poss. 1939	Phthalocyanine Green (PG7)
				1934	Barite, Calcite
Blue	Shower	4C	Wall painting	1950	Barite, Ultramarine blue
				1934	Barite, Prussian Blue
Blue	Hairdresser	C11	Wood painting	Und.	Calcite, Barite, Zinc white
				1950	Ultramarine blue, Calcite, Barite, Zinc white, Aluminosilicates
				Und.	Barite, Zinc white, Calcite
				Und.	Zinc white
				Und.	Barite, Zinc white, Carbon black, Quartz, Calcite
				Und.	Titanium white, calcium salts, Zinc white
				Und.	Barite, Zinc white, Calcite, Carbon black, Titanium white (anatase)
				Und.	Titanium white (anatase), calcium salts, Zinc white
Blue	Outdoor walls (Balcony)	C2	Wall painting	Poss. Post 1965	Aluminosilicates, Titanium white (rutile), Calcite
				Poss. 1960–1965	Barite, Phthalocyanine Blue (PB15:4)
				Und.	Barite, Zinc white
				Und.	Aluminosilicates, Calcium salts
				Und.	Barite, Zinc white, Aluminosilicates, Iron oxides, Carbon black
				Und.	Barite, Zinc white, Carbon black
Blue	Bedroom	C12	Painting on metal	Poss. 1950	Ultramarine blue, Calcium salts, Barite, Zinc white, Potassium & aluminium silicates

[a]Und.: Undetermined date; Poss.: Most probable date, Post: Later than.

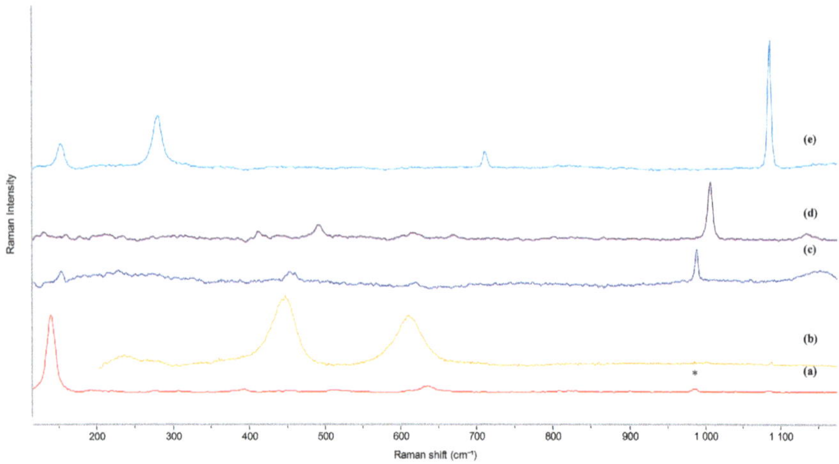

Figure 8.12 Raman spectra of white and yellowish pigments detected in Le Corbusier's apartment. Raman bands reported in Table 8.2; (a) Anatase (#C13), (b) Rutile (#C1), (c) Barite (#C13), (d) Gypsum (#3A) and (e) Calcite (#2L"). * Refers to other minor phases, *e.g.* barite, gypsum or calcite.

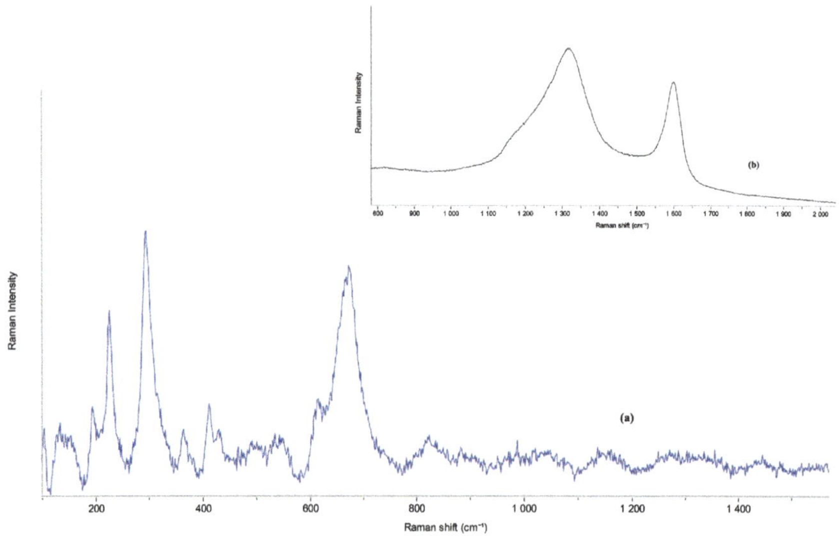

Figure 8.13 Raman spectra of black and brown pigments detected in Le Corbusier's apartment. Raman bands reported in Table 8.2; (a) Anatase (#C13), (b) Rutile (#C1), (c) Barite (#C13), (d) Gypsum (#3A) and (e) Calcite (#2L"). * Refers to other minor phases, *e.g.* barite, gypsum or calcite.

8.7.3.3 Red Pigments

The red of the wall (2K) is particularly well characterized and differentiated during the examinations of the cross-section (Figure 8.14).

The first red corresponds to the 1946 "Matroil" and is based on molybdenum red, PR104 (C.I. 77605).

The presumed 1950 red repaint is based on an organic synthetic pigment belonging to the mono-azo family, PR4 (Permanent Red R, C.I. 12085) (as with Matroil paints, this paint is based on an oil-emulsion, its final aspect is matt and friable).

PR3 (Naphthol Red AS, C.I. 12370) and PR112 (Toluidine Red, C.I. 12120) correspond to a paint layer applied in 1965 and are also observed on several sections of the wall on the balcony (C1) (see Figure 8.15).

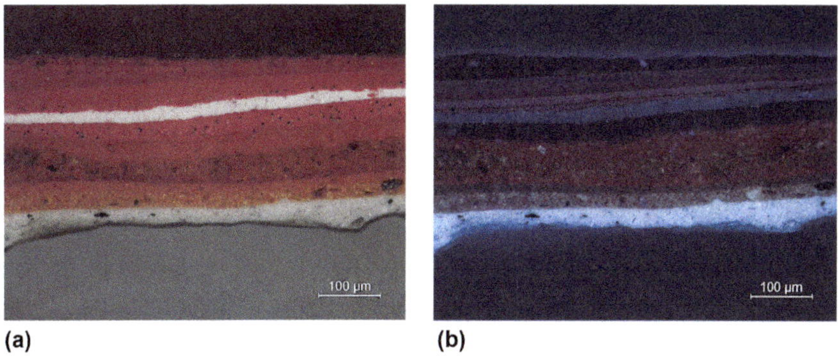

Figure 8.14 Cross-sections of sample 2K from the sounding 2N1 of Le Corbusier's apartment under visible (a) and UV (b) light.

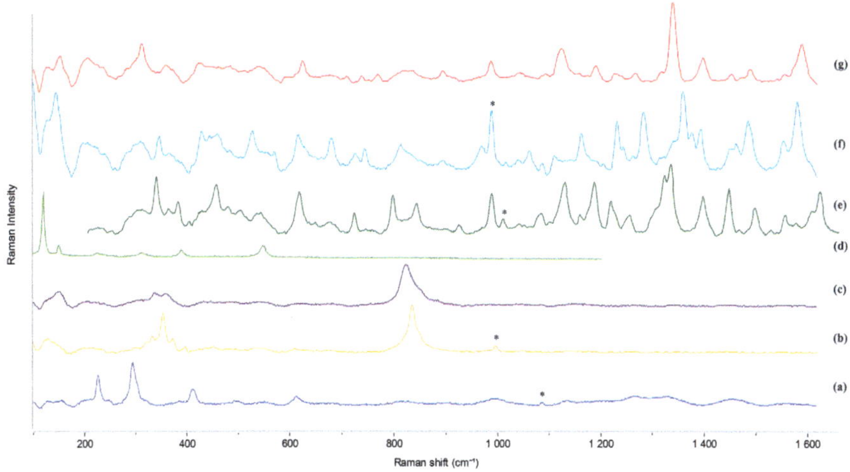

Figure 8.15 Raman spectra of red pigments detected in Le Corbusier's apartment. Raman bands reported in Table 8.2; (a) Haematite (#3A), (b) Chrome orange (#P6b), (c) Molybdate orange (#2K), (d) Minium (#C5), (e) PR3 (#2K), (f) PR112 (#2K) and (g) PR4 (#2K). * Refers to other minor phases, *e.g.* barite, gypsum or calcite.

8.7.3.4 Green Pigments

As for the greens, first layers of phthalocyanine green (PG7) are detected both in the salon niche (C10) and on the desk office (C7). This green is similar in composition to the Matroil colours analysed on other projects of Le Corbusier and in particular his Cabanon from the 1950s.[63]

The upper layer of these two green samples contains, in addition to PG7, minor amounts of phthalocyanine blue (PB15:3). These layers correspond to 1965 repainting campaigns but are not further documented.

8.7.3.5 Blue Pigments

The dark blues in the room (aerator, C12, and hairdresser, C11) are made of ultramarine blue (PB29, C.I. 77007). This colour is also very characteristic in Le Corbusier projects. These paints were found on very small painted surfaces in the apartment and they echo the documented use of Salubra blue wallpapers of the same hue for a wall of the room.

These ultramarine blues are part of the 1934 and 1950 painting campaigns.

Another similar blue is identified on a very altered part of the balcony wall (C2), but this paint is covered by a more recent paint made of phthalocyanine blue and attributed to the early 1960s campaign (see Figure 8.16).

The clear blue of the shower room and the bathroom (4C), despite their yellowish appearance during sampling, are real blue pigments (not a mixture of yellow and green as presumed). They were characterized as a succession of ultramarine blue and Prussian blue corresponding to two different painting campaigns (see Table 8.3).

8.7.4 Conservation Issues

As part of the complete restoration (2014–2018) of the apartment-studio of Le Corbusier, the Foundation, the owner of this place, wished to restore the interior paintings at the last state wanted by Le Corbusier in 1965. The very exhaustive work done by the restorers and the scientists was therefore essential in order to understand the different layers of paintings on walls and all other objects of the apartment (bed, table de nuit, wooden curtains, placards, *etc.*).

The first objective was to determine each layer of paint chosen by Le Corbusier between 1934 and 1965, and also all later on layers applied after 1965 during the Foundation intervention. This phase was made possible by conducting survey stratigraphies, which were confronted

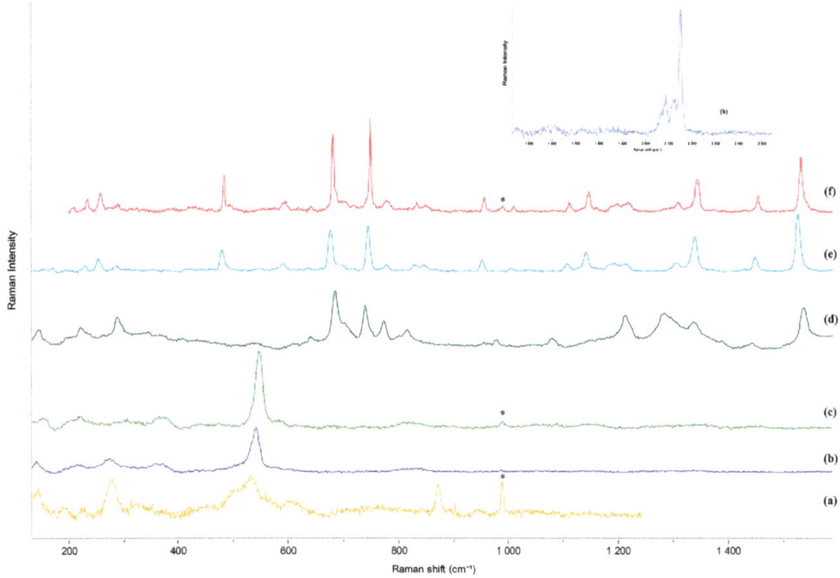

Figure 8.16 Raman spectra of green and blue pigments detected in Le Corbusier's apartment. Raman bands reported in Table 8.2; (a) Zinc chromate (#Ep2), (b) Prussian blue (#P2-2), (c) Ultramarine blue (#7A), (d) PG7 (#1A), (e) PB15-4 (#2L'') and (f) PB15-3 (#C10). The inset shows the high wavenumber bands of (b) Prussian blue (#P2-2). * Refers to other minor phases, *e.g.* barite, gypsum or calcite.

in particular with the archival documents, in order to be able to correlate colours and dates.

From this first understanding of the "historical" layers and "restoration" layers, the objective was therefore to identify and restore the layer of 1965. The choice of this date is due to the fact that the state of the interior masonry did not allow preserving the earlier historical layers. Nevertheless, all the surveys carried out by the conservators on the site remain visible to the public after the restoration of the apartment.

Once the different layers were identified, it was important to be able to characterize them using physicochemical analyses, in order to understand precisely the different industrial materials used (pigments, fillers and binders) and obtain the closest restitution to the original appearance in 1965.

As an example, the blue hues of the bathroom and the shower appeared indeed to be blues and not greens; the white hue on the outside wall of the shower is in fact a white that has yellowed and not a pale ochre.

New colours will be referenced to be added to the colour chart of Le Corbusier for the planned restoration, in order to get closer to the

Table 8.3 Raman wavenumbers of the different pigments reported in Figures 8.10 to 8.14 and detected in the architecture and on the furniture of Le Corbusier's apartment in Paris.

CI Common Name	CI Generic Name	CI Constitution Number	Chemical or mineral Name	Chemical or mineral Composition	Fig.	Raman wavenumbers (cm^{-1})	References
PW6	Titanium White	77891	Anatase	TiO_2	10	144, 396, 517, 639	108–121
PW6	Titanium White	77891	Rutile	TiO_2	10	233, 445, 610	108–121
PW21/22	Barium Sulfate	77120	Barite	$BaSO_4$	10	153, 455, 619, 987	108–121
PW25	Gypsum	77231	Gypsum	$CaSO_4 \cdot 2H_2O$	10	130, 412, 481, 617, 668, 1007, 1136	108–121
PW18	Chalk	77220	Calcium carbonate (Calcite)	$CaCO_3$	10	151, 278, 712, 1085	108–121
n/a	Magnetite	77491 77492 77499	Ferric oxide	Fe_3O_4	11	193, 306, 538, 670	108
PBk6	Carbon Black	77266	Carbon	C	11	~1318, 1593	121
PR101	Iron Oxide Red	77491	Haematite	Fe_2O_3	13	224, 291, 299, 411, 496, 610	108–121
PR103	Chrome Orange	77601	Lead chromate	$PbCrO_4 \cdot PbO$	13	336, 358, 374, 835	121
PR104	Molybdate Orange	77605	Molybdated lead chromate	$PbCrO_4$, $PbMoO_4$, $PbSO_4$	13	340, 824	106
PR105	Red Lead	77578	Minium	Pb_3O_4	13	60, 84, 120, 150, 223, 313, 391, 480, 549	121
PR3	Toluidine Red	12120	Beta naphthol toluidine	$C_{17}H_{13}N_3O_3$	13	225, 340, 383, 455, 504, 542, 617, 676, 723, 759, 77, 842, 924, 985, 1040, 1057, 1076, 1127, 1186, 1217, 1256, 1333, 1395, 1445, 1470, 1496, 1526, 1554, 1604, 1621	122
PR112	Naphthol Red AS-D	12370	Monoazo; naphthol	$C_{24}H_{16}C_{13}N_3O_2$	13	148, 191, 346, 389, 429, 442, 467, 527, 571, 616, 680, 725, 745, 813, 865, 893, 967, 1015, 1039, 1061,	108–122

						1109, 1161, 1205, 1230, 1243, 1260, 1282, 1314, 1332, 1358, 1375, 1392, 1448, 1461, 1484, 1526, 1552, 1579	
PR4	Permanent Red R	12085	Chlorinated para-nitroaniline (Beta naphthol)	$C_{16}H_{10}ClN_3O_3$	13	208, 313, 340, 356, 403, 420, 466, 514, 593, 624, 654, 709, 737, 769, 838, 893, 942, 986, 1041, 1093, 1124, 1156, 1189, 1224, 1239, 1266, 1288, 1338, 1396, 1451, 1487, 1554, 1587	122
PY36	Zinc Yellow	77955	Zinc chromate	$ZnCrO_4$	14	872, 941	123
PB27	Prussian Blue	77510	Iron(III) hexacyanoferrate(II)	$C_{18}Fe_7N_{18}$	14	277, 539, 2102, 2151	108–123
PB29	Ultramarine Blue	77007	Polysulfide of sodium, potassium or lithium alumino-silicate	$Na_{8-10}Al_6Si_6O_{24}S_{2-4}$	14	548	108–123
PG7	Phthalocyanine Green BS	74260	Polychlorinated copper phthalocyanine	$C_{32}H_3Cl_{13}CuN_8$ to $C_{32}HCl_{15}CuN_8$	14	642, 685, 740, 772, 820, 979, 1082, 1187, 1200, 1214, 1283, 1338, 1445, 1538	108–122
PB15-4	Phthalocyanine Blue NCF	74160	Beta Copper phthalocyanine	$C_{32}H_{16}CuN_8$	14	257, 287, 382, 421, 482, 547, 594, 639, 679, 746, 776, 829, 848, 951, 1006, 1037, 1107, 1142, 1190, 1216, 1305, 1336, 1426, 1448, 1525	108–122
PB15-3	Phthalocyanine Blue BGS	74160	Beta Copper phthalocyanine	$C_{32}H_{16}CuN_8$	14	258, 483, 594, 641, 681, 719, 747, 781, 832, 849, 954, 1008, 1109, 1144, 1195, 1217, 1307, 1341, 1429, 1451, 1530	108–122

"historic" shades. The characterization of the pigments and binders has led the project management team to restore oil emulsions for all paintings in the apartment.[††]

Acknowledgements

The practical study is done with the support and participation of Jean-François Dedomenici (Nuances minerals; Switzerland).

The analyses were principally completed in the ITODYS (UMR7086) infrastructure in Paris. The authors would also like to thank Mary Inglis and David C. Smith for their support.

Recommended Reading

M. De Keijzer, The History of Modern Synthetic Inorganic and Organic Artists' Pigments, in *Contributions to Conservation: Research in Conservation at the Netherlands Institute for Cultural Heritage*, ed. A. Jaap, ICN Instituut Collectie Nederland, Amserdam, 2002, p. 42.

N. Eastaugh, W. Valentine, T. Chaplin and R. Siddall, *Pigments Compendium*, Routledge, London, 2008.

R. J. Gettens and G. L. Stout, *Painting Materials: A Short Encyclopedia*, Dover Publications Inc., New York, 2000.

W. Herbst and K. Hunger, *Industrial Organic Pigments: Production, Properties, Applications*, Wiley-VCH Verlag GmbH, Weinheim, 2004.

T. Jesionowski and F. Ciesielczyk, Pigment, Inorganic, in *Encyclopedia of Color Science and Technology*, ed. M. R. Luo, Springer, New York, 2014, p. 1.

S. Q. Lomax and T. Learner, A Review of the Classes, Structures, and Methods of Analysis of Synthetic Organic Pigments, *J. Am. Inst. Conserv.*, 2006, **45**, 107.

K. Venkataraman, *Chemistry of Synthetic Dyes*, Academic Press Inc., New York, 1971.

References

1. W. Herbst and K. Hunger, *Industrial Organic Pigments: Production, Properties, Applications*, Wiley-VCH Verlag GmbH, 2004.
2. N. Eastaugh, W. Valentine, C. Tracey and S. Ruth, *Pigment Compendium*, Routledge, 2008.

[††]The new pigments (paints) are made by M.J.F. Dedominici, Nuances Minérales, Switzerland.

3. R. J. Gettens and G. L. Stout, *Painting Materials: A Short Encyclopedia*, New York, 2000: Dover Publications Inc.
4. K. Hunger, *Industrial Dyes: Chemistry, Properties, Applications*, John Wiley & Sons, 2003.
5. S. Quillen Lomax and T. Learner, A Review of the Classes, Structures, and Methods of Analysis of Synthetic Organic Pigments, *J. Am. Inst. Conserv.*, 2006, **45**(2), 107–125.
6. M. Pastoureau, La couleur et l'historien, in *Pigments et colorants de l'Antiquité et du Moyen âge: Teinture, peinture*, CNRS Editions, Paris, 2002, pp. 21–40.
7. R. Halleux, Pigments et colorants dans la Mappae Clavicula, in *Pigments et colorants de l'Antiquité et du Moyen âge: Teinture, peinture*, CNRS Editions, Paris, 2002, pp. 173–180.
8. J. A. Chaptal, *Ann. Chim.*, 1809, **22**, 70.
9. H. Davy, *Phil. Trans.*, 1815, **105**, 97–124.
10. M. Farnsworth, *J. Chem.*, 1951, **28**, 72.
11. G. Agricola, *De re metallica*, 1556.
12. P. Ball, *Bright Earth: The Invention of Colour*, University of Chicago Press, 2008.
13. R. Fuchs and D. Oltrogge, *Pigments et colorants de l'Antiquité et du Moyen âge: Teinture, peinture*, CNRS Edition, Paris, 2002, pp. 309–323.
14. C.-E. Guignet, *Fabrication des Couleurs*, Vve Ch. DUNOD, Paris, 1888, vol. Tome X.
15. Winston-&-Newton, Personal Communication, 2009.
16. *Society of Dyers and Colourists, Color Index*, 3rd edn, Bradford, UK, 1971, vol. 4.
17. E. E. Jaffe, Pigments, Organic, in *Encyclopedia of Chemical Technology, Inc.*, John Wiley & Sons, 1996, vol. 19.
18. C. Graebe and C. Liebermann, Ueber Alizarin and Anthracen, *Berichte der Deutschen Cemischen Gesellschaft, II*, 1868, pp. 49–51.
19. R. Pfister, *Teinture et alchimie dans l'Orient hellénistique*, Institut Kondakov, Prague, 1935, p. 59.
20. V. Guichard and B. Guineau, Identification de colorants organiques naturels dans des fragments de peintures murales de l'Antiquité, in *Pigments et colorants de l'Antiquité et du Moyen âge: Teinture, peinture*, CNRS edn, 2002, Paris, pp. 21–40: pp. 245–254.
21. M. Balter, Clothes Make the (Hu) Man, *Science*, 2009, **325**, 1329.
22. E. J. W. Barber, *Prehistoric Textiles*, 1993, 504.
23. J. A. Goodwin, *Dyer's Manual*, Pelham, 1982.
24. H. C. Bhardwaj and K. K. Jain, Indian Dyes and Industry During 18th-19th Century, *Indian J. History Sci.*, 1982, **17**, 70–81.
25. M. Dubrunfaut, *Bulletin des sciences technologiques-Arts Chimiques*, Soc. pour la Propagation des Connaissances Sc. et Ind., Paris, 1828, vol. IX.
26. W. H. Perkin, On mauveine and allied colouring matters, *J. Chem. Soc., Trans.*, 1879, 717–732.
27. H. Zollinger, *Color Chemistry: Syntheses, Properties, and Applications of Organic Dyes and Pigments*, Wiley VCH, 3rd edn, 2003.
28. J. Kirby, C. Cooksey, A. Quye and M. R. van Bommel, *Dyes in History and Archaeology: Including Papers Presented at the 20th Meeting*, Held in the Instituut Collectie Nederland, Amsterdam, 2001, vol. 20, p. 200.
29. N. N. Mahapatra, *Textile Dyes*, WPI Publishing, 2016.
30. *International Agency for Research on Cancer*, General Introduction to the Chemistry of Dyes, Lyon, France, 2010, vol. 99.
31. M. Saito, M. Uehara and K. Yamazaki, Components of mordants and the silk fabrics treated with them, *Dyes in History and Archaeology 2.0*, 2005, 95–101.

32. A. G. Perkin and A. E. Everest, *The Natural Organic Coloring Matters*, London, 1918.
33. Pliny, *Natural History*, XXXV.
34. Westlake Knits & Fibres, (last access April 2019) *Westlake Knits & Fibres*, www.westlakeknits.ca.
35. R. J. H. Clark, C. J. Cooksey, M. A. M. Daniels and R. Withnall, Indigo, Woad, and Tyrian Purple: Important vat Dyes from Antiquity to the Present, *Endeavour*, 1993, **17**, 191–199.
36. C. J. Cooksey and R. S. Sinclair, *Dyes in History and Archaeology, 2.0*, Archetype Publications, London, 2005, vol. 20, pp. 127–135.
37. T. Jesionowski and F. Ciesielczyk, Pigment, Inorganic, *Encyclopedia of Color Science and Technology*, 2014, 1–21.
38. F. Delamare, Les pigments à base d'oxyde de fer et leur utilisation en peinture pariétale et mural, *PACT*, 1987, **17**, 333–344.
39. Vitruve, De Architectura, p. Book IX.
40. E. B. Faulkner and R. J. Schwartz, *High Performance Pigments*, 2nd edn, 2009, Wiley-VCH.
41. K. Venkataraman, *Chemistry of Synthetic Dyes*, Academic Press Inc, 1972, vol. 5.
42. M. De Keijzer, The history of modern synthetic inorganic and organic artists' pigments, *Contributions to Conservation: Research in Conservation at the Netherlands Institute for Cultural Heritage (ICN Instituut Collectie Nederland)*, 2002, pp. 42–54.
43. M. Doerner, *The Materials of the Artist and Their Use in Painting*, Harcourt Brace International, 1984.
44. M. R. Christie, *Pigments: Structures and Synthetic Procedures*, Oil and Colour Chemists Association, London, 1993.
45. A. P. Lewis, *Pigment Handbook, Properties and Economics*, 2nd edn, 1988, John Wiley & Sons Inc, New York, vol. 1.
46. C. James and M. B. Cohn, *Old Master Prints and Drawings: A Guide to Preservation and Conservation*, Amsterdam University Press, Amsterdam, 1997.
47. É. Delange, M. Grange, B. Kusko and E. Menei, Apparition de l'encre métallogallique en Égypte à partir de la collection de papyrus du Louvre, *Revue d'égyptologie*, 1990, **41**, 213–217.
48. NIIR Board of Consultants & Engineers, *Modern Technology of Printing & Writing Inks*, Asia Pacific Business Press Inc., India, 2nd edn, 2015.
49. T. Padfield and S. Landi, The light fastness of the natural dyes, *Stud. Conserv.*, 1966, **11**, 181–196.
50. N. Kohara *et al.*,, Degradation and Color Fading of Cotton, in *Historic Textiles, Papers, and Polymers in Museums*, American Chemical Society, 2001, vol. 779, ch. 6, pp. 74–85.
51. G. Field, *Chromatography; or, A Treatise on Colours and Pigments, and of Their Powers in Painting*, London, C. Tilt, 1835.
52. C. Cennini, *Il libro dell'Arte*, 1390–1437.
53. A. Verhecken, A concise history of dye analysis, *yes in History and Archaeology 2.0*, 2005, 1–22.
54. J. Sbriglio, Immeuble, 24 N. C. et appartement Le Corbusier, Birkhauser, Paris-Basel, 1996.
55. G. Marino, *Les multiples vies de l'appartement-atelier du 24 NC: histoire, projet, méthode*, Rencontre FLC, 2017.
56. M.-J. Dumont, Article Immeuble Molitor, *Le Corbusier Plans*, Echelle-1/FLC, 2005.
57. S. Jumsai, *24 NC. Appartement de Le Corbusier à Paris*, Dessins de Sumet Jumsai, 2011.

58. J. Fillacier, *Recoloration (La). Essai sur la portée de la couleur dans le monde moderne*, Entreprise de peinture Esgépé d.l., Paris, 1951.
59. R. Papillault, *L'éblouissement et le contre-jour dans l'atelier du 24 N.C.* Rencontres FLC, 2005.
60. L. Corbusier, *Claviers de Couleurs*, Salubra, Bâle, 1931.
61. Le Corbusier, *Claviers de Couleurs*, Ed. Zürich, Salubra, 1959.
62. F. Graf and G. Marino, *Les multiples vies de l'appartement-atelier Le Corbusier*, ed. Cahiers du Tsam, PPUR, 2017, vol. 2.
63. CICRP, Analyses Le Corbusier, Internal Report ref DI06-27, 10 Sept. 2010.
64. M. de Keijzer, A microchemical analysis on synthetic organic pigments discovered in the 20th century, in *ICOM Committee for Conservation. 9th Triennial Meeting*, Dresden, Germany, 1990, pp. 220–225.
65. B. V. Kharbade and O. P. Agrawal, Analysis of Natural Dyes in Indian Historic Textiles, *Stud. Conserv.*, 1988, **33**, 1–8.
66. Z. C. Koren, Non-destructive vs. Microchemical Analyses: The Case of Dyes and Pigments, in *9th International Conference on NDT of Art*, Jerusalem, Israel2008, 25–30 May , pp. 1–10.
67. W. Mc Crone, Polarized Light Microscopy in Conservation: A Personal Perspective, *JAIC*, 1994, **33**, 101–114.
68. C. S. Palenik and S. Palenik, Seeing Color: Practical Methods in Pigment Microscopy, *Microscope*, 2014, **6**, 51–61.
69. J. Nissen, SEM/EDX-Analysis in Art History: A Portrait by Lukas Cranach the Younger, *Imaging Microsc.*, 2010, 1–2.
70. M. Schreiner, M. Melcher and K. Uhlir, Scanning electron microscopy and energy dispersive analysis: applications in the field of cultural heritage, *Anal. Bioanal. Chem.*, 2007, **387**, 737–747.
71. M. Home, R. Dudley and D. K. Laing, The discrimination of modern household paints using thin layer chromatography, *J. Forensic Sci. Soc.*, 1982, **22**, 147–154.
72. A. Tsukui, K. Kuwano and T. Mitamura, Anthocyanin Pigment Isolated from Purple Root of Sweet Potato, *J. Home Econ. Japan*, 1983, 153–159.
73. V. Aginsky, Forensic Examination of "Slightly Soluble" Ink Pigments Using Thin-Layer Chromatography, *J. Forensic Sci.*, 1993, **38**, 1131–1133.
74. M. F. Striegel and J. Hill, *Thin-Layer Chromatography for Binding Media Analysis*, Getty Conservation Institute, Los Angeles, CA, USA, 1996.
75. W. Wegener, W. C. Klamer, H. Govers and U. Brinkman, Determination of organic colorants in cosmetic products by high-performance liquid chromatography, *Chromatographia*, 1987, **24**, 865–875.
76. J. Wouters, C. M. Grzywacz and A. Claro, A Comparative Investigation of Hydrolysis Methods to Analyze Natural Organic Dyes by HPLC-PDA – Nine Methods, Twelve Biological Sources, Ten Dye Classes, Dyed Yarns, Pigments and Paints, *Stud. Conserv.*, 2011, **56**((3)), 231–249.
77. I. Karapanagiotis, S. Daniilia, A. Tsakalof and Y. Chryssoulakis, Identification of Red Natural Dyes in Post-Byzantine Icons by HPLC, *J. Liq. Chromatogr. Relat. Technol.*, 2005, **28**, 39–749.
78. M. R. van Bommel, I. VandenBerghe, A. M. Wallert, R. Boitelle and J. Wouters, High-performance liquid chromatography and non-destructive three-dimensional fluorescence analysis of early synthetic dyes, *J. Chromatogr. A.*, 2007, **1157**, 260–272.
79. R. E. de la Rie, Fluorescence of Paint and Varnish Layers (Part III), *Stud. Conserv.*, 1982, **27**, 102–108.
80. A. Pelagotti, L. Pezzati, A. Piva and A. Del Mastio, Multispectral UV fluorescence analysis of painted surfaces, in *14th European Signal Processing Conference*, Florence, 2006, pp. 1–5.

81. V. Vascotto and A. Pelagotti, A study of uv fluorescence emission of painting materials, in *8th International Conference on Non-destructive Investigations and Microanalysis for the Diagnostics and Conservation of the Cultural and Environmental Heritage*, Lecce, Italy, 2005, pp. Art 05.
82. N. Sonoda, J. P. Rioux and A. Duval, Identification des materiaux synthetiques dans les peintures modernes.2. Pigments organique et matiere picturale, *Stud. Conserv.*, 1993, **38**, 99–127.
83. N. Sanoda, Application des methods chromatographiques a la caracterisation des peintures alkyds pour artistes, *Techne*, 1998, **8**, 33–43.
84. N. Sanoda, *Stud. Conserv.*, 1999, **44**, 195–208.
85. T. Learner, The characterisation of acrylic painting materials and implications for their use, conservation and stability, in *The Characterisation of Acrylic Painting Materials and Implications for Their Use, Conservation and Stability*, PhD thesis, University of London, 1996.
86. T. Learner, *Analysis of Modern Paints*, Getty Conservation Institute, Los Angeles, 2004.
87. C. A. Menke, R. Rivenc and T. Learner, The use of direct temperature-resolved mass spectrometry (DTMS) in the detection of organic pigments found in acrylic paints used by Sam Francis, *Int. J. Mass Spectrom.*, 2009, **284**, 2–11.
88. N. Wyplosz, *Laser Desorption Mass Spectrometric Studies of Artists' Organic Pigments*, PhD diss., University of Amsterdam, 2003.
89. F.-D. Munteanu, N. Dinca and A. Cavaco-Paulo, Maldi-TOF Mass Spectrometry in Textile Industry, in *Applications of Mass Spectrometry in Life Safety*, Springer, Dordrecht, 2008, pp. 193–203.
90. J. J. Soltzberg, A. Hagar, S. Kridaratikorn and A. Mattson, MALDI-TOF mass spectrometric identification of dyes and pigments, *J. Am. Soc. Mass Spetrom.*, 2001, **2007**, 18.
91. B. Marino, T. Learner, J. Van der horst and J. Boon, A mass spectrometry study of the red Naphtol PR10 as pigment and as Acrylic emulsion Paint, *Dyes History Archaeol.*, 2005, **20**, 187.
92. C. A. da Silva Junior and P. A. Segura, Identification of Dyes in Pen Ink by High Resolution Mass Spectrometry: A Laboratory Experiment, *J. Lab. Chem. Educat.*, 2017, 108–115.
93. P. I. Humphrey, The Use of Mass Spectrometry to Differentiate Blue Dyes From Indigo and Woad, *McNair Schol. Res. J.*, 2017, **10**, 7.
94. A. Whitaker, X-ray powder diffraction, in *The Analytical Chemistry of Synthetic Dyes*, ed. K. Venkataraman, Academic Press, New York, 1977, pp. 280–285.
95. B. Honigmann, Modifikationen und physikalische formen organischer Pigmente, *Farbe und Lack*, 1964, **70**, 787.
96. P. Rioux, Pigments organiques, liants et vernis du XXe siecle, *Techne*, 1995, **2**, 82.
97. M. C. Corbeil, E. Moffatt and D. Miller, A study of the materials and techniques of Alfred Pellan, CCI, *CCI Newsletter*, 1994, **14**, 13.
98. H. E. Shurvell and L. Pinzuti, Sur les spectres infrarouges des phthalocyanines, *Can. J. Chem.*, 1966, **44**, 125–136.
99. T. Learner, The use of a diamond cell for the FTIR characterisation of paints and varnishes available to twentieth century artists, in *Proceedings of the 2nd Infrared and Raman Users Group Meeting*, London, 1998, pp. 7–20.
100. A. M. Jonsson, Separation of acrylic paint components and their identification with FTIR spectroscopy, in *Proceedings of the 6th Infrared and Raman Users Group (IRUG 6)*, ed. M. Picollo, II Prato, Florence, 2004, pp. 58–63.
101. R. Davey, D. J. Gardiner, B. W. Singer and M. Spokes, Examples of analysis of pigments from fine art objects by Raman microscopy, *J. Raman Microsc.*, 1994, **25**, 53–57.

102. D. C. Smith, M. Bouchard and M. Lorblanchet, An initial Raman microscopic investigation of prehistoric rock art in caves of the Quercy District, *J. Raman Spectrosc.*, 1999, **30**, 347–354.
103. P. Vandenabeele, A. Hardy, H. G. M. Edwards and L. Moens, Evaluation of a principal componentsbased searching algorithm for Raman spectroscopic identification of organic pigments in 20th century artwork, *Appl. Spectrosc.*, 2001, **55**, 525–532.
104. P. Vandenabeele, L. Moens, H. M. G. Edwards and R. Dams, Raman spectroscopic database of azo pigments and application to modern art studies, *J. Raman Spectrosc.*, 2000, **31**, 509–517.
105. W. E. Smith and B. N. Rospendowski, Raman spectra of phthalocyanines, in *Phthalocyanines: Properties and Applications*, ed. A. C. C. Leznoff and A. B. P. Lever, Wiley, New York, 1992, vol. 3, pp. 171–217.
106. G. Massonnet and W. Stoecklein, Identification of organic pigments in coatings: applications to red automotive topcoats, *Sci. Justice*, 1999, **39**, 181–187.
107. P. Ropret, S. A. Centano and P. Bukovec, Raman identification of yellow synthetic organic pigments in modern and contemporary paintings: reference spectra and case studies, *Spectrochem. Acta A*, 2008, **69**, 486.
108. M. Bouchard, R. Rivenc, C. Menke and T. Learner, FTIR and Raman Microscopy study of paints used by Sam Francis, *e-Preserv. Sci.*, 2009, **6**, 27–37.
109. M. Bouchard and A. Gambardella, Raman microscopy study of synthetic cobalt blue spinels used in the field of art, *J. Raman Spectrosc.*, 2010, **41**, 1477–1485.
110. F. Pozzi, Development Of Innovative Analytical Procedures, For The Identification Of Organic Colorants Of Interest In Art And Archaeology, Thesis, 2011, 169 pp.
111. C. L. Brosseau, K. S. Rayner, F. Casadio, C. M. Crzywacz and R. P. Van Duyne, Surface-Enhanced Raman Spectroscopy: A Direct Method to Identify Colorants in Various Artist Media, *Anal. Chem.*, 2009, **81**, 7443–7447.
112. S. G. Harroun, J. Bergman, E. Jablonski and C. L. Brosseau, Surface-enhanced Raman spectroscopy analysis of house paint and wallpaper samples from an 18th century historic property, *Analyst*, 2011, **17**, 136.
113. M. Leona and J. R. Lombardi, Identification of berberine in ancient and historical textiles by surface-enhanced Raman scattering, *J. Raman Spectrosc.*, 2007, **38**, 853–858.
114. M. V. Cañamares, M. Leona, M. Bouchard, C. M. Grzywacz and J. Wouters, Evaluation of Raman and SERS analytical protocols in the analysis of Cape Jasmine dye (Gardenia augusta L.), *J. Raman Spectrosc.*, 2010, **41**, 391–397.
115. I. T. Shadi, B. Z. Chowdhry, M. J. Snowden and R. Withnall, Semi-quantitative analysis of alizarin and purpurin by surface-enhanced resonance Raman spectroscopy (SERRS) using silver colloids, *J. Raman Spectrosc.*, 2004, **35**, 800–807.
116. F. Pozzi, K. J. van den Berg, I. Fiedler and F. Casadio, A systematic analysis of red lake pigments in French Impressionist and Post-Impressionist paintings by surface-enhanced Raman spectroscopy (SERS), *J. Raman Spectrosc.*, 2014, **45**, 1119–1126.
117. J.-J. Colin and P. J. Robiquet, Recherche sur la matière colorante de la garance, *Annelles de Chimie et de Physique*, 1827, 225–253.
118. M. Garfield, Mauve, Faber & Faber, London, W. W. Norton & Company; Reprint edition, 2002, p. 240.
119. F. Perego, *Dictionnaire des matériaux du peintre*, Belin, Paris, 2005.
120. K. Venkataraman, *Chemistry of Synthetic Dyes*, Academic Press Inc, 1972, vol. V, p. 5.

121. M. Bouchard and D. C. Smith, Database of 74 Raman Spectra of, Standard Minerals of Relevance, to Metal Corrosion, Stained Glass, or Prehistoric Rock Art, in *Raman Spectroscopy in Archaeology and Art History*, ed. H. Edwards and J. Chalmers, Royal Society of Chemistry, U.K., 2005, pp. 429–464.
122. N. C. Scherrer, S. Zumbuehl, F. Delavy, A. Fritsch and R. Kuehnen, Synthetic organic pigments of the 20th and 21st century relevant to artist's paints: Raman spectra reference collection, *Spectrochim. Acta, A*, 2009, **73**, 505–524.
123. L. Burgio and R. J. Clark, Library of FT-Raman Spectra of Pigments, Minerals, Pigment Media and Varnishes, and Supplement to Existing Library of Raman Spectra of Pigments with Visible Excitation, *Spectrochim. Acta A*, 2001, **57**, 1491–1521.

9 Paints and Binding Media

A. Nevin

Department of Conservation, Courtauld Institute of Art, Somerset House, Strand, WC2R 0RN, UK
Email: austin.nevin@courtauld.ac.uk

9.1 Introduction

Organic binding media are widely found in paint and their analysis represents a significant challenge.[1–4] Primary sources – in the form of treatises and recipes – suggest the use of organic materials as binding media for paint, for retouching, for gilding, for preparation of walls, for the preparation of grounds and for stuccos. Thus, media have complex roles and a heterogeneous spatial and stratigraphic distribution within a painting, a serious complication for their analysis, identification and quantification. Questions by conservators are key in the study of binding media and may relate to the distribution of a particular binder on a surface or in a paint stratigraphy, to the classification of a group of binders (*i.e.* oil or protein) or the identification of a specific binder (*i.e.* glair or egg white, PVAc or acrylic), and the determination of how a binder was prepared (cooked heat-bodied oil or cold pressed linseed oil). Analysis may instead seek to understand degradation processes – the chemical reactions that follow ageing and interaction between pigments, binders and the environment – a newer frontier in the study of oils and proteins.

The analytical challenge for the identification of binding media in paint depends on the type of question posed, intrinsic sample restrictions that allow only microscopic fragments to be taken from

Conservation Science: Heritage Materials, 2nd Edition
Edited by Paul Garside and Emma Richardson

Published by the Royal Society of Chemistry, www.rsc.org

paintings, and the difficulty in combining spatial resolution with chemical specificity. Methodological developments and unique expertise in museums and research institutions continue to contribute to our understanding of binding media and their use in paint.[5–7] Different natural and, since the 19th C., synthetic materials have been used as paint, and the same materials have also been used for conservation (for example consolidation, fixing of flaking) or treatment, adding a layer of complexity to the identification and analysis of organic binding media.

The main differences in commercial paints that are used today can be related to their formulation and, critically, the binding media mixed together with pigments and other additives: whether house paint (water-based acrylic), alkyd-based oil paints, linseed oil, acrylic paint tubes, gum-based watercolours, or ill-defined "tempera" paint. Each binder or paint mixture will allow different forms of expression, and have different working or drying properties: in some cases, paint can be diluted with water, or turpentine, and may dry rapidly in minutes for acrylic and alkyd paints, or over days for many drying oils. All binders have varying adhesive properties and serve to distribute and adhere paint made of pigments and colourants to a surface. For painters and for conservation scientists it is useful to classify binders based on their origin or chemical composition as: animal proteins, natural oils, waxes, polysaccharides and terpenoids (used as varnishes) and the very broad class of synthetic modern materials.

This chapter aims to provide a broad overview of the context, chemistry and analysis of common binding media in paint and presents selected case studies of the application of analytical methods to identify media in paint samples and to map their distribution. Rather than focus on all media or specific techniques, many of which are described comprehensively in dedicated publications,[2,8–13] this chapter aims to provide the reader with an appreciation of the use of binding media, basic chemistry of media with a principal focus on protein-based and oil-based media, and case studies applied for the analysis of paint containing oils and proteins. Indeed while binding media are vastly different, and their chemistry and reactions are fundamental for conservation, analytical strategies applied for the identification of media are generally far less specific, and only in some cases are tailored to the detection of proteins (for example MALDI-MS[14] or ELISA[15,16]). Before considering how media can be identified it is useful to consider where, when and why they are used in the first place.

9.1.1 The Use of Proteins as Binding Media

Protein-based media made from animal products (including eggs, milk, bones and skins) have exceptional working properties as binders, are strong adhesives, and have been used in paint since prehistory. The preparation and use of glues and binding media from protein-based materials are described in detail in artist treatises and technical literature:[1,17] by Vitruvius (probably *ca.* 30–20 BCE),[18] Theophilus (early 12th C.),[19] Cennini (late 14th C.),[20] Vasari (16th C.), Sorte (late 16th C.), de Mayerne (mid 17th C.), Pacheco (mid 17th C.) and Baldinucci (late 17th C.).

Treatises cite different sources of proteins for paint that include glues extracted from mammal skins and fish bones/tissues, as well as media deriving from eggs, milk and cheese. For example, Vitruvius mentions 'glue for walls' as well as the preparation of colours from wine residues and charcoal black with 'glue.'[18] In his *De diversus artibus*, Theophilus mentions the application of metal foils with hide glue, gilding with egg white, egg yolk as a paint medium, and the preparation of various glues from cheese casein, sturgeon bladder and animal hides (see Chapter 5).[19]

Protein-based media may be found in every stage of the production of paintings. Grounds may be applied to a range of supports such as parchment (animal skin), wood or metal, stone, lime or earth-based surfaces. Wooden panels and canvases are often prepared with multiple layers of gypsum and/or chalk mixed with glue to give a smooth and reflective surface for painting.[21] Similar grounds containing lead white and protein have been identified in the preparation of stone surfaces for the execution of oil-based medieval wall paintings.[22] Further treatments of un-painted stone with glue (as associated with *scialbo*) or the lime-washing of surfaces with lime-protein paint are often found. Although the most typical traditional varnishes include resins (mastic and dammar, for instance), Cennini also recommends the use of animal glue. Egg white (glair) has also been used as a varnish or temporary coating for oil paintings, as described by Vincent Van Gogh.[23] Proteins have also been implicated in the preparation of some synthetic pigments, as in the case of artificial copper blues and greens, described in 18th-century Russian documents.[24] While natural materials have been partially replaced by artificial polymers such as acrylics and alkyds (see Chapter 11), naturally found proteins are still used by modern and contemporary painters: Mark Rothko used emulsions of egg in the execution of his works of the 1950s and '60s[25] and egg-based paint media continue to be popular with painters.

9.1.2 On the Use of Drying Oils

While proteins are the most ancient and ubiquitous binding medium (for example in wall paintings worldwide), Western paintings have been dominated for the last 500 years by media based on drying, or siccative, oils.[4,26–29] Indeed, drying oils have been used for millennia, by Ancient Egyptian and Greek civilisations: Theophilus, Cennini and Vasari all cite drying oils for painting. It is understood that while oils may have been used as sealants and for glazes, they only became the predominant medium for panel painting during the 16th C.

Common drying oils deriving from linseeds, walnut and poppy seeds are eloquently described by Mills and White[2] and in seminal work on the application of gas chromatography to the analysis of paint from the National Gallery of Art in London.[30] Gettens and Stout describe some of the many processes which can be used during the extraction and treatment of oils.[9] Following extraction from ground seeds and nuts (that also include safflower and tung seeds) drying oils can be manipulated by heating or cooking (as in the case of thicker heat-bodied stand oil) or through mixing with reactive pigments (for example with the addition of PbO or litharge, or other metal ions), all of which may change the working properties of paint, and influence drying time and the chemical stability of oil paints.[31–33] Modern paints contain significant additives that will modify their rheological, chemical and physical properties – and these additives may be detected during analysis.

9.1.3 20th C. Binding Media

Major changes to paint media occurred during the 20th C.[11] with the introduction of vinyl acrylic and later alkyd artists' paint,[34–37] and there is a growing interest in understanding modern paint binders and conservation implications associated with their ageing.[38] Modern adhesives may also be found as paint binders and include cellulose to vinyl acetates. A range of additives are found in modern paints and include extenders, bulking agents and driers.[39]

9.2 Chemistry, Structure and Microstructure of Binding Media

Binding media allow the preparation of paint from a dry pigment (see Chapter 8) and will dry or cure to form a solid film. The function of binding media is therefore both as vehicle and adhesive. For dispersed and dissolved media like proteins, gums and many modern polymers, water is either naturally present (for example in milk or egg) or is used to dissolve a glue or to disperse the emulsion of fats and proteins in egg, and the evaporation of water leads to the formation of a soluble film. For other media containing reactive fatty acids in oils and egg yolk (lipid and protein), the medium oxidises and polymerises to form a solid film. In many cases, aged binders chemically degrade and their solubility in water will decrease with cross-linking reactions. The chemistry of natural and synthetic binding media – the stability and reactivity and the physical properties of the resultant films – differs greatly and is of staggering complexity. For simplicity, the chemistry of binding media is briefly summarised for each class of materials below.

9.2.1 The Structure of Protein-based Binding Media

Protein-based paint media contain macromolecules based on long chains made of 20 different amino acids that are linked by peptide bonds. The peptide bond between a carbonyl and an amide group is key to the properties of proteins because it allows the formation of endless chains. The peptide bond is also central in analytical approaches: peptide bonds are detected by vibrational spectroscopy, and they are broken either selectively (*e.g.* enzymatic digestion) or completely (*e.g.* acid hydrolysis) to release amino acids for analysis using chromatographic techniques. Typical proteins contain many hundred peptide bonds and different concentrations of amino acids. The chemistry of amino acids is dependent on the functional groups present in each, and these are described in detail elsewhere (for example see Chapter 5 and also Leather in the 1st Edition[40]). Large differences in concentrations of amino acids in paint media are due to the sources of proteins of different media (Table 9.1). For example, the amino acid hydroxyproline (OH-Pro) is found only in collagen, making it a marker for animal glue while tryptophan (Trp) (absent in collagen) is indicative of a non-collagenous protein.[2]

The main groups of proteins found in binding media are classified on the basis of structure as serpins (ovalbumin and lactalbumin), casein and collagens as listed in Table 9.2. For a discussion of the chemistry of collagen see ref. 40.

Proteins present different secondary structures. Collagen, the source of animal glues, is found with multiple coils of amino acids,

Table 9.1 Average amino acid weight percentages of common sources of binding media.[2,41] (Data reference – USDA Food Composition Databases https://ndb.nal.usda.gov/ndb/.)

Amino acid, weight %	Whole egg	Egg white	Egg yolk	Lactic casein	α-Lactalbumin	Calcium caseinate	Milk	Collagen
Alanine	6%	6%	5%	3%	6%	3%	3%	10%
Arginine	6%	5%	7%	4%	3%	4%	3%	8%
Aspartic acid	10%	10%	10%	7%	11%	7%	7%	7%
Cystine	2%	3%	2%	0%	3%	0%	1%	0%
Glutamic acid	13%	13%	13%	23%	16%	23%	20%	11%
Glycine	3%	3%	3%	2%	2%	2%	2%	27%
Histidine	2%	2%	3%	4%	3%	4%	3%	1%
Isoleucine	5%	6%	5%	5%	5%	6%	6%	2%
Leucine	9%	8%	9%	10%	13%	10%	9%	4%
Lysine	7%	7%	8%	9%	10%	8%	8%	4%
Methionine	3%	3%	2%	3%	3%	3%	2%	1%
Phenylalanine	5%	6%	4%	6%	4%	6%	5%	2%
Proline	4%	4%	4%	0%	0%	0%	9%	14%
Serine	7%	7%	9%	7%	5%	6%	5%	4%
Threonine	5%	5%	5%	4%	5%	4%	4%	2%
Tryptophan	1%	1%	1%	0%	0%	0%	1%	0%
Tyrosine	4%	4%	4%	6%	4%	6%	5%	1%
Valine	6%	6%	6%	7%	6%	7%	6%	3%

Table 9.2 Proteins in binding media.[41]

Sepins: Ovalbumin and Lactalbumin (Egg and Milk)	Casein	Collagen (Animal tissue)
Group of proteins thought to inhibit peptidases (enzymes which break down proteins)	Predominant phosphoprotein found in milk, bound to calcium in a micellular suspension	Main protein found in connective tissue, component of ligaments and tendons, with high tensile strength and elasticity
Ovalbumin: metal-coordinating carriage protein surrounding the egg yolk Lactalbumin: present in milk of mammals Often contains three beta sheets and eight or nine alpha helices	Structurally, many inactive (random coil) areas and no disulfide bridges lead to little secondary or tertiary structure; hydrophobic $\alpha s1:\alpha s2:\beta:\kappa$ in ratio 4 : 1 : 4 : 1	Large concentration of amino acids of Gly, Pro (and OHP) which allow formation of tightly coiled helices that twist with other similar helices to form a triple helix

which are formed due to the large presence of Gly, Pro, OHP and Ala which are small amino acid molecules. Every third amino acid in collagen is Gly that is the basis for chain formation. Alpha helices constitute the building block for collagen, which is formed by the twisting of three alpha helices together to make a triple-helix.[42] Gels of refined animal glue extracted from collagen form a collagen-like helical structure upon loss of water which is strongly dependent on the setting, film-forming conditions (especially temperature and solvent).[43] In contrast to collagen, the milk protein casein has little tertiary structure: no di-sulfide bridges are associated with the macromolecule and the exposure of hydrophobic residues accounts for its insolubility in water at neutral pH.

Although the tertiary structure of biologically active proteins may seem far removed from the study of binding media in paintings, this is not the case. Aspects of the behaviour of natural proteins used as binding media, especially their preparation (solubility), stability, and their sensitivity to water and response to changes in humidity are largely dependent on the tertiary and secondary structure of proteins, as can be understood by the following example. Film formation of collagen, hence its suitability as a paint medium, is due to the extraction of long alpha helices from animal tissue following heating in water. As the liquid cools, the alpha helices rearrange to form a semi-organised lattice which accounts for the formation of a gel; upon final loss of water, the gel dries to form a strong transparent solid. This solid glue can be rehydrated in water to form another gel, and this is due to the retention of secondary structure in the solid glue.

9.2.2 The Structure of Drying Oils

Dryings oils are mixtures of triglycerides, esters of trihydric alcohol glycerol (1,2,3-propanetriol) and a limited range of saturated fatty acids, the most important being C-16 and C-18 chains, summarised in Table 9.3. Egg yolk also contains approximately 33% fatty acids together with proteins, and water.[2]

Longer chain fatty acids, phytosterols and steroid alcohols are found in small concentrations in drying oils. Linseed, poppy seed, walnut oils and egg yolk have similar composition but small differences in the concentrations of palmitic and stearic acids that account for their drying properties and are the basis of their differentiation by separation techniques like GC–MS[44,45] and are summarised in Table 9.4.

Drying oils react rapidly because the unsaturated fatty acids linoleic and lineolenic acids oxidise to form a polymeric network and film or solid matrix.[2,47,48] The reactions in oil drying continue to be of major interest in conservation science. For example, recent research has modelled 70 chemical reactions that occur during the first stages of drying of oil paints, highlighting the complexity of competing reactions.[49]

Chain free radical oxidation reactions occur following radical abstraction from methyl groups in fatty acids, and leads to fragmentation and cyclisation reactions, and the generation of a network of bonds and molecules of massive molecular weight.[47]

Table 9.3 The most abundant fatty acids found in drying oils.[2]

IUPAC nomenclature	Common name	Formula
Nonanoic	Nonanoic	$C_9H_{18}O_2$
Dodecanoic	Lauric	$C_{12}H_{24}O_2$
Tetradecanoic	Myristic	$C_{14}H_{28}O_2$
Hexadecanoic	Palmitic	$C_{16}H_{32}O_2$
9-Hexadecanoic	Palmitoleic	$C_{16}H_{30}O_2$
Octadecanoic	Stearic	$C_{18}H_{36}O_2$
9-octadecenoic	Oleic	$C_{18}H_{34}O_2$
(*cis*) 9,12-octadecadienoic	Linoleic	$C_{18}H_{32}O_2$
(*cis*) 9,12,15-octadecatrienoic	Linolenic	$C_{18}H_{30}O_2$
Cis,*trans*,trans-9,11,13-octadecatrienoic	α-eleostearic	$C_{18}H_{30}O_2$
12-hydroxy-9-octadecenoic	Ricinoleic	$C_{18}H_{34}O_3$

Table 9.4 Fatty acid composition (%) of common drying oils used in paint.[2,46]

	Palmitic	Stearic	Oleic	Linoleic	Linolenic
Linseed	4–10	2–8	10–24	14–19	48–60
Poppyseed	9–11	1–2	11–18	69–77	3–5
Walnut	3–8	0.5–3	9–30	57–76	2–16
Egg Tempera	27	9	44	13	0.5

Another important reaction of oils is aldehyde and ketone oxidation, which generates dicarboxylic acids with the commonly found azelaic acid (nonanedioic acid), with ratios of azelaic to palmitic acid providing an indication of the extent of oxidation and polymerization in a drying oil.[50]

9.2.3 A Brief Note on Synthetic Paints

During the 20th C., binding media in paint changed with the introduction of synthetic polymers (see Chapter 11), which include acrylics and their dispersions in water, polyvinyl acetates (PVAc), nitrocellulose, and polyesters modified with the addition of oils, fast drying alkyd paints, as described in detail by others (on Alkyds: ref. 35, on PVAc in Paint: refs. 51,52). In conservation, acrylics have also been widely used as adhesives and varnishes, and many 20th C. artists employ commercially available formulations of paint (including house paint or their own mixtures of paint).[6] Additives in paint, including drying agents, preservatives, plasticisers and bulking agents are common and may have significant implications for the conservation of modern paints.[38]

9.2.4 Stratigraphy and Distribution of Binding Media

When used as paint media, regardless of their composition, binders provide a cohesive matrix in which pigments or colourants are dispersed that must also adhere to a support or surface. In other instances, binders may function only as an adhesive (which include for gilding and for varnish), or as a cohesive barrier (when used as a sealant for stone). The pigment to binder ratio and the pigment volume concentration in paint depends on paint media – while in pre-20th C. easel paintings, oil-based binders may dominate the paint composition, in wall paintings binders are often much less concentrated.[53]

Paint samples removed for analysis are necessarily minute and within such samples the binders can be found in trace concentrations, thus requiring sensitive analytical techniques for their identification. Paintings are characteristically multi-layered, and typically contain a protein-based ground covered by multiple paint layers of media and pigment, with each layer in the order of tens of micrometres in thickness and paint sometimes applied using different media. Organic glazes, varnish and multiple coatings are common. Hence paintings are multi-material systems in which binding media are one of many components, and commonly more than one binder will be present in a painting. The conservation of paintings and the introduction of new adhesives that may be based on proteins or synthetic binders further complicates the identification of original binders and their stratigraphic distribution in paintings.

9.3 Analytical Techniques Appropriate for Binding Media

Because of the intrinsic complexity of paint different analytical approaches should be adopted for the analysis of binding media based on questions posed during conservation and, critically, the availability of both instrumentation and samples. *In situ* and non-invasive techniques (that require no sampling or contact) may provide reliable information regarding the presence of particular classes of binding media, but it is noted here that most routine analysis of proteins, oils and modern paint is performed with destructive chromatographic techniques, summarised in Table 9.5. In some instances, if binding media is sufficiently concentrated on a surface, and if a painting is not varnished with a coating, non-invasive methods can provide indications about the type of binder used for paint.[54,55] In other instances, in the analysis of modern polymers used in binding media, non-invasive methods based on the detection of molecular vibrations are particularly promising[56] but may be limited to the detection of broad classes of media.

9.3.1 *In Situ*, Non-invasive Techniques

In situ analysis may reveal the presence of proteins or oils in an unvarnished painting[57] or in paintings with a minute concentration of binder with respect to inorganic material as is the case for many wall

Table 9.5 A summary of commonly applied analytical techniques for the analysis of paint binding media.

Technique	Type of sampling	Size of sample/ area of analysis	Principle of the technique	Information provided	Basis for identification	Level of identification and comments
Near Infrared Imaging and Spectroscopy	*In situ*	Depends on focus/spot size	Detection of reflected IR which is modified by absorptions by molecular vibrations	A spectrum of the intensity of reflected radiation as a function of wavelength in the NIR range	Comparison with database of NIR spectra and the attribution of overtones to specific combinations of vibrations, often from O–H and N–H vibrations	In-depth interpretation of spectral bands and the adoption of statistical methods have greatly improved the utility of NIR, and imaging systems based on NIR have great potential for the detection of materials on a surface (*i.e.* if paint and binding media are not covered by a varnish)
FTIR reflectance (portable)	*In situ*	4 mm diameter	Fibre optic bundle or portable instrument brought into near contact with the surface of an object to deliver a beam of IR radiation and other fibres in the bundle are used to detect reflected IR	An IR spectrum (Reflectance *vs.* wavenumber/ cm^{-1}) is indicative of molecular vibrations which occur following a change in the dipole moment	Oils and many synthetic media can be identified by C–O and C–H bonds and spectra matching with reference databases. Proteins can be identified by amide-I and amide-II vibrations at approximately 1650 cm^{-1} and 1580 cm^{-1} respectively. Other C–H stretching between	General protein, lipid and broad class of synthetic media: the presence of inorganic materials can limit observation of signals from trace materials. Insensitive technique. Detection limits are greater than those for FTIR (see below)

Table 9.5 *(Continued)*

Technique	Type of sampling	Size of sample/area of analysis	Principle of the technique	Information provided	Basis for identification	Level of identification and comments
					2700–3200 cm^{-1} may be indicative of organic materials	
Micro-FTIR (transmission, reflectance, imaging, ATR) on solid or liquid (including cross-section, powder or surface)	*In situ*, or contact	Semi-destructive (sample integrity may be compromised but the sample can be re-analysed), size of sampling area depends on signal:noise, but is generally not below the diffraction limit 10×10 μm	Using an IR-objective (Cassegrain) or ATR crystal, the IR beam is delivered to a selected area of a sample, and is collected either in transmission or reflectance	Spectra are similar to FTIR reflectance (see above)	See above; FTIR may also be able to resolve rgw secondary structure of proteins on the basis of the position of Amide-I/II	General class of binding medium: FTIR may provide sufficient information for the detection of the presence of proteins and lipids but quantification is not straightforward
Micro-Raman and Micro FT-Raman spectroscopy	*In situ* if portable, or non-destructive	Spot size (micro) depends on the objective used and signal: noise ratio, generally 5–10 μm diameter.	A laser beam is directed onto the surface of a sample selected using a microscope and Raman scattering is detected	A Raman spectrum (Raman intensity *vs.* wavenumber or Raman shift, cm^{-1}) is produced and indicates molecular vibrations	Protein may be identified by amide-I and other peaks from amino acids	General class of binding medium: Raman may provide qualitative analysis of a surface provided fluorescence does not dominate the spectrum. The identification of a specific protein or synthetic polymer may be possible depending on the purity of a sample

Micro-chemical stain tests on sections or fragments	Destructive	Staining with antibodies, fluorophores, or dyes which may label or react with compounds within proteins on the surface of a sample, size of sample: not less than 10 µm, usually much larger	Dyes/antibodies react with amino acids or fragments of protein-chains	Positive/negative stain or colour	Protein identification following a specific colour change of sample following treatment	General protein and lipid: solubility of proteins may be a problem and lead to false-negative results Tests are not always reliable for porous samples where false positives are common or where interferences with other organic materials limit access to proteins
Immunological-based methods, including Enzyme-Linked Immunosorbent Assay (ELISA)	Destructive	>10 ng	Proteins are extracted from a sample and are then detected using stained protein-specific antibodies	Positive/negative stain or colour of labelled antibody	Protein identification following detected fluorescence or absorption of colour from stain following reaction with antibody	Class: Due to the specificity of antibodies this is specific to particular proteins which are found in the original materials used for paint media. The extraction of proteins of suitable chain length may indeed be a serious limitation and interference with pigments is often a problem.

Table 9.5 (*Continued*)

Technique	Type of sampling	Size of sample/area of analysis	Principle of the technique	Information provided	Basis for identification	Level of identification and comments
FTIR transmission	Destructive	KBr or diamond cell to yield a transparent sample, μg samples	The IR absorption of a sample is measured using a transparent sample (dispersed in KBr) or flattened with a diamond cell	An FTIR transmission spectrum [Transmittance *vs.* wavenumber (cm^{-1})] is recorded and can be compared with a database of reference materials	See above	General class of binder: FTIR may provide sufficient information for the detection of the presence of proteins, but quantification is not straightforward and protein identification will depend on the purity of a sample
Pyrolysis-Mass Spectrometry, Curie Point Py–MS and DT–MS	Destructive	Sample size ~μm	1. Decomposition of a sample (which can be mixed with a derivitising agent) over a temperature gradient 2. Detection of fragments using MS	A chromatogram is a plot of intensity *vs.* Retention time/s and each peak is associated with a mass spectrum (intensity *vs.* mass/charge)	Peaks can be compared with a database of materials for identification based on mass-spectra of markers which include degradation products of amino acids	Class: Minimal sample preparation, but potentially difficult interpretation usually based on specific markers, complex reactions and byproducts may interfere with the interpretation of results

Gas Chromatography–Mass Spectrometry	Destructive	Sample size ~μg, with detection of individual amino acids ~100 pg	1. Extraction of proteins/lipids/other fractions from a sample 2. Hydrolysis of protein/lipids 3. Derivitisation 4. Column separation 5. Mass detection	As above. A chromatogram is produced and peaks correspond to different amino acids/fatty acids and other markers found in binding media	Amino acid concentrations can be compared with a database for identification of a particular binding medium but rigorous instrumental calibration is usually required for corrections of derivitisation and detection efficiency	Class: Quantitative and qualitative analysis of stable amino acids or fatty acids and markers can be used to determine media following statistical data analysis and comparison with reference materials for proteins, or P : S ratios for oils and the identification of markers. The extraction of proteins may indeed be a serious limitation of liquid-derivitisation and interference with pigments is often a problem Reproducibility and repeatability are a problem due to multiple wet-chemical steps
Reverse-phase Liquid Chromatography	Destructive	Sample size ~μg, with detection of individual amino acids ~100 pg	1. Extraction of proteins from a sample 2. Hydrolysis of protein 3. Derivitisation 4. Column separation 5. Fluorescence detection			

Table 9.5 *(Continued)*

Technique	Type of sampling	Size of sample/area of analysis	Principle of the technique	Information provided	Basis for identification	Level of identification and comments
MALDI-MS	Destructive	Sample size ~10 μg	1. Extraction of proteins from a sample 2. Enzyme hydrolysis of protein 3. Electrophoresis for separation of amino acids or protein fragments 4. Laser Mass spectrometric analysis with Time of Flight detection 5. Peptide sequencing can be achieved with nanochromatography following electrophoresis	A mass spectrum of intensity *vs.* mass/charge is produced and can be compared with databases of known proteins and protein-fragments	Identification of proteins on the basis of a database of fragments. Can be used for the understanding of potential changes in proteins as a result of ageing and is particularly powerful due to the large fragments which are analysed which give mechanistic insights into degradation	Class: Quantitative and qualitative analysis of peptides and their sequences to determine specific protein (*e.g.* lysozyme) following statistical data analysis and comparison with reference databases. The extraction of proteins may indeed be a serious limitation of liquid-derivitisation and interference with pigments is often a problem

paintings.[58] UV-imaging has often been cited for the detection of organic materials, including oils, waxes and proteins, that may account for fluorescence,[53] but detection usually takes place in the visible, beyond the UV fluorescence emissions of amino acids, and signals from spectra of oil paintings are often broad and non-specific.[59,60]

Fibre-optic and Reflectance Fourier Transform Infrared (FTIR) Spectroscopy have been used for analysing the surface of paintings and may be able to detect the presence of proteinaceous materials, lipids and polymers based on characteristic spectral bands and comparison with reference databases.[61,62] Depending on the concentration of inorganic and other organic materials, absorptions from the amide-stretching and C–H stretching vibrations may be indicative of the presence of peptides or proteins in paintings, as has been shown convincingly in NIR spectroscopic imaging,[54] and in other examples of the analysis of a painting by Jackson Pollock.[63] Other examples of FTIR in reflectance include the detection of oil-based media[55] and synthetic media.[62,64]

9.3.2 Sampling Techniques

Binding media analysis usually requires samples of paint that are often first analysed using Fourier Transform Infrared (FTIR) spectroscopy for the detection of pigments and binders (see Chapter 2). FTIR may be used for the identification of the presence of different classes of binding media, but detection limits of the technique and the lack of specificity to a particular natural binding medium may compromise its utility. A consolidated technique, FTIR is ideal for the analysis of micro-samples because only limited sample preparation is required and results are often sufficient to answer most questions. Examples of FTIR applied in conservation science are numerous.[8,65] While FTIR spectra can rarely be used to distinguish between different proteins, or drying oils, that are molecularly almost identical, it is ideal for the estimation of significant chemical changes like the oxidation of oils or reactions between oils and pigments (the formation of lead soaps, for example).[66] FTIR is also a key technique for the analysis of modern materials and inorganic pigments. A major advantage of FTIR is that it can be coupled to a microscope for surface analysis of cross-sections or fragments of paintings[67–70] and in transmission for the analysis of powders, films or thin sections[71] (see the Case Study in Section 9.5.2).

Micro Raman spectroscopy is ideal for the analysis of samples of pigments and has been reported for the analysis of binding media.[72] The utility of Raman analysis is often intrinsically limited by the luminescence of media and signals from organic material are often masked by stronger scattering from inorganic pigments (see Chapter 2). Modern polymers are routinely analysed by Raman spectroscopy. When Raman scattering is not masked by luminescence it may be possible to distinguish between different binding media and, with statistical methods, even specific proteins.[72,73] Spectral databases using different laser excitation sources ranging from visible to infrared have been published and examples of the application of Raman spectroscopy to the analysis of pigment composition are numerous.[74] Technical advances promise more widespread applications of portable Raman spectroscopy.[75]

Micro Raman has rarely been applied to the analysis of the organic binding media found in paintings. However, the analysis of proteinaceous materials from archaeological remains including degraded collagen has been carried out using Fourier Transform Raman spectroscopy.[76,77]

9.3.3 Case Study of the Application of FTIR and Raman for the Analysis of Binding Media

A useful case study is the assignment of FTIR and Raman spectra for the identification of media. A careful assessment of bands in acquired spectra allows the identification of different classes of pigments, as seen in Figure 9.1. Spectral assignments, rather than automated methods, are particularly important when bands from pigment and binder may be present in a spectrum and do not require a large database. Casanova *et al.* use a combination of FTIR and Raman spectroscopy to classify 23 natural and synthetic organic compounds from 11 different chemical classes: resins, proteins, sugars, waxes, fats, acrylics, vinylics, epoxies, amides, polyesters and cellulosic derivatives.[78] The compounds analysed within each family were selected based on the most common organic products used as adhesives or coatings. Different regions of the spectra are examined sequentially for the differentiation of media in the following steps:

A. Three doublets at 2920–2850, 1470–1460 and 730–720 cm^{-1} in the FTIR spectrum that are assigned respectively to the stretching, deformation and torsion of C–H bonds typical of waxes.
B. The carbonyl C=O functional group, characteristic of esters, ketones and carboxylic acids is identified in infrared and

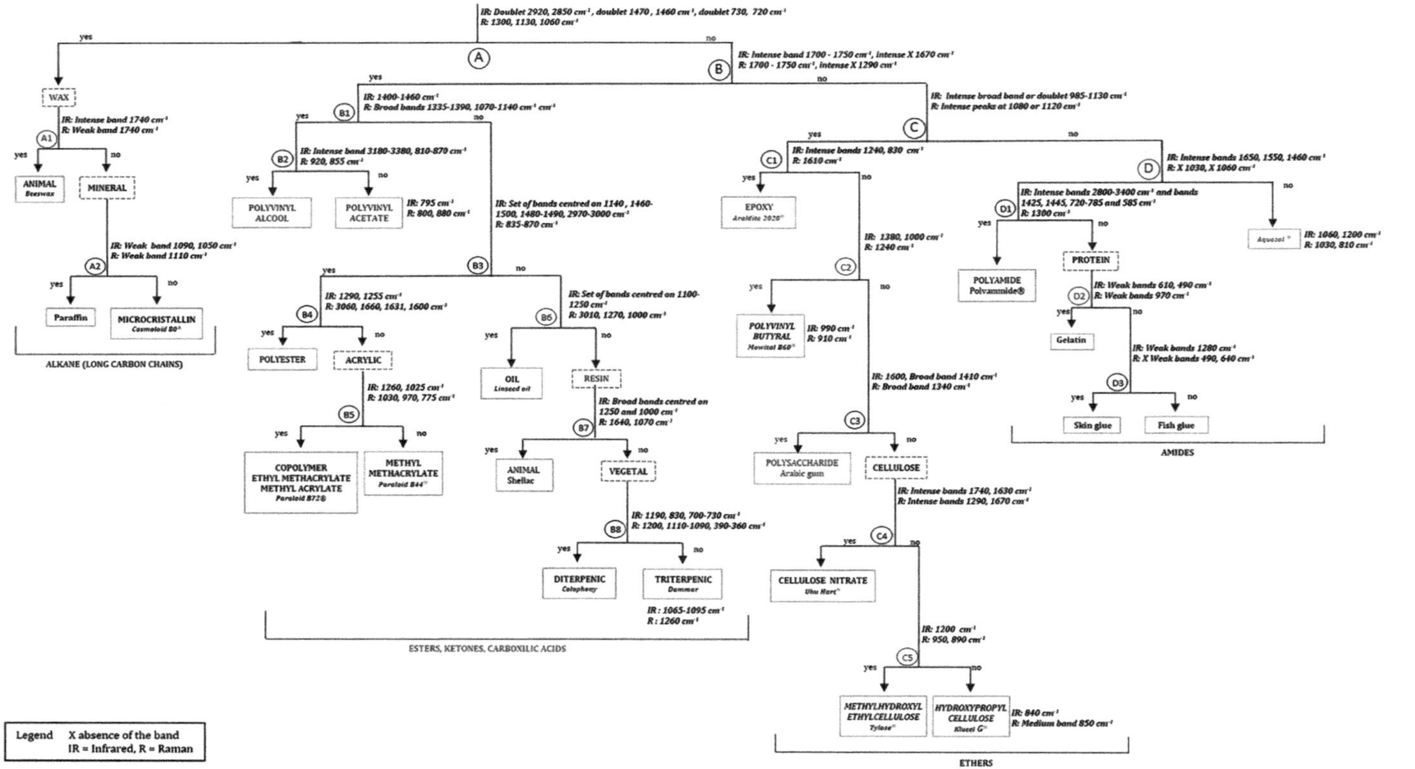

Figure 9.1 Flowchart enabling the discrimination of synthetic and natural organic compounds widely used in restoration work.
Reproduced from ref. 78 with permission from the Royal Society of Chemistry.

Raman spectroscopy by a C=O stretching band between 1700–1750 cm^{-1} (very intense in infrared and medium-weak in Raman).

C. Ether groups present a strong and broad infrared band between 985–1130 cm^{-1} and a medium, broad Raman band between 1080–1120 cm^{-1}.

D. Amide bonds typical of proteins are assigned to the C=O stretching vibration below 1700 cm^{-1}.

9.3.4 Destructive Techniques Based on Mass Spectrometry

The majority of techniques for binding media analysis requires a sample, and usually the sample is destroyed during analysis – therefore it is important to choose reliable techniques and balance the need for specific detection of a particular binder with sample restrictions. Sometimes it is possible to separate paint samples into different paint layers, necessary for the stratigraphically resolved analysis of each layer. Despite sample limitations, the information gained from destructive analysis can be extremely specific and may also lead to the understanding of some of the chemical reactions that occur as a result of ageing and reactions between binders and pigments.

Gas chromatography for separation and mass spectrometry for analysis are the most effective and commonly used methods for the identification of binding media in samples from paintings (see Chapter 2). Matrix Assisted Laser Desorption Ionisation coupled with Time of Flight (MALDI–TOF) has recently demonstrated that much of the primary structures from proteins found in eggs are retained even after ageing of both model and real paintings painted in egg-based binding media.[12,79]

Simply, with all solution-based (see ELISA for example[80]) analysis and chromatographic techniques, a critical step is the solubilisation or extraction of organic material from a sample.[81] For chromatographic techniques, extraction is followed by the hydrolysis of proteins or oils into components (amino acids or fatty acids) using a variety of methods which include high temperature decomposition of the macromolecules (pyrolysis), and hydrolysis of the protein chain into fragments.[82,83] Mass spectrometry (MS), usually following gas chromatography (GC) for the separation of materials, is then used to measure the mass of protein or fatty acid fragments (or amino acids), and can determine the concentration of certain amino acids, or identify specific markers associated with amino acid degradation. Recent advances in the analysis of protein-based binding using a variety of techniques include HPLC, GC–MS, Pyrolysis–GC–MS, Direct Temperature-Mass spectroscopy and MALDI–TOF for the determination of the concentration of different amino acids, peptides or proteins, and oils for the analysis of binding media in microgram samples from paintings.[84,85]

Markers are molecules that can be used as tracers for the presence of a specific material and can be intrinsic (*i.e.* present prior to analysis) or form with reactions during separation and/or preparation. For example, phosphate ions may be indicative of the phosphoproteins casein from milk or ovalbumin from egg, and cholesterol is typically found in egg yolk. Hydroxyproline is found only in collagen. Other drying oils have specific markers and include, for example, 9,11,13-octadecatrienoic acid, α-eleosteric acid a marker for tung oil.

Although markers are good indicators and can sometimes be used for the identification of media, it is more common to consider the concentrations of selected stable amino acids, ratios of fatty acids, or longer peptide fragments to confirm the presence of a particular protein or a specific drying oil.[85]

9.3.4.1 Case Study: GC–MS for the Analysis of Proteinaceous Binding Media

GC–MS is a sensitive invasive technique that allows the qualitative and quantitative identification of amino acids, fatty acids and polysaccharides, and can resolve the presence of different binding media. Due to the extreme sensitivity of detectors, samples of less than 100 μg can be analysed.[86] The quantitative analysis of selected amino acids is achieved during GC–MS of paint samples containing proteins but requires significant sample preparation.[13,87] In addition to protein-based binding media, lipidic and resinous compounds are commonly found, and it is necessary to extract proteins from a sample and remove pigments, compounds and molecules that can interfere during subsequent steps of preparation and detection. Following extraction, long protein chains are separated into individual amino acids; this can be achieved in a variety of ways using acids, bases, or enzymes for the hydrolysis of peptide bonds. There is an intrinsic compromise between efficient hydrolysis, destruction of sensitive amino acids and the formation of condensation products between trace sugars and free amino acids as a result of the long exposure to high temperatures (above boiling point) during hydrolysis. After hydrolysis, and further extraction, the free amino acids, in the presence of an internal standard, are reacted with specific derivitisating agents, resulting in the formation of a mixture of volatile amino acid derivatives. After derivitisation, the liquid mixture of volatile compounds is injected into the GC column coupled to the mass spectrometer. As the gases travel through the column at different rates, they are separated and then subsequently detected by MS as they reach the end of the column.

The result of GC–MS analysis is a chromatogram, the spectrum of detected mass (intensity) over time. Qualitative data are provided since each peak in a chromatogram is associated with a mass spectrum and hence peaks in the chromatogram can be identified with different amino acids for proteins (or fatty acids for oils), ions or other markers. In addition, quantitative results can be derived from the analysis since the area under each peak is associated with the quantity of the mass fraction. The interpretation of GC–MS data for the identification of binding media and spectral data processing can seriously affect the outcome of analysis.

In GC–MS analysis of protein-based media the markers that are separated and quantified are selected derivatives of amino acids as well as phosphates, oxalates and glycerol. Binding media are differentiated based on their amino acid composition. Due to the analytical procedure, which commonly involves acid hydrolysis of peptide bonds, amino acids tryptophan (Trp) and cysteine (Cys) are destroyed, and dehydration of serine (Ser) and threonine (Thr) have also been reported.[88]

The preparation of samples for the GC–MS analysis of proteins involves the following steps.

1. Extraction of proteins and free amino acids from a sample
2. Hydrolysis of the extracted protein
3. Derivitisation of amino acids
4. GC–MS of derivitised sample

(1) Proteins can be extracted from small samples with highly polar solutions. Water is unable to dissolve most solid proteins used as binding media which are aged, cross-linked (for example egg yolk that contains lipids) and often form chemical complexes with pigments. It is common to employ reagents like 2.5 M ammonia (aq) in ultrasound and at moderate temperatures (60 °C) to extract free amino acids and proteins from samples, that are later analysed for fatty acids.[88]

(2) Significant differences in acid hydrolysis conditions are found across laboratories for protein analysis with GC–MS; these range from milder conditions of 1 hour microwave hydrolysis with 6 M HCl[87,89] to 24–36 hours acid hydrolysis using HCl.[44]

(3) Following the hydrolysis of peptide bonds, and extraction of fatty acids, the free amino acids are derivitised to yield volatile molecules for each amino acid. Derivitising agents employed for the detection of amino acids in binding media are based on the following compounds:

- *N*-trifluoroacetic acid (CF_3CO_2H) (to yield *N*-trifluoroacetyl-0-2-propyl esters)[44]
- Ethylchloroformate (CH_3CH_2OCOCl) derivatives[90]
- *N*-(tert-butyldimethylsilyl)-*N*-methyltrifluoroacetamide (MTBSTFA, $CF_3CCON(CH_3)Si(CH_3)_2C(CH_3)_3$)[87]

The determination of the concentration of amino acids requires both careful calibration and rigorous experimental protocol. The presence of internal standards in known concentration in analysed samples allows an assessment of the derivitisation procedure and calculation of the yield of derivitisation that is dependent on the sample as well as systematic error and allows for fluctuations in the GC–MS instrument (column, ioniser, and detector). Further, the calibration of the detection of derivitised standard amino acid solutions at known concentrations corrects for non-linearity in the detection of amino acids.

The stability of amino acids to ageing and in the presence of pigments is of key interest for their identification, and has been investigated with standard references of commonly found protein-based binding media exposed to artificial ageing and with different pigments.[91,92] Amino acids vary significantly in concentration following chromatographic separation, particularly due to losses from oxidation, cross-linking and other degradation reactions (*i.e.* Tyr and Lys, and sulfur containing Met and Cys),[93] and are often excluded from datasets for comparison with standards. Further, Trp, which is also sensitive to oxidation reactions prior to analysis, is completely destroyed during acid hydrolysis which is the first step during GC–MS sample preparation. Therefore, Trp is not considered in the determination of protein-based paint media. Indicative amino acids are often selected for comparison with reference materials: Alanine (Ala),

Glycine (Gly), Valine (Val), Leucine (Leu), Isoleucine (Ile), Proline (Pro) and Hydroxyproline (OHP or Hyp). Ratios between amino acids have been reported for the differentiation between proteins in GC–MS[85] and Liquid Chromatography (LC).[94] Correlation coefficients used in the past may provide an indication of the quality of a match between data sets of a known and unknown sample and have been employed for the comparison of detected amino acids with reference binding media,[90] but correlation coefficients may not give an indication of the deviation of specific amino acids from a standard reference or allow the detection of mixtures of protein-based media.

Identification of mixtures of proteins with GC–MS is not straightforward through amino acid ratios alone. Instead, statistical treatment of data, which allows the evaluation of multiple components (multivariate analysis) including Principal Component Analysis (PCA) has been used. With such analysis, data from many samples are grouped into clusters graphically (in score plots with specific loading for each Principal Component) based on contributions from the seven different stable amino acids. Although PCA score plots bear no direct physical significance, the separation of a set containing the seven variables (amino acids) from different chromatograms into different clusters is especially useful as it takes into account contributions from each variable and allows the rapid, unbiased assignment of a particular sample to a cluster of reference materials or standards.

Specific markers retrieved during GC–MS make the identification of collagen-based proteins especially straightforward: Hyp is present only in samples of animal glues and Gly is also found in large concentrations (>35). Further differences between the samples can be appreciated with reference to the concentration of Glu which is much higher in milk and casein than in the egg proteins and it is found in even lower quantities in the collagen samples. By contrast, in the egg and milk proteins, amino acids Phe, Leu and Ile are present in three to four times the concentrations of the non-collagen proteins. Furthermore, high levels of Asp and Ser are associated with the egg proteins.

To compare the calculated concentrations of amino acids with reference standards PCA was applied to a matrix of 7 amino acid percentages of over 80 different protein-based paint media (from eggs, milk or collagen-based tissues) which were naturally and artificially aged. The score plot of the two first Principal Components (PCs) of the database of standards from the University of Pisa, together with the protein-based media analysed is given in Figure 9.2.

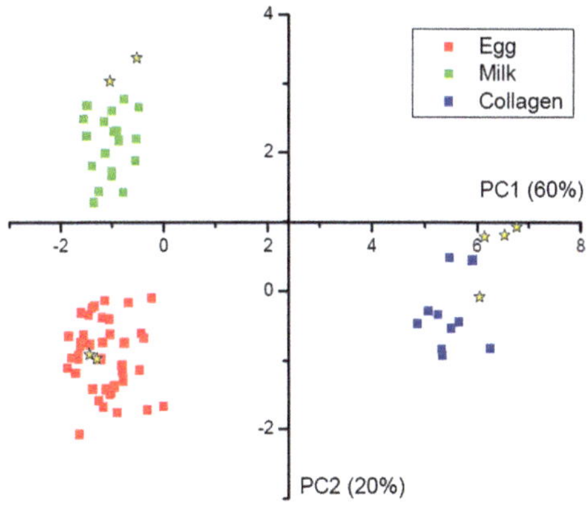

Figure 9.2 PCA of the amino acid profiles of 8 reference samples (stars) compared with standard references of binding media. Samples from materials used in this work are grouped according to the concentration of amino acids and cluster with the reference samples from the University of Pisa.

Table 9.6 Mass percent of ten amino acids (those not included in PCA analysis are marked with an asterisk) calculated on the basis of GC-MS analysis following procedure described in ref. 95.

Sample/Amino acid	Ala	Gly	Val	Leu	Ile	Ser*	Pro	Phe*	Asp*	Glu*	Hyp
Source: Milk											
Casein	4.6	3.4	7.4	10.1	4.9	9.4	17.3	4.0	7.2	31.8	0.0
Milk	5.8	3.7	8.9	11.9	6.7	8.8	18.6	5.3	9.2	21.0	0.0
Source: Eggs											
Egg white	12.3	8.3	10.2	12.0	7.1	10.6	5.8	6.0	12.9	14.8	0.0
Egg yolk	11.2	7.5	8.8	12.4	7.2	13.7	6.3	4.3	14.5	14.1	0.0
Source: Animal collagen											
Isinglass	13.6	38.5	2.1	2.3	1.4	5.5	12.8	1.7	5.4	7.5	9.2
Ox bone glue	12.5	36.0	2.2	2.7	1.3	3.8	16.5	1.5	4.6	7.0	11.9
Parchment glue	12.0	36.2	2.4	3.0	1.5	3.9	15.6	1.6	5.2	7.8	10.8
Rabbit skin glue	12.5	36.2	2.0	2.5	1.2	3.7	17.4	1.3	4.3	6.6	12.3

Amino acid concentrations (given in weight per cent) of seven stable amino acids in binding media are listed in Table 9.6. PCA of these concentrations can separate the profiles of different binding media reference data into three distinct clusters shown in Figure 9.2. Following analysis, 80% of the variance in concentration is described by components PC1 and PC2.

9.3.4.2 A Note on Complications in the Analysis of Protein-based Media

Although quantitative data are provided with MS, losses of material are incurred through side-reactions within the protein, between amino acids and with derivitising agents commonly used in destructive analysis. Free sugars are present in most sources of binding media (*i.e.* lactose in milk) and will react with amino acids in the presence of strong acids (used for hydrolysis) to form humins (coloured compounds), especially when heat is employed during the hydrolysis of proteins.[96] Variables affecting the interpretation of GC–MS results include sample preparation methods, differences in column setup, eluents and detection methods, calibration and data processing. Therefore, the comparison of results between laboratories is quite difficult.

Another complication arises from the interaction between amino acids and ions that may be dissolved during hydrolysis and include Cu^{+2}. In order to remove interferences from copper complexes, filtration techniques have been employed for the purification of extracted material prior to derivitisation.[97–99]

9.3.5 Analysis of Oil-based Binding Media

The analysis of oil-based binding media relies on the reaction of lipids in samples with a suitable reagent, followed by GC–MS detection of different fatty acids. A common agent for derivisation is (*m*-trifluoromethylphenyl)trimethylammonium hydroxide, for the saponification of triglycerides to methyl esters, and is also useful for the production of methyl esters from resins and waxes.[100–102] In other approaches *N*,*O*-bis-trimetilsilil-trifluoroacetamide is used.[13] The identification of oils is based on the comparison of fatty acid profiles, and most commonly the P : S ratio in oils, to differentiate drying oils egg yolk, linseed, walnut and poppy oil, and the detection of molecular markers for some siccative, and semi-siccative oils.[2]

9.3.5.1 Case Study: Analysis of Multiple Binding Media Using GC–MS

The analysis of proteins and oils and other organic media in the same sample is also possible and requires suitable combinations of extraction and separation, and the separate GC–MS analysis of different

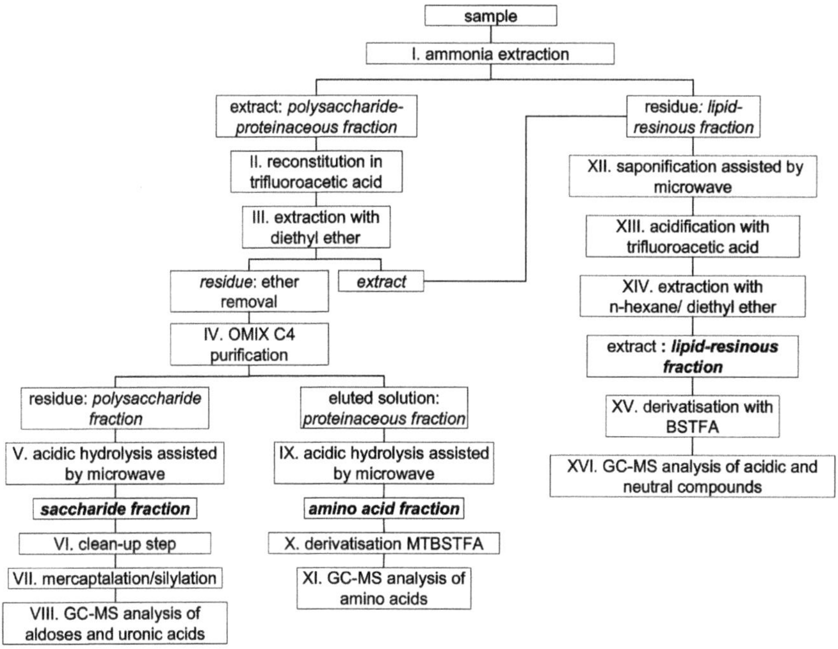

Figure 9.3 The GC–MS combined analytical procedure applied for the analysis of different organic binding media from a single sample.
Reproduced from ref. 13 with permission from American Chemical Society, Copyright 2010.

extracts. One approach that has been successfully adopted for the analysis of samples of paintings allows the sequential analysis of proteins, lipids, resins, polysaccharides and waxes, and is shown schematically in Figure 9.3.[13]

This approach has been successfully applied for the detection of egg-based binders, mixed with drying oils in samples from the Bamayan Buddha,[103] which was also studied using alternative approaches based on FTIR coupled with Synchrotron Radiation[104] and proteomics.[103]

9.4 Impact of Chemistry and Analysis on Conservation, Preservation and Scholarship

There are various reasons for the analysis of paint media. These include interest in artists' materials and historical painting techniques.

Another serious motivation for analysing organic material is the conservation of paintings. The behaviour of binding media in the past, the reactions between media and pigments and potential future changes in their stability may have a direct impact on the condition of a painting. The identification of media will also influence risk assessment that is key to preventive conservation as well as for the choice and application of remedial and passive conservation treatments.

Organic additives and binders used for the execution of wall paintings have often been ignored. A recent international project coordinated by the Getty Conservation Institute on the Organic Materials in Wall paintings was motivated by the damage past conservation treatments caused because they were chosen with the belief that many wall paintings were painted in true *fresco* and did not contain added organic materials, thus inorganic and organic materials present on the surface of paintings were due to past restorations.[58] The project highlighted the need to acknowledge the presence of organic materials, but also the critical problem associated with their analysis that most often requires both sampling and significant expertise.

The organic binding media found in paint will respond to the environment. Proteinaceous paint is often hygroscopic and will contract in dry conditions, but may also be a nutrient source for microorganisms. An important case of a flour-based coating on a ceiling painting that is particularly sensitive to fluctuations in relative humidity is reported by Blauer *et al.*[105] Binding media in canvas paintings are semi-rigid at room temperature, but may become more brittle with ageing reactions, or glass like at lower temperatures. Therefore, the transportation and storage of paintings require careful selection of the most appropriate temperature and relative humidity.[106]

9.5 Selected Case Studies

Three case studies are presented to showcase different applications of analytical methods for the analysis of paint binding media. In the first, FTIR is applied to cross-sections of paintings to map the presence of different binding media (see also ref. 70). The second case study reviews the applications of MS to the study of paintings by Lucio Fontana (see Kokkori *et al.* for a comparison of MS techniques for the analysis of paint[107]). The third case study reports the combination of

MS and proteomics for the analysis of binder in an archaeological sample (see Lluveras *et al.* for another case study combining GC–MS and proteomics on wall paintings[103]).

9.5.1 Mapping and Imaging Cross-sections with FTIR

The analysis of paint fragments is often the only way to understand stratigraphy and the complex inorganic and organic materials present in paint.[67,108] Recent research on the application of FTIR microscopy to the analysis of paint samples is particularly promising for the detection of organic binding media, and degradation products like metal soaps.[66] When considering porous wall paintings the embedding resin used in the preparation of cross-sections may penetrate samples and prevent the analysis of media. Different approaches to overcome this constraint have been published, including the encapsulation of samples and the analysis of thin sections prepared by compressing the sample between faces of a diamond cell.[109,110,111] In a striking example, Cotte *et al.* demonstrate the spatial distribution of binding media throughout a specifically prepared sample in cross section that included lipids, proteins and resins, as well as copper oxalates and lead and copper soaps in an 8th C. paint sample from the Fodali cave in Bamayan.[104] While FTIR cannot here be used to distinguish between specific proteins, as was possible with GC–MS of a sample from the 6th C. Eastern Buddha,[13,103] vibrations detected with FTIR are assigned to amides, lipids and polysaccharides (Figure 9.4).

In other examples of FTIR applied to the analysis of cross-sections using ATR mapping,[70] and ATR Imaging at the National Gallery London allow the analysis of individual layers in traditionally-embedded cross-sections.[5] This may be particularly useful for pigments, but can also lead to insights on the use and distribution of binding media. For example, an FTIR–ATR imaging setup was used, and maps reveal the presence of individual layers from an embedded paint sample.

In Figure 9.5a paint sample from the National Gallery London is shown, with different layers revealed in the paint sample, and false colour images are generated to represent the signal from different ranges in the FTIR spectrum. FTIR spectra extracted from the green paint layer (Figure 9.5c) have typical absorption bands at 1660 and 1540 cm^{-1} that may be ascribed to protein-based media, while bands at 1535 and 1400 cm^{-1} ascribed to lead carboxylates are from areas at the bottom of the spectral image, seen in red in Figure 9.5b.

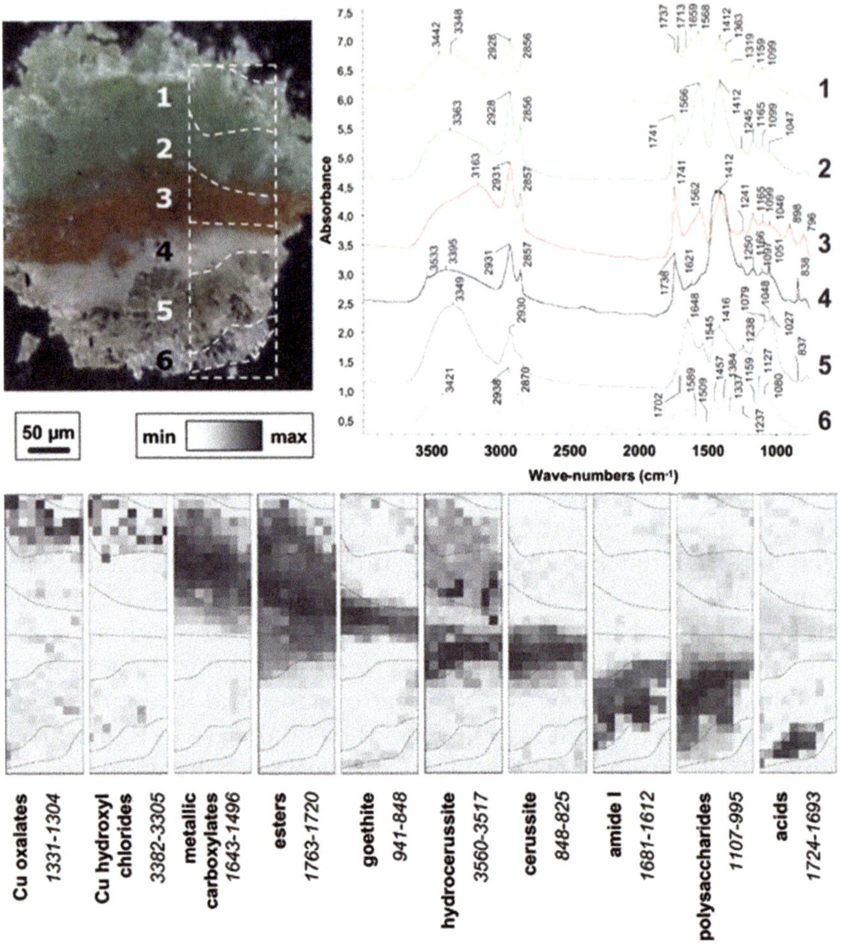

Figure 9.4 μFTIR analysis of a pressed painting fragment from Foladi. Visible light picture, average FTIR spectra and chemical mappings. Regions of interest considered to calculate maps are indicated in wave-numbers. Map size: 144×384 μm^2; step and beam size: 12×12 μm^2.
Reproduced from ref. 104 with permission from the Royal Society of Chemistry.

9.5.2 Analysis of Paintings by Lucio Fontana

Recent technical studies of paintings by the highly experimental 20th C. artist Lucio Fontana have revealed a range of binding media employed by the artist and are presented here to give an example of the use of different analytical approaches that are increasingly useful for the analysis of modern paints.[34,102,107,112] Analysis of samples without derivitisation using pyrolysis coupled with GC–MS is useful for the

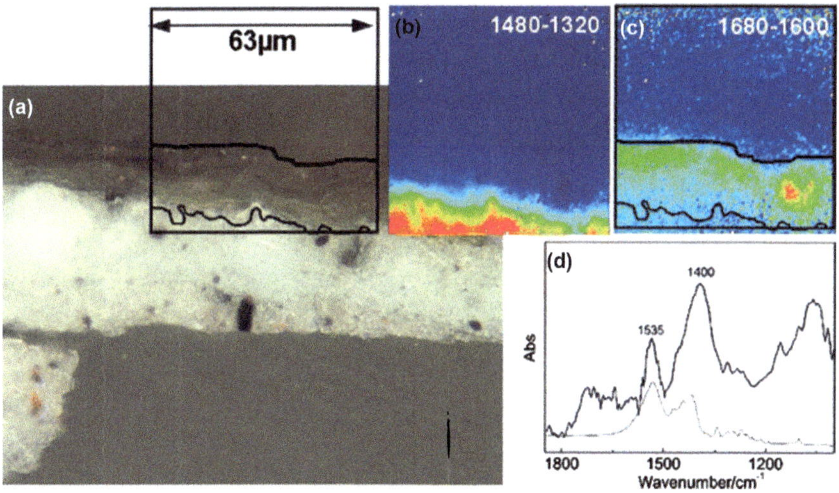

Figure 9.5 Cross section from the Portrait of Paul Pennington by Peter Romney: (a) visible microscopic image with a black box marking the area imaged by ATR-FTIR ($63 \times 63\ \mu m^2$), the line of the top of the lead white layer and the line at the top of the surface coatings; (b) FTIR image representing the integrated absorbance of the band between 1480 and 1320 cm^{-1}; (c) FTIR image representing the integrated absorbance of the band between 1680 and 1600 cm^{-1}; (d) FTIR spectrum extracted from the area below the green zone in the FTIR image shown in (c). The lower dotted trace is a lead azelate reference spectrum.
Reproduced from ref. 5 with permission from Springer Nature, Copyright 2008.

detection of synthetic media while transesterification using (trifluoromethylphenyl)trimethylammonium hydroxide is powerful for the identification of complex mixtures of drying oils[102] (see Table 9.7). It is striking that different analytical methods yield complementary, rather than contradictory, data.

PY–GC–MS of paints and primers revealed of mixtures of oils and synthetic media): EA, MMA, EMA in Emulsion paints that also contained BMA, as well as oil and short chain fatty acids from alkyd-based paints and widespread use of polyvinyl acetate (PVAc), a glue that Fontana used as an additive and adhesive, and also found in commercial paints, mixed with acrylics.[34] Ideally suited to the analysis of modern paint, Py–GC–MS reveals alkyd-based house paints and mixtures of PVAc glue and oil paints were found in different paintings. These media allowed Fontana to experiment and achieve matte paintings with remarkable 3D effects, holes, cuts and adhesions.

Through the combination of thermal methylation and direct Py–GC–MS Chiantore *et al.* were able to determine the presence of

Table 9.7 Summary table of the analysis of some paint samples from paintings by Lucio Fontana from the 1960s.[34,102,112]

Painting	Analytical Technique	Derivitisation	Molecules identified with Mass Spectrometry	Palmitic: stearic acid ratio	Attribution	Reference
61 O 117	THM–PY–GC–MS	TMAH	Glycerol, hexanoic acid, octanoic acid, azelaic acid, palmitic acid, oleic acid, stearic acid, and linoleic acid	1,21	Suggests Linseed	34
64 FD 4	THM–PY–GC–MS	TMAH	Palmitic acid, stearic acid	2,14	Semi -siccative oil	34
60 O 81	GC–MS	TMAH	Palmitic acid, stearic acid, suberic acid, azelaic acid, sebacic acid, oleic acid Linoleic acid, arachidic acid, behenic acid,ricinoleic acid, glycerol	1	Linseed oil with the addition of ricinoleic acid (castor oil), $A/P = 3.2$	102
62 O 66	GC–MS	TMAH	Llauric acid, myristic acid, erucic acid and gondoic acid and their oxidation products 13,14-dihydroxydocosanoic and 11,12-dihydroxyeicosanoi acids.	3,5	Oil-based paint containing rapeseed oil, $A/P = 1.6$	102
64 FD 5	GC–MS	TMAH	Llauric acid, myristic acid, erucic acid and gondoic acid and their oxidation products 13,14-dihydroxydocosanoic and 11,12-dihydroxyeicosanoi acids, glycerol	1,7	Polyvinyl acetate and rapeseed oil mixed with linseed oil	102

61 O 41	Py–GC–MS	None	Phtalic anhydride, styrene, diethylene glycol		Alkyd	112
64 T 15	Py–GC–MS	None		Polyvinyl acetate, polymethylmethacrylate, polyethylmethacrylate	Emulsion paint containing Polyvinylacetate and Acrylic	112
63 FD 9	Py–GC–MS	None	Pentanoic acid, hexanoic acid, heptanoic acid, octanoic acid nonanoic acid, decanoic acid,dodecanoic acid, hexadecanoic acid	Even:odd predominance in short chain	Alkyd	112
64 T 149	Py–GC–MS	none	Ethyl acrylate (AE), ethyl methacrylate EMA, methyl methacrylate (MMA); sequimers, dimers and trimers of EA/MAA		Polyvinyl acetate and acrylic paints	34

oil-based binders or synthetic paints respectively. Alkyd paints, PVAc-based adhesives, acrylic and vinyl-acrylate paints were identified. While ratios of palmitic to stearic acid are sometimes reported as compatible with linseed oil, in other cases Chiantore *et al.* suggest a mixture of semi-siccative oils. More specific analysis of oils in samples is possible through esterification and GC–MS for the detection of fatty acids and their oxidation products. For example, non-siccative castor oil was detected through the marker ricinoleic acid (12-hydroxy-9-cis-octadecenoic), and poorly drying rapeseed oil, detected by markers erucic (13-docosenoic) and gondoic (11-eicosenoic) acids and their oxidation products (13,14-dihydroxydocosanoic acid and 11,12-dihydroxyeicosanoicacid respectively).[102] It is not clear if Fontana or paint manufacturers added these oils to paint, but it is noted that the conservation problems in the paintings are in part due to incomplete drying of the paint due to the non-drying oils.

9.5.3 Proteomics for the Analysis of Paintings from 6th C. B.C.E.

Aegean civilisations are known for the remains of beautiful wall paintings that often contain the early synthetic pigment Egyptian blue. Based on suspicions that paint was applied with a binder, analysis of minute samples of the blue paint was carried with a direct temperature-mass spectroscopy which did not require extraction of proteins from samples that is usually necessary for the alternative GC–MS. Peaks in the total ion thermogram [fragment ions (m/z 67, 91, 107, 117, 131) are assigned to pyrrole and alkylpyrrole, indole and alkylindole, phenol and alkylphenol, toluene, styrene and ethylcyanobenzene], typically attributed to combustion products from egg-based binding media were identified in mass spectra (Figure 9.6).

Proteomics-based analysis was then carried out with the aim of detecting peptide fragments from binding media following enzymatic digestion of a small sample. Analysis of peptides using MALDI-MS then allowed the detection of a peptides sequence of amino acids that can be compared with references, and the identification of egg white Lysozyme C, one of the main proteins found in egg. This finding is of specific relevance to the technology of ancient paintings and to the conservation of similar paintings. Extreme care is required in the handling of paintings that, until recently, were believed to be painted in fresco on wet plaster without organic binding media.[113,114]

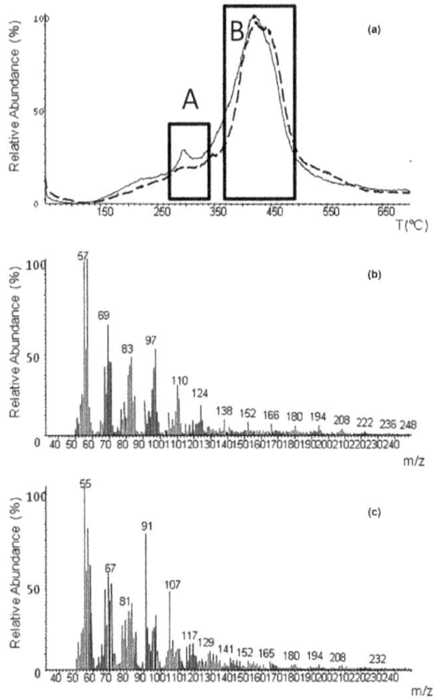

Figure 9.6 (a) Total ion thermogram of sample K1 from Tel Kabri showing Egyptian blue paint layer (straight line) and a preparation layer (dotted line). (b) Average mass spectra of area A. (c) Average mass spectra of area B.

9.6 Conclusions

Two distinct trends can be delineated in the analysis of binding media. The first is the increasing use of non-invasive portable devices for analysis, and it is likely that more sophisticated and accessible instrumentation will become available for the analysis of many painting surfaces, even if the identification of trace media may remain elusive. A second trend is the use of increasingly sophisticated, albeit sometimes costly, techniques for the study of media, their stratigraphic distribution, and their aging or degradation -and in this chapter many of these techniques have been highlighted. It is likely that our understanding of binders and their interaction with pigments will improve, and that new methods for their identification will be developed, both enabling conservation in the future.

Recommended Reading

J. S. Mills and R. White, *Organic Chemistry of Museum Objects, 2nd edn*, Routledge, London, 2011.

S. Dallongeville, N. Garnier, C. Rolando and C. Tokarski, Proteins in Art, Archaeology, and Paleontology: From Detection to Identification, *Chem. Rev.*, 2016, **116**, 2.

T. Learner, *Stud. Conserv.*, 2001, **46**, 225.

F. Piqué and G. Verri, *Organic Materials in Wall Paintings: Project Report*, Getty Conservation Institute, Los Angeles, 2015.

A. Liuveras, I. Bonaduce, A. Andreotti and M. P. Colombini, GC/MS Analytical Procedure for the Characterization of Glycerolipids, Natural Waxes, Terpenoid Resins, Proteinaceous and Polysaccharide Materials in the Same Paint Microsample Avoiding Interferences from Inorganic Media, *Anal. Chem.*, 2010, **82**, 376.

References

1. M. P. Merrifield, *Original Treatises on the Arts of Painting*, Dover Publications, Inc., Mineola, New York, United States, 1967.
2. J. S. Mills and R. White, *Organic Chemistry of Museum Objects*, Routledge, London; New York, 2nd edn., 2011.
3. J. Mills and R. White, *Natl. Gallery Tech. Bull.*, 1989, **13**, 69–71.
4. J. Mills and R. White, *Natl. Gallery Tech. Bull.*, 1985, **9**, 70–72.
5. M. Spring, C. Ricci, D. A. Peggie and S. G. Kazarian, *Anal. Bioanal. Chem.*, 2008, **392**, 37–45.
6. P. Dredge, M. R. Schilling, G. Gautier, J. Mazurek, T. Learner and R. Wuhrer, *J. Am. Inst. Conserv.*, 2013, **52**, 213–226.
7. S. Sotiropoulou, G. Sciutto, A. L. Tenorio, J. Mazurek, I. Bonaduce, S. Prati, R. Mazzeo, M. Schilling and M. P. Colombini, *Microchem. J.*, 2018, **139**, 278–294.
8. M. R. Derrick, D. Stulik and J. M. Landry, *Infrared Spectroscopy in Conservation Science,* Getty Conservation Institute, Los Angeles, 1999.
9. R. J. Gettens and G. L. Stout, *Painting Materials: A Short Encyclopaedia*, Dover Publications, New York, 1966.
10. R. Mazzeo, M. Alfeld, W. Anaf, L. Bellot-Gurlet, U. Bergmann, S. Bernard, L. C. Bertrand, I. Bonacini, I. Bonaduce, B. Brunetti, D. Caramelli, L. Cartechini, F. Casadio, C. L. Daher, I. Degano, B. Doherty, A. L. Gourrier, M. Guardigli, K. H. A. Janssens, M. Lari, S. Legrand, F. Marone, C. Miliani, M. Mirasoli, F. Modugno, L. Monico, J. L. Nasa, W. D. Nolf, G. Nuyts, M. Palmieri, M. P. Colombini, L. Pitzurra, S. Prati, I. Reiche, E. Ribechini, A. Roda, A. Romani, F. Rosi, P. Sciau, G. Sciutto, A. Sgamellotti, M. Thoury, M. Vagnini, S. Vai, G. v. d. Snickt, M. v. Strydonck, F. Vanmeert, J. Verbeeck, M. Vermeulen, K. d. Wael and M. Zangheri, *Analytical Chemistry for Cultural Heritage*, Springer, Cham, Switzerland, 2017.
11. T. Learner and Getty Conservation Institute, *Analysis of Modern Paints*, Getty Conservation Institute, Los Angeles, 2004.
12. S. Dallongeville, N. Garnier, C. Rolando and C. Tokarski, *Chem. Rev.*, 2016, **116**, 2–79.

13. A. Lluveras, I. Bonaduce, A. Andreotti and M. P. Colombini, *Anal. Chem.*, 2010, **82**, 376–386.
14. S. Dallongeville, M. Koperska, N. Garnier, G. Reille-Tailefert, C. Rolando and C. Tokarski, *Anal. Chem.*, 2011, **83**, 9431–9437.
15. H. Y. Lee, N. Atlasevich, C. Granzotto, J. Schultz, J. Loike and J. Arslanoglu, *Anal. Methods*, 2015, **7**, 187–196.
16. E. Barberis, S. Baiocco, E. Conte, F. Gosetti, A. Rava, G. Zilberstein, P. G. Righetti, E. Marengo and M. Manfredi, *Microchem. J.*, 2018, **139**, 450–457.
17. F. Casadio, I. Giangualano and F. Piqué, *Rev. Conserv.*, 2004, 63–80.
18. P. Vitruvius, *Vitruvius: The Ten Books on Architecture*, Dover Publications Inc., London, 1960.
19. *Theophilus, in De diversis artibus = The various arts*, 1961, ed. C. R. Dodwell.
20. C. D. A. Cennini, *The Craftsman's Handbook: Translation of Il libro dell'arte*, Dover Publications Inc., Mineola, 1960.
21. J. Wouters, M. Van Bos and K. Lamens, *Stud. Conserv.*, 2000, **45**, 106–116.
22. M.-P. Subes-Picot, *Revue de l'art*, Presses universitaires de France. Departement des revues, 1992, pp. 85–93.
23. C. Stringari, in *A Closer Look: Technical and Art-historical Studies on Works by Van Gogh and Gauguin*, 1991, ed. C. Peres, M. Hoyle and L. v. Tilborgh, pp. 61–74.
24. M. M. Naumova and S. A. Pisareva, *Stud. Conserv.*, 1994, **39**, 277–283.
25. C. Mancusi-Ungaro, in *Personal Viewpoints: Thoughts About Paintings Conservation: A Seminar Organized by the J. Paul Getty Museum, the Getty Conservation Institute, and the Getty Research Institute at the Getty Center, Los Angeles, June 21–22, 2001*, ed. M. Leonard, The Getty Conservation Institute, Los Angeles,California, United States, 2003, pp. 83–94.
26. R. J. Gettens and E. W. FitzHugh, *Artists' Pigments: A Handbook of Their History and Characteristics*, National Gallery of Art, Washington, 1993, vol. 2, pp. 23–35.
27. S. Georgiou, V. Zafiropulos, D. Anglos, C. Balas, V. Tornari and C. Fotakis, *Appl. Surf. Sci.*, 1998, **129**, 738–745.
28. J. S. Mills and R. White, in *Application of Science in Examination of Works of Art: Proceedings of the Seminar, 7–9 September 1983*, ed. P. England and L. van Zelst, Museum of Fine Arts (Boston), Boston, 1985, pp. 29–34.
29. J. Mills and R. White, *Natl. Gallery Tech. Bull.*, 1983, **7**, 65–67.
30. J. S. Mills and R. White, *Natl. Gallery Tech. Bull.*, 1982, **6**, 3–18.
31. C. S. Tumosa and M. F. Mecklenburg, in *New Insights into the Cleaning of Paintings. Proceedings from the Cleaning 2010 International Conference. Universidad Politécnica de Valencia and Museum Conservation Institute*, ed. M. F. Mecklenburg, A. E. Charola and R. J. Koestler, Smithsonian Institution Scholarly Press, Washington DC, United States, 2013, pp. 51–58.
32. M. F. Mecklenburg, C. S. Tumosa and E. P. Vicenzi, in *New Insights into the Cleaning of Paintings. Proceedings from the Cleaning 2010 International Conference. Universidad Politécnica de Valencia and Museum Conservation Institute*, ed. M. F. Mecklenburg, A. E. Charola and R. J. Koestler, Smithsonian Institution Scholarly Press, Washington DC, United States, 2013, pp. 59–67.
33. M. F. Mecklenburg, in *Museum Microclimates: Contributions to the Copenhagen Conference, 19–23 November 2007*, ed. T. Padfield, K. Borchersen and M. C. Christensen, Nationalmuseet (Denmark), Copenhagen, Denmark, 2007, pp. 19–25.
34. O. Chiantore, R. Ploeger, T. Poli and B. Ferriani, *Stud. Conserv.*, 2012, **57**, 92–105.
35. R. Ploeger, D. Scalarone and O. Chiantore, *J. Cult. Heritage*, 2008, **9**, 412–419.
36. T. Learner, O. Chiantore and D. Scalarone, in *13th Triennial Meeting, Rio de Janeiro, 22–27 September 2002: Preprints*, ed. R. Vontobel, Earthscan Ltd., London,United Kingdom, 2002, pp. 911–919.

37. F. Cappitelli, *The Chemical Characterisation of Binding Media in 20th Century Art*, Royal College of Art, 2002.
38. *Issues in Contemporary Oil Paint*, ed. K. J. van den Berg, A. Burnstock, M. de Keijzer, J. Krueger, T. Learner, A. de Tagle and G. Heydenreich, Springer, New York, 2014.
39. T. Learner, *Stud. Conserv.*, 2001, **46**, 225–241.
40. E. May and M. Jones, *Conservation Science: Heritage Materials*, Royal Society of Chemistry, Cambridge, 2006.
41. H. Berman, K. Henrick and H. Nakamura, *Nat. Struct. Biol.*, 2003, **10**, 980.
42. B. M. Haines, in *Leather: its Composition and Changes with Time*, 1991, ed. C. Calnan and B. Haines, pp. 5–9.
43. P. Koslov and G. Burdygina, *Polymer*, 1983, **24**, 651–666.
44. A. Casoli, P. C. Musini and G. Palla, *J. Chromatogr. A*, 1996, **731**, 237–246.
45. M. R. Schilling and H. P. Khanjian, in *11th Triennial Meeting, Edinburgh, Scotland, 1–6 September, 1996: Preprints (ICOM Committee for Conservation)*, ed.J. Bridgland, James & James (Science Publishers) Ltd., 1996, pp. 220–227.
46. J. D. van den Berg, K. J. van den Berg and J. J. Boon, *J. Chromatogr. A*, 2002, **950**, 195–211.
47. J. D. van den Berg, K. J. van den Berg and J. J. Boon, *J. Chromatogr. A*, 2002, **950**, 195–211.
48. M. Lazzari and O. Chiantore, *Polym. Degrad. Stab.*, 1999, **65**, 303–313.
49. P. D. Iedema, J. J. Hermans, K. Keune, A. van Loon and M. J. N. Stols-Witlox, in *ICOM-CC 17th Triennial Conference Preprints, Melbourne, 15–19 September 2014*, ed. J. Bridgland, ICOM Committee for Conservation, Paris, France, 2014.
50. F. Modugno, F. Di Gianvincenzo, I. Degano, I. D. van der Werf, I. Bonaduce and K. J. van den Berg, *Sci. Rep.*, 2019, **9**, 5533.
51. S. Y. Wei, V. Pintus and M. Schreiner, *J. Anal. Appl. Pyrol.*, 2012, **97**, 158–163.
52. T. Learner, *Conservator*, 2000, **24**, 96–103.
53. F. Piqué, in *Safeguarded Cultural Heritage: Understanding & Viability for the Enlarged Europe: Proceedings of the 7th European Conference "Saveur": 31st May-3rd June, 2006, Prague, Czech Republic*, ed. M. Drdácký and M. Chapuis, Ceská akademia ved. Ústav teoretické a aplikované mechaniky, Prague, Czech Republic, 2007, pp. 31–39.
54. K. A. Dooley, S. Lomax, J. G. Zeibel, C. Miliani, P. Ricciardi, A. Hoenigswald, M. Loew and J. K. Delaney, *Analyst*, 2013, **138**, 4838–4848.
55. L. Giorgi, A. Nevin, L. Nodari, D. Comelli, R. Alberti, M. Gironda, S. Mosca, E. Zendri, M. Piccolo and F. C. Izzo, *Spectrochim. Acta, Part A*, 2019, **219**, 530–538.
56. F. Gabrieli, F. Rosi, A. Vichi, L. Cartechini, L. P. Buene, S. G. Kazarian and C. MilianiT, *Anal. Chem.*, 2017, **89**, 1283–1289.
57. L. Nodari and P. Ricciardi, *Heritage Science*, 2019, **7**, 7.
58. F. Piqué, and G. Verri, *Organic Materials in Wall Paintings: Project Report*, Getty Conservation Institute, Los Angeles, 2015.
59. E. R. de la Rie, *Stud. Conserv.*, 1982, **27**, 65–69.
60. A. Nevin, D. Comelli, G. Valentini and R. Cubeddu, *Anal. Chem.*, 2009, **81**, 1784–1791.
61. C. Miliani, F. Rosi, I. Borgia, P. Benedetti, B. G. Brunetti and A. Sgamellotti, *Appl. Spectrosc.*, 2007, **61**, 293–299.
62. F. Rosi, A. Daveri, C. Miliani, G. Verri, P. Benedetti, F. Pique, B. G. Brunetti and A. Sgamellotti, *Anal. Bioanal. Chem.*, 2009, **395**, 2097–2106.
63. F. Rosi, C. Grazia, R. Fontana, F. Gabrieli, L. P. Buemi, E. Pampaloni, A. Romani, C. Stringari and C. Miliani, *Heritage Sci.*, 2016, **4**.
64. F. Rosi, C. Miliani, C. Clementi, K. Kahrim, F. Presciutti, M. Vagnini, V. Manuali, A. Daveri, L. Cartechini, B. G. Brunetti and A. Sgamellotti, *Appl. Phys. A: Mater. Sci. Process.*, 2010, **100**, 613–624.
65. F. Casadio and L. Toniolo, *J. Cult. Heritage*, 2001, **2**, 71–78.

66. M. Cotte, E. Checroun, W. De Nolf, Y. Taniguchi, L. De Viguerie, M. Burghammer, P. Walter, C. Rivard, M. Salome, K. Janssens and J. Susini, *Stud. Conserv.*, 2017, **62**, 2–23.
67. R. M. A. Heeren, J. J. Boon, P. Noble and J. Wadum, in *12th Triennial Meeting, Lyon, 29 August–3 September 1999: preprints (ICOM Committee for Conservation)*, ed. J. Bridgland, James & James (Science Publishers) Ltd., 1999, pp. 228–233.
68. J. Van Der Weerd, H. Brammer, J. J. Boon and R. M. A. Heeren, *Appl. Spectrosc.*, 2002, **56**, 275–283.
69. C. Martin de Fonjaudran, A. Nevin, F. Pique and S. Cather, *Anal. Bioanal. Chem.*, 2008, **392**, 77–86.
70. E. Joseph, C. Ricci, S. G. Kazarian, R. Mazzeo, S. Prati and M. Ioele, *Vib. Spectrosc.*, 2010, **53**, 274–278.
71. J. Pilc and R. White, *Natl. Gallery Tech. Bull.*, 1995, **16**, 73–84.
72. P. Vandenabeele, B. Wehling, L. Moens, H. Edwards, M. De Reu and G. Van Hooydonk, *Anal. Chim. Acta*, 2000, **407**, 261–274.
73. I. Osticioli, A. Nevin, D. Anglos, A. Burnstock, S. Cather, M. Becucci, C. Fotakis and E. Castellucci, *J. Raman Spectrosc.*, 2008, **39**, 307–313.
74. L. Burgio and R. J. H. Clark, *Spectrochim. Acta, Part A*, 2001, **57**, 1491–1521.
75. P. Vandenabeele, *J. Raman Spectrosc.*, 2004, **35**, 607–609.
76. A. C. Williams, H. G. M. Edwards and B. W. Barry, *Biochim. Biophys. Acta, Int. J. Biochem. Biophys.*, 1995, **1246**, 98–105.
77. H. G. M. Edwards, A. C. Williams and B. W. Barry, *Spectrosc. Eur.*, 1996, **8**, 10–18.
78. E. Casanova, C. Pele-Meziani, E. Guilminot, J. Y. Mevellec, C. Riquier-Bouclet, A. Vincotte and G. Lemoine, *Anal. Methods*, 2016, **8**, 8514–8527.
79. R. Hynek, S. Kuckova, J. Hradilova and M. Kodicek, *Rapid Commun. Mass Spectrom.*, 2004, **18**, 1896–1900.
80. J. Arslanoglu, N. Atlasevich and C. Tokarski, *Abstr. Pap. Am. Chem. Soc.*, 2017, **254**.
81. S. Vallence, *Analyst*, 1997, **122**, 75R–81R.
82. G. Chiavari, P. Bocchini and G. C. Galletti, *Sci. Technol. Cult. Heritage*, 1992, **1**, 153–158.
83. S. M. Harrison, I. Kaml, V. Prokoratova, M. Mazanek and E. Kenndler, *Anal. Bioanal. Chem.*, 2005, **382**, 1520–1526.
84. I. Bonaduce, M. Cito and M. P. Colombini, *J. Chromatogr. A*, 2009, **1216**, 5931–5939.
85. A. Casoli, G. Palla and J. Tavlaridis, *Stud. Conserv.*, 1998, **43**, 150–158.
86. A. Derieux, S. Rochut, M. Papillon and C. Pepe, *Chemie*, 2001, **4**, 295–300.
87. A. Andreotti, I. Bonaduce, M. P. Colombini, G. Gautier, F. Modugno and E. Ribechini, *Anal. Chem.*, 2006, **78**, 4490–4500.
88. M. P. Colombini, F. Modugno, E. Menicagli, R. Fuoco and A. Giacomelli, *Microchem. J.*, 2000, **67**, 291–300.
89. M. P. Colombini and F. Modugno, *J. Sep. Sci.*, 2004, **27**, 147–160.
90. M. R. Schilling and H. P. Khanjian, *J. Am. Inst. Conserv.*, 1996, **35**, 123–144.
91. M. R. Schilling and H. P. Khanjian, in *11th Triennial Meeting, Edinburgh, Scotland, 1–6 September, 1996: preprints (ICOM Committee for Conservation)*, ed. J. Bridgland, James & James (Science Publishers) Ltd., 1996, pp. 211–219.
92. R. Arbizzani, U. Casellato, E. Fiorin, L. Nodari, U. Russo and P. A. Vigato, *J. Cult. Heritage*, 2004, **5**, 167–182.
93. A. Karpowicz, *Stud. Conserv.*, 1981, **26**, 153–160.
94. C. M. Grzywacz, *J. Chromatogr. A*, 1994, **676**, 177–183.
95. A. Liuveras, I. Bonaduce, A. Andreotti and M. P. Colombini, *Anal. Chem.*, 2010, **82**, 376–386.
96. S. L. Vallance, *Analyst*, 1997, **122**, R75–R81.

97. G. Gautier, PhD Dissertation, *A reliable analytical procedure for the characterization and identification of proteinaceous binding media in wall paintings*, University of Pisa, Pisa, 2006.
98. A. Nevin, J. L. Melia, I. Osticioli, G. Gautier and M. P. Colombini, *J. Cult. Heritage*, 2008, **9**, 154–161.
99. G. Gautier and M. P. Colombini, *Talanta*, 2007, **73**, 95–102.
100. K. Sutherland, *J. Chromatogr. A*, 2007, **1149**, 30–37.
101. V. Pitthard, S. Stanek, M. Griesser and T. Muxeneder, *Chromatographia*, 2005, **62**, 175–182.
102. F. C. Izzo, B. Ferriani, K. J. Van den Berg, H. Van Keulen and E. Zendri, *J. Cult. Heritage*, 2014, **15**, 557–563.
103. A. Lluveras-Tenorio, R. Vinciguerra, E. Galano, C. Blaensdorf, E. Emmerling, M. P. Colombini, L. Birolo and I. Bonaduce, *PLoS One*, 2017, **12**, e0172990.
104. M. Cotte, J. Susini, V. A. Sole, Y. Taniguchi, J. Chillida, E. Checroun and P. Walter, *J. Anal. At. Spectrom.*, 2008, **23**, 820–828.
105. C. Bläuer-Böhm, K. Zehnder, H. Domeisen and A. Arnold, *Stud. Conserv.*, 2001, **46**, 251–268.
106. S. Michalski, *Art in Transit*, 1991, 223–248.
107. M. Kokkori, K. Sutherland, J. Boon, F. Casadio and M. Vermeulen, *Heritage Sci.*, 2015, **3**, 30.
108. K. Keune and J. J. Boon, *Anal. Chem.*, 2004, **76**, 1374–1385.
109. M. Derrick, L. Souza, T. Kieslich, H. Florsheim and D. Stulik, *J. Am. Inst. Conserv.*, 1994, **33**, 227–245.
110. C. M. de Fonjaudran, A. Nevin, F. Pique and S. Cather, *Anal. Bioanal. Chem.*, 2008, **392**, 77–86.
111. E. Pouyet, A. Lluveras-Tenorio, A. Nevin, D. Saviello, F. Sette and M. Cotte, *Anal. Chim. Acta*, 2014, **822**, 51–59.
112. P. Gottschaller, N. Khandekar, L. F. Lee and D. P. Kirby, *Stud. Conserv.*, 2012, **57**, 76–91.
113. R. Linn, I. Bonaduce, G. Ntasi, L. Birolo, A. Yasur-Landau, E. H. Cline, A. Nevin and A. Lluveras-Tenorio, *Angew. Chem., Int. Ed.*, 2018, **57**, 13257–13260.
114. H. Brecoulaki, A. Andreotti, I. Bonaduce, M. P. Colombini and A. Lluveras, *J. Archaeol. Sci.*, 2012, **39**, 2866–2876.

10 Textiles

P. Garside

The British Library, 96 Euston Road, London NW1 2DB, UK
Email: dr.paul.garside@gmail.com

10.1 Introduction

Textiles form a vital part of our cultural heritage and material legacy, and represent objects ranging from commonplace items of clothing and household goods, to high status and high fashion costume, decoration and furniture. They are ubiquitous items, and their changing use, composition and construction reflect the ways in which cultures and technologies have developed. As a result, they provide insights into social history, international trade, agricultural development, artistic trends and technological progress. However, textiles are ephemeral objects, and so they – along with the stories they can tell – are easily lost. In the main, they are composed of organic materials which decompose naturally under the influence of the environment and the action of pests and microorganisms. Furthermore, due to the common perception of many textiles as inconsequential day-to-day items, they are often discarded, repurposed or recycled in ways which other cultural artefacts may not be.

The role of the textile conservator, then, is to ensure that as far as possible these objects will survive for future generations, and the role of the conservation scientist is to provide the knowledge about composition, chemistry, microstructure, deterioration and physical state which will facilitate this. Historically this has often been achieved through analyses which require invasive or destructive

Conservation Science: Heritage Materials, 2nd Edition
Edited by Paul Garside and Emma Richardson

Published by the Royal Society of Chemistry, www.rsc.org

sampling. These technologies are now being widely replaced by modern analytical methods that can be used non-invasively, and also often provide more rapid and reproducible results. But however the information is gathered, it must also be informed by an understanding of the materials and their uses in order to be of greatest benefit.

This chapter will broadly describe the different classes of textile materials in terms of their chemistry, microstructure and mechanisms of degradation. The use and suitability of different analytical methods will also be briefly discussed, as will the application of this knowledge to common conservation procedures, highlighting the benefits of a close working relationship between scientist and conservator.

10.1.1 Textile Materials

Stem (or 'bast') fibres such as flax and seed fibres like cotton were probably the first materials to be commonly used to create textiles. Linen, derived from flax fibres, was the general fabric of ancient Egypt, and the plant was the first cultivated source of textile fibres in Europe. Although cotton only achieved widespread use in Europe within the past few centuries, its history in Asia and Central America dates back over 7000 years. The spinning of animal-derived fibres to form threads and yarns seems to have occurred somewhat later. In Mesopotamia, woven wool cloth was being produced just over 6000 years ago, and the used of silk ('sericulture') originates in China, probably in the early 3rd millennium BC. These fibres and fabrics form the mainstay of most historic textile collections, and due to their long history of use there is a large body of existing literature concerning their properties and deterioration, but they nonetheless still provide a wealth of opportunities for new research.

Although subsequently there were many significant developments in textile technology, perhaps the most dramatic change occurred in the 20th century with the revolution in synthetic polymer chemistry by companies such as Courtaulds, DuPont and ICI. This led to the production of materials like regenerated natural cellulose (rayon), modified cellulose (acetate) and fully synthetic polymers including polyamides (nylon), polyesters, acrylics and elastomers. Many of these materials were originally developed in the search for inexpensive substitutes for luxurious but costly fibres such as silk, or to compensate for shortages of more commonplace materials, particularly those encountered during the Second World War. A century of usage has ensured that these materials are now found in increasing numbers in collections, and they form an important (if sometimes underappreciated) part of the story of many technological and social revolutions, including warfare, the space race and lifestyles. In the past, conservation of these materials may have been hampered by the perception of synthetic fibres as indestructible (hence their reputation for being 'environmentally unfriendly'), therefore requiring little attention, although there in now an increasing understanding that many are prone to rapid deterioration and can also be attacked by microorganisms and other pests.

More recently still, the specific and adaptable properties of synthetic polymers have been exploited to fulfil particular roles, including those designed to cope with extreme conditions or to incorporate responsive or 'smart' functionalities. The long-term behaviours of such materials are not yet well understood, and so will present novel challenges to conservators and conservation scientists for generations to come.

In addition to the range of fibre and fabric types which can be encountered, textile artefacts may incorporate a wide range of other materials. These include dyes, mordants, optical brighteners, weatherproofing and fireproofing treatments, paints, adhesives, glazes, waxes, sizing, residues from processing methods (such as bleaching), functional or decorative elements (for example, buttons, zips, beads or sequins, which may include such materials as metals, glass, bone, ivory, feathers, plastics, wood or minerals) and structural components (frequently wood, metal or plastic). A knowledge of the ways these materials may interact with the surrounding textiles and the local environment, and in which they will deteriorate and change, is vital to properly care for such artefacts, and science has a crucial role to play in providing this understanding. Many of the other chapters of this book deal with the specific properties of these materials in greater detail.

Whereas the textiles chapter in the first edition of this book concentrated on natural fibres (linen, silk and wool), this chapter will endeavour to provide a more wide-reaching overview of the wide range of materials which may be encountered in a modern textile collection – natural, synthetic and semi-synthetic.

A case study is provided by the analysis and conservation work carried out on the fore topsail of the *HMS Victory*, Nelson's flagship at the Battle of Trafalgar in 1805, demonstrating the way in which an understanding the chemistry, microstructure and properties of a textile artefact can inform appropriate and sympathetic treatment choices.

10.2 Chemistry, Microstructure and Structure of Textiles and Fibres

The overall performance of a fibre or textile is dependent on a wide range of factors, including chemistry, microstructure, fibre structure, yarn structure and textile structure. Textile fibres are principally polymeric materials, composed in the main of long chain molecules which are ultimately derived from simple precursor monomers (or oligomers). The length of the polymer chain is described by 'degree of polymerisation' (DP), defined as the average number of repeat units per polymer molecule in the material. In general, longer chains give better physical properties in the bulk material. Most textile fibres are highly crystalline (for example, the cellulose of cotton is typically 70–80% crystalline), other otherwise exhibits significant long-range ordering (as is the case with wool); this imparts many of the properties necessary for the material to be used in such a role.

The microstructure describes the way in which these polymer chains aggregate, orient, order and relate to one another. Microstructures are often hierarchical, with several different levels of ordering of increasing scales. A particularly important factor is the extent to which the polymer adopts crystallinity or long-range ordering. In crystalline regions (crystallites), the polymer chains are constrained to specific arrangements through inter- and intramolecular interactions; these crystalline regions often also have a specific orientation with respect to the bulk fibre axis. These crystallites are typically set within an amorphous matrix, comprised of the remainder of the polymer (and sometimes additional components as well); the amorphous matrix lacks short- or long-range ordering. There may also regions of poorly ordered crystallinity, in which crystallites are present, but do not possess a consistent long-range orientation; this is common in less highly crystalline materials, and may also arise as the result of damage (break-down of the amorphous region can allow the ordering of the embedded crystallites to 'drift'). The relative properties of the crystalline and amorphous regions give fibres their unique properties, with their balance between strength and flexibility, resilience and elasticity. The strength of the material is dictated by the interlinked, inextensible crystalline component, whilst any viscoelastic and plastic behaviour is determined by the amorphous regions. Most fibres will exhibit elastic recovery if stretched, but only within limits; as loading on the fibre increases, the chains in the amorphous zones will slip past each other effecting first recoverable and then irrecoverable creep. In the extreme, of course, the aggregate will disrupt and the fibre will fracture.

This may be modelled in a number of ways, but the simple model of the 'fringed micelle' is a useful way of understanding these functional properties. This model presents a picture of a semi-crystalline polymer aggregate, in which each polymer chain passes through several ordered regions (crystallites), which result from the close packing of neighbouring chains, and through several amorphous regions, in which it adopts an unordered, random coil arrangement (Figure 10.1).

Increasing in scale from the microstructural level to that of complete fibres, the length of the resultant fibre dictates how it is then employed. Longer natural fibres require little spinning to form coherent yarns. This is particularly the case with good quality silk, in which a single fibre may be several hundred metres long; flax fibres (up to 70 cm long) similarly do not require tight spinning. Synthetic fibres may also be treated in this way, when used as continuous filaments. Shorter fibres, or cut or broken sections of longer fibres, may be treated as 'staples', and these generally require tighter spinning in order to form a yarn with sufficient strength; some fibres, such as wool, may be used with shorter staple lengths, due to their propensity to inter-tangle. Spinning may be either in the 'S' or 'Z' sense – see Figure 10.2.

Textiles can then be produced by a variety of techniques. Yarns may be woven or knitted, and different methods will yield textiles with particular physical properties and appearances. Yarns or individual fibres can be used to create non-woven textiles; traditionally this was commonly employed with wool to produce felt, using the tendency of the fibres to tangle under particular conditions, and barkcloth is also created in this way. It can also be achieved with synthetic fibres which possess thermoplastic properties, using sufficient heat to cause the material to soften and fuse. Synthetic materials may also be produced as continuous sheets.

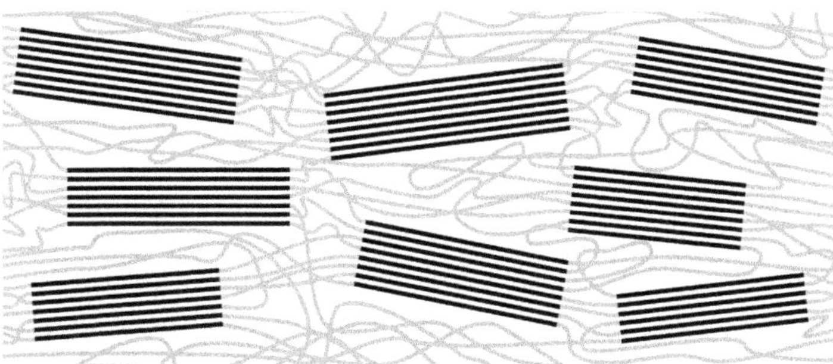

Figure 10.1 The fringed micelle of fibre polymers.

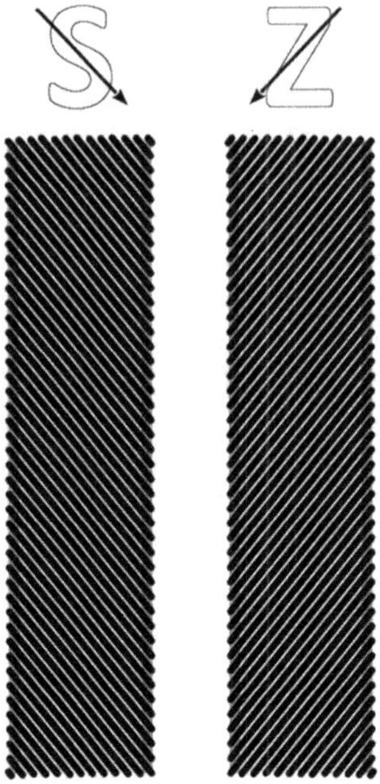

Figure 10.2 S and Z twist.

10.3 Cellulosic Fibres

Cellulosic fibres are derived from plant cells, and generally fall within three types: seed fibres including cotton, kapok and coir (taken from fibres surrounding the seed-head), bast fibres such as flax, hemp, jute and ramie (derived from the stem of the plant) and leaf fibres like sisal and abaca. The fibres are cellular in nature, and these cells are predominantly composed of cellulosic cell walls around a hollow lumen. Fibres from some species (such as cotton) are found as individual cells; in other species (for example, bast fibres such as flax and hemp), bundles of cells are cemented together by the material of the middle lamella.

Cotton is the most widely used of the seed fibres, and is the plant fibre in greatest current commercial production. The fibres are commercially cultivated from four species: *Gossypium hirsutum*, *G. barbadense*, *G. arboreum* and *G. herbaceum*. Cotton fibres are almost entirely composed of cellulose (up to 99% after processing), which is highly crystalline (usually in the range 70–80%). The fibres are allowed to dry after harvesting, and on doing so they become flattened and adopt their characteristic random spiral convolutions, alternating between clockwise and anti-clockwise, which are readily observed under the microscope (Figure 10.3).

Bast fibres are derived from the long structural cells of plant stems, and include such species as flax, hemp and ramie. Like cotton, these fibres are soft, and typically have a relatively low lignin content (jute being an exception). Unlike seed fibres, which are found as individual elongated cells, bast fibres are composed of bundles of elongated cells (ultimates) cemented together by hemicelluloses, pectins, lignin and other materials (which comprise the middle lamella).

Leaf fibres, such as sisal and abaca, tend to be hard and coarse, and have a relatively high lignin content.

Regardless of the source, the cell walls of all plant fibres are principally composed of polysaccharides, with a smaller proportion of lignin, proteins, pigments, waxes and minerals. The polysaccharide component is primarily cellulose (in the form α-cellulose or cellulose I), along with hemicelluloses and pectins. Cellulose itself is composed of 1,4-β-(D-glucose) units; the repeat unit is cellobiose, a glucose dimer (Figure 10.4(a) and (b)).

For undegraded cotton cellulose the degree of polymerisation is typically 6000–8000. The polymer is linear, but hydrogen bonding creates a strong intermolecular attraction between adjacent chains (Figure 10.5), which leads to a high degree of crystallinity (typically 60–90% for textile fibres). This crystallinity gives the fibres physical strength and chemical resistivity. The unit cell is presented in Figure 10.6.

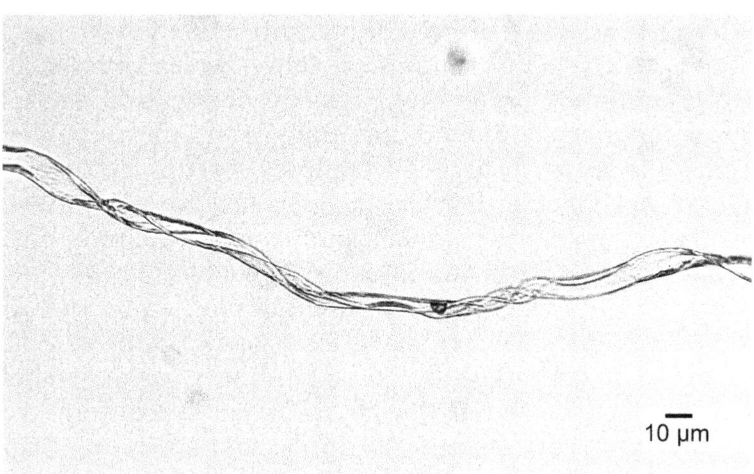

Figure 10.3 A typical micrograph of cotton fibres.

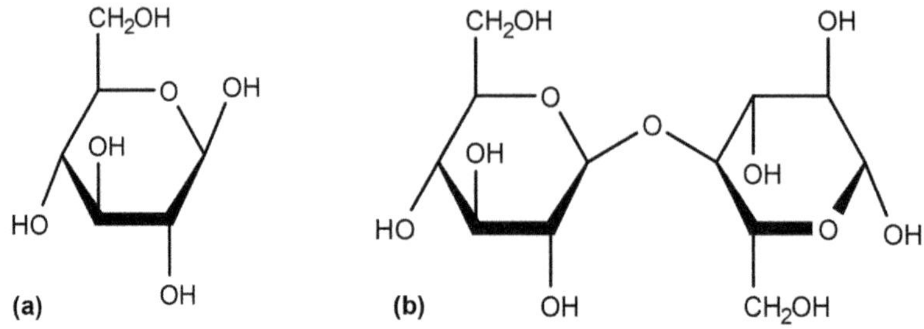

Figure 10.4 (a) The structure of glucose. (b) The structure of cellobiose.

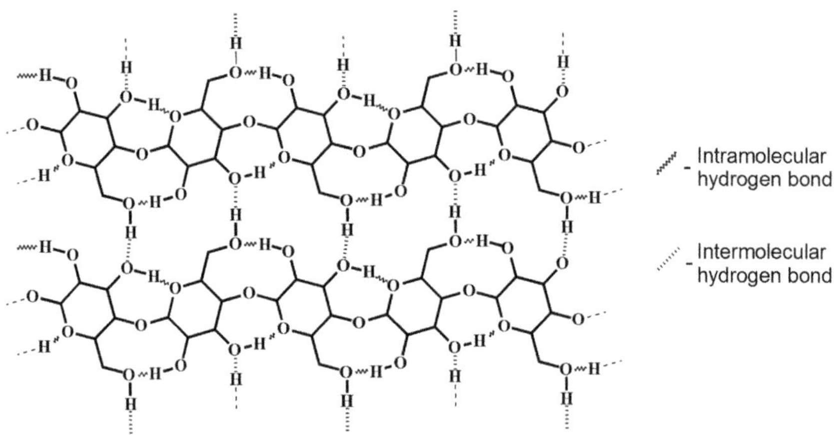

Figure 10.5 Hydrogen bonding between cellulose chains.

The fibres possess a hierarchical microstructure in which regions of well-ordered crystalline cellulose are interspersed with sections of random crystallinity, lacking long-range ordering, and amorphous regions, without crystallinity or ordering; within these components additional species such as hemicelluloses, pectins and lignin are also found. At the molecular level, aggregates of roughly 100 cellulose chains organise into parallel – and hence crystalline – bundles, which associate to form micelles or elementary fibrils (5–6 nm diameter). In turn, roughly 15 of these micelles, embedded within the amorphous intermicellar space, aggregate to form microfibrils (75–90 nm diameter). Bundles of microfibrils, again embedded within amorphous material (the interfibrillar space) form macrofibrils (also referred to simply as fibrils, and typically 0.5 μm in diameter). Macrofibrils are found as lamellae in the cell walls. The intermicellar and interfibrillar matrices (as well as the intercellular

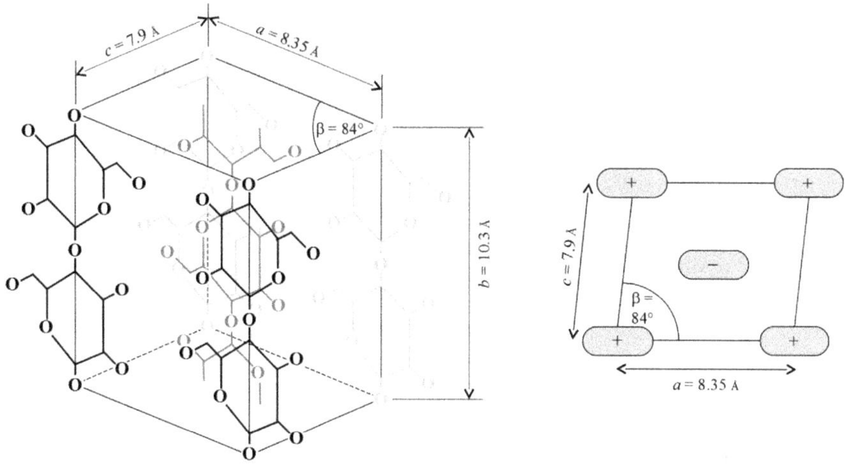

Figure 10.6 The unit cell of crystalline cellulose.

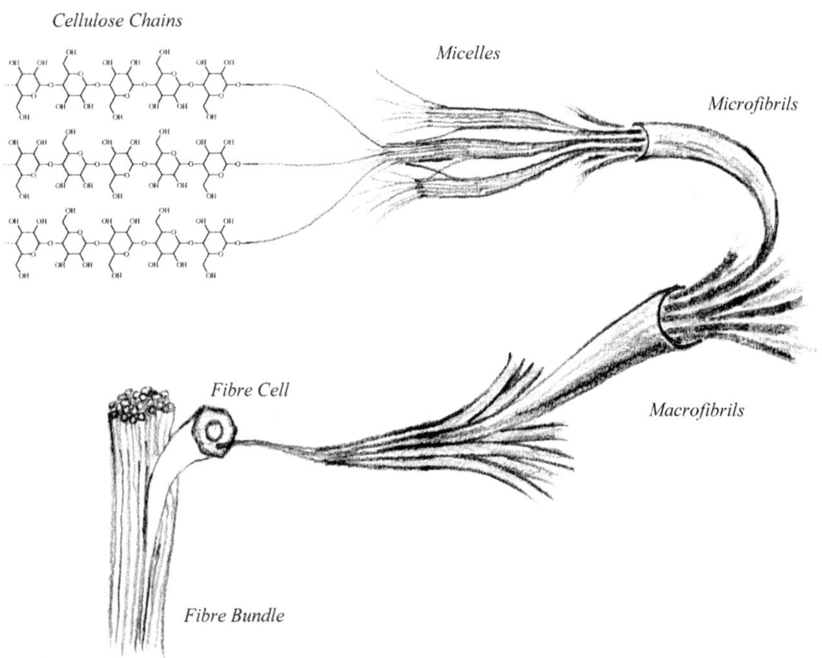

Figure 10.7 Hierarchical microstructure of a cellulosic bast fibre.

middle lamella, in the case of fibres found as ultimates) are composed of the non-crystalline components – amorphous cellulose, hemicelluloses, pectin and lignin; within these regions there are extensive networks of capillaries, which play a role in the material's response to water. See Figure 10.7.

The cell walls are organised in layers; the secondary wall is thicker than the primary (several microns, as opposed to several tenths of a micron, respectively, depending on the species), possesses a greater degree of crystallinity, and makes the dominant contribution to the physical properties of the fibre. In the cell walls, macrofibrils are wound with an angle and sense (direction) characteristic of the fibre species – for example, the secondary cell wall is S wound at ~6.5° for flax, or Z wound at ~7.5° for hemp – except for cotton, which whilst wound at a consistent angle of 12°, has a sense of wind that varies along the length of the fibre and leads to the characteristic convolutions of the material (see Table 10.1).

The other main polysaccharide components are hemicellulose and pectin. Hemicelluloses (primarily composed of xylose and mannose) are much shorter in length than cellulose and are extensively branched. In plant fibres, the cellulose microfibrils are embedded in an almost entirely amorphous hemicellulose matrix. These species may also act to cross-link cellulose and other non-cellulosic polymers. Pectins are jelly-like, acidic polymers composed of 1,4-β-(D-galacturonic acid) units, along with some galactose and arabinose. The acid groups are generally present as either methyl esters or metal salts (primarily those of magnesium, calcium and iron).

Lignin is a heterogeneous phenylpropanoid polymer with a complex amorphous structure; it is aromatic and acidic. It is found in the hemicellulose matrix, to which it may be covalently bound, and is also a major component of the intercellular matrix (middle lamella). It is highly light sensitive (particularly to ultraviolet radiation), with prolonged exposure causing yellow-brown discolouration and an increase in acidity. Like cellulose, it has a structural role, and helps resist swelling. In higher concentrations it gives some fibres a 'woody' character, but its exact nature varies with species and location within

Table 10.1 Composition and properties of cellulosic fibres.

		Cotton	Flax	Hemp	Ramie	Sisal
Composition, %	Cellulose	83	64	67	67	66
	Hemicellulose	6	17	16	13	12
	Pectin		2	1	2	1
	Lignin	0	2	3	1	10
	Other	11	15	13	17	11
Crystallinity, %		70–80	70	70	70	65
Winding (2° cell wall)		*Variable*	S	Z	S	Z
Angle of wind (2° wall)		12°	6.5°	7.5°	7°	20–25°
Fibre length, mm		10–60	400–900	20–50	200–400	800–1200

the plant; as a result, some relatively highly lignified fibres are soft to the touch (for example, kapok has a lignin content of about 15%).

The remainder of the fibre is composed of: waxes, generally found on the fibre surface; pigments, primarily chlorophyll, xanthophyll and carotene (imparting green, yellow and red/orange colours, respectively); residual protein; and minerals, such as oxalate crystals, silica and calcium and potassium salts.

10.3.1 Deterioration of Cellulosic Fibres

The degradation of cellulosic fibres is largely related to the chemistry of cellulose, although the behaviour of the other components also plays an important role. They are susceptible to a range of degradative processes, though in general most deterioration is ultimately attributable to oxidative or hydrolytic reactions, which can lead to both scission and cross-linking of the cellulose polymer. Chemical degradation tends to result in a polymer assembly with a reduced degree of polymerisation, increased apparent crystallinity due to aggregation of smaller chain fragments, and the development of some networked character, which leads to weakening and embrittlement.

Water is crucial to the normal behaviour of cellulosic fibres. Under moderate conditions (RH 45–65%) water is readily absorbed through pores and capillaries in the fibres, and acts as a plasticiser in the amorphous regions, modifying hydrogen bonding and giving flexibility to the bulk polymer; in the absence of water the fibre would be permanently brittle with an effective glass transition point well above room temperature. Higher humidities, or exposure to liquid water, leads to swelling which may cause substantial physical damage, particularly in aged fibres in which the polymer is already compromised, as well as opening the polymer structure which can lead to ingress by pollutants and other reagents. These conditions also favour mould growth and microbiological attack. Desiccation, on the other hand (occurring with RH less than 30%) can lead to increased cross-linking in the amorphous regions through secondary bonding, ultimately causing shrinkage, brittleness and loss of flexibility. Lignin and hemicellulose will moderate interactions with water, with the former helping to resist swelling due to its hydrophobicity and structural function, and the latter in retaining moisture through its branched, highly amorphous and hydrophilic polymer structure.

Thermal oxidation is driven by free radicals and results in both chain scission and cross-linking, with damage initiating in the accessible amorphous regions. The process also leads to discolouration.

These reactions occur slowly at room temperature, but are accelerated by the presence of pre-existing free radicals, such as those created by photlytic processes. Lignin and, to a lesser extent, hemicellulose are both more susceptible than cellulose.

Cellulose is vulnerable to both acid and alkaline hydrolysis, with the amorphous regions once again being the initial sites of attack. Acid hydrolysis occurs readily and randomly throughout the polymer chain, leading to the scission of the glycosidic ether bond (Figure 10.8). The resulting material, hydrocellulose, tends to be highly crystalline as the shorter chain fragments are readily mobile and can reorient themselves to permit extensive interchain hydrogen bonding, but the material is mechanically weak and lacks flexibility. Cellulose displays a greater resistance to alkaline attack, which, unlike acid hydrolysis, only occurs at the 'C1' end of the polymer chains (Figure 10.9) and proceeds *via* ring opening, reorganisation and β-elimination, leading to a progressive depolymerisation occurring one unit at a time. As a result, this will not cause significant deterioration at the bulk level unless the polymer is already compromised by pre-existing damage – the short chains of hydrocellulose and oxycellulose (the products of acid hydrolysis and photo-oxidation) will degrade much more rapidly.

Light, and particularly ultraviolet radiation, will damage cellulosic fibres by various photodegradative reactions, including direct photolysis, photochemical and radical oxidation, and photosensitised degradation (Figures 10.10 and 10.11). These processes are accelerated by moisture and catalysts such as transition metal-based dyes and mordants. Extensive photodegradation leads to depolymerisation and the formation of variety of small, water soluble, chromophoric acidic species; although they can be washed out of degraded fabric to reduce acidity and discolouration, this also risks further physical weakening of the material.

As with most natural organic materials, cellulose is prone to biological degradation by moulds, microbes and pests. Microorganisms which feed on cellulose break down the material into metabolites using cellulase enzymes, of which there are three main types: endo-β-1,4-glucanase, which breaks β-1,4 bonds; exo-β-1,4-glucanase, which removes cellobiose units from the chain ends; and β-glucosidase, which cleaves the glycosidic link in cellobiose to form glucose. Other metabolic by-products of these organisms, such as hydrogen peroxide and organic acids, may cause additional damage. Biological attack will be facilitated by improved access to the polymer, whether in the naturally vulnerable amorphous regions, through the opening of the structure

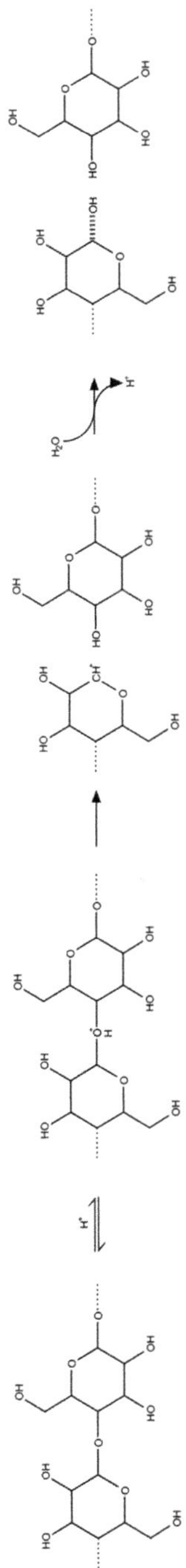

Figure 10.8 Acid hydrolysis of cellulose.

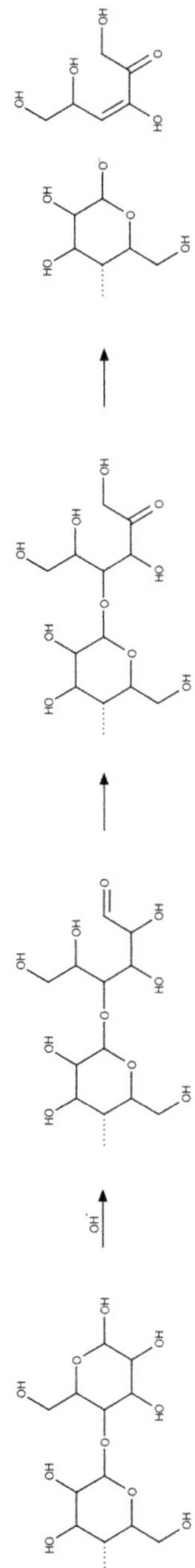

Figure 10.9 Alkaline hydrolysis of cellulose.

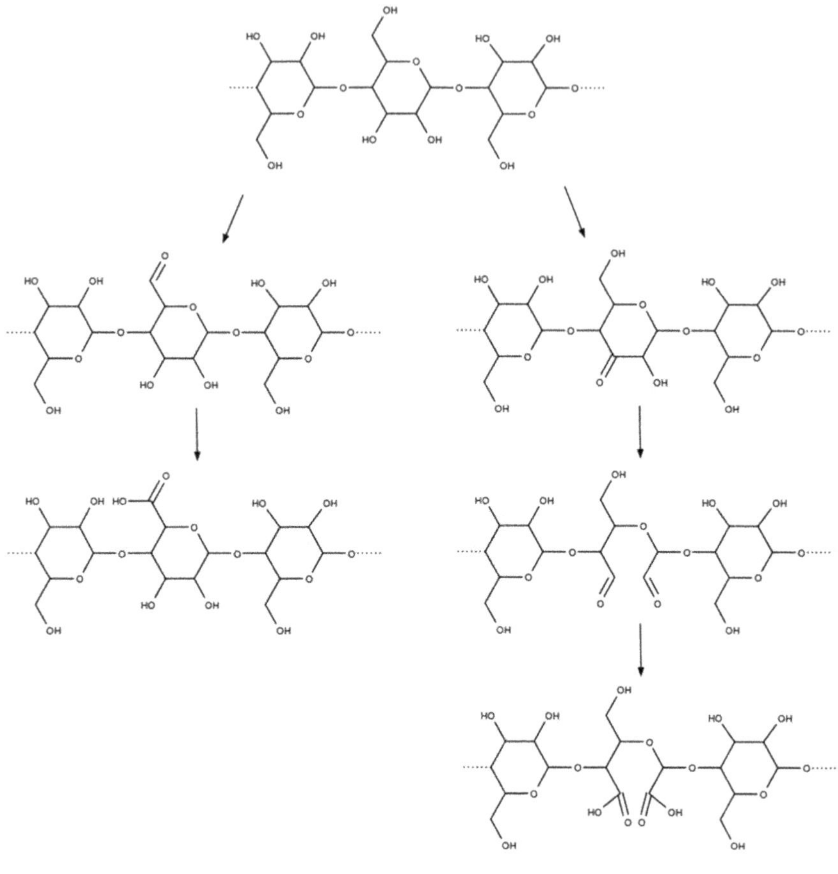

Figure 10.10 Photodegradation of cellulose.

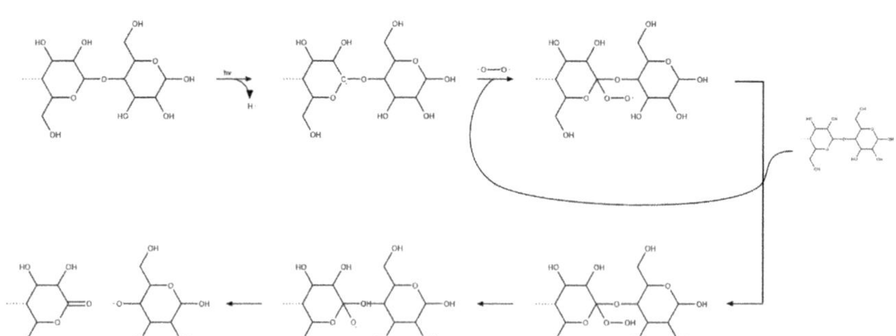

Figure 10.11 Radical photolysis of cellulose.

resulting from swelling or due to pre-existing damage. Hemicellulose and pectin are particularly prone to this type of damage due to their lack of ordering, but lignin exhibits a greater resistance.

10.4 Silk

Silks are produced by a number of insect and spider species, but the major sources of commercial textile silks are the silk moths of the family *Bombycidae*, of which the most important is the domesticated silk moth, *Bombyx mori*. Moth larvae (silkworms) produce the fibre to form a protective cocoon prior to pupation. The first known references to the cultivation of silkworms (sericulture) come from China in the second or third millennium BC, from where this practice spread through Asia, then on to the Middle East and eventually Europe.

The polymeric material is produced by the silkworm from a liquid crystalline phase, and the fibre is formed by the extrusion of this from the larva's spinneret as a 'bave', consisting of two roughly triangular fibroin filaments ('brins') bound together by a second protein, sericin (Figure 10.12). In commercial production, the chrysalid is killed before it can hatch (which would damage the thread), and a continuous filament can then be unreeled from each cocoon, giving typically 500–800 m of useable fibre.

Fibroin is largely composed of three amino acids, glycine (45%), alanine (29%) and serine (12%); it can be divided into two main components, the crystalline and the amorphous regions, which are in the approximate ratio 3 : 2. The crystalline regions are formed of a hexapeptide motif (Figure 10.13), folded into an anti-parallel β-sheet secondary structure, held together be extensive hydrogen bonding (Figure 10.14) and strongly aligned with the fibre axis:

-Gly-Ala-Gly-Ala-Gly-Ser-

The crystalline close packing of this region is facilitated by the small, non-bulky side-chains of the residues (–H, –Me or $-CH_2OH$); see Figure 10.15. The unit cell is built from four protein chains, with

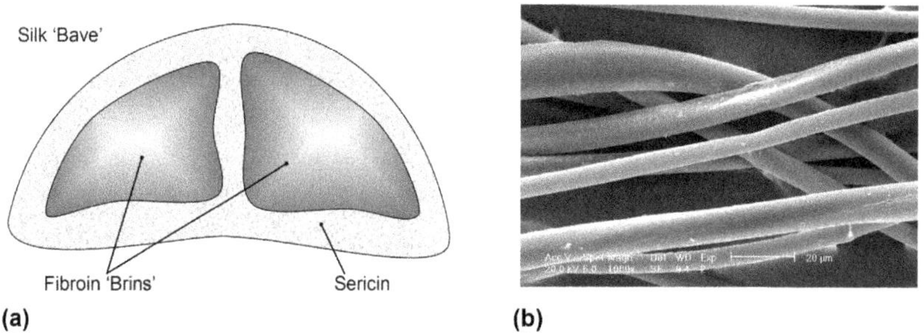

Figure 10.12 (a) Diagram of the cross-section of a native silk fibre. (b) Scanning electron micrograph of silk fibres.

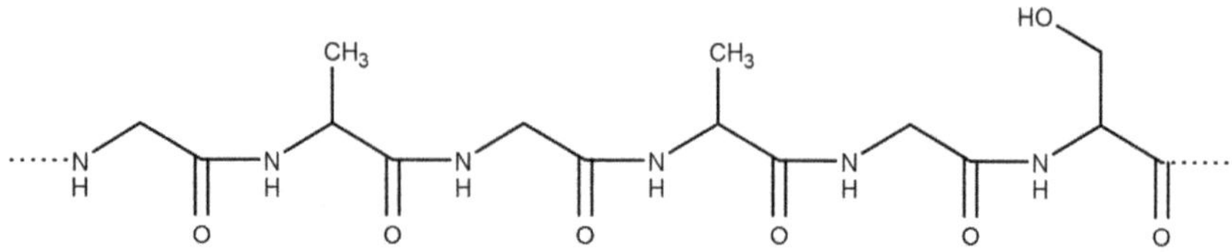

Figure 10.13 Silk hexapeptide motif.

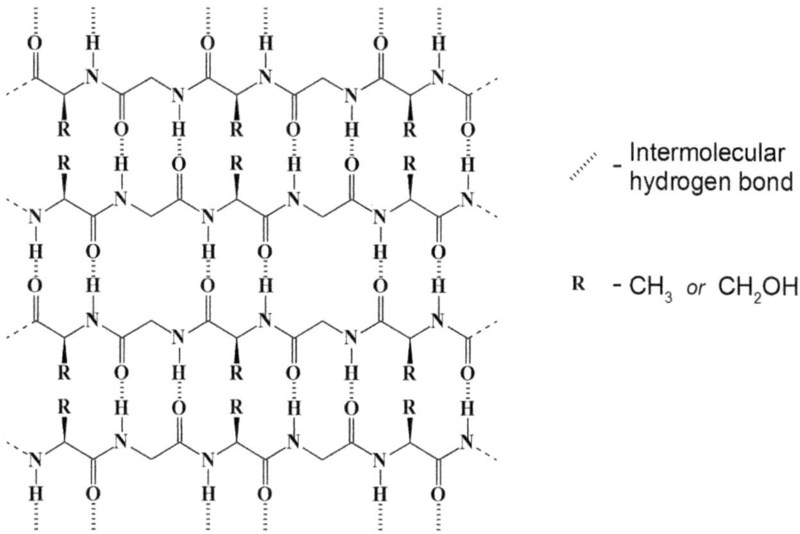

Figure 10.14 Secondary structure hydrogen bonding in silk β-sheets.

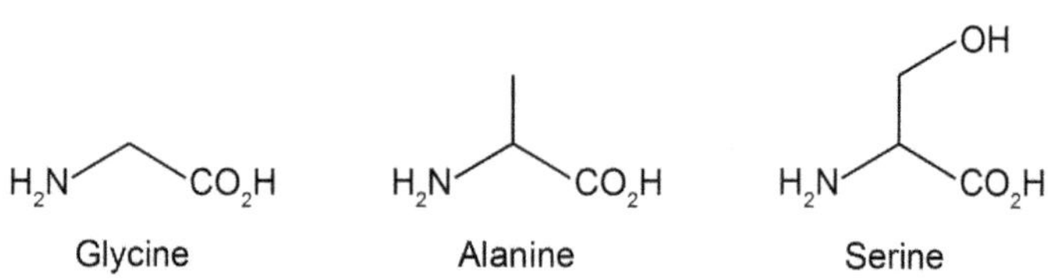

Figure 10.15 The structure of glycine, alanine and serine.

ab being in the plane of the β-sheets and *b* lying along the fibre axis (Figure 10.16). The high crystallinity of fibroin leads to the characteristic properties of the fibre: good mechanical strength and limited extensibility, arising from the highly inter-bonding and full extension of the protein chains, and a high degree of chemical resistivity.

The amorphous and semi-amorphous regions that comprise the remainder of the fibre contain a high proportion of residues

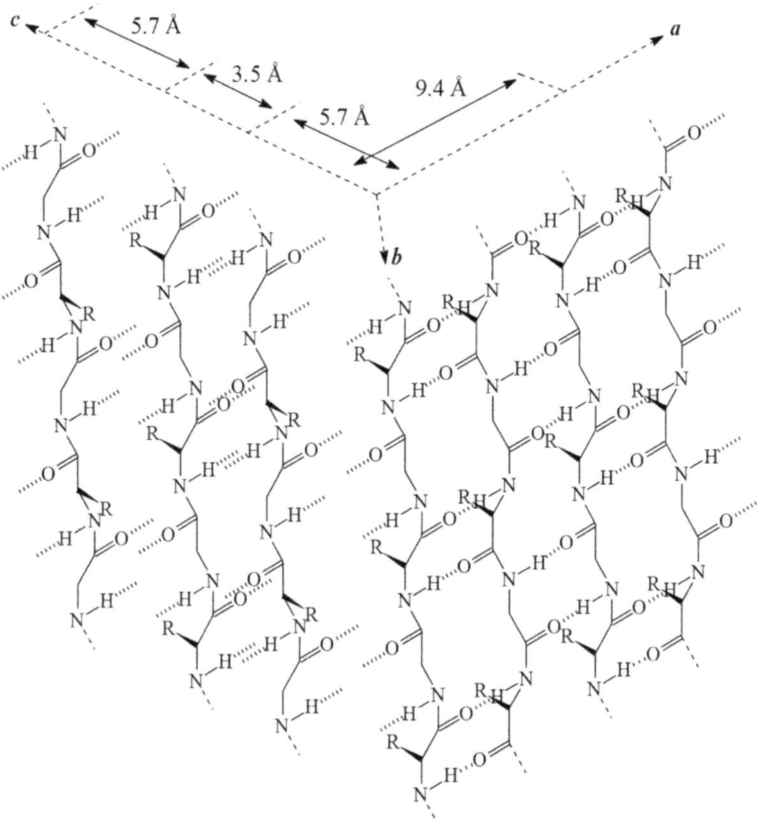

Figure 10.16 The silk unit cell in silk β-crystallites; *b* lies along the fibre axis.

possessing bulky side-groups, which dictates a lower degree of ordering and greater susceptibility to chemical attack.

As with other natural fibres, silk has a hierarchical microstructure – about five anti-parallel β-sheets (each with around 12 chains) aggregate to form parallel, crystalline microfibrils (approximate diameter 10 nm), bundles of which make up fibrillar elements (approximate diameter 1 μm), which in turn associate to comprise the individual fibroin filaments (7–12 μm); at each level of organisation, the ordered elements are embedded within amorphous matrices derived from the non-crystalline components (Figure 10.17). The behaviour of the structural composite can be understood in terms of the semi-crystalline array of its component parts.

Sericin, the protein that binds the pairs of fibroin filaments, has a markedly different composition and structure. In the native fibre, it may have a role in dehydrating the fibroin and encouraging its

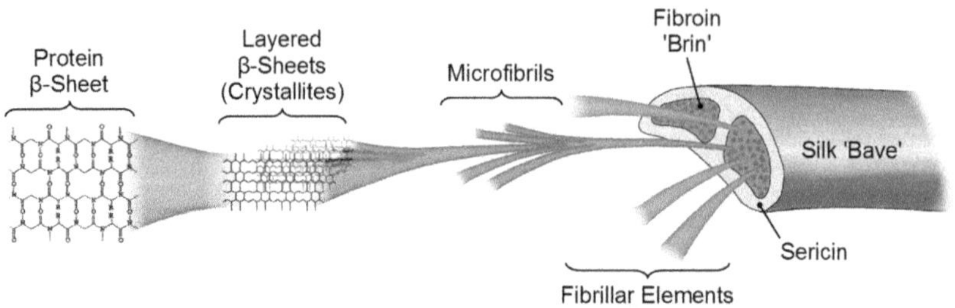

Figure 10.17 The hierarchical microstructure structure of silk.

crystallisation. The material is largely amorphous and rich in serine (~32%), aspartic acid (~14%) and glycine (~13%); there is a much greater proportion of residues with polar and/or bulky side-chains. The predominance of these polar, hydrophilic groups means that sericin is readily soluble in hot water.

Higher quality silk yarns are 'thrown' as lightly twisted bundles of long, continuous filaments. Shorter broken sections of fibre may instead be spun to produce lower quality yarns. The sericin may be left intact, to give rougher textured, more readily dyed 'raw' silk, or removed through 'degumming' to leave individual fibroin filaments. Degumming can be achieved by the use of boiling water or steam, alkaline (soap) washes, fermentation or, more recently, with acid treatments. Degummed fibres are smoother and more lustrous, but may have reduced physical strength and a lower affinity for water, dyes and weighting agents.

Following degumming silk can be bleached to remove residual colour. Historically sulphur 'stoving' (exposure to the fumes of burning sulphur) was common in Europe, until superseded by modern 'wet' methods, including the use of hydrogen peroxide, sodium peroxide, various bisulfites and 'blueing' agents. Incomplete removal of bleaching agents can lead to significant damage to the material over the longer term.

A particularly important aspect of European silk processing is 'weighting'. Historically silk was sold by weight, up to 25% of which may be lost through degumming; to restore this lost weight, and hence maintain the value of the material, weighting agents were added. This ultimately became an accepted processing method which altered the feel (crisper and stiffer) and drape of the fabric. In extreme cases the weighting agents may represent several times the original weight of the fibre. Weighting agents have included gum Arabic, salts of tannic, gallic and formic acids, catechu and logwood ('vegetable weighting'), protein glues, waxes, sugar and various metal salts. The use of metal salts was especially widespread from the late 18th to early

20th centuries, and these materials may also act as mordants, especially for darker dyes. They include compounds of iron, lead, tin, aluminium, sodium and zinc, amongst others; the most common are salts of lead and tin (as lead$^{(II)}$ acetate and tin$^{(IV)}$ chloride). They are readily absorbed into the amorphous regions of the fibre. All weighting agents, and especially metal salts, can promote degradative reactions in silk, causing long-term preservation issues, and this may be exacerbated by the harsh conditions encountered during the weighting process itself (such as hot or acidic aqueous treatments). Consequently, many historic silks, especially European silks of the 19th and early 20th century, are in a relatively poor state.

10.4.1 Deterioration of Silk

The high crystallinity of silk gives it a generally good chemical resistance, but it is prone to damage under certain conditions. The majority of degradative processes are hydrolytic or oxidative, and attack tends to start in the amorphous regions which are both more open to ingress by reagents and contain a higher proportion of reactive amino acids (particularly histidine, lysine, phenylalanine, proline, threonine, tryptophan, tyrosine and valine).

Water bonds strongly to silk due to the high proportion of polar groups in the amorphous region, and acts as a plasticiser. At 65% RH, bound water can represent up to 30% of the weight of the fibre, and can lead to swelling. Pre-existing damage and the presence of salts will alter the pattern of swelling, and if there is significant deterioration this can lead to a catastrophic breakdown of the fibre structure. Silk retains good flexibility at humidities as low as 40%, but below this the fibre will become desiccated, rigid and brittle with prolonged exposure. Higher temperatures will encourage dehydration and can also promote free-radical oxidation, primarily of amino acids with side-groups that readily lose hydrogen (Figure 10.18), accompanied by discolouration.

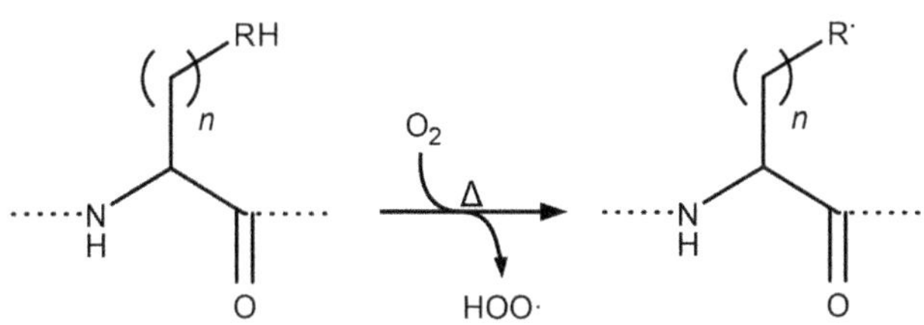

Figure 10.18 Thermal oxidation of silk.

Silk is generally resistant to acid and alkaline hydrolysis under moderate conditions. However high concentrations of acids will attack the amorphous regions rapidly and randomly along the polymer chain, leading to the hydrolysis of peptide bonds (Figure 10.19). Acids may also induce the rearrangement of peptides to esters and ethers. Acid treatments may additionally be used for certain commercial processes: weak acids (particularly acetic and tartaric) can impart 'scroop', a characteristic rustling thought to be caused by the accumulated microcrystalline deposits, whilst concentrated sulphuric acid causes the contraction of the polymer and modification of side-chains, altering lustre and softness to yield silk crêpe. Reaction with alkalis is generally much slower and proceeds from the chain ends (Figure 10.20). Alkalis can additionally cause cross-linking, for example by forming lysinoalanine through the creation of bridges between lysine and serine residues. Hot, concentrated alkaline solutions can cause the complete dissolution of the material. The presence of both acids and alkalines may also disrupt the hydrogen bonding and salt linkages which define the secondary and tertiary structure of the polymer, leading to microstructural changes, brittleness and loss of strength. Weighting agents may be implicated in hydrolytic degradation – for example, tin salts can be hydrolysed to amphoteric species which can effect both acid and alkaline attack.

Strong oxidants, such as those used to bleach the fibres, also have the capacity to cause serious deterioration. The most significant reactions occur at tyrosine and threonine side-chains, resulting in their conversion to acids and the disruption of surrounding peptide bonds; cross-links may also be created, for example between lysine and tyrosine residues.

Silk is particularly vulnerable to damage by light, especially by ultraviolet radiation; this is typified by weakening, embrittlement and yellowing. This sensitivity arises from the relatively high proportion of aromatic residues, which are prone to photo-oxidation. The degradation products are generally not well characterised, but include chromophoric derivatives of tryptophan and tyrosine. In addition to side-chain modification, light will induce the formation of radicals which can cleave adjacent peptide bonds (Figure 10.21). A variety of other photochemical reactions can also occur, including cross-linking in the amorphous regions which leads to the loss of flexibility. Moisture will influence these reactions, and the presence of certain dyes, particularly anthraquinones, will tend to accelerate photodegradation. Weighting agents are also implicated – metals ions, such as tin, can play a catalytic role, and tannins may reduce pH to the

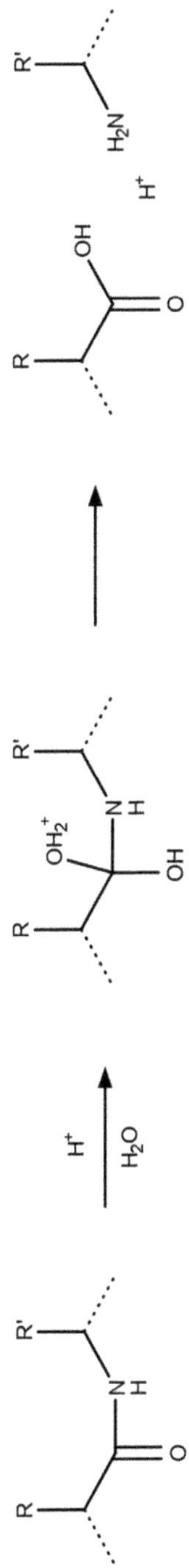

Figure 10.19 Acid hydrolysis of silk.

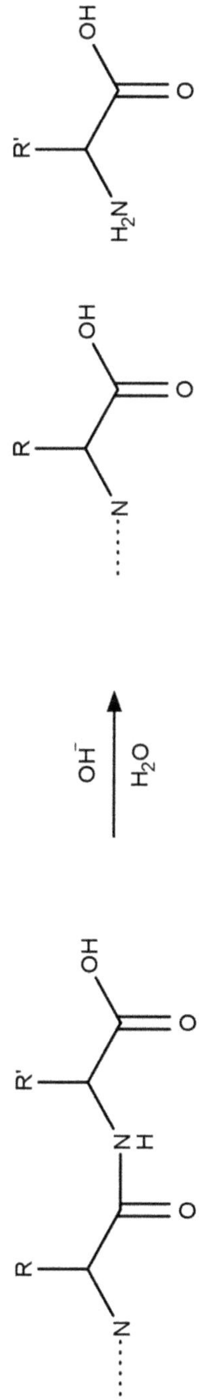

Figure 10.20 Alkaline hydrolysis of silk.

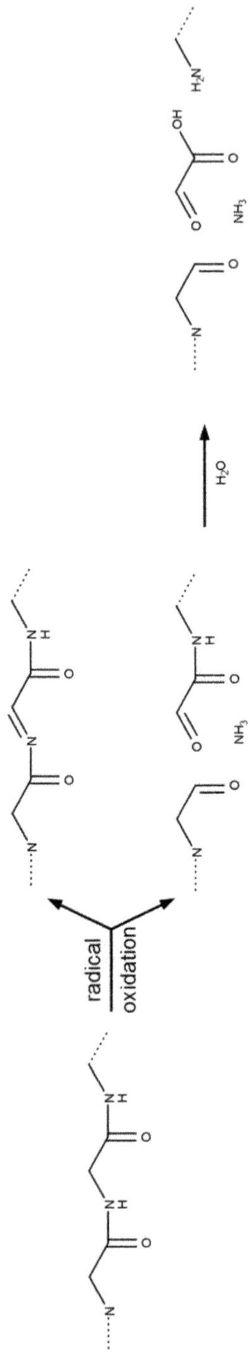

Figure 10.21 Photodegradation of silk.

point where the material becomes particularly susceptible to light damage.

Sericin is especially vulnerable to microbial (enzymatic) attack, a property which is exploited in degumming, but fibroin is relatively resistant to this form of damage.

10.5 Wool

Wools and other mammalian hairs are largely composed of keratin proteins. Wool fibres are cellular in nature, unlike the other natural proteinaceous fibre, silk. They consist of relatively hard, flattened, overlapping cuticle cells, which surround the central cortical cells; in some fibres, these may in turn surround a hollow medulla (Figure 10.22). Wool and hair fibres can be taken from a wide range of species including sheep, goats, camels, yak, musk ox, horse, llamas and related species, and 'fur' animals (rabbits, foxes, beavers and the like). The source will determine fibre properties such as length, diameter and fineness, mechanical characteristics and durability, colour and dyeability. Compared to silk, wool has a lower mechanical strength but greater extensibility and elasticity, due to its cellular nature and lower degree of crystallinity.

Depending on the origin of the fibre, the cuticle can vary in thickness from a single cell, for the finest wools, up to several dozen. These cells contain amorphous keratin arranged in layers, with the resistant exocuticle overlying the inner endocuticle. The resilience of the exocuticle substantially derives from its sulphur (cysteine) rich nature, which allows the formation of disulphide bridges that reinforce the polymer; the endocuticle has a much lower sulphur content and so is more susceptible to deterioration. The exposed surface of the cells is covered by a protein membrane, the epicuticle, along with waxes and fatty acids (such as lanolin in the case of wool), which impart further chemical and mechanical resistance. The overlapping cuticle cells give wool fibres their characteristic scaly appearance (Figure 10.23). A continuous band of water resistance intercellular material is found between the cuticle and the cortex.

The bulk of the fibre consists of the cortex. The elongated cortical cells are composed of crystalline proteins within a sulphur rich amorphous protein matrix. The keratin adopts two principal secondary structural motifs, the α- and γ-helices, which form due to a combination of extensive hydrogen bonding and steric factors, with large side-groups arranged around the spiralling peptide

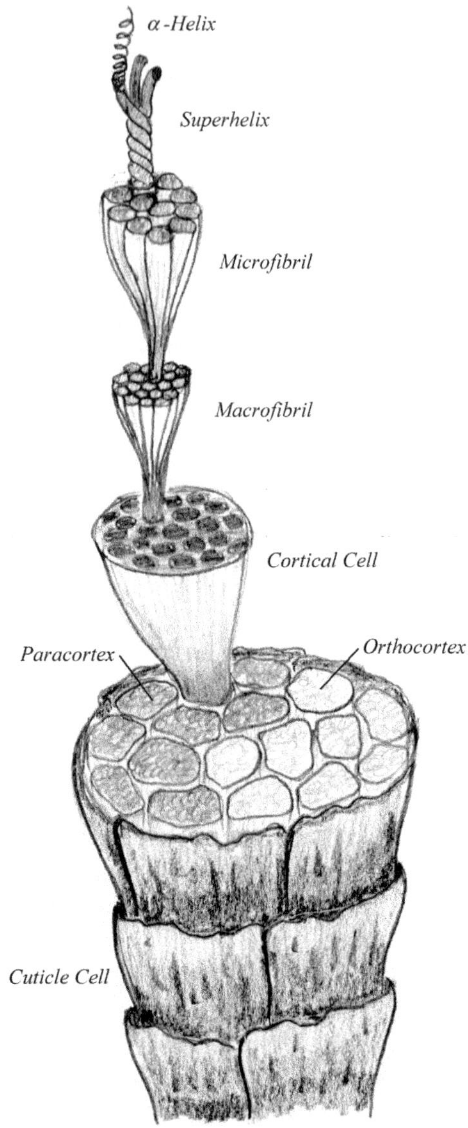

Figure 10.22 The hierarchical microstructure structure of wool.

backbone. In the crystalline regions of the fibre, superhelices or protofibrils are formed from thee α-helices (Figure 10.24) which are themselves arranged in a helical fashion. These then aggregate as microfibrils, which in turn form macrofibrils, and are located within the amorphous bulk of the keratin. The amorphous matrix itself is principally stabilised by the large number of

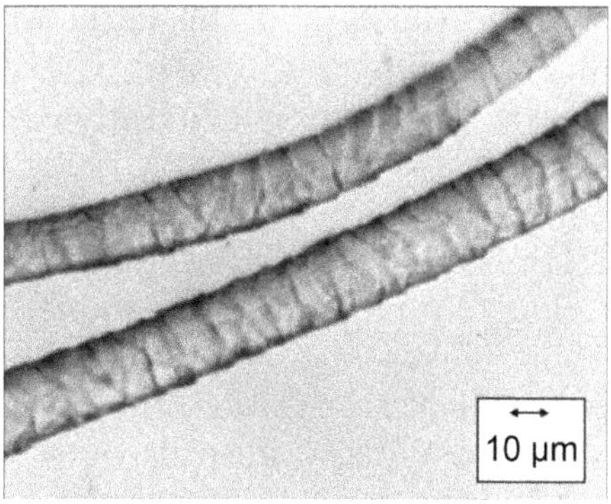

Figure 10.23 The scales of a wool fibre.

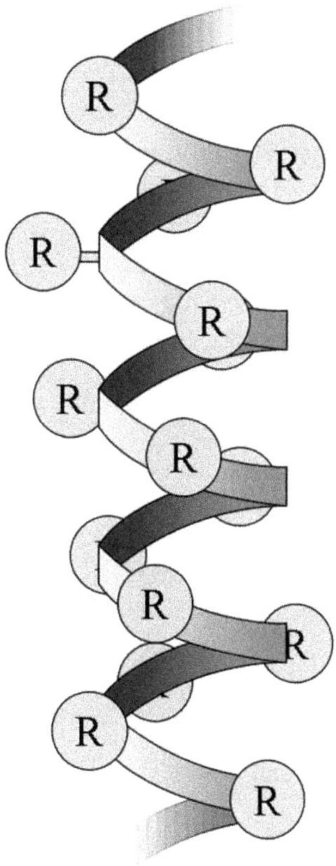

Figure 10.24 Keratin α-helices.

disulphide bridges which form between cysteine residues. The cortex can be further divided into two general regions, the orthocortex and the paracortex, the former containing a greater proportion of crystalline fibrils and the latter more largely composed of amorphous material.

Finally, a medulla – a hollow, air-filled cavity – may be present at the centre of the fibre. This is generally only found in coarser fibres, and may take a number of different forms, ranging from a continuous unbroken channel (for example, wool from many species of sheep) to interrupted, lattice-like structures (for example, rabbit fur); see Figure 10.25.

Wool is particularly valued as a textile fibre for its desirable properties in terms of comfort. When it absorbs water, heat is liberated, making woollen fabrics ideal for cold, wet conditions; this ability to trap and retain water also keeps the relative humidity at the skin comparatively high, making it suitable for hot, dry environments. However, wetting weakens the fibres and may promote bulk dimensional changes in fabrics. The bulky, crimped nature of wool fibres ensures that the textiles contain a high proportion of air space, giving them good insulative properties. In its natural state wool repels water well, due to the presence of lanolin and other oily or waxy materials; if the fibres are degreased to remove these materials, to give a more pleasant sensation against the skin, this water resistance is impaired.

Wool fibres may be subjected to fullering – the fibres are treated with an agent which causes the individual surface scales of the fibre to lift, allowing them to act like hooks and encouraging adjacent fibres to

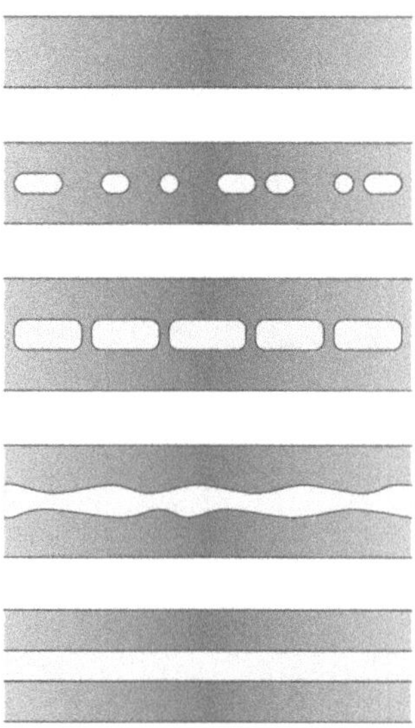

Figure 10.25 Various different possible medulla structures, from absent to continuous.

entangle (traditionally mercury compounds were employed in this role). This increases the cohesion and bulk of the material. The process can be employed to make non-woven woollen textiles (*i.e.*, felt), but can also give a greater durability and bulk to woven materials.

10.5.1 Deterioration of Wool

Like other natural fibres, wool is subject to attack by heat, light, hydrolysis, the action of microorganisms and through mechanical damage (wear-and-tear); however, wool is also especially prone to attack by insects. The degradation of wool, like silk, is largely dictated by protein chemistry, but wool generally has a greater vulnerability due to a more amorphous structure and a higher proportion of residues with reactive side-chains.

Wool will normally bind about 16–18% water by weight, but soaking the fibres can increase this to as much as 200%; as a result, degraded fibres are particularly prone to damaging swelling, and are much more susceptible to mechanical damage in this state. Both swelling and desiccation may lead to significant rearrangement of the polymers at the microstructural level, and the formation of new bonds, causing changes in physical properties and bulk dimensions.

Wool is vulnerable to a similar range of hydrolytic processes as silk, though these reactions tend to occur more rapidly, particularly if the resilient outer layers of the fibre have already been compromised. The presence of a greater number of cysteine residues leads the material to be more prone to cross-linking, through the formation of lanthionine and lysinoalanine bonds. Acids and alkalis are also effective in stripping the outer protective layers of waxes and fatty acids, facilitating further attack.

Wool fibres are generally more resistant to photodegradation than silk. However exposure to light, particularly below 380 nm and in the presence of water, will initially cause yellowing due to reactions of amino acid side-chains, and will eventually lead to chemical modification, mechanical weakness, embrittlement and loss of flexibility through various free-radical reactions and the scission of disulphide bonds. Histidine, tryptophan, tyrosine, methionine and cysteine residues are especially prone to deterioration. Exposure to longer wavelengths of light may cause bleaching.

The material is particularly vulnerable to attack by various species of moths and beetles (such as the well-known clothes moth), which can rapidly cause substantial damage and losses. Wool may also be targeted by a variety of keratinophilic bacteria and fungi, with proteins being broken down *via* enzymatic oxidation, reduction or hydrolysis; this attack predominantly occurs in the amorphous regions, especially in the cuticle.

10.6 Synthetic and Semi-synthetic Fibres

The class of synthetic and semi-synthetic fibres encompasses the wide range of novel textile materials developed from the late 19th century onwards. These materials represent a very broad spectrum of chemistries, properties and uses. Some were intended as inexpensive replacements for natural fibres, whereas others were developed with properties to suit specific – and often very specialised – applications. A short list of these materials might include acrylics and modacrylics, polyamides (nylons and aramids), viscose, cellulose acetate and triacetate, chlorofibres and fluorofibres (such as polyvinylchloride and Teflon), polyesters, polyolefins, regenerated protein fibres and glass fibres.

Many synthetic fibres are derived from a single species of monomer, the 'building blocks' of polymers. Co-polymers of various kinds are also common; the majority of these will employ alternating or block copolymer systems, as the regular structures allow close packing and crystallinity or long-range ordering, whereas random co-polymers are more likely to adopt amorphous structures which are less likely to provide the physical properties necessary for useable fibres (Figure 10.26). Graft co-polymers, comprising of a linear backbone with pendant side-chains or cross-linking, are also found. Polymerisation itself may occur through addition reactions (which normally occur when a double bond in the monomer opens, allowing new linking bonds to form) or condensation reactions (in which a linking bond is formed with the elimination of a small molecule such as water or ammonia).

Synthetic fibres generally lack the hierarchical microstructures found in natural fibres, and are usually filamentary in nature. They can be manufactured with a wide range of cross-sectional morphologies, determined by the shape of the drawplate through which the

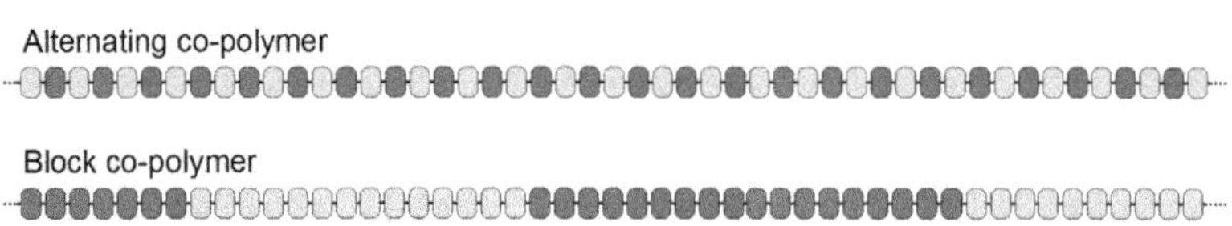

Figure 10.26 Schematics of alternating and block co-polymers.

fibre is extruded during manufacture; this property, along with the chemistry of the material, will influence lustre, insulative quality, feel and drape. Although certain polymer types are often manufactured with particular morphologies, these shapes cannot really be considered characteristic as they are purely a property of the drawplate rather than an intrinsic quality of the polymer itself. There are broadly three methods of manufacture of synthetic textile fibre: wet spinning, in which the polymer solution is extruded into an aqueous bath, and hardens as a result of physical or chemical changes; dry spinning, in which the polymer solution is extruded into a drying chamber (often with the aid of a hot air stream) which allows the solvent to evaporate and leave the solid filament; and melt spinning, in which a thermoplastic polymer is melted, extruded and then allowed to cool and harden.

Although most synthetic fibres are produced as filaments, they are frequently then cut into shorter staple lengths for use as textile fibres. These shorter lengths require spinning to form yarns, and generally have greater extensibility and lower strength than the original filament, but are usually also more pleasant to the touch.

Many synthetic polymers used for textiles fibres will also contain plasticisers, which effectively lower their glass transition point and ensure that the material remains flexible over a useable range of temperatures. If the plasticiser is lost (through reaction or migration), or the polymer is exposed to temperatures below the glass transition point, the material will become rigid and brittle.

The dyability of synthetic fibres relies on the availability of suitable reaction sites in the material to which dyes can bond. Not all materials possess this property (polyolefins are a notable example of this), and in this case colouration is usually achieved through the addition of pigment particles during synthesis. Delustrants may also be added, to reduce the typical glossiness of these materials (titanium dioxide particles are commonly used, although this compound is implicated in certain degradative reactions).

A variety of common synthetic fibres are briefly described below (although this is not exhaustive list and a variety of other synthetic polymers may also be found as textiles in specific roles). The summaries include the basic chemical composition of the materials, their typical microstructure, the most common production methods, methods of modification, and general physical and chemical properties, including vulnerability to factors which may commonly be encountered by textile artefacts (such as hydrolytic and oxidative reactivity, response to water and solvents, light sensitivity and susceptibility to attack by microorganisms). However, these materials may undergo a much wider range of reactions under specific circumstances or when exposed to less common reagents, and their properties may also vary considerably depending on manufacturing methods (including alternative synthetic routes, co-polymerisation or blending), chemical or physical modifications, and the presence of additives.

Many of these materials are also described in a different context in Chapter 11.

10.6.1 Viscose

The earliest successful man-made fibre was viscose, also commonly referred to as rayon, developed in late 19th century. The material is a semi-synthetic or regenerated fibre, as it uses natural cellulose as the base polymer; this is generally obtained from wood pulp, and occasionally from cotton linters. Cellulose is derivatised to the soluble cellulose xanthate, and then extruded and coagulated in an acid bath (*i.e.*, *via* a wet spinning process); stretching the resultant fibre imparts some long-range ordering and strength. Viscose lacks the hierarchical microstructure, extensive crystallinity and non-cellulosic components (such as hemicellulose, pectin and lignin) of natural cellulosic fibres; it also has a much shorter chain length, with a typical DP of about 200–700 (by comparison the DP of native cotton cellulose is typically 6000–8000). The coagulation process leads to differences between the outer skin and the core, with the rapidly crystallised skin containing numerous small, imperfectly formed crystals and the core containing larger, more regular crystals; modifying this process to form a thicker skin on one side of the filament produces crimped rayon due to differential swelling, giving a material with improved resilience and bulk. Viscose generally has a poor wet strength and abrasion resistance, and is prone to creasing, though appropriate finishing can address some of these issues. It is readily subject to acid and alkaline hydrolysis, and will dissolve in alkaline solutions unless cross-linked. It undergoes similar photodegradation processes to natural cellulosic fibres, although these reactions may be accelerated by the presence of titanium dioxide, a commonly used delustrant. Other production methods can create viscose fibres with modified properties. Deriving the material from cellulose acetate *via* saponification creates a fibre with greater polymer orientation. High tenacity fibres, with greater long-range ordering and crystallinity, can be produced by subjecting the filaments to a much higher degree of stretch after extrusion, thereby imposing longitudinal orientation on the polymer chains. Cupraommonium rayon, polynosic rayon and solvent spun rayon all use modified chemical processing methods to generate fibres with improved physical and chemical properties.

10.6.2 Cellulose Esters: Acetate and Triacetate

Unlike viscose, cellulose esters use a chemically modified cellulose derivative, rather than the 'pure' raw material, resulting in fibres with markedly different chemical and structural properties.

Cellulose triacetate is created by the esterification of cellulose to replace the three hydroxyl groups with acetate groups (Figure 10.27). The polymer is extruded by dry, wet or melt spinning from the solution (in methylene chloride), and the fibres are employed in both filament and staple forms. Triacetate fibres are soft, silk-like and possess good drape properties, but lack durability. In its initial state the material has low crystallinity, but heat setting can significantly increase this, leading to greater heat resilience, lower moisture regain and excellent dimensional stability. The polymer is resistant to dilute

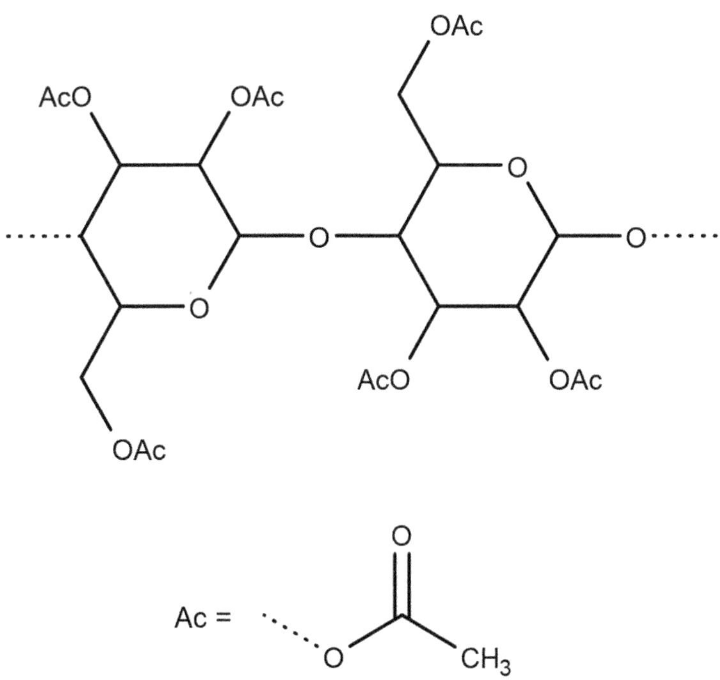

Figure 10.27 Cellulose triacetate.

acids, but is attacked by strong, concentrated acids; it is also resistant to de-acetylation (saponification) by alkali. It is highly resistant to light damage; if damage does occur the processes are similar to those of acetate (see below). Triacetate fibres are highly resistant to moulds, mildew and insects.

Cellulose diacetate (commonly referred to as acetate) is produced by limited re-hydroxylation of cellulose triacetate, and can be dry, wet or melt spun. The fibre is generally used as a filament, has a silky appearance and is soft to the touch. It has low strength (both wet and dry) and poor abrasion resistance, but good dimensional stability and low water absorption. The random distribution of acetyl and hydroxyl groups limit the crystallinity of the material. The material has a good heat resistance (softening and becoming tacky at roughly 190–205 °C). It is not affected by dilute weak acids, but will dissolve in strong, concentrated acids; organic acids, such as acetic and formic acid, cause the fibres to swell. Strong alkaline solutions lead to saponification, the loss of acetate groups and re-conversion to cellulose. Prolonged light exposure causes loss of strength, but no yellowing; this may be accelerated by delustrants such as titanium dioxide. Short wave ultraviolet light leads to the direct photolysis of bonds within the material, whereas longer wave ultraviolet and visible light causes degradation *via* photosensitisation mediated by environmental conditions. It is resistant to insects and microorganisms.

10.6.3 Acrylic and Modacrylic

Acrylic is formed through the addition polymerisation of acrylonitrile (Figure 10.28). The presence of the bulky nitrile (–C≡N) groups causes the polymer backbone to adopt a helical conformation with three repeats units per turn, in which the pendant nitrile groups are

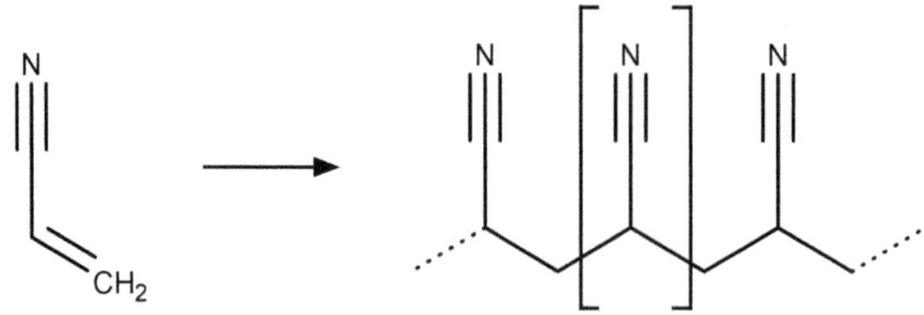

Figure 10.28 Acrylonitrile.

positioned on the exterior of the helix; these roughly cylindrical fibrillar structures are bound together *via* interactions between the exposed nitrile groups, leading to a relatively stiff, pseudo-crystalline ordering, which accounts for about half the material, with the remainder being randomly ordered, although it cannot readily be described using the standard two phase 'crystallites in amorphous matrix' model. A typical DP for the material is 40 000–60 000, amongst the greatest for artificial fibres. Modacrylic (modified acrylic) consists of 35–85% acrylic, with the remainder being co-polymers; this allows properties such as dyability, strength and flame resistance to be adjusted.

The fibres have a good resistance to abrasion and ageing, but poor dimensional stability. Generally, the material is broken into short lengths (staple fibres) which can then be spun into yarns; the fibres have a soft touch, and wool-like properties which means that they are often blended with wool fibres.

The bulk polymer is heat sensitive due to the weakness of the intermolecular van der Waals forces, and even body heat can cause wrinkling or distortion, but it has good thermal stability at the molecular level up to about 230 °C; at higher temperatures the material discolours and decomposes, rather than melting. It is exceptionally resistant to damage by light, making the material particularly suited for outdoor applications. Moisture does not detrimentally affect the polymer to any great extent, though water will act as a plasticiser. Acrylics are resistant to the action of weak acids and bases, and exhibit a reasonable resistance to strong acids. Alkaline hydrolysis will only occur in the presence of strong bases, but alkalis may lead to surface saponification, so washing can often have the effect of yellowing the material. It is not subject to damage by moulds or microorganisms. Acrylics have a good chemical resistance, and are generally not damaged by solvents (including dry-cleaning solvents), the exception being acetone which may lead to the swelling and dissolution of some types of the material.

10.6.4 Polyamides: Nylon and Aramid

Polyamides are defined by the presence of an amide bond formed in the polymerisation process, and fall into two broad categories: nylon, in which less than 85% of the amide linkages are attached to two aromatic rings (and are predominantly derived from aliphatic monomers); and aramids, in which at least 85% of the amide groups are directly adjacent to aromatic groups.

Nylon fibres are formed either as a polymer of an amino acid or as the copolymer of a diamine and a diacid. The nature of the polymer, and the lengths of the monomer carbon chains, is indicated in the nomenclature, so nylon 6 is formed from a 6 carbon amino acid (6-aminohexanoic acid or caprolactam; Figure 10.29), nylon 11 from an 11 carbon amino acid, and nylon 6,6 from a diacid and diamine each with a 6 carbon length (Figure 10.30); these three are the most common forms of the polymer found as textile fibres. The fibres are formed by rapid extrusion and cooling of the molten polymer, followed by drawing to improve orientational properties, and are used as multifilament yarns, monofilaments, staples and tow. Tenacity, strength and elongation vary with the manufacturing conditions. The different forms of nylon may be used to create bicomponent fibres with specific properties; for example, fibres with a nylon 6,6 core and nylon 6 skin can be used to create non-woven fabrics which benefit from the superior physical properties of the former, but also exploit the lower melting point of the latter to allow the fibres to be readily fused together. Nylon generally has high impact strength, toughness, elasticity, flexibility and abrasion resistance; the strength, elasticity and toughness of nylon 6,6 are superior to those of silk. Its thermosplasticity allows it to be heat set, which makes it ideal for shaped garments such as stockings. At room temperature the

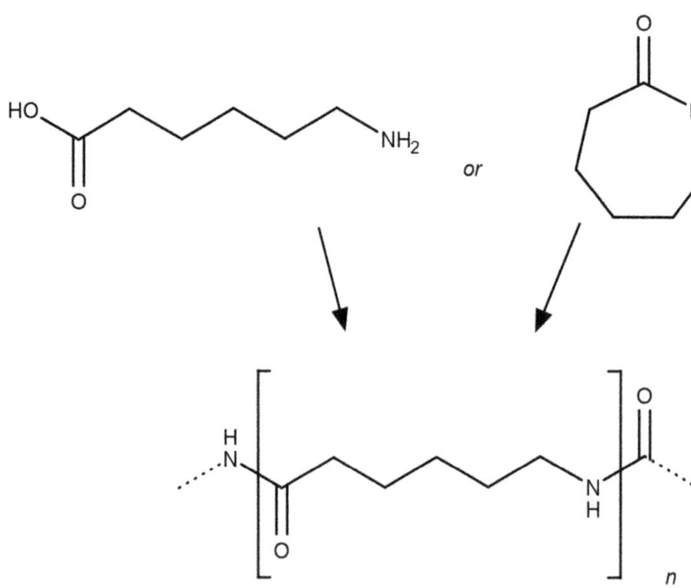

Figure 10.29 Nylon 6.

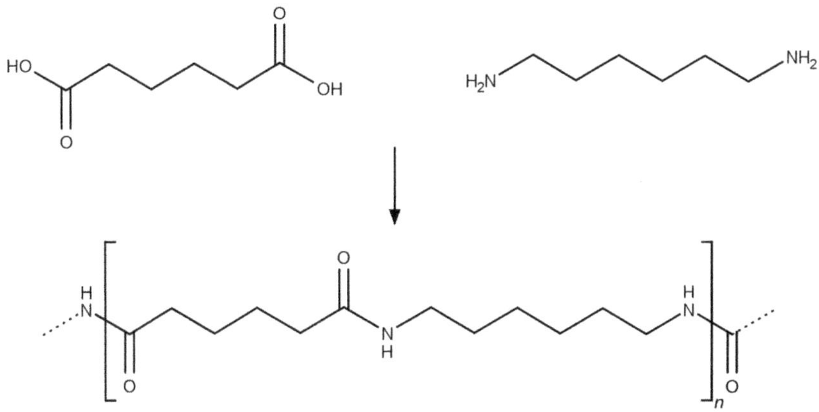

Figure 10.30 Nylon 11.

extensive $-CH_2-$ sequences are flexible and only weakly attracted by van der Waals forces; the majority of interchain linking occurs *via* hydrogen bonding between amide groups. Photodegradation occurs through the oxidation of imino hydrogens, creating various reactive groups which then lead to chain scission; this causes a progressive weakening but is not generally accompanied by discolouration. Elevated temperatures can result in both yellowing and loss of strength. Acids will attack the amide bond through hydrolysis, and the material is soluble in concentrated acid; prolonged exposure to alkalis cases similar damage. Oligomers may migrate to the fibre surface over time, causing brittleness. In general, nylon will undergo a typical range of amide reactions, and unlike many synthetic fibres, it is vulnerable to attack by mildew and bacteria.

Aramids, such as Kevlar and Nomex, are polymers in which aromatic groups are linked by amide or imide bonds (Figure 10.31), which allows very extensive and regular interchain ordering (Figure 10.32). These materials are produced by extrusion from solutions in powerful solvents or strong acids. They are highly flame retardant and will withstand high temperatures. The extensive crystallinity and strong interchain bonds impart resistance to most forms of chemical attack. They have a very high strength to weight ratio. These properties mean that the fibres are employed in flame retardant clothing, bulletproof vests, cables and belts intended for use in harsh environments, and as the polymer component of polymer-matrix composites where both lightness and strength are required.

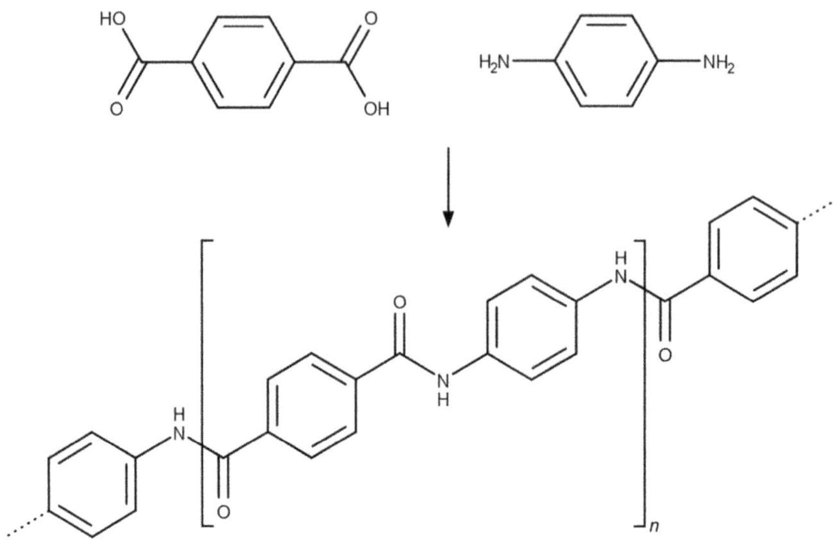

Figure 10.31 Imide bond.

10.6.5 Polyesters

Polyesters represent a very broad class of polymers, but have the common feature of an ester linkage; in general, they are the result of a polycondensation reaction between a dibasic acid and dihydric alcohol. The most common form used for textiles is poly(ethylene terephthalate) (PET; Figure 10.33); other forms find more minor usage, particularly for specialist roles. The polymer is generally produced by melt spinning, and forms a smooth, featureless fibre. Manufacturing and processing options will influence crystallinity and long-range ordering – polyesters used for textiles are highly crystalline (65–85%), resulting in relatively stiff, hydrophobic fibres. PET is strong, with good dimensional stability and a high elastic modulus. It is thermoplastic, and burns readily. It has a low affinity for water, but its hydrophobic nature means it will readily absorb fats, greases and similar oily stains. As with nylon, oligomers may migrate over time, leading to brittleness. The material is potentially vulnerable to both acid and alkaline hydrolysis, but its hydrophobicity limits the extent to which this can readily occur; it is most likely with prolonged exposure to high temperature moisture (boiling water or steam). It has a good thermal stability below the melting point. It is relatively resistant to damage from light, although prolonged exposure will lead to a loss of strength; the majority of photodegradation proceeds by photolysis, due to the strong absorption of ultraviolet radiation by

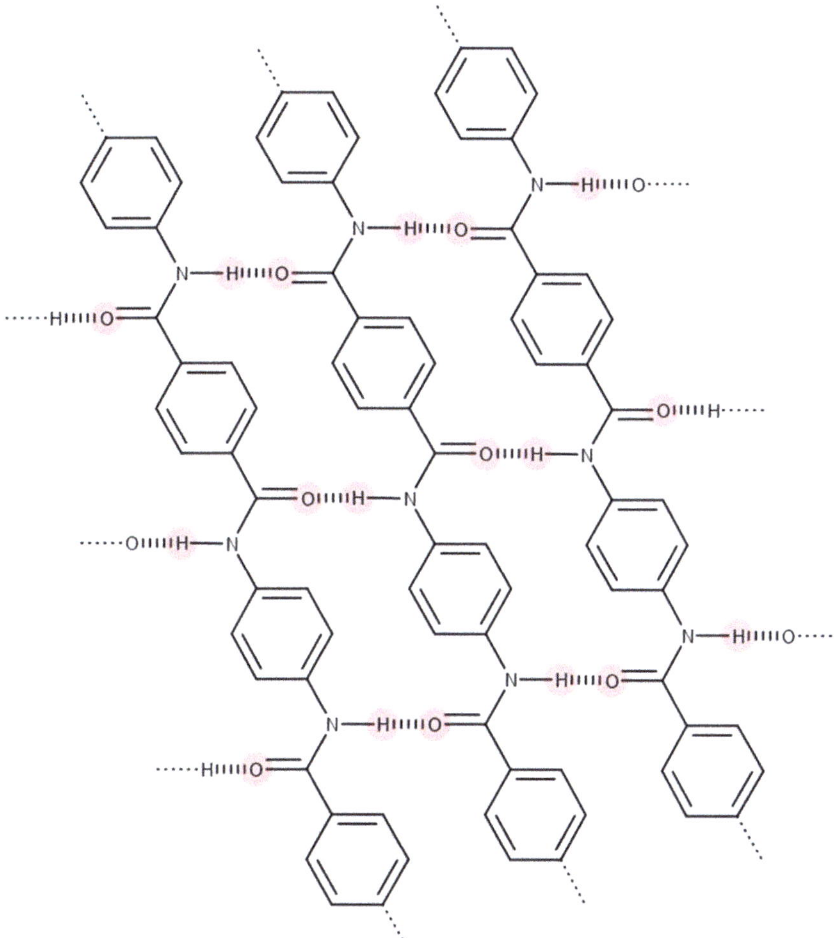

Figure 10.32 Aramid structure.

terephthaloyl groups, which also tends to limit damage to the fibre surface. Chain scission is accompanied by cross-linking (in the absence of oxygen) or hydroxylation (in the presence of oxygen); both processes lead to discolouration. PET exhibits a generally high chemical resistance, and is not affected by pollutants, except for nitrogen dioxide. Moulds, mildew and fungi will not readily attack the polymer.

10.6.6 Polyolefins: Polyethene and Polypropene

Polyolefins are simple petrochemical-based polymers with a non-alternating repeat, formed by addition polymerisation.

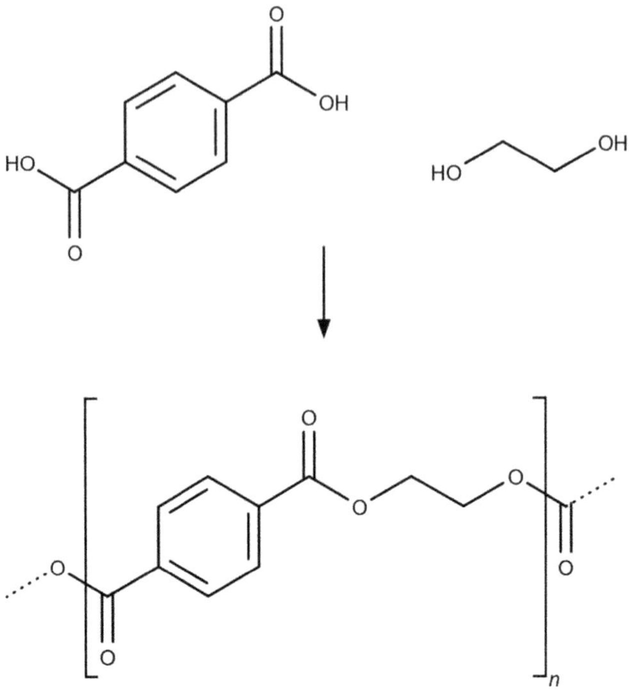

Figure 10.33 Poly(ethylene terephthalate).

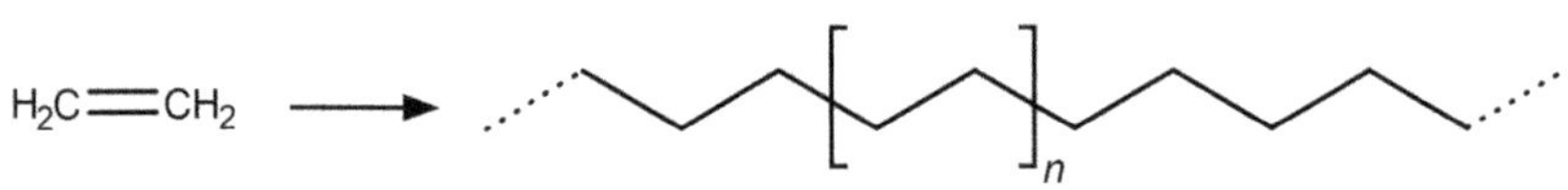

Figure 10.34 Polyethene.

Polyethene (also called polythene or polyethylene) is one of the simplest polymers, formed from ethene and consisting of a $-[CH_2CH_2]-$ repeat (Figure 10.34). In general, it is not a particularly good textile material, due to its softness and extensibility. Its low melting point allows it to be used for fusible interlinings, and it also finds some use as an upholstery fibre.

Polypropene (polypropylene) is formed from propene. Polymerisation may occur by one of three routes, depending on the manufacturing process: isotactic polypropylene results from head-to-tail polymerisation, so has a regular, alternating arrangement of side-chains; syndiotactic polypropylene results from head-to-head polymerisation; and atactic polypropylene results from random polymerisation, so the orientation of the side-chains is also random (Figure 10.35). For use as a textile, the polymer is typically 90–95%

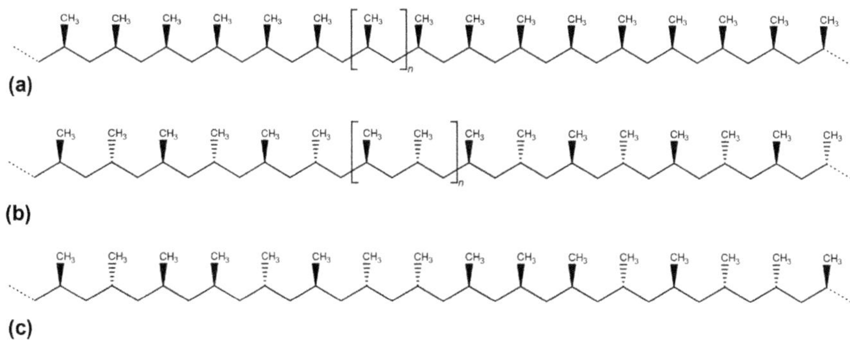

Figure 10.35 Polypropylene, in (a) isotactic, (b) syndiotactic and (c) atatic forms.

isotactic, with a small amount of atactic material; in this form the polymer can be best regarded as consisting of lamellar crystals interspersed by amorphous, non-crystalline zones. The polymer is relatively strong, and due to its resistance to water retains its strength when wet, but it also has a relatively low resistance to extension, low elastic recovery and poor creep behaviour. It is primarily employed as an upholstery fabric, and is found in many of the roles formerly taken by jute; it is only of limited use as a clothing fibre. Generally, it is not directly dyeable as it lacks appropriate dye reaction sites. The material is susceptible to oxidation, even at room temperature, so anti-oxidants are needed to stabilise the material during production and over the longer term. Exposure to sunlight (specifically UV) in the presence of oxygen will lead to photolytic oxidation unless some form of UV stabiliser is present. The lack of reaction sites means that the polymer is generally resistant to other forms of chemical attack.

10.6.7 Polyurethane

Polyurethanes are generally polyfunctional, but all contain the urethane group (–CO_2NH–) and are formed from isocyanate and polyol monomers. There are two main classes, depending on the nature of the polyol – polyester polyurethanes and polyether polyurethanes. These materials are widely used as foams and are found widely in furnishings and upholstery, despite their flammability and tendency to produce toxic fumes on burning. They also find use as fibres (particularly elastomeric fibres or elastanes), adhesives, consolidants and textured surface foams in synthetic suedes and leathers, textile/foam laminates and water repellent finishes. The elastic modulus is relatively low (comparable to that of nylon or cotton); the elastic recovery is not quite as complete as that of natural rubber, with stretching leading to a slightly permanent elongation. The material is thermoplastic and hydrophobic; its tensile properties are only slightly affected by water. On degradation, polyurethanes tend to become discoloured, followed by embrittlement and stickiness; this is eventually followed by the complete loss of elasticity and compressive strength. Due to their extensive surface area, foams degrade more rapidly than the corresponding fibres. All polyurethanes will degrade by oxidative processes, although polyether polyurethanes are more susceptible *via* the oxidation of the ether groups in the polyol segments. Oxidation products tend to be volatile polar species (often acidic) of low molecular weight (which may lead to a pungent odour in the degraded material), and will further contribute to degradation, particularly in the presence of moisture. The material yellows with prolonged light exposure although the exact processes are dependent on the constituent monomers. They show a generally good resistance to alkalis. Hydrolysis leads to chain scission, and polyester polyurethanes are markedly more susceptible to this process. Plasticiser migration is a common problem over time.

10.6.8 Regenerated Protein Fibres

Regenerated protein fibres use natural protein as a base material, and are analogous the regenerated celluloses such as viscose; however, they generally exhibit poor physical and chemical properties, and so have been much less successful commercially. Various protein sources have been employed, including peanuts (arachin, produced commercially as *Ardil*), soya beans, maize (zein) and milk (casein); the last is the only one produced in significant quantities. Casein fibres possess wool-like properties; they are slightly weaker than wool when dry, but are exceptionally weak when wet, a common property of most regenerated proteins. They are occasionally used in blends with cotton, viscose or wool.

10.7 Analytical Techniques Appropriate for Textiles and Fibres

General information about the theory and application of various analytical techniques can be found in Chapter 2. The use and suitability of these techniques to the study of historic textiles are discussed below.

When choosing the most appropriate method of studying these materials, it is important to bear in mind how the information will be used, what kind of samples (if any) can be taken and whether or not a particular technique will provide useful information given the specific nature of the specimen. For example, all cellulosic fibres have similar chemistries but distinctive morphologies, so microscopy is generally the best method of distinguishing them. Conversely the morphologies of synthetic fibres are largely dictated by the drawplate used to create them, and so cannot generally be considered characteristic of their composition; as a result, a technique which interrogates their underlying chemistry is typically more useful to identify them, such as vibrational spectroscopy.

The most widespread and useful techniques for the analysis of textile artefacts are generally microscopy, elemental analysis, vibrational spectroscopy, mechanical testing, chromatography and chemical testing. Other methods may also find use in specific circumstances.

10.7.1 Light Microscopy

Light microscopy is useful for studying physical, morphological and dimensional features of fibres and fabrics. Depending on the technique chosen, it is possible to examine a range of scales from single fibres up to bulk weave structures.

Low magnification stereomicroscopy is useful for studying weave structures, relatively large-scale surface features, bulk damage, discolouration and structural or decorative components, and has the advantage that it can often be carried out *in situ*, without requiring samples to be taken.

If representative fibres can be extracted from an object, they can be identified by transmitted light microscopy which can allow them to be identified by their characteristic morphologies, structures and dimensions, such as the hollow lumen of plant fibres (as well as the surface dislocations of bast fibres and the characteristic twisting of cotton), the scales and medulla or hair and wool fibres, and the cross-sectional shape of synthetic fibres which indicate the type of die plate used in their manufacture. Furthermore, it is often possible to identify regions of damage or alteration. If it is acceptable to permanently alter the sample, areas of interest may be enhanced with suitable dyes, or it may be set in resin and cross-sectioned (Figure 10.36). In addition, the presence of pigment particles, contaminants or surface treatments may be identified, although other techniques are generally necessary to identify them more fully.

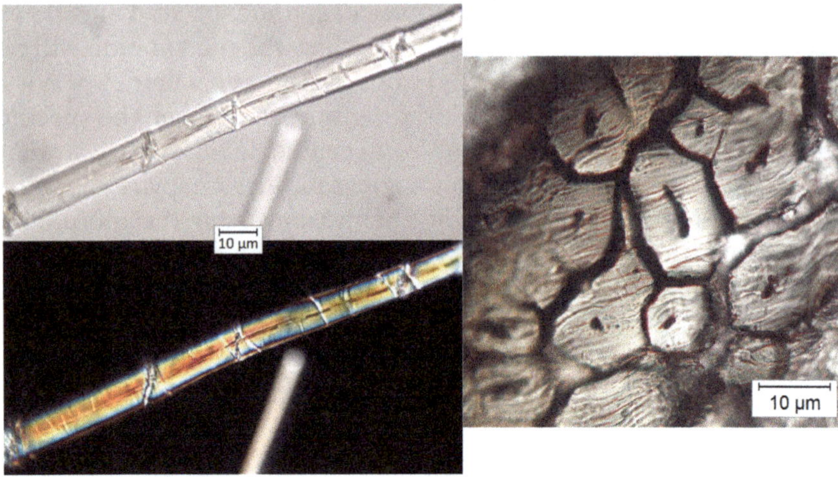

Figure 10.36 Micrographs of flax fibres, revealing different features of interest: conventional and polarised light images of a transverse fibre, and a cross-sectional image of a bundle of resin set fibres.

The use of polarised light microscopy is particularly useful to highlight structural features in fibres, such as the hollow lumen at the centre of cellulosic materials (see Figure 10.36).

Light microscopy has the advantages that it is inexpensive, rapid and readily available, but it is often necessary to take samples from the object, particularly if fibre identification is required. At higher levels of magnification, even individual fibres have a highly three-dimensional nature which means it is difficult to keep an entire sample in focus due to the limits of depth of field. As a result, although magnifications of up to ×1000 may be achievable in theory, for most work with fibres and textiles the usefulness of magnifications greater than ×400 is limited.

10.7.2 Electron Microscopy

Scanning electron microscopy (SEM) can allow the study of the surface of a sample at much higher magnifications than light microscopy, and can also provide information on composition. It requires samples to be taken, and to achieve the highest quality images, these must then be coated with a conductive material such as carbon or gold; environmental SEM does not require this coating, but is a more limited technique. Generally, samples have to be held under vacuum as they are examined, which can potentially cause dehydration,

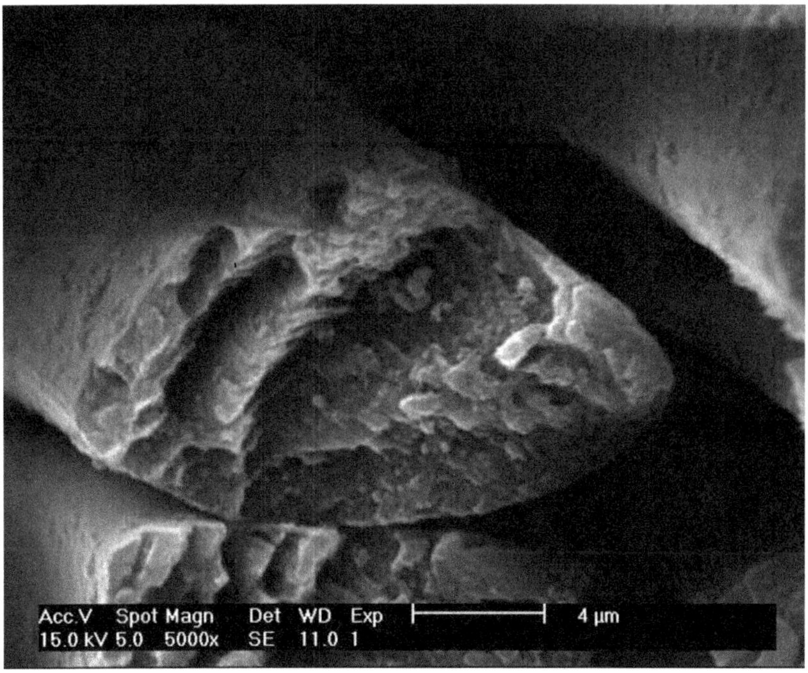

Figure 10.37 Scanning electron micrograph of damage to silk fibres.

morphological changes or other damage. The equipment is limited to larger institutions, and is relatively expensive to operate. Samples may also suffer direct damage from the high-energy electron beam.

It is particularly useful for looking at features such as surface damage, fracturing or defibrillation – various forms of damage to silk fibres can be seen in Figure 10.37.

10.7.3 Elemental Analysis

Elemental analysis encompasses a range of related techniques, which reveal information about the elemental composition of a sample and are particularly useful when examining inorganic materials. In the context of textiles artefacts, these methods generally do not yield much useful information when considering organic fibres themselves, but can be especially useful when studying components such as metal threads, buttons, sequins, decorative elements, pigments, silk weighting agents, optical brighteners, fire retardants and similar treatments, insecticides, corrosion products, dirt and soiling.

Elemental analysis can be achieved by techniques such as energy dispersive X-ray spectroscopy (EDS or EDX, usually performed in an electron microscope) or X-ray fluorescence (XRF), which may be found either as benchtop devices or as the more compact, but more limited, portable units (p-XRF). In general, all of these methods will only detect only heavier elements (typically sodium or above).

Elemental mapping can be used to investigate the distribution of inorganic materials across surfaces or through cross-sections. A particular application of this, for example, would be to determine the way in which weighting agents permeate silk fibres.

10.7.4 Vibrational Spectroscopy

Vibrational spectroscopy includes techniques such as Fourier transform infra-red (FTIR), near infra-red (NIR) and Raman spectroscopy. These methods are particularly useful for identifying organic materials, such as polymers (natural and synthetic fibres, plastics, foams, waxes, *etc.*), degradation products, binding media, adhesives and some dyes and pigments; changes in such materials (through degradation, modification or environmental response) may also be assessed, and it is possible to gain insights into their crystallinity and microstructure.

FTIR spectroscopy can be used either to identify specific chemical functionalities within a sample or to provide a spectral 'finger-print' of the sample which can then be used to characterise it by comparison with a library of spectra of known materials. For fibre analysis a particularly useful implementation of this technique is attenuated total reflectance (ATR) spectroscopy, which can be used to record spectra from single fibres taken from an object. NIR spectroscopy is generally less sensitive than mid infra-red spectroscopy, but is good for investigating water and hydroxyl content, and has the particular advantage that is can be used with an optical fibre probe, thus allowing *in situ* analysis. Raman spectroscopy is especially suited to dyes and pigments, but care must be taken not to damage the sample with the incident laser light. Specific refinements of these techniques may allow spectral 'maps' to be created, with each 'pixel' of the map being composed of a complete spectrum, or permit microspectroscopic analysis, with measurements takes from areas as small as several tens of square microns.

10.7.5 Mechanical Testing

Mechanical testing reveals information about the physical properties of a fibre or textile sample, allowing the determination of parameters such as strength, elastic modulus, elastic recovery, long-term creep or deformation. These properties are dependent both on the original nature of the material and on its subsequent history, including factors such as treatment, usage and deterioration. Mechanical testing allows the effect that structural, microstructural and chemical changes have had on the bulk material to be better understood, and in turn this will help to inform decisions about the appropriate handling, conservation, display and storage of the item. Such results may, for example, indicate that a textile is no longer capable of supporting its own weight, or that particular regions require special support or consolidation to relieve local fragility or brittleness.

10.7.6 Chromatography

Chromatographic techniques separate out components of a soluble or volatile sample in a solvent or gas stream (mobile phase) by exploiting the differing affinities of these components for another material (stationary phase). These affinities may arise from the chemistry, electrical charge or size of the component. The techniques are useful for looking at dyes and other soluble components, and if the fibre polymer itself can be solubilised it is also possible to measure properties such as DP.

Liquid chromatography includes methods which range from very simple thin layer chromatography (TLC) through to sophisticated high-performance liquid chromatography (HPLC).

10.7.7 Chemical Tests

Chemical testing relies on the way in which particular fibres behave when subjected to specific chemical reactions or treatments, including staining, dissolution and burning. These tests were once widespread, but are now less commonly used, due to their drawbacks and limitations. Particular disadvantages with these methods are that specific tests are required for specific fibres (unlike infra-red spectroscopy, for instance, where the procedure is the same regardless of the fibre type), they are destructive and relatively large sample sizes are needed. The less is known about the sample initially the greater are the number and range of tests required, which can become extremely time consuming and wasteful of material.

Staining tests rely on the reactivities of particular chemical and microstructural features in a fibre, and are highly specific. For example, some typical chemical test methods for plant fibres include:

the phloroglucinol test for lignin; the Herzberg reagent (zinc chlor-oiodide), which highlights structures within fibre cells; and Schweitzer's reagent (cuprammonium hydroxide) which causes the swelling and dissolution of cellulose, in patterns dependent on cellulose content and distribution, and the presence of other components. Some commercial analytical stains are also available, composed of mixtures of dyes designed to stain certain fibre types specific colours. All stain and dye tests are of limited use when dealing with dyed, aged or heavily degraded samples. In general, age will tend to darken fibre specimens, thus obscuring the results of these techniques, and the accurate identification of colour changes will vary between observers.

It is also possible to devise a series of solubility tests that will systematically categorise a fibre based on its chemical nature, using readily available reagents. With the appropriate scheme it is possible, for example, to distinguish secondary acetate, triacetate, silk, nylon, wool, polyester and natural and regenerated cellulose. However, as all plant fibres are chemically similar (*i.e.*, primarily composed of cellulose), as are wool and hair samples (composed of keratin proteins), tests of this type will generally not allow a more detailed level of identification, and modifications or other treatments can alter the way in which a fibre responds to a solvent.

Burn tests rely on the differing ways in which fibres behave when subjected to a flame. Various factors are taken into consideration, including whether or not the specimen melts, whether it continues burning when removed from the flame, the resultant smell, the final appearance of the material, *etc.* In general, synthetic fibres melt whilst natural ones do not. Cellulosic fibres generally burn with a smell akin to burning grass or paper, protein fibres (wool or silk) are similar to the smell of burning hair, whilst synthetic fibres have an acrid or 'chemical' smell.

One area in which chemical tests are particularly useful is in the assessment of damage to fibres, in combination with microscopy; reagents which stain regions of damage will highlight features which may not be apparent by simple examination under the microscope. Examples specific to cellulosic fibres include: Fehling's solution, which precipitates red copper I oxide in regions of acid damage; the Clibbens and Geake test, which causes damaged regions to darken; the Turnbull Blue test, which stains regions of acid or oxidation damage dark blue; and the Congo Red test, which will stain the interior of the fibre more strongly than the exterior, thus revealing areas where the fibre had swollen or split.

10.8 Conservation

The conservation of an object generally aims to ameliorate the effects of pre-existing damage, provide physical and chemical stabilisation and limit future deterioration, ideally with the minimum of intervention and often with the requirement the object is in a suitable state for display or further investigation. Conservation treatments should not cause further damage or excessively modify the object, and as far as possible should be reversible and allow for future treatment. Although an important outcome of conservation is to aid in the interpretation of the original object and its history, in general it is not appropriate to intervene unduly to restore lost function, renew faded or fugitive dyes or otherwise attempt to recreate the object in its assumed original state. The techniques involved in conservation commonly include cleaning, reshaping, support, pest control and mounting, as briefly summarised below. In order to apply these techniques to best effect, and ensure that any intervention is appropriate, sympathetic and compatible, it is important to appreciate the material nature of the object – composition, chemistry, microstructure and structure – and to be able to investigate properties of the object which may not be immediately apparent. An understanding of the principles of conservation science, or close collaboration with a conservation scientist, is invaluable in this regard.

Cleaning allows the removal of dirt, soiling, stains, frass, dye bleeding or other material such as mould residues and spores. It may be desirable to remove other substances as well, such as adhesives, consolidants or other treatments, if these are leading to distortion or damaging stresses. Cleaning may simply involve the removal of surface dirt through low-powered suction or, for more robust fabrics, gentle mechanical action with a suitable brush or cloth. It may also be carried out using an aqueous or organic solvent wash, in combination with detergents, sequestrants and other cleaning agents; in this case it is necessary to understand the object sufficiently to choose a wash system of appropriate solvency and pH to remove dirt without causing damage, including bleeding or colour changes in dyes. As with all aspects of conservation, it is preferable to err on the side of caution and leave soiling rather than risk damaging the underlying object by removing it. It should also be borne in mind that in some cases soiling is intrinsic to the history of the artefact, and it is not appropriate to remove it.

Reshaping is important as many objects are encountered in a heavily creased or crumpled state, which not only means that their original form is lost but they may also be under undue stress. Humidification is often involved in this process, as water acts as a plasticiser for most natural and some synthetic fibres, restoring flexibility to these materials and allowing them to be reshaped with minimal risk of causing damage. However, some materials do not respond well to humidification (heavily weighted, 'shattered' silk, for example), so correct identification of materials is important.

Support involves affixing a fragile or fragmentary fabric to a suitable inert backing or overlying material (often a silk, nylon or polyester net, dyed to ensure that it is as unobtrusive as possible), using either a suitable, reversible adhesive or careful stitching. This supports the item, thus minimising further physical damage. Cosmetic in-fills may also be used, to make areas of damage less obvious or to aid the intelligibility of an incomplete design or feature, but in these cases the intent is to not to make the object appear undamaged (and the infill should be recognisable as such on close examination), but to give a general impression of the missing area, to better to understand the item as a whole.

Pest control is an important consideration in those artefacts that may harbour mould spores or insect eggs, capable of remaining dormant on a fabric for extended periods; to some extent cleaning will remove these contaminants, but other measures might also be necessary. The most common procedures are freezing and anoxia, and again a knowledge of the composition of the object is important – on freezing, some materials may be taken below their glass transition temperature and thus be at risk of physical damage on subsequent handling, whilst anoxia can induce unwanted reactions, particularly colour changes in certain classes of dye. Pesticides are to be avoided as they may interact unfavourably with the artefact and are likely to leave residues.

Mounting is usually the final stage of conservation, and provides the object with appropriate structure to minimise stresses whilst, usually, allowing it to be presented in a manner suitable for display. The structures should be purpose built for the artefact in question, and a well-designed support may minimise or avoid the need for interventive conservation. In all cases it is important to ensure that the materials used are compatible with those of the object itself and will not interact in a deleterious manner.

10.9 Case Study: The *Victory* Sail

The fore topsail of *HMS Victory* is housed at Portsmouth Historic Dockyard, and is an object of major national maritime importance.

The *HMS Victory* was Admiral Lord Nelson's flagship at the Battle of Trafalgar, fought on 21st October 1805, and the sail itself was extensively damaged during the battle, being holed by musket and cannon fire, and suffering an extensive tear when it was struck by a falling mast. The sail's origin can be traced back to Baxter Brothers Linen and Jute Manufacturers of Dundee; it was in use by 1803 and was probably employed at sea for roughly 18 months prior to the battle. Not only does it represent a link to one of the most crucial sea battles in British history, it is also believed to be the only surviving sail from the period. So, in addition to being one of the largest and most significant naval textile artefacts, it bears key witness to the early 19th century sailmakers' art.

The sail is 24 m wide at the foot, 17 m in height and weighs approximately 360 kg. It is constructed from bolts of linen cloth (each roughly two feet wide), running from head to foot, and is reinforced with similar material running across the width of the sail in bands (Figure 10.38); the edge is strengthened with hemp rope. The sailcloth

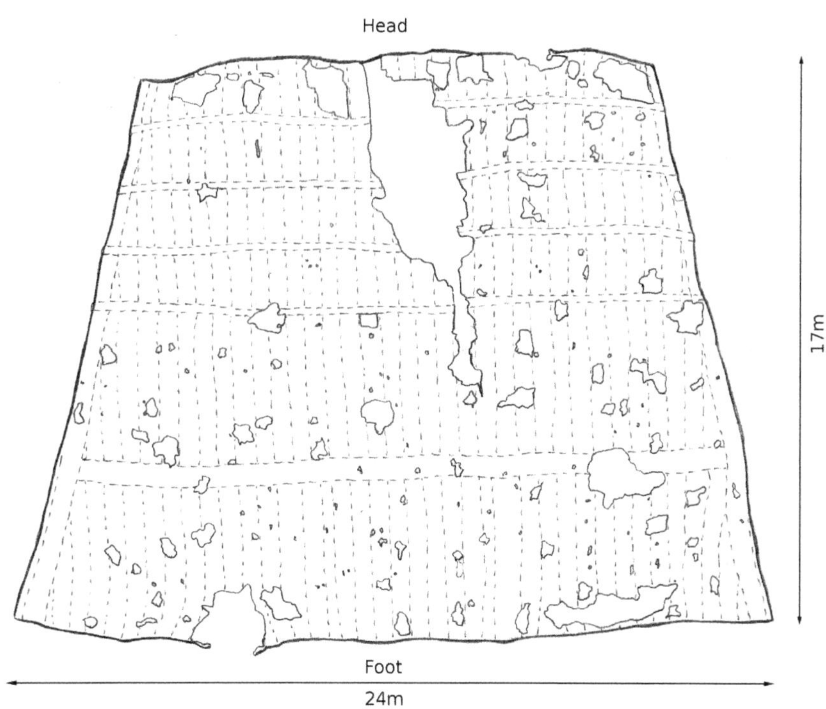

Figure 10.38 A sketch of the fore topsail of *HMS Victory*, showing the areas of loss.

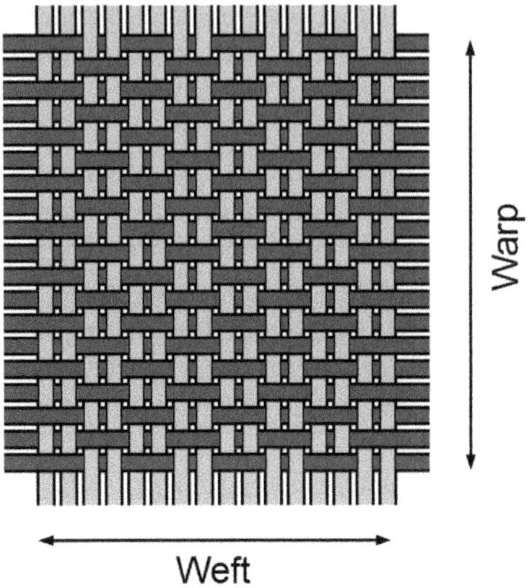

Figure 10.39 Weave structure of the sailcloth, with paired warp yarns.

is of a plain weave (Figure 10.39), with paired warp yarns (10.8 ± 0.4 pairs per cm) and single wefts (7.9 ± 0.9 yarns per cm).

In 2003 the conservators at the Textile Conservation Centre were charged with carrying out first stage conservation, which involved cleaning the sail; at the same time, the Centre's conservation scientists were asked to assess the physical state of the sail, in relation to its handling and stability, and to determine if the sailcloth was at immediate risk. To be able to provide complete answers and ultimately advise on the most appropriate approaches to conservation and display, it was essential to have an understanding of the microstructure, chemistry and degradation mechanisms of the constituent linen fibres. Further knowledge of the sail's life also proved critical – the mechanical stresses during its brief use and its exposure to the marine environment, the damage it suffered during the battle, and the more subtle deterioration experienced over the subsequent two centuries of storage and occasional display have all played a role in determining its current condition. Soaking in sea spray, of particular relevance to the *Victory* sail, will have led to increased swelling of the fibres, due to the penetration of hydrated sodium and chloride ions. On drying, swollen fibres will suffer dimensional changes, which in turn can lead to the distortion of the fabric. For the brine-laden sailcloth, further disruption of the linen would have ensued as the salt crystallised within the fibres. A scanning electron micrograph of fibres from the sail clearly shows extensive damage, most notably in the form of defibrillation (Figure 10.40).

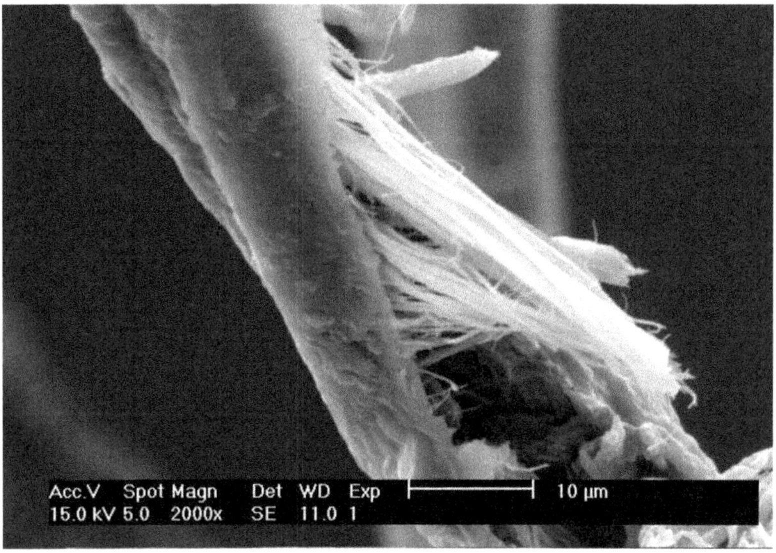

Figure 10.40 SEM image of fibres from the *Victory* sail, showing extensive fracturing and defibrillation.

To provide this information, several methods of analysis were employed, using sample from both the sail itself and from modern surrogates. Light and electron microscopy allowed the composition, structure and condition of the sail and its components to be visually assessed, informing the choices and sampling strategies for more detailed subsequent investigations. Mechanical measurements were carried out to establish the current physical condition of the cloth, and differences in performance between the historic and modern fibres were related to the disruption of the fibres consequent upon ageing. As oxidative deterioration of cellulosic materials leads to the generation of carboxylic acids, with potentially deleterious consequences for the long-term stability of the cellulose polymer chains, acidity measurements were also performed. The results of these studies made it possible to make recommendations concerning the use and longer-term preservation of the sail.

10.9.1 The Mechanical Performance of the *Victory* Sail

Linen is particularly suited to the production of sailcloth due to its good mechanical properties, including a high strength and low extensibility; furthermore, the strength of new linen fibres actually increases on wetting. Inevitably, though, the material will gradually deteriorate through use and interaction with its environment, and its performance will reduce.

Therefore, properties of the *Victory* sailcloth are bound to have changed in the two centuries since its manufacture, and these changes will be expressed at the chemical, microstructural and bulk levels. This could have progressed to the point at which the sail cannot even support its own weight, which would have implications both for any immediate conservation treatments and also for longer-term considerations. An appreciation of the sail's current condition was therefore crucial in deciding handling and display methods, and this could be informed by carrying out mechanical testing of the cloth. However, such tests are normally destructive and require samples to be taken; therefore, after consideration of the risks and benefits of sampling, and in consultation with the other stakeholders, it was deemed acceptable to take small specimens from areas of pre-existing damage, such as the holes in the cloth, where loose yarns could easily be removed without further damaging the bulk sail.

The following characteristics were measured: linear density ($g\,m^{-1}$), the mass per unit length of the yarns; breaking strength (N), the force required to break the sample; elongation, the percentage of original length by which the sample is stretched; and tenacity ($cN\,tex^{-1}$), the breaking strength per unit linear density. By calculating the tenacities of the yarns, the problem of comparing yarns of different linear densities is avoided. A typical mechanical test curve for a yarn from the sail, illustrating the percentage extension against load, is shown in Figure 10.41. Failure results from a combination of fracture (breaks forming in cells and fibres) and slippage (cells and fibres pulling past one another and losing cohesion).

From these data, average tenacities of the yarns were calculated for the various sampling locations; it became evident that these tenacities were all quite close in value, and what variation was observed did not occur in a systematic manner across the sail. Some areas are more heavily degraded and weaker, but the bulk appears to have retained about 30% of its strength when new (derived by comparison with data from appropriate modern equivalents).

Whilst tests of this sort reveal the mechanical properties of individual yarns, they do not necessarily reflect the characteristics of the cloth as a whole – the structure of the weave also plays a role in the load-bearing behaviour of the fabric. Since it was obviously not possible to remove whole pieces of the *Victory* sail to investigate such behaviour, an appropriate surrogate material was used instead, namely a modern linen sailcloth of a comparable weave, which was

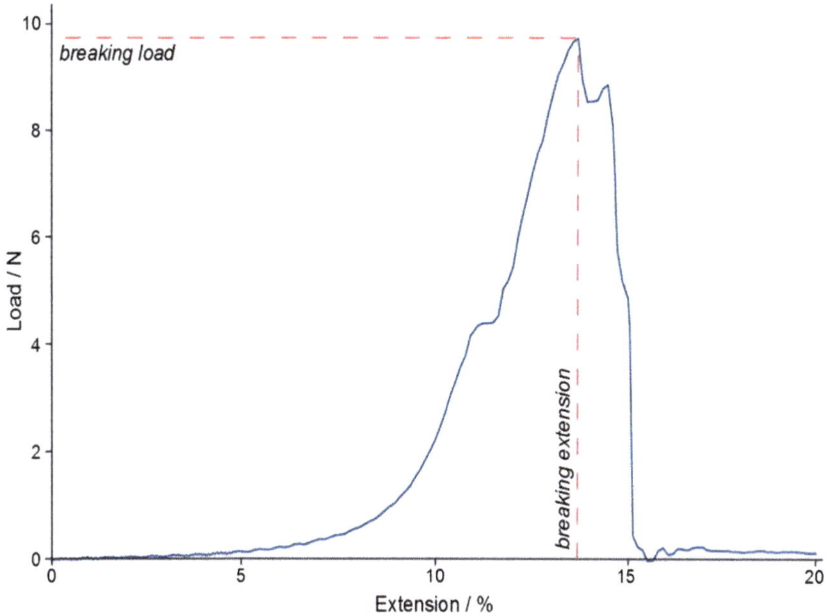

Figure 10.41 A typical load/extension (stress *versus* strain) curve for a yarn from the *Victory* sail, using a gauge length of 2.5 cm.

subjected to an artificial ageing regime based on the known history of the sail. With this material, a good correlation was found between the breaking load of individual yarns and that of whole sailcloth pieces. The only proviso was that the value for the cloth should be taken as 80% that of the fibre. The reason for this is that there is an additional failure mechanism when dealing with long specimens. For very short samples, fracture of the fibres dominates the failure behaviour of the material. At slightly greater sample lengths, slippage of individual cells (around 7 cm in length) within the ultimate will also play a role. And once the length of a sample approaches or exceeds the length of the staple (*i.e.*, the length of the fibre, which is generally a few tens of centimetres), then slippage of the fibres themselves is also important.

This information now made it possible to calculate the load-bearing capacity of the *Victory* sail, and, taking account of the holes and losses at the head of the sail, permitted a conservative estimate that the sail would still be able to bear twice its own weight, despite its aged condition. However, in addition to the danger of catastrophic failure of individual yarns and the bulk fabric, the more subtle risk of permanent stretching that could result from the sail supporting its own weight was also of concern. Deformation of this kind is an additional effect of slippage mechanisms at a variety of scales: it occurs when individual fibres in a yarn slip past each other, when the ultimates within a single fibre slide over one other, and when the secondary bonds that bind polymer chains together break, allowing the chains to creep past each other before new bonds are formed. From further direct investigations it was possible to determine that the sail would undergo a permanent deformation long before it failed. Therefore, it was possible to offer advice concerning the handling of the sail during its conservation to avoid undue and prolonged stress, and make suggestions for its display, specifically that it should be either flat or resting at a shallow angle on a support platform. These recommendations were necessary in light of the reduced strength and, in particular, the likelihood of deformation resulting from the aged condition of the sailcloth and the degraded state of the constituent fibres.

10.9.2 Acidity Measurements on the *Victory* Sail

As noted in Section 10.3.1, in the absence of active biological attack, the two dominant processes leading to natural deterioration of cellulose are acid hydrolysis and oxidation. The latter is quite slow, and cellulosic artefacts that are above neutrality and kept in a stable environment might be expected to remain in relatively good condition for hundreds of years. However, those that are acidic can show significant signs of deterioration within a few decades. The ageing of fibres like flax often results in the formation of acidic degradation products, and these, together with adsorbed acidic atmospheric pollutants, can continue to promote further damage through the hydrolytic depolymerisation of cellulose. Therefore, to better understand the likely future stability of the *Victory* sailcloth, it was important to estimate the acid content of the material. These measurements were carried out in two ways: surface measurement on the bulk fabric and acid extractions from individual fibres.

Surface pH was recorded by pressing a moistened non-bleeding indicator paper to the fabric for a short period, after which the resultant colour change was interpreted. These revealed pH values between 3.3 and 4.0 in the various locations tested, whereas for new linen sailcloth the value was around 6.0. Whilst it is difficult to deduce the true acid content from this type of assessment, nevertheless values below pH 4 are generally regarded as a cause for some concern over the longer term, leading to recommendations for remedial action.

More detailed total acidity assessments were performed on short sections of the isolated yarns by extracting mobile and fibre-bound acid with a degassed saline solution, before measuring the acidity with a microelectrode and pH meter. These saline extractions allow a better understanding of acid distribution, with the presence of salt ensuring the equilibration of immobile acid within the solution. Such acids may be strong (for example, those arising from atmospheric pollutants) or weak (such as the carboxylic acids that are natural components of plant fibres or generated through oxidative deterioration). The total acid content deduced in this manner represents a lower estimate since the weak acids may not be fully ionised, and this underestimate will be most marked for the more acidic samples.

Yarns from the modern, unaged sailcloth had a small acid content, probably just reflecting residual weak acid constituents of the plant fibres, such as hemicelluloses and lignin. The *Victory* sail samples had a higher acid content; for some yarns this increase was only marginal, whilst others had more than a ten-fold raised acid content, and in the case of a few heavily deteriorated specimens, a much-increased acidity was observed. These results contrast with the surface pH data, which suggested a similarly raised acidity across the whole sail. One interpretation of this anomaly is that in cases where the yarn acid content is lower, the acid is localised in the outer regions of the fibres, and that as the acid content rises it is found deeper within the fibres. This would be consistent with an initial surface oxidation, followed by the progressive deterioration of the secondary cell walls of the fibres. Titration implies that the majority of this acid is weak, derived from the oxidation of cellulose and other fibre constituents, which would fit with this hypothesis.

10.9.3 Conservation and Display of the *Victory* Fore Topsail

Before any treatment, the sail was placed in an environmentally controlled chamber and allowed to acclimatise under conditions of moderate humidity, where the fabric would be the most pliable. In the first phase, the conservators then surface cleaned the *Victory* sail with low suction vacuum cleaners and static cloths, removing a considerable quantity of particulate soiling. This not only led to an aesthetic improvement but also reduced the risk of abrasive wear, and removed material that might act as either a source of deleterious reagents and or a foodstuff for pests. Taking account of the need to handle the fabric gently and prevent undue stress, the sail was laid flat (over Tyvek®, an inert high-density polyethylene sheet, on a suitable platform), and a specially constructed inflatable roller was used to help to move the large artefact within a relatively confined space. There was then negligible risk of uneven short-term loading which could have ruptured the fabric.

The mechanical tests suggested that it would be inadvisable to hang the sail from a yard, in the manner originally used onboard ship, which was one of the display possibilities originally considered. Instead, it was determined that it would be more suitable to display the artefact either flat or at a shallow angle on a suitable frame; this would support the weight of the fabric and prevent fibre slippage and permanent deformation. In terms of the sail's continued preservation, the apparent undue acidity of the fabric also needed to be borne in mind. Since over the longer term, acid deterioration products can become mobile and may even be evolved in gaseous form, adjacent materials could become affected. This suggested that it is best to keep the sail unfolded and in an air-conditioned environment to prevent any build-up of acidic gases, though for storage it might be appropriate to include acid sorbents between the layers of the cloth. Since some fibres are already quite acidic, deacidification also needs to be considered in the future. After weighing up the relative risks though, non-intervention may be deemed the more appropriate course at present, avoiding the structural damage that might be caused by such an intensive process.

Acknowledgements

The author would like to thank The Mary Rose Trust and Portsmouth Historic Dockyard for permission to publish the *Victory* sail case study.

Recommended Reading

Fibre Structure, Chemistry and Properties

A. Tímár-Balászy and D. Eastop, Chemical Principles of Textile Conservation, Butterworth-Heinemann, Oxford, 1998.

K. L. Hatch, Textile Science, West Publishing Company, Minneapolis, 1993.

M. Lewin, Handbook of Fibre Chemistry, 3rd edn, CRC Press, Boca Raton, 2006.

W. E. Morton and J. W. S. Hearle, Physical Properties of Textile Fibres, The Textile Institute, Manchester, UK, 1993.

M. A. Taylor, *Technology of Textile Properties*, 3rd edn, Forbes Publications, London, 1990.
R. R. Mather and R. H. Wardman, The Chemistry of Textile Fibres, Royal Society of Chemistry, Cambridge, 2011.
P. Garside, Textiles, in Cultural Heritage Microbiology, ed. R. Mitchell and C. J. McNamara, ASM Press, Washington, DC, 2010, p. 97.

Analysis

P. H. Greaves and B. P. Saville, *Microscopy of Textile Fibres*, Bios, Oxford, 1995.
M. Sawbridge and J. E. Ford, *Textile Fibres under the Microscope*, Shirley Institute, Manchester, 1987.
D. M. Catling and J. E. Grayson, *Identification of Vegetable Fibres*, Archetype Publications, London, 1998.
B. P. Saville, *Physical Testing of Textiles*, Woodhead Publishing Ltd, Cambridge, 1999.
Chemical Testing of Textiles, ed. Q. Fan, Woodhead Publishing Ltd, Cambridge, 2005.

Conservation

J. Robinson and T. Pardoe, *An Illustrated Guide to the Care of Costume and Textile Collections*, Museums and Galleries Commission, London, 2000.
Textile Conservation: Advances in Practice, ed. F. Lennard and P. Ewer, Butterworth-Heinemann, Oxford, 2010.
B. Powell, *Collection Care: An Illustrated Handbook for the Care and Handling of Cultural Objects*, Rowman & Littlefield Publishers, Lanham, 2016.

The *Victory* Sail

K. Gill and P. Garside, The HMS Victory Fore-Topsail, in *'Big Issues': Forum of the* UKIC *Textiles Section, 2005*, ed. F. Nuttgens and M. Jordan, ICON, London, 2006, p. 36.
P. Garside and P. Wyeth, Assessing the Physical State of The Fore-Topsail of the HMS Victory, in *Scientific Analysis of Ancient and Historic Textiles: Informing Preservation, Display and Interpretation*, ed. R. Janaway and P. Wyeth, Archetype Publications, London, 2005, p. 118.

11 Plastics

E. Richardson*[a] and C. Flack[b]

[a] Image Permanence Institute, Rochester Institute of Technology, 70 Lomb Memorial Drive, Rochester, NY 14623-5604, USA; [b] Tate, Department of Conservation, Millbank, SW1P 4RG, London, UK
*Email: ejrpph@rit.edu

11.1 Introduction

Plastics are some of the most varied and challenging classes of materials the museum conservator will face. The degradation of plastic objects in collections has been a growing problem for museums and galleries for more than 25 years,[1,2] yet the problem still confounds conservators and collectors due to the potential for rapid and often spectacular disintegration of objects with little advance warning. Without extensive research into this area, such problems are set to escalate.

Since their development in the late 19th Century, synthetic polymers have moved steadily into almost every area of life, and as a consequence, into a growing number of museum collections. Plastics are found everywhere in modern and contemporary cultural heritage, from sculpture to design objects, architectural models to decorative arts. Many classes of plastic have become household names: polythene, polyester, polyurethane, acrylic, polyvinyl chloride (PVC), and cellulose acetate, to name just a few. They can take many forms, including solid sheet plastics, foams, fabrics, and 3D objects. They are typically cheap, lightweight, readily moulded and shaped into all kinds of forms and structures and, since their introduction, they have opened up incredible new design

Conservation Science: Heritage Materials, 2nd Edition
Edited by Paul Garside and Emma Richardson

Published by the Royal Society of Chemistry, www.rsc.org

possibilities for artists, architects, and designers, owing to novel properties not found in natural materials.

The general perception of plastics is one of persistence, materials that do not readily biodegrade and pose significant environmental problems. Whilst this may well be the case, the original form and condition of these plastics will likely alter over time even though their presence remains. It is these alterations in the physical nature of plastics that are of concern to the heritage profession, with many types of plastic already exhibiting serious signs of deterioration that often appear with little or no warning. Common signs of deterioration include discolouration (yellowing or opacifying), crazing and cracking, warping, becoming sticky (as plasticisers migrate out of the bulk material onto their surfaces), and in extreme cases, turning completely to powder (Figure 11.1). The combination of the potential for catastrophic changes in certain classes of plastic and the sheer range of plastics available means that their preservation constitutes a significant challenge for the conservation profession.

One key problem when dealing with such materials is that the term ‘plastic’ covers such a wide variety of different polymers, which can exhibit highly varied properties and behaviours. A significant body of research is therefore needed before it will be possible to decide on the optimum ways in which plastic objects/materials can be exhibited, treated, and stored in order to lower their deterioration rate. Due to the complex composition of modern synthetic polymers, deterioration processes are complicated to understand and predict, and unfortunately at this time impossible to reverse. Therefore, ongoing research is

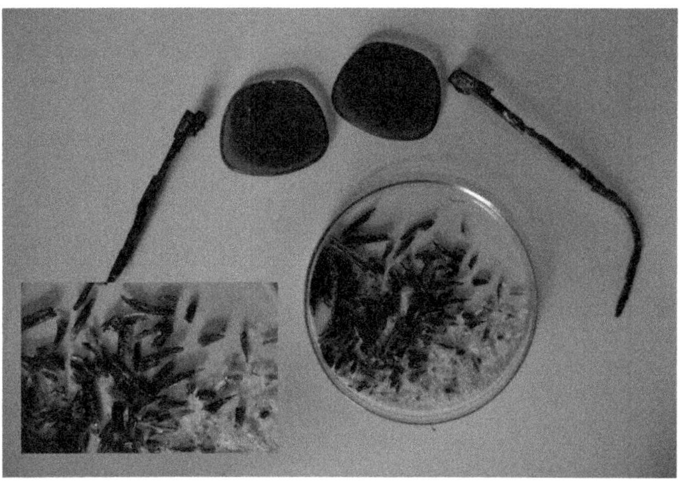

Figure 11.1 Degraded cellulose acetate sunglasses displaying complete loss of physical integrity and loss of triphenyl phosphate additive.

currently being carried out at various institutions to find ways to preserve and treat plastic artworks for future generations.

The 21st century would be unrecognisable if plastics did not exist. This modern material, "the stuff of alchemy" as Roland Barthes puts it in his essay *Mythologies*,[3] represents consumer driven mass culture; it is the material of the everyday. The development of plastics has helped to raise the standard of living for many people through making products available that otherwise would have proven too expensive for the masses. Plastics have become ubiquitous materials that have steadily moved into every walk of life, from car manufacturing to computers, clothing to medical components.

The term plastic is the general name given to a variety of natural, synthetic and semi-synthetic materials, which, as the name suggests, exhibit plastic behaviour and can be moulded or shaped into a range of forms. The earliest plastics were naturally occurring and can be dated back to prehistoric times. For example, horn, amber and tortoiseshell are natural plastics that can be reshaped using heat. It was not until the 19th century that the 'world of synthetic chemistry began to boom'[4] as scientists such as Alexander Parkes in Britain and Leo H. Baekeland in America began experimenting to create new commercial materials. Originally these were based on natural materials, such as cellulose, derived from wood pulp, or casein, derived from milk products, which were chemically modified in the laboratory and used to create semi-synthetic plastics (Figure 11.2) One of the most notable

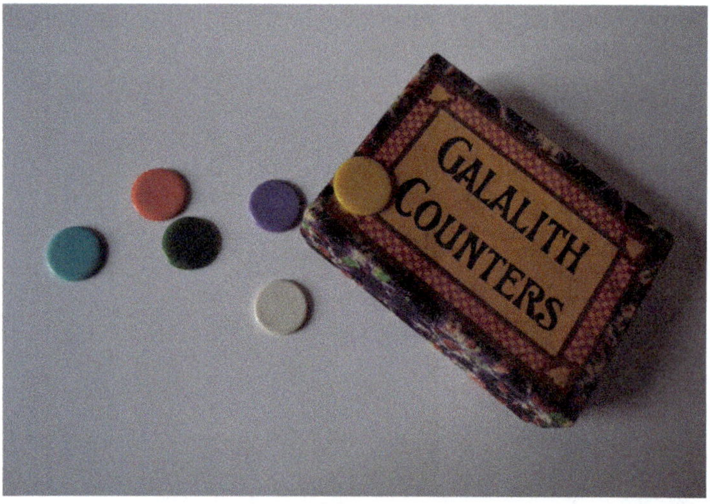

Figure 11.2 Box of casein formaldehyde counters, a thermosetting semi-synthetic plastic sold commercially under the name of Galalith, derived from the Greek words gala (milk) and lithos (stone).

of these is cellulose nitrate, synonymous with the 20th-century film industry and most commonly known as celluloid.

As synthetic chemistry developed, in some cases driven by the onset of the Second World War and a need to push material boundaries, entirely synthetic polymers emerged onto the market. Polymer scientists saw the possibilities offered by developing completely synthetic polymers derived from petrochemicals, the earliest originating from the gas ethylene, and used to form plastics such as PVC, polystyrene and polyethylene.[5]

11.2 Chemistry, Structure and Microstructure of Plastics

Much like metal alloys, plastics are composite materials that are made from two or more constituents that have different chemical and physical properties. These constitutions include the base polymer or polymers that are combined with materials such as plasticisers and pigments. In the case of plastic, the bulk material is comprised of one or more organic polymers (the term coming from the Greek for having *many parts, polumerēs*) in combination with various secondary additives that impart particular favourable properties on the plastic. The additive materials may be materials such as plasticisers, pigments, UV stabilisers amongst others. Organic polymeric materials will all degrade over time, whether natural, semi-synthetic or synthetic. However, semi-synthetic and synthetic types tend to pose significant stabilisation problems due to production methods, additives, and treatments, which can introduce varying and complex routes of degradation.

Polymers are formed by the linking of many smaller molecules known as monomers (Figure 11.3). Organic polymers can be natural, semi-synthetic or synthetic in origin. Natural polymers are those derived from plant and animal sources, such as wood, cotton, and leather and parchment. These materials are composites of different cellulosic or proteinaceous polymers and have been discussed in detail in Chapters 3, 5 and 10.

Manmade polymers are grouped into two distinct categories of materials, owing to their origin. Semi-synthetic materials are those that have been derived from naturally occurring polymers and chemically modified within the laboratory. In contrast, synthetic polymers are formed through induced polymerisation reactions of small molecular weight hydrocarbons derived from petroleum fractions.

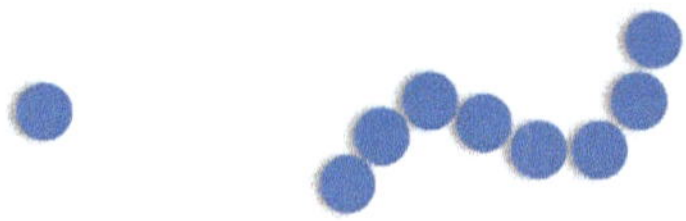

Figure 11.3 Schematic representation of monomer and polymer units.

The physical characteristics of all polymers are based on their high molecular weights and the underlying crystalline and non-crystalline structures. Therefore, processes of deterioration such as hydrolysis and thermal degradation that alter the structure of these polymer chains are naturally going to affect their material properties and stability.

11.2.1 Polymerisation

The formation of synthetic polymers requires that the initial monomer molecules have a functionality of two or more. This allows for linkages to be formed at these activation sites. Many types of reactions are possible, but most can be classified as either step polymerisation reactions or chain polymerisation reactions.

11.2.1.1 Step Polymerisation

Step polymerisation occurs when two difunctional molecules (monomers) react with each other to form a product that also contains two functional end groups. The molecular mass of the polymer increases as a function of reaction time.[5] This relationship is, however, governed by factors such as the reactivity of the functional groups and the degree of intramolecular reactions competing with intermolecular reactions.[6] This process can be divided into two categories: polycondensation and polyaddition. The former are those reactions where small molecules are eliminated, with the resultant polymer units containing fewer atoms than the original monomer units.[7] The formation of synthetic polyamides, the most commonly known being nylon, is formed through polycondensation reactions (Figure 11.4).

By contract, polyaddition reactions do not eliminate molecules during polymer formation. Any side products formed during the reaction are retained in the final polymer material. The formation of polyurethanes is *via* a polyaddition reaction (Figure 11.5).[8]

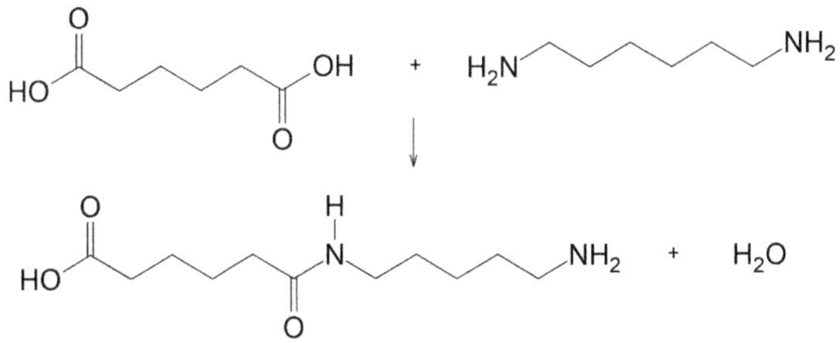

Figure 11.4 Schematic of the formation of Nylon 6,6.

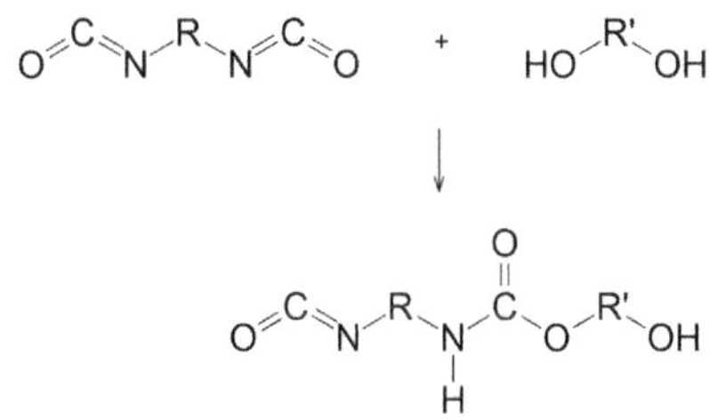

Figure 11.5 Schematic of the formation of polyurethane.

11.2.1.2 Chain Polymerisation

Chain polymerisation can follow a number of reaction routes. These include ionic polymerisation, Ziegler–Natta coordination polymerisation, ring opening polymerisation and free radical polymerisation.[7] The latter is the most common reaction for the production of polyalkene vinyl polymers, such as polythene (Figure 11.6), polystyrene and PVC. Their formation is reliant on the presence of unsaturation in the monomer unit enabling a chain polymerisation.

As the name suggests, chain polymers are built up by the addition of one monomer unit at a time, in a radical initiated reaction, without the formation of other products. During chain growth, a monomer is activated and polymerisation propagates by activating neighbouring monomer units. It follows three key steps during the reaction, namely initiation, propagation and termination.

During the *initiation* step, free radicals called initiators are formed under catalytic conditions. Due to their highly reactive nature, these often break down further into smaller radicals with the elimination of a molecule. The free radical is then available to react with the double bond of an alkene. The new carbon radical can then go on to react with a monomer unit. This *propagation* step is repeated many times to form the polymer chains.[9] The polymerisation reaches *termination*

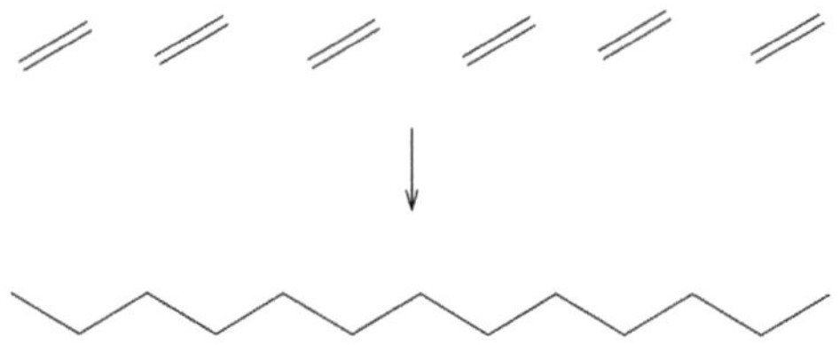

Figure 11.6 Schematic of the formation of vinyl polymers.

when the radicals are all consumed. The most common of these reactions are combination and disproportionation.[7]

During the combination mechanism, two carbon radicals come together to form a bond and a single polymer chain. Disproportionation occurs when a hydrogen atom is removed from one carbon radical to another. The result is the retardation of one radical, with the other gaining a second unpaired electron. The molecule is then stabilised by the formation of a double bond. If the disproportionation radical attack occurs along the polymer backbone, the energy cannot be dissipated. Therefore, the newly formed radical will go on to react with further alkene molecules forming a branched or network polymer structure.[10]

11.2.2 Degree of Polymerisation and Molecular Weight

The degree of polymerisation (DP) is a measure of the average number of monomer units per polymer chain. It is determined statistically and is based on the stoichiometric balance of the reaction. The reader is referred to the literature for an in-depth discussion on the subject,[5,7,11] but it suffices to say that the type of monomer units and the degree of polymerisation will govern the molecular weight of the material. If reactant quantities are not stoichiometric then the degree of polymerisation is reduced, because the reaction is terminated by the monomer in deficit. This can often be used to the advantage of manufacturing, enabling the control of the length of the polymer chains, and therefore manipulating the physical properties of the material.

DP can be misleading since it refers to the number of links and does not convey anything about the size of the starting molecules. It is clear from Figure 11.7 that the polymer on the left is larger than that on the right, owing to the size of the monomer units. However, the degree of polymerisation of the polymer on the left is lower than that on the right.

Due to this, it is sometimes more helpful to discuss alternative distributions of the molecules to give us a better indication of the

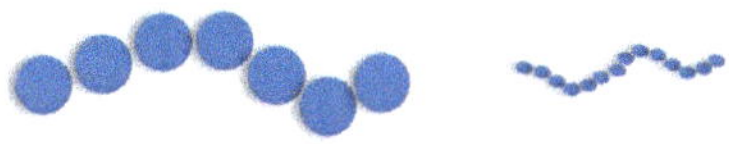

Figure 11.7 Schematic representation of the degree of polymerisation.

weight and number of the constituents. Distributions might include the number average molar mass (Mn) or the weight average molar mass (Mw). The former refers to the average number of molecules of a particular molar mass and the latter is the average number of molecules of a particular size. These distributions are determined experimentally through osmotic pressure methods and viscometry, respectively.[11]

11.2.3 Microstructure

Organic polymeric materials are long chain macromolecules made up of many repeat units of smaller organic monomers. Depending on the stereochemistry and functionality of the initial reactants the resulting polymer will have a linear, branched or three-dimensional network (Figure 11.8).

Polymerisation can occur between monomer units of the same species, or from two or more different reactants. The former polymerisation route derives homopolymers with the latter producing copolymers of differing repeat units.[7] In the production of copolymers the beneficial characteristics of the individual molecules can be employed within one material.

The microstructure of polymers and plastic materials shows a mixed phase morphology known as a fringed micelle structure; low order, randomly oriented amorphous fringes and well ordered, closely packed crystalline micelles (Figure 11.9). The crystalline regions are formed by folded polymer lengths, called lamellae, with a high degree of intermolecular bonding between adjacent polar groups along the polymer. The amorphous regions contain disoriented polymer chains, many having large side groups, branches or random stereochemistry that reduce packing, orientation and forces of attraction.

The polymerisation of monosubstituted alkenes often produces highly amorphous polymers, where the chains possess varying stereochemistry dependent on the asymmetry of the carbon group. When the substituted group is positioned in a random manner along the polymer backbone the polymer is said to be atactic (Figure 11.10).

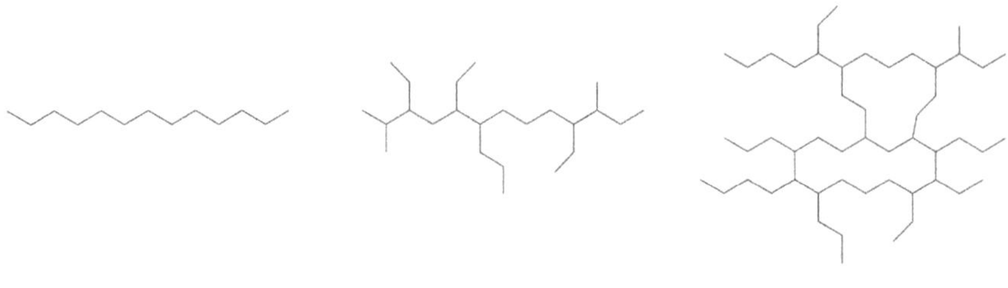

Figure 11.8 Schematic of polymer microstructure, namely linear, branched and networked.

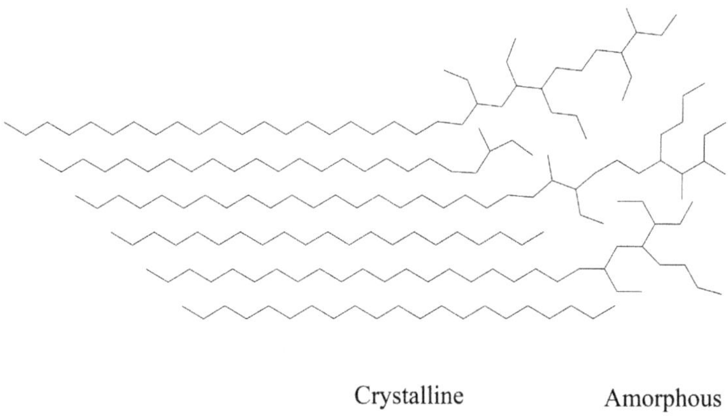

Figure 11.9 Schematic of the semi-crystalline polymer microstructure.

Polymers possessing group conformations on the same side of the polymer backbone are isotactic and on alternating sides syndiotactic. The atactic conformation prevents the chains from closely aligning with adjacent lengths producing polymers of low crystallinity.[9] Free radical polymerisation of substituted alkenes often produces polymer chains that possess random, atactic stereochemistry.[9] Polymers that possess isotactic and syndiotactic conformations are formed through a process known as Ziegler–Natta polymerisation.

Crystallinity, orientation and forces of attraction all have an effect on the physical characteristics of polymers, such as strength, melting point (T_m) and the glass transition temperature (T_g). A high proportion of crystalline regions within a plastic will lead to increased polymer strength. However, this also produces a rigid polymer backbone, reducing flexibility and increasing brittleness. For these polymers, plasticisers are often added to intersperse within the matrix, thus enabling increased freedom of the molecular chains.[6]

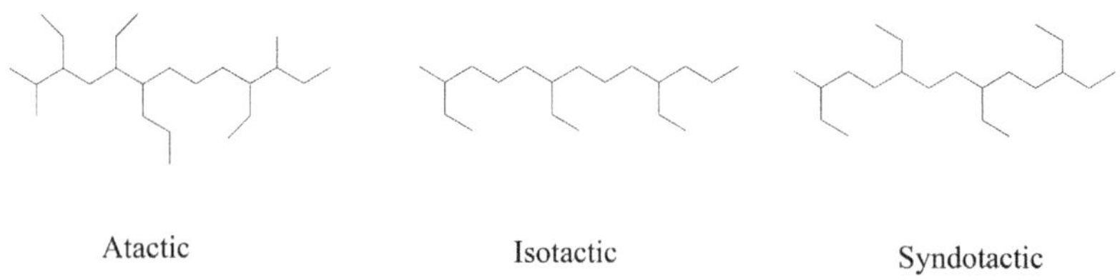

Figure 11.10 Schematic of the stereochemistry of monosubstituted alkenes, namely atactic, isotactic and syndotactic.

11.2.4 Crystalline and Amorphous Fractions

The crystalline regions will exhibit long range order, whilst the amorphous regions will show some short-range ordering but will be disordered over greater lengths. The degree of crystallinity is dependent on the ability of the polymer chains to establish close packing and to form strong intermolecular bonds. The thermal behaviour of the crystalline regions is characterised by the melting temperature (T_m) and exceeding this results in the complete breakdown of the ordered regions. It is due to the presence of the intermolecular bonds that polyamides melt at a higher temperature than polyolefin materials, for example,[12] with a number of amide groups reportedly remaining hydrogen bonded even in the polymer melt.[5,13] Tm will increase with increasing intermolecular bond density due to mobility restrictions on polymer chains. The melting temperatures of a number of synthetic polymers are presented in Table 11.1. The melting temperatures of polyamides will decrease as the number of methylene groups along the polymer backbone (per amide group) increases. For example, polyamide 2,20 has a melting temperature of 230 °C compared to a melting temperature of 170 °C for polyamide 12,20[14] (see Chapter 10). In the case of polypropylene, the isotactic form cannot crystallise in a planar zig-zag due to steric hindrance and will form a tightly packed helix,[15] with interactions between neighbouring methylene groups. The presence of the amorphous, atactic form within polypropylene disrupts the packing and reduces entanglements. As such, the melting temperature of polypropylene is dependent on the percentage isotactic content.[16]

Many of the synthetic polymers found within museum and heritage collections are semi-crystalline in nature, where the regions outside of the crystal lattice show some short-range order and intermolecular bonding, but are disordered over longer distances.[17] The non-crystalline, amorphous regions are characterised by segmental motion affecting the final flexibility of the material. Where the thermal behaviour of the crystalline regions are characterised by the melting

Table 11.1 The melting temperatures of selected synthetic polymers.

Polymer	Melting temperature, °C
Polyamide 6	215
Polyamide 6,6	264
Polypropylene (isotactic)	160–170
Polypropylene (syndotactic)	125–135
Cellulose Triacetate	300
Cellulose Diacetate	230

temperature (Tm), the bulk amorphous regions are characterised by a number of thermal transitions linked to local vibrations and rotations and longer range chain mobility.

Fully crystalline materials or low molecular weight materials experience defined thermal transitions accompanied by a phase change from solid to liquid and liquid to gas. Whereas high molecular weight polymeric materials experience a number of second order transitions over discrete temperature ranges, which do not exhibit a phase change. The secondary transitions are associated with main chain and side chain motions within the polymer matrix. The temperatures at which these motions are free to occur are dependent on, and influenced by, a number of factors relating directly to conformational changes in the macromolecule, which have a significant effect on the physical properties of a material.

The primary relaxation of the amorphous regions is known as the glass transition (occurring at T_g). This is the most important transition with respect to the physical behaviour of polymers. It covers a narrow range at which the non-crystalline regions move from a rigid, glassy state to a rubbery state caused by an increase in energy.[18] The glass transition temperature is a viscoelastic transition, governed by the amorphous regions, where free volume begins to increase resulting in increased molecular motion of the long chain segments and rupturing of the intermolecular bonds.[19]

11.2.5 Thermal Properties

Temperature is a key factor in the working properties of a material. When considering the thermal properties of plastics it is common practice to group materials into two different types, namely thermoplastics and thermosets.

Thermoplastic materials refer to those materials that can be softened by the application of heat but become rigid on cooling.

In thermoplastic materials this is a reversible, physical process whereby the further application of heat enables the re-softening of the plastic, enabling it to be reshaped and moulded numerous times. It is this class of materials that often lend themselves to household recycling processes. Thermoplastics are characterised by high molecular weight polymers modified with various chemicals to alter the elements along the chain lengths, which induce forces of attracting between the chains. The presence of functional groups and intermolecular bonding allow for short-range, crystalline regions that impart rigidity and strength, whilst longer range disordered regions allow for a degree of flexibility and extensibility.

In semi-crystalline materials the room temperature working properties are governed by the glass transition temperature. In high molecular weight linear polymers, changes of state are not clearly defined and can occur over a finite temperature range. The glass transition temperature is characterised by the transition from a glassy state to a rubbery one. The relaxation behaviour can be separated into two parts: a solid-rubbery transition, known as the glass transition (T_g) and a rubber-liquid transition at higher temperatures that produces a material capable of viscous flow. Raising the temperature above the T_g results in greater molecular motion of the long chain segments and eventual rupturing of the intermolecular bonds. Below the temperature the molecules have reduced energy, becoming increasingly restricted preventing deformation. At this point the material becomes brittle and will readily fracture under external forces.

By contrast, thermosetting types of plastics offer an altogether different set of physical properties. In thermosetting materials, the starting polymer is in the form of a viscous fluid or powder that can be readily cast into a mould, but solidification occurs through the addition of heat, pressure or a catalyst. The solidification process is known as curing and on a molecular level induces bridges or cross-links to be formed between the different polymer chain lengths. The cross-linking causes a three-dimensional network of chemical bonds to occur throughout the material rendering it a rigid solid, which on the application of heat cannot be readily softened or manipulated further.[5]

11.3 Preparation

The process of forming the bulk physical shape of plastics is dependent on whether the plastic is a thermoplastic or thermoset material, with the former type affording the possibility of remoulding and reshaping and the latter producing temperature stable, rigid structures.

11.3.1 Injection Moulding

As the name suggests, the process of injection moulding relies on the injection of the molten polymer into a mould known as a die. Preformed polymer pellets are introduced into the system *via* a funnel called a hopper, which gravity-feeds the pellets into the heating chamber. The molten polymer is then forced through an injection nozzle into the mould under pressure caused by a plunger at the rear of the chamber. The temperature of the die is controlled *via* cooling channels and as the molten polymer enters the mould solidifies into its three-dimensional form (Figure 11.11).

11.3.2 Extrusion Moulding

The term extrusion refers to the expulsion of material in a continuous form due to the application of pressure, the most common example being the extrusion of toothpaste from a tube. In a similar fashion to injection moulding, the polymer is introduced into a heating chamber *via* a hopper and forced through a narrow die, which ensures a constant cross-section. As the extrudate exits the die it solidifies and is wound around a series of wheels to ensure that the exit of the mould remains clear for further extrusion (Figure 11.12). Synthetic fibre formation is generally *via* the process of melt spinning, which is usually followed by fibre drawing in order to improve mechanical properties (see Chapter 10). Depending on the conditions, this has the effect of increasing the percentage crystallinity, chain orientation and crystal size, as well as inducing some orientation of the amorphous regions.

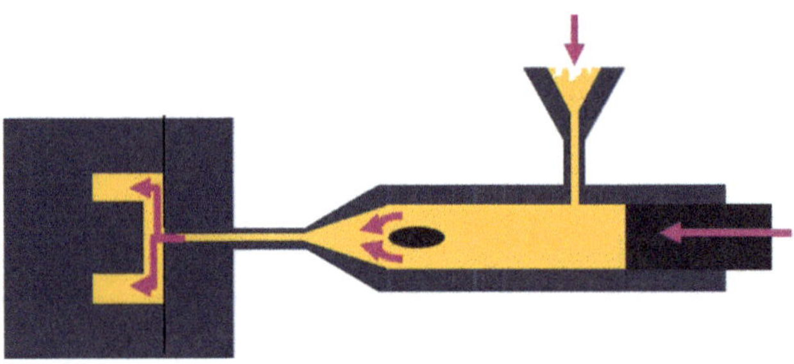

Figure 11.11 Schematic of injection moulding of thermoplastic (after Fry).

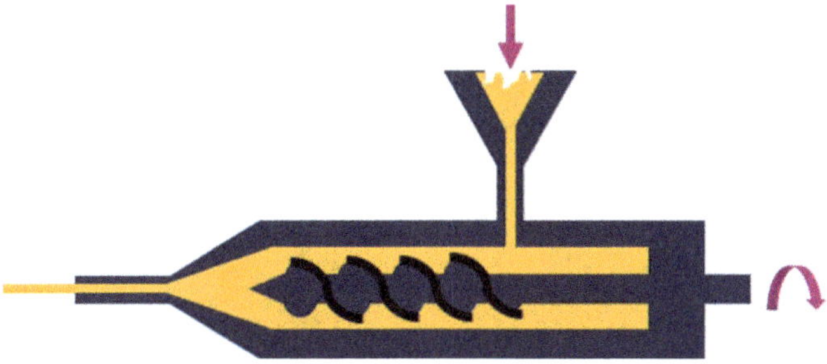

Figure 11.12 Schematic of extrusion moulding of thermoplastic (after Fry).

The drawing of polymer fibres is normally carried out above the glass transition temperature as this allows for greater chain mobility and rearrangement.[20] The draw ratio (DR) is a measure of the degree of extension of a fibre after polymer drawing. It is the ratio of the cross-sectional area of the undrawn fibre to that of the drawn fibre. Vasanthan *et al.*[20] showed that the size of the crystallites in polyamide 6,6 fibres increases rapidly with an increase in DR from one to two, but this tails off with higher draw ratios. They also demonstrated that the degree of orientation develops more slowly in the amorphous regions than in the crystalline regions.

11.3.3 Calender Forming

Calendering machines are used in the formation of polymer films and sheets and are reliant on the control of a series of polished cylinders between which molten material is fed (Figure 11.13). Opposing cylinders are rotated in opposite directions to feed through the molten polymer. The thickness of the sheet material is controlled by adjusting the distance between neighbouring cylinders and the degree of crystallinity adjusted by carefully controlling the cooling rate during the rolling process.

A significant proportion of vinyl polymers are formed into sheet materials, as will be discussed in the Case Study in Section 11.6.

11.3.4 Compression Moulding

Historically, the use of compression moulding is usually reserved for thermosetting plastics, where the use of heat cannot be readily used to alter the physical form of the plastic material. In compression moulding the compounding material is introduced into the cavity of a die and the upper mould placed on top. The compression forces cause

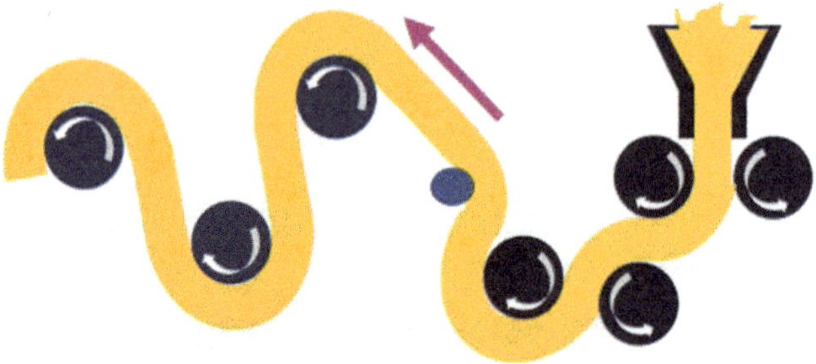

Figure 11.13 Schematic of calendar forming of thermoplastic sheets (after Fry).

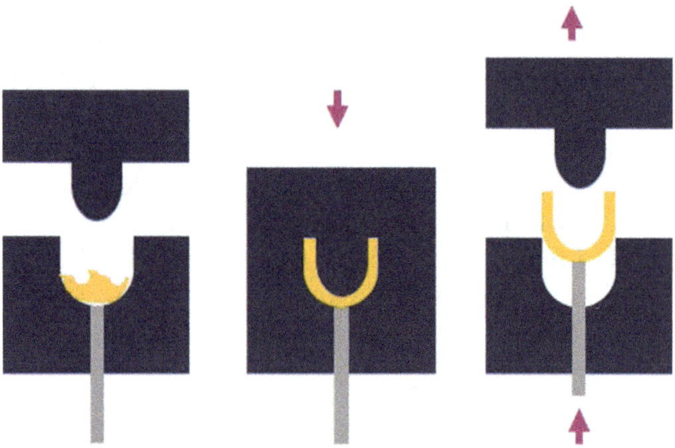

Figure 11.14 Schematic of compression moulding of thermosetting plastic (after Fry).

the polymer to move through the cavity and the thermosetting plastic is catalysed to form a networked solid through the application of heat and pressure (Figure 11.14).

11.4 Additives

In terms of plastics manufacturing, synthetic and semi-synthetic polymers are rarely used in isolation and are combined with other materials to form composites (see Chapter 9). Although the polymer components generally represent the greatest proportion of a plastic, a range of different materials known as additives are incorporated into the polymer matrix to impart different chemical and physical properties. There is an extensive range of additive types that can be loosely classified into i) those actively involved in the polymerisation process, ii) those which aid the manufacturing process, and iii) those influencing the final bulk properties of the polymer. The type and application of these additive components are too extensive to cover in this text, but some of the primary additives conservators are likely to encounter within collections or during treatment are introduced below.

11.4.1 Plasticisers

The primary role of a plasticiser is to soften a material and increase its malleability. Although contemporary plastic manufacturing generally relies on synthetic plasticiser compounds to improve the original properties of the polymer, natural plasticisers have been used for millennia in various forms to manipulate material properties. A familiar example is the addition of water to clay mixtures to form a highly pliable material for moulding and forming (see Chapter 7). The early semi-synthetic plastic cellulose nitrate used the natural plasticiser camphor, which is a terpene distillation from the wood of camphor trees, but the use of natural oils can be readily traced back to Ancient Egypt when cedar oil was used to prevent the formation of cracks in skin during mummification processes (see Chapter 5).

Plasticisation occurs through the addition of relatively low molecular weight molecules into the polymer matrix, which causes the separation of neighbouring polymer chains. Plasticisers primarily reside in amorphous regions, resulting in the interruption of the interchain hydrogen bonds, reducing the steric hindrance of the polymer chains and increasing mobility.[13] This has the effect of significantly reducing the modulus of elasticity[21] and the materials softening temperature (T_g).

11.4.2 Fire Retardants

The majority of polymers are combustible and therefore their safe use often relies on the incorporation of fire retardants into the plastic composite. It is not uncommon to find fire retardants across modern collections, such as furniture collections where polyurethane foams are often present as padding and filling materials. Since the 1980s, regulations have required retardants to be incorporated into polyurethane foam because its large surface area and air permeability causes it to be highly flammable, whilst the combustion of the polyurethane is known to release carbon monoxide and hydrogen cyanide.[22]

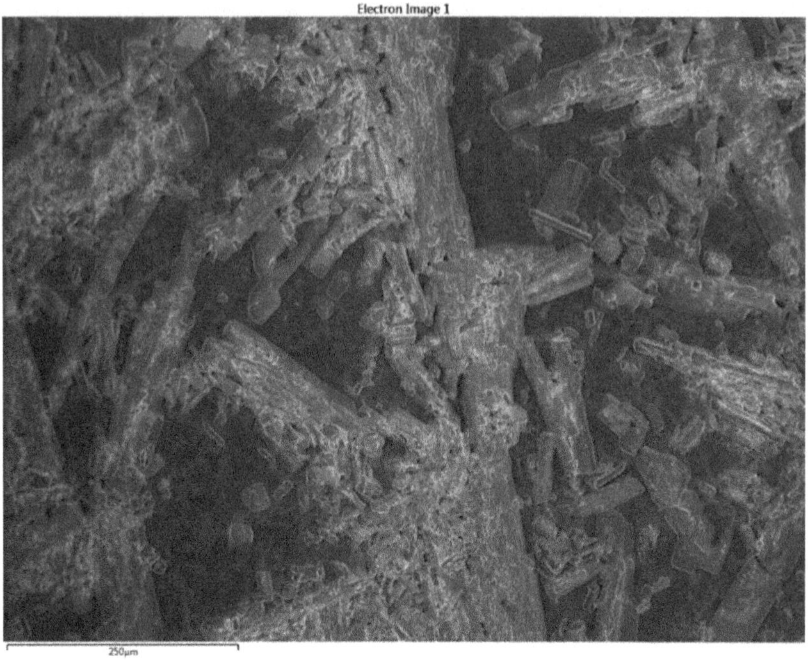

Figure 11.15 Scanning electron micrograph of triphenyl phosphate crystals following migration to the surface of a cellulose acetate film, x250 magnification.

The largest share of the fire-retardant market is the phosphate compounds, which not only have the effect of retarding fire by scavenging free radicals in the flame, but also impart plasticising properties. Triphenyl phosphate (TPP) is historically important in film materials since it was widely incorporated into cellulose acetate,[23] marketed as Safety Film, in response to highly flammable cellulose nitrate film. TPP is a crystalline solid that displays low compatibility with cellulose acetate and therefore often found in the presence of liquid plasticisers to help reduce the type of recrystallisation and blooming seen in Figure 11.15.

11.4.3 Surfactants

Depending on the polarity of the polymer, and the nature of the additive components to be incorporated into the bulk plastic, surfactants are often required to ensure compatibility. A surfactant is a long chain organic molecule that has non-polar and polar groups at opposing ends of the molecule. These are used to reduce the surface tension of a liquid by bridging the gap between a polar substance and a non-polar substance, such as water and polyvinyl acetate (PVAc) respectively (see Chapter 9).

PVAc emulsion paints are reliant on the presence of surfactants to disperse the non-polar polymer in water, thus enabling water to be used as a paint vehicle rather than organic solvents such as turpentine.[24] The presence of surfactants also has the benefit of reducing surface tension and increasing wettability which helps prevent clumping of pigment particles.

Over time surfactants are often seen to migrate through the bulk of a plastic to the surface due to changing environments and incompatibility, in a similar manner to triphenyl phosphate discussed above. Once the water has evaporated from an acrylic paint film the surfactants are no longer stable within the non-polar PVAc film, which causes their movement to the surface. This behaviour is analogous to the efflorescence of fatty acid metal soaps in oil paints.[25]

11.4.4 Fillers

Fillers are usually in the form of solid additives, incorporated into plastic materials to improve a range of physical properties, from mechanical strength to opacity. Air can be considered a filler in foaming polymers due to the creation of voids and cells, but more generally they refer to solid materials. These solids are generally categorised as particulate, rubbery, resinous and fibrous.[5]

Particulate fillers are insoluble in the polymer compound and are often used to reduce cost as they act as bulking agents. These particulates are often described as inert, in so far as their action is passive, but they will inevitably have an impact on the overarching properties of the final polymer or influence processing. Depending on the polymer in question, the incorporation of inorganic particles can improve electrical insulation, reduce dimensional change during solidification and, depending on the size of the particles, increase hardness by reducing crack propagation. As many fillers are mineral derived they may have hydroxyl groups on the surface, which necessities the incorporation of surfactants to allow wetting and compatibility with non-polar polymers.

Fibrous fillers are amongst the oldest type to be incorporated into plastic materials and range from organic wood flour to inorganic fibres such as asbestos. The use of wood flour or shavings was an integral component of the 19th-century thermosetting plastic Bois Durci.[26] This patented compound combined wood flour, pigs' blood and egg whites, which on heating under pressure would cross-link to form a dark brown, rigid polymer. Bois Durci was superseded by thermosetting plastics such as Bakelite, but objects in collections remain in good physical form owing to the structural stability afforded by the filler. Contemporary fillers include glass fibres and carbon fibres, often incorporated in unsaturated polyester resin to provide mechanical cohesion and to reduce cracking due to shrinkage or thermal shock.

11.5 Analytical Techniques to Determine Thermal and Mechanical Properties of Plastics

Due to the wide range of plastic materials and their composite nature, there are numerous instruments available for their characterisation. Analysis of plastics usually relies on the coupling of techniques to fully characterise the chemical and physical characteristics. However, sampling can often prove challenging due to characteristics such as smooth morphologies of injection moulded thermoplastics and hard, brittle thermosets. This has led to the application of non-invasive or minimally-invasive methods for the characterisation of plastics in heritage collections, such as Fourier-transform near-infrared spectroscopy (FT-NIR),[27] mid-infrared spectroscopy (FT-MIR),[28] and Raman spectroscopy (FT-Raman).[29]

Where sampling is possible, the mechanical and thermal properties are some of the most important characteristics to determine. As their name suggests, plastics undergo plastic and elastic deformation. The physical properties of polymeric materials are dependent on the underlying chemical and physical structure. Their properties are essentially governed by the length of the long chain backbone, the presence of functional groups, the polymer processing method and the overall chain length and homogeneity. These factors govern their response to external stresses and consequently their final application.

The significance of the glass transition temperature essentially arises from the change in physical properties experienced as a polymer is exposed to temperatures above or below this transition. Characteristic changes on increasing temperature include a decrease in hardness and modulus, and an increase in volume and percentage elongation at break.[13] It is these changes in physical characteristics that are analytically exploited to detect the temperature range of the transitions and the working properties of a material. Table 11.2

Table 11.2 Analytical techniques appropriate for thermal and mechanical characterisation.

Characteristic	Technique
Elastic and plastic deformation	Mechanical analysis
Thermomechanical deformation and thermal coefficients of expansion	Thermomechanical analysis and dynamic mechanical analysis
Thermal stability and decomposition	Thermogravimetric analysis
Degree of crystallinity	Differential scanning calorimetry
Thermal transition temperatures	Differential scanning calorimetry, Thermomechanical analysis and dynamic mechanical analysis

indicates the techniques available for interrogating these characteristics. The close relationship between thermal properties and a material's mechanical response means that these properties are often conducted concurrently using techniques such as dynamic mechanical analysis (DMA) and thermomechanical analysis (TMA).

11.5.1 Elastic and Plastic Properties – Mechanical Analysis

The elasticity of polymeric materials is the ability of a material to return to its original dimensions once an exerted load has been removed;[30] elastic behaviour is reversible. If the stress continues to be applied, then eventually the material will reach a point where it can no longer return to its original conformation and is said to have reached its plastic limit; any further strain will cause inelastic, permanent deformation.

For an ideal elastic solid, the strain (ε) or deformation applied to a material has a linear relationship with the applied stress (σ) or force (eqn (11.1)):

$$\sigma = E\varepsilon \tag{11.1}$$

Where E is the modulus of elasticity. This behaviour is known as Hooke's law, which states that the applied stress is proportional to the resultant strain, but independent of the strain rate.

However, in the case of polymers the response is not entirely elastic, as they will exhibit characteristics that are intermediate between an elastic solid and a liquid capable of viscous flow.

This liquid component is described by Newton's law, where the stress is independent of strain, but proportional to the strain rate. This means the response of a polymer to an external load is dependent on both magnitude and time. This characteristic dual-component affords polymers the term viscoelastic. The viscoelastic nature of polymers means that the mechanical properties are time (or frequency) dependent and temperature dependent (see Chapter 10).

Key Features: When a plastic is subjected to a tensile force there are several key features that can be derived from the stress–strain curve (Figure 11.16). The initial slope of the curve is referred to as the modulus of elasticity (E) and is defined as $E = \Delta\sigma/\Delta\varepsilon$. Given that this corresponds to the rate of change of strain as a function of stress it is seen as a measure of the material's response to the applied stress. The yield stress, σY, refers to the region of the stress–strain curve where a maximum is reached. This is the point at which the material begins to plastically deform, *i.e.* irreversible deformation of the sample. The ultimate tensile strength, σS, is the final stress at the point of break. For materials such as thermoplastics, the point of rupture usually follows a process known as cold-drawing. After the yield point the sample forms a neck, which propagates along the length of the sample. The load remains stable as the amorphous regions become oriented in the direction of drawing, until ultimately the neck reaches the tensile clamps, the load increases, and the neck fracture.

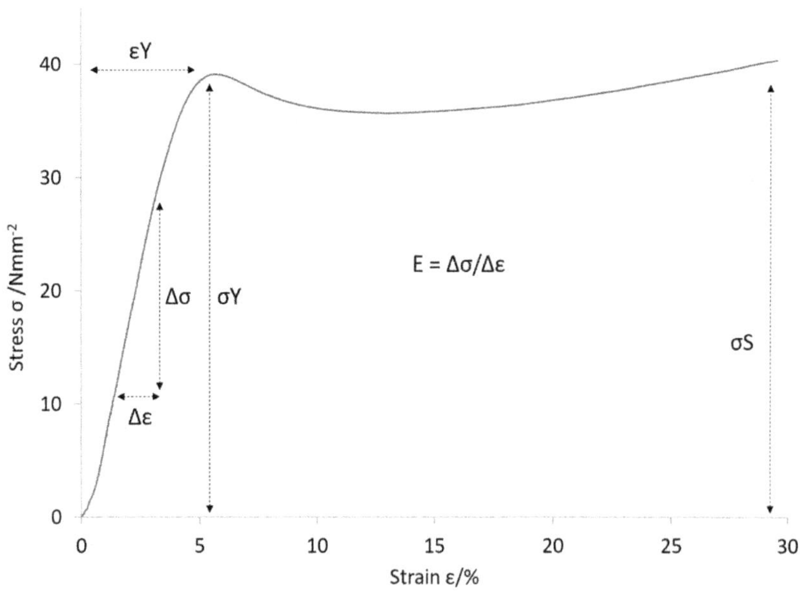

Figure 11.16 Stress *versus* strain curve of cellulose acetate sheet.

11.5.2 Thermomechanical - TMA/DMA

As the temperature of a material is lowered or raised, the amount of force required to cause elastic and plastic deformation is altered. At temperatures above the glass transition molecules increasingly move freely and deform readily to external forces allowing permanent deformation by a combination of elastic and viscous flow. Lowering a material to well below its T_g will produce a substance resistant to deformation, which is stiff and rigid in nature.

Thermal mechanical analysis (TMA) measures changes in dimensions of or the mechanical properties of a material whilst it is subject to a temperature regime, therefore enabling the glass transition temperature to be determined. Depending on the geometry of the sample and its final application there are different approaches to analysis, but samples are usually testing in tensile or bending modes. The sample is held between two clamps, one of which is stationary and the other mobile, and a specified static load is applied to the sample. The analysis chamber is heated, and the expansion of the material is measured as a function of time or temperature.

Key Features: There are two primary pieces of information that can be derived from a TMA heating curve, namely the glass transition temperature and the coefficient of thermal expansion. The former can be readily determined by extrapolating the tangents of the heating curve before and after the rapid change in slope. The midpoint of the glass transition temperature is at the inflection point *i.e.* where these tangents intersect. This corresponds to a temperature of 143 °C in Figure 11.17.

The coefficient of thermal expansion is calculated by dividing the slope of the heating curve by the original length of the sample *i.e* the rate of change relative to the original sample length. Given that polymers display distinct molecular changes over a range of temperatures, a given coefficient of expansion only holds true over small temperature ranges and is therefore calculated across small temperature regions. The values change significantly depending on whether the sample is above or below its glass transition temperature. For the cellulose diacetate sample analysed in Figure 11.17, the coefficients of expansion below and above the glass transition temperature are 0.00009/°C and 0.0275/°C respectively.

Dynamic Mechanical Analysis (DMA) is a similar technique to TMA, but rather than applying a static load or displacement, the sample is subject to a cyclical (sinusoidal) load or displacement. The sample's measured response is in the form of the amplitude and phase

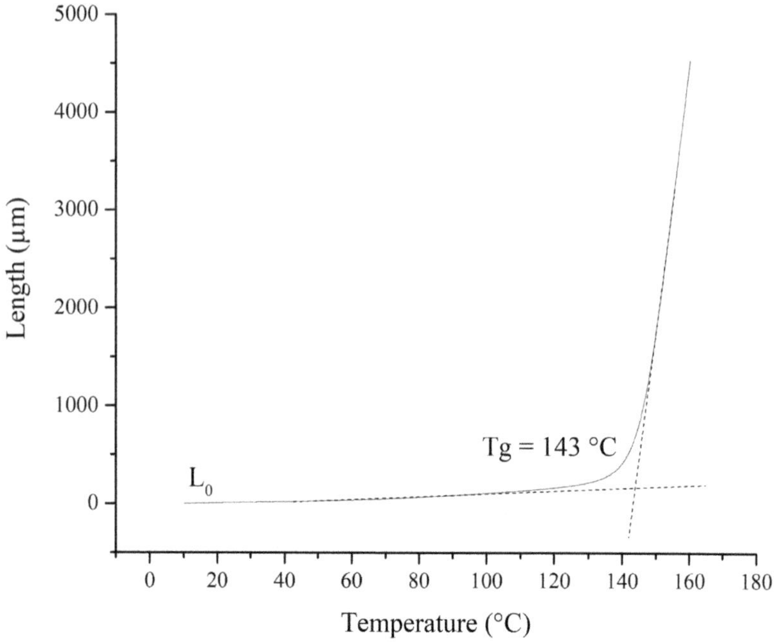

Figure 11.17 Thermomechanical analysis (TMA) heating curve of a cellulose acetate film.

difference, also known as the phase lag, δ. For a completely elastic material no lag would be experienced between the applied displacement and the sample's response, therefore the phase lag is equal to zero. By comparison, for a completely viscous material a phase lag is experienced between the applied displacement and displacement response. As such, the displacement is separated into two parts, the relative proportion and amplitude of both signals allowing various physical properties to be derived.

The storage modulus (E') is the in-phase, elastic response of the material to the applied force or displacement. E' measures the stored energy like a spring and is seen to decrease with increasing temperature. By comparison, the loss modulus (E'') is the out-of-phase, viscous response where energy is lost through internal motion. It measures the mechanical energy converted to heat through viscous forces of the material and in the first instance is seen to increases with increasing temperature.

The ratio of these two components is known as the loss factor and is used as a measure of the material's damping properties *i.e.* its capacity to dissipate mechanical energy from the impact force *via* conversion to heat energy. The loss factor is calculated as the difference in the tangents of the storage and loss moduli (eqn (11.2)) and is often termed tan delta. Tan delta will increase with increasing temp as the loss modulus (E'') is greater than the storage modulus (E').

$$\text{Tan Delta} = E''/E' \tag{11.2}$$

Key Features: There are several pieces of information that can be derived from a DMA heating curve, but the information most useful for conservation purposes relates to the material's mechanical properties and thermal transitions. When the strain is kept constant, the storage modulus provides information on the force required to displace the material by a given amount at each temperature (Figure 11.18). This is a measure of the material's resistance to deformation and can alter by several orders of magnitude as the material is heated. In Figure 11.18, the onset of the glass transition temperature is seen as a sudden decrease in the storage modulus at temperatures nearing 100 °C.

Figure 11.18 also shows the tan delta curve, which is seen to increase with increasing temperature as the material more efficiently absorbs energy. The glass transition temperature is determined as the peak of the tan delta curve and is centred some 25 °C higher than that of the storage modulus. This is because the tan delta peak is derived

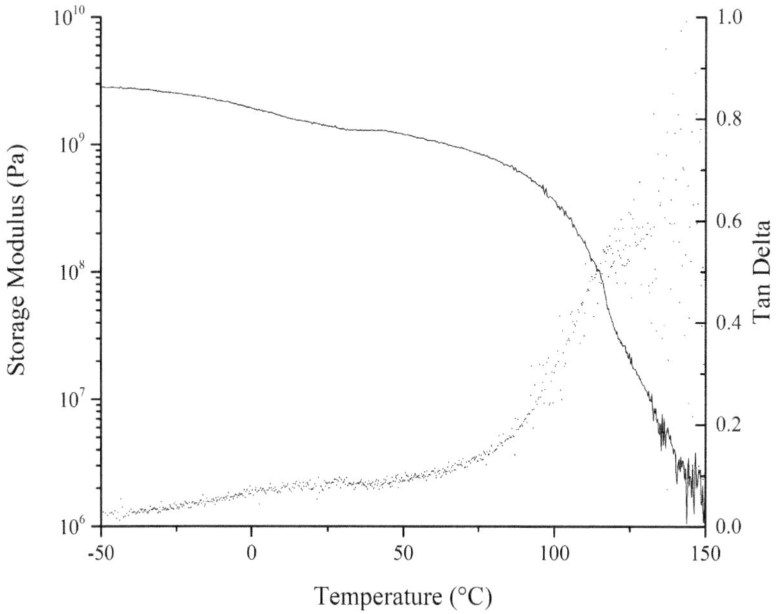

Figure 11.18 Dynamic mechanical analysis (DMA) heating curve of a cellulose acetate film.

from the ratio of both the loss and storage moduli. Therefore, when quoting glass transition temperatures, it is critical to state how the T_g was derived as it will indicate whether the material is at its lower or higher working temperature.

DMA can also be employed to monitor the creep and creep recovery behaviour of polymers at isothermal temperatures with a static load. During a creep experiment a load is applied to the sample and held constant, whilst the resultant deformation is measured as a function of time. After a specified time the load is removed and recovery of the displacement is measured.

Key Features: There are three main features of a creep/recovery curve, namely the initial strain caused by the loading of the sample, the final strain following the creep period, which holds the sample at a constant load, and the final recovery after the load has been removed.

Figure 11.19 shows the creep/recovery of two plain woven nylon textiles subject to a tensile force of 0.3 MPa and held isothermally at 35 °C and −15 °C; 35 °C being the midpoint of the glass transition temperature for both samples. At the beginning of the experiment the

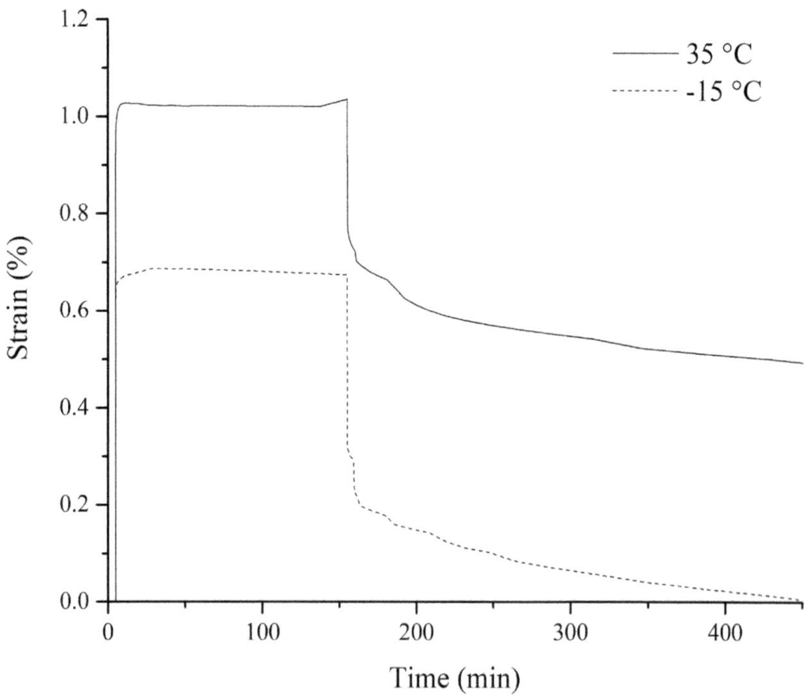

Figure 11.19 Dynamic mechanical analysis (DMA) isothermal creep and creep recovery curve of polyamide 6,6 woven textile.

samples were allowed to equilibrate at the respective temperatures for 5 minutes, known as the soak time. The force was subsequently applied with both samples displaying an initial, instantaneous extension. After the initial extension the samples were held constant and any time dependant changes in strain measured over a creep period of 150 minutes. For the sample held at 35 °C there is a slight increase in strain after 145 minutes of creep, suggesting the drawing of the fibres. After the creep period the load was removed displaying an initial and immediate contraction in each sample. Following 300 minutes of recovery, only the sample at −15 °C was seen to fully recover to its original length. The sample loaded at 35 °C, the glass transition temperature of the samples, did not fully recover to its original length indicating permanent deformation.

11.5.3 Thermal Stability – TGA

Thermogravimetric Analysis (TGA) is employed to determine the change in mass of a sample relative to temperature. Accurately weighed samples are introduced into the furnace, and the change in mass is measured as a function of temperature and time. TGA can be used for a number of polymer applications, such as determining the quantity of volatiles within a polymer and the temperatures of volatilisation, as well as providing information on oxidative decomposition or thermal decomposition.[31] The mass of inert aluminium oxide crucibles is accurately weighed prior to introducing a sample. The initial mass of both sample and crucible is recorded and then further measured as a function of increasing temperature. Samples are heated either in an inert gas or oxygen, depending on whether thermal or oxidative decomposition respectively are of interest.

Key Features: Depending on the composition of the sample the TGA weight loss curve will often show an initial change in mass as small volatile fractions, such as moisture, are lost from the material. In Figure 11.20 this can be seen between 85 °C and 165 °C, followed by a second weight loss step between 165 °C and 200 °C. The weight loss curve is of a cellulose acetate film plasticised with diethyl phthalate (DEP) and the two-step volatilisation is believed to be caused by two types of plasticisation. Friedman and Sorokina[32] reported that at concentrations below 25 wt%, DEP is sorbed by the active sites of the cellulose acetate polymer. However, above this level, it was shown that the DEP vapour pressure above the plasticised samples increased and was attributed to a portion of the plasticiser existing as an individual phase. The latter being more readily lost from the polymer matrix as

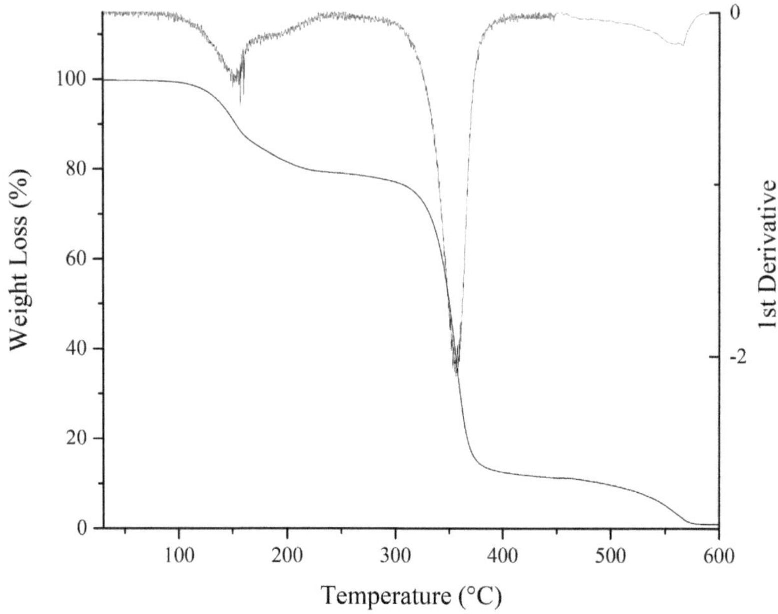

Figure 11.20 Thermogravimetric analysis (TGA) weight loss curve of a cellulose acetate film.

the sample is heated *i.e.* at lower temperatures. Here the second weight loss step corresponds to a total volatile weight loss of 23 wt%.

The final and most rapid weight loss shown in Figure 11.20 occurs at 240 °C and corresponds to the thermal decomposition of the cellulose acetate polymer, with a smaller, secondary step occurring at 550 °C.

By calculating the first derivative of the weight loss curve it is possible to highlight the midpoint of each of these main thermal steps.

11.5.4 Degree of Crystallinity – DSC

Differential scanning calorimetry (DSC) is one of a number of thermal analysis techniques that can be employed to monitor thermal transitions, such as the glass transition temperature and the melting temperature, with the latter allowing for the degree of crystallinity to be determined. DSC provides a measure of the heat flow into the sample as the temperature is raised and lowered, and thus provides information on thermal transitions. Accurately weighed samples are presented into the instrument. The difference in heat flow between an empty, control crucible and that of the sample and crucible is determined.

Two different DSC methods are available. The first is a heat flux DSC, which records the temperature difference between the sample and control crucible as they are both heated at the same rate using one furnace. The temperature difference in the sample is caused by changes in heat capacity and enthalpy of the specimen under investigation. The second type is a power compensated DSC, which monitors the difference in power as the temperature of the sample and the reference are kept the same using different heating furnaces. It is the energy difference that is a measure of the change in heat capacity and enthalpy.[33]

Key Features: Figure 11.21 shows the DSC heating curve for a polyamide 6,6 woven textile. As the sample reaches its glass transition temperature it displays a change in the heat capacity, which is seen here as an inflection in the heating curve between 20 °C and 60 °C. In this heating curve the midpoint of the glass transition is impacted by the presence of a peak, which is due to the relaxation of internal stresses that remain in the oriented fibres as a result of processing. As the sample is heated to the glass transition region the previously stressed molecules relax and absorption of energy.

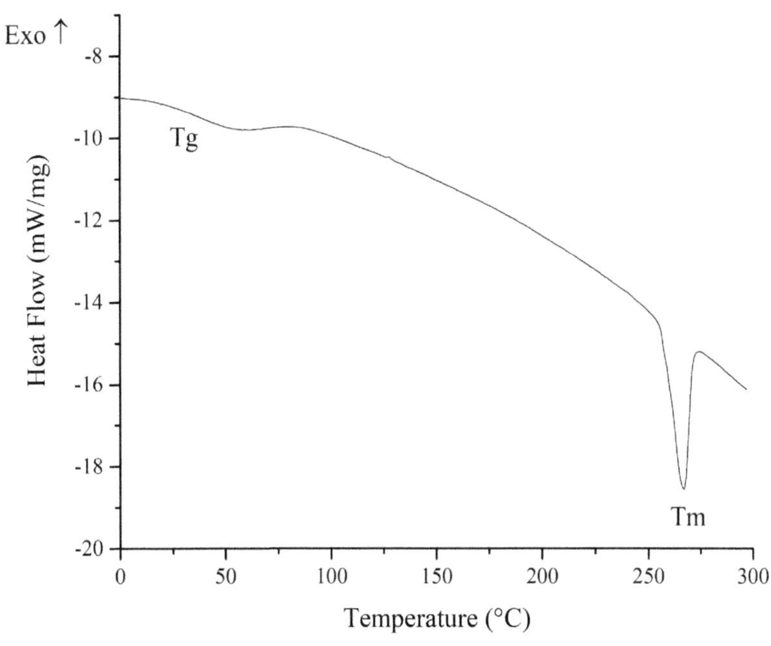

Figure 11.21 Differential scanning calorimetry (DSC) heating curve of polyamide 6,6 woven textile.

The DSC heating curve can provide information on the glass transition temperature through monitoring changes in the heat capacity, but it can also be used to provide information on the crystallinity of a material. The melting temperature is characterised by an endothermic response (Figure 11.21), where heat energy is taken into the sample and intermolecular bonds are broken down. The extent of the response is directly proportional to the percentage of crystallites within the sample. Assuming a two-phase, semi-crystalline morphology, the percentage crystallinity can be determined by the heat of melting derived from the melting endotherm (eqn (11.3)):

$$\%\,\text{Crystallinity} = \frac{[\Delta\text{Hm}]}{\Delta\text{Hm}^{\circ}} \times 100 \tag{11.3}$$

Where ΔHm° is the heat of melting of a 100% crystalline reference of the same material, and ΔHm refers to the sample of interest, in J g^{-1}.[34]

The melting endotherm shown in Figure 11.21 shows a broad, ill-defined endotherm with evidence of a shoulder at 259 °C and a trough minimum at 267 °C. Simal and Martin[35] suggest that the presence of a double endotherm in polyamide 6,6 fibres relates to differences in size and perfection of the crystalline regions. In their work they show that on increasing the temperature above 70 °C, the growth and perfection of the crystallites increased. Thanki *et al.*[36] also showed a similar double melting peak, which was attributed to the melting of the bulk crystal structure at the higher temperature, and melting of the smaller crystallites, which occurs at a lower temperature.

11.6 Case Study: *Burn Hole*, Henk Peeters, 1961, Tate

Despite the earliest manmade plastics being developed in the 19th century these materials have remained firmly associated with contemporary living, which has been mirrored by their use as artists' materials. As early as the 1920s artists first started to experiment with plastics, most notably a group of artists whose work are collectively known as the Constructivists. At the forefront of this experimentation were the Russian artists Naum Gabo and his brother Antoine Pevsner (Figure 11.22), who rejected traditional figurative sculpture for more abstract, geometrical forms. Gabo stated in 1920 that 'space and time are the only forms on which life is built – and hence – art must be constructed'.[37] Making use of cellulose nitrate, cellulose acetate and polymethyl methacrylate (Perspex) they had discovered contemporary

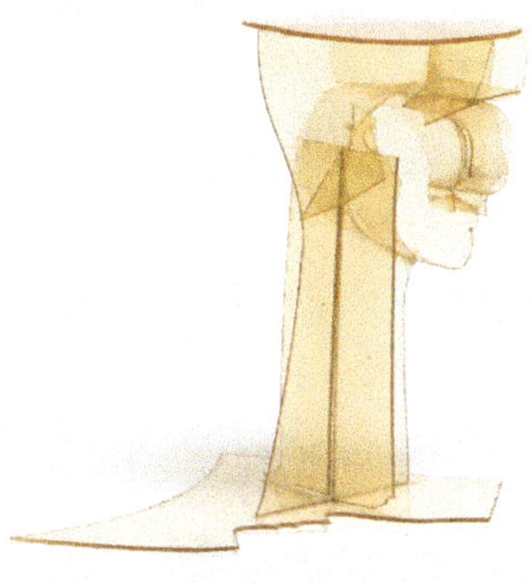

Figure 11.22 Antoine Pevsner, T02241, *Head* (c.1923-4) Cellulose Nitrate. Reproduced with permission from © Tate © ADAGP, Paris and DACS, London 2019.

materials that enabled them to express concepts of space through exploiting properties such as rigidity and transparency. At this time they also firmly believed that these materials would have longevity and last well into the future, if not forever. It is now well documented that this is not the case.[23,38,39]

Throughout the 20th Century innovations in plastic technologies have been quickly followed by their application as art materials, with plastics developed for military and aerospace applications, and growth in car manufacturing, one of the most notable iterations was seen during the 1960s and 70s. In post Second World War Britain the New Generation artists, including Phillip King and Tim Scott, saw the possibilities that industrial plastics offered and employed them to create large scale, innovative and colourful works (Figure 11.23).

Concurrently the Arte Povera, ZERO and Nul groups (Italy, Germany, and The Netherlands, respectively) found plastics to be perfect materials for conveying many of their ideas, with these artists rejecting '…the revival of traditional painting in the fifties and striving to eliminate all mythical elements from their work whilst using materials in a depersonalised way'.[40] However, the use of plastics as art materials was quite controversial, meeting resistance from gallerists and collectors, due to its links to mass production and industry; properties and associations that the New Generation and Nul artists particularly championed (Figure 11.24).

As previously discussed, plastic materials can be classified in many ways based on aspects such as structure, origin, application and thermal properties. It is this latter classification that is of particular

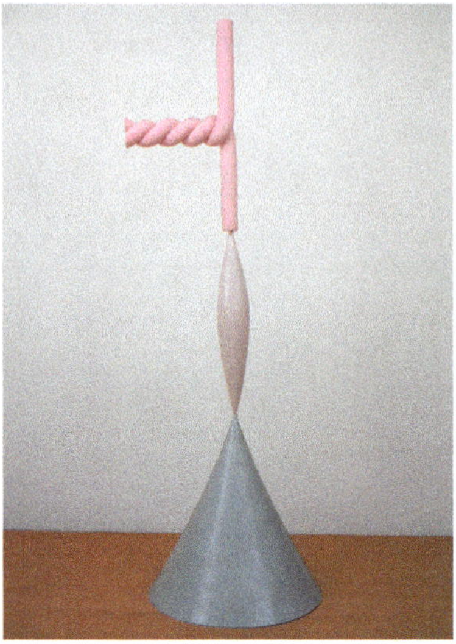

Figure 11.23 Philip King, T01206, *Tr-La-La* (1963) Fibreglass. Reproduced with permission from © Tate © Phillip King. Courtesy of the artist and Thomas Dane Gallery.

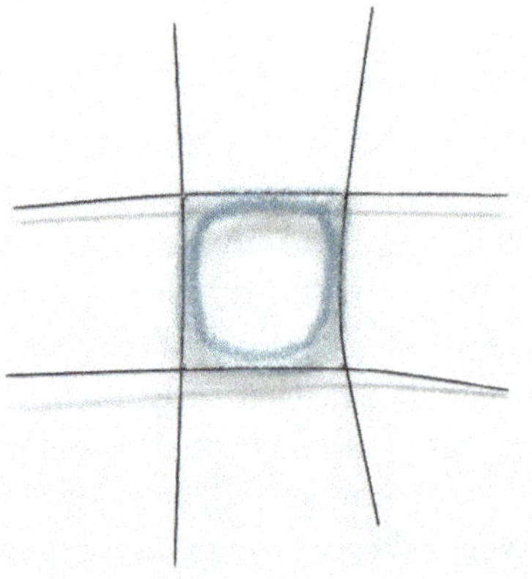

Figure 11.24 Marissa Merz, T13031, *Untitled* (1969) Steel and Nylon. ©Tate.

pertinence when considering the *pyrographies* works by Nul artist Henk Peeters, the focus of this case study. The term pyrography is the art of decorating materials with burn marks through the controlled application of heat and as such Peeters' works rely on the exploitation of a material's thermal properties.

Henk Peeters' *pyrographies* date from the late 1950s, placing his works at the forefront of plastics experimentation. Peeters used a range of off the shelf materials to create this series of works including

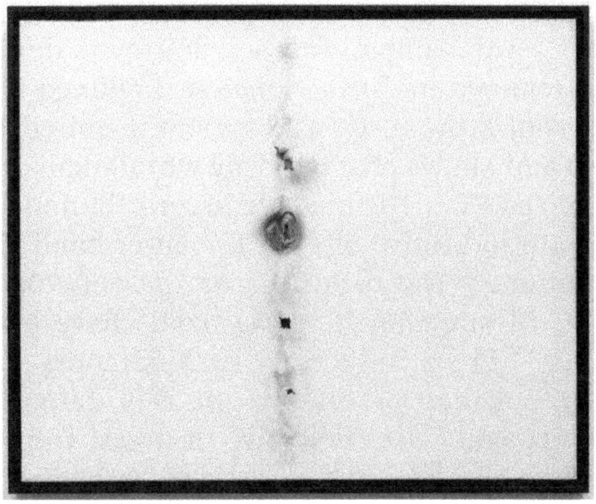

Figure 11.25 Henk Peeters, T13407, *Burn Hole* (1961). Reproduced with permission from © Tate © Henk Peeters, c/o Pictoright. DACS 2019.

canvas, polyurethane (PUR) foam and polyvinyl chloride (PVC) film.[40] The Mayor Galley describes the properties of Peeters' works as having 'a very tactile appeal, whilst he simultaneously created a certain untouchability.... He (also) used fire on canvases, leaving behind traces of thick smoke, or he burnt holes into plastic, the so-called *pyrographies*'.[41] At their time of production plastic was still a relatively new material, particularly amongst artists, and such works were often ill-received by the museum going public.

Looking at *Burn Hole* (Figure 11.25) it is possible to focus on the conservation challenges it brings as well as the conservation techniques that can be used to keep it looking 'fresh'. Each polymer has its own characteristics that can be further modified with the use of additives, such as plasticisers and pigments, which are often added to increase flexibility, durability and stability. Therefore, it is crucial for the conservator to characterise the materials of construction before an understanding of an artist's techniques can be determined or a conservation strategy put in place. The degradation and stabilisation of organic polymers vary through the classes, therefore illustrating the need for positive identification within collections, which is not always easy with limited sampling of artefacts.

In order to identify the film material a small sample was removed from the reverse of the work for laboratory analysis. The sample was analysed using a Bruker Alpha Fourier-transform infrared spectrometer fitted with an attenuated total reflectance (ATR) accessory with a Zn–Se crystal (see Chapter 2). Acquisition was across a wavenumber region of

400–4000 cm^{-1}, with 32 scan accumulations and a wavenumber resolution of 2 cm^{-1}. The sample was not destroyed during characterisation and was retained for further analysis (outlined below).

Figure 11.26 shows the ATR-FTIR spectra acquired from a small physical sample and shows characteristic vibrational signatures for a phthalate plasticised PVC. The methylene contributions at 1436 cm^{-1} and 1424 cm^{-1} are indicative of PVC, the latter band relating to the CH_2–Cl deformation.[42] The bands across the 600–700 cm^{-1} region correspond to C–Cl stretches.[43] These bands have been shown by Tabb and Koenig[44] to be influenced by differences in crystallinity induced during polymer processing; in this case the bands at 603 cm^{-1} and 631 cm^{-1} are indicative of a cast film. The doublet between 1580–1600 cm^{-1} is characteristic of phthalate plasticisers,[43] as is the C–H benzene absorption at 740 cm^{-1}.[45] The broad band superimposed across the 700–400 cm^{-1} region indicates the PVC is pigmented with titanium dioxide,[46] most likely in its rutile form.

PVC is one of the most widely manufactured and used classes of synthetic plastics. It was first seen in the 1920s,[26] but its popularity grew after the Second World War when its durable and flame-resistant nature were exploited for uses such as insulating electrical

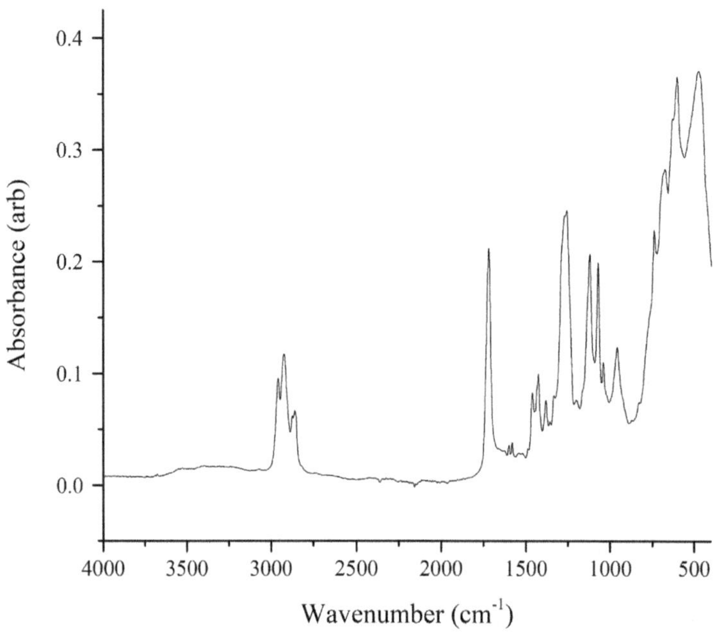

Figure 11.26 ATR-FTIR spectrum of a sample removed from the reverse of Henk Peeters' *Burn Hole*.

wire. Some of its many varied uses have included sheeting, flooring, shoes and rainwear, toys, adhesives, and guttering and drainpipes, each having quite different characteristics and appearances. Many of these characteristics and appearances are created from the use of additives, in particular stabilisers and plasticisers, making it a very adaptable material. This also makes it an attractive artists' material, however it also means it is an unpredictable and complicated material for conservation and preservation.

PVC is a thermoplastic material and has the propensity to undergo deformation at raised temperatures under the application of force. By increasing the temperature of PVC, the non-crystalline, disordered regions of the polymer are provided with enough energy to move and rearrange. The temperature at which this rearrangement happens is termed the glass transition temperature and is the point at which a material moves from a glassy, rigid solid to a flexible, malleable solid. Raising a material above its glass transition temperature enables it to be readily manipulated into various forms, in the case of this work through extrusion as sheet material, whilst retaining its form on the removal of heat. By including additives, such as plasticisers, the glass transition temperature and the temperature at which the material can be deformed are significantly lowered, thus producing a plastic that is flexible at ambient temperatures. The role of a plasticiser is to increase the spaces between the PVC molecules, rendering the plastic in its rubbery state at ambient conditions, producing a soft and flexible material. For PVC sheet materials plasticisers often constitute as much as 40% by weight of the total mass. This can be determined using thermal analysis methods, such as thermogravimetric analysis (TGA), which monitors the percentage weight loss of volatile components over a given temperature range.

Henk Peeters' use of the word *pyrographies* clearly indicates the production methods utilised by the artist. That is to say, the act of decorating the film material with burn marks through the controlled application of a heated object, such as a poker or a cigarette. In producing *Burn Hole,* Peeters exploited the temperature dependent properties of the polyvinyl chloride film. Therefore, to fully understand the methodologies of making, the precise thermal properties of the plasticised PVC film were established using laboratory techniques. All analyses were undertaken in an air atmosphere, as opposed to a nitrogen atmosphere, given this more closely represented the conditions of combustion used by Henk Peeters.

Utilising the small sample that was removed for characterisation, dynamic mechanical analysis (DMA) was employed. This technique measures the modulus of the sample as a function of temperature, and thus offers information on molecular motion and thermal transitions, such as the glass transition temperature. Analysis was undertaken using a PerkinElmer DMA8000 with Pyris thermal analysis software. The sample was run in tension mode, with a gauge length of 4.81 mm, a width of 4.08 mm and a thickness of 0.18 mm. The chamber was manually temperature cooled with liquid nitrogen and analysed at a heating rate of 2 K per min over a temperature range of −100 °C to 100 °C. The sample was subjected to a displacement of 0.01 mm, equating to a strain of 0.2, at an oscillation frequency of 1 Hz.

The storage modulus (Figure 11.27) clearly shows that the onset of the glass transition temperature occurs at approximately −20 °C, below ambient temperatures. Using a material with a glass transition temperature below ambient provided Peeters with a polymer film that was soft and flexible. This enabled him to stretch the polymer taught over the work's wooden frame (Figure 11.28), with the folds of the film stapled in place having a malleability that would not have been possible with a stiffer plastic.

Figure 11.29 shows the thermogravimetric weight loss curve acquired from a 3.8 mg sample of the plasticised PVC. As previously mentioned, TGA can provide information on the loss of volatile components over a given temperature range, in addition to information on temperatures of decomposition. The sample was analysed using a Netzsch STA DSC/TGA in an aluminium oxide crucible, a heating rate of 5 K per min and a purged atmosphere of air at a flow rate of 2 ml per min.

The data clearly show that the plasticised PVC undergoes four stages of thermal degradation.

i. The slight increase in mass up to 200 °C suggests the formation of oxidative products. This results from the formation of non-volatile compounds through thermal oxidation reactions with the purge gas.[47] Although phthalate plasticisers show a range of volatilities within this temperature region, which would result in weight loss, it was not possible to determine the percentage of volatile due to this concurrent weight gain up to 200 °C.
ii. TGA/DSC detail [Figure 11.30] shows a slight weight loss and concurrent endothermic response between 70–90 °C, which is likely due to the loss of moisture.[39]
iii. About 200–300 °C is the most rapid change and corresponds to the elimination of HCl through free radical mechanisms and the concurrent formation of polyene structures.[48]

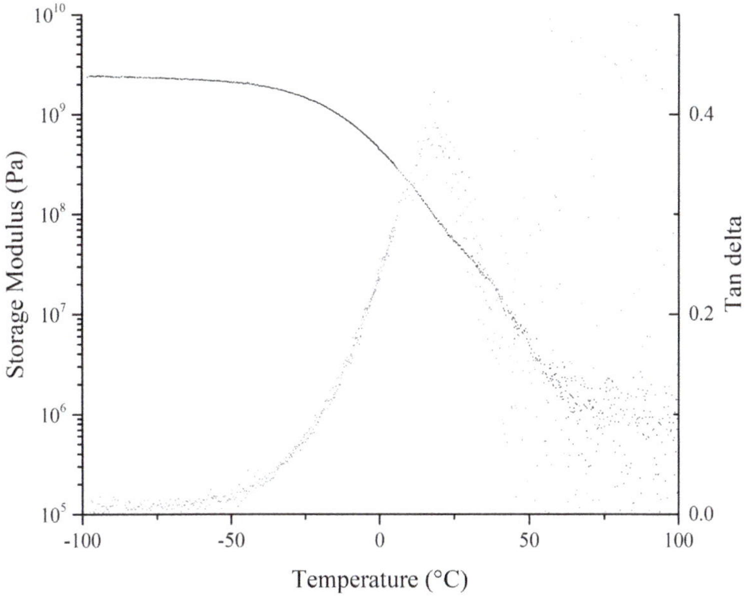

Figure 11.27 Storage modulus acquired by dynamic mechanical analysis.

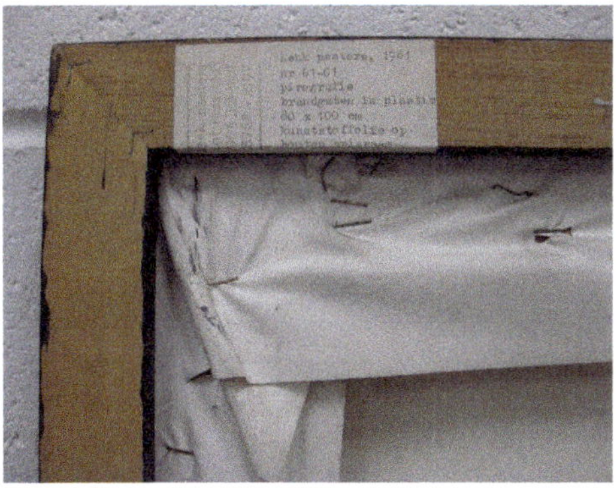

Figure 11.28 Rear view of Henk Peeters' *Burn Hole* (1961), showing the PVC film pulled and stapled over the wooden stretcher. Reproduced with permission from ©Tate. ©Henk Peeters, c/o Pictoright. DACS 2019.

iv. These new polyene compounds are transformed to aromatic compounds between 300–420 °C.[42]

v. The final weight loss is seen to occur above 420 °C, which is attributed to the degradation of complex structures resulting from aromatisation.[48]

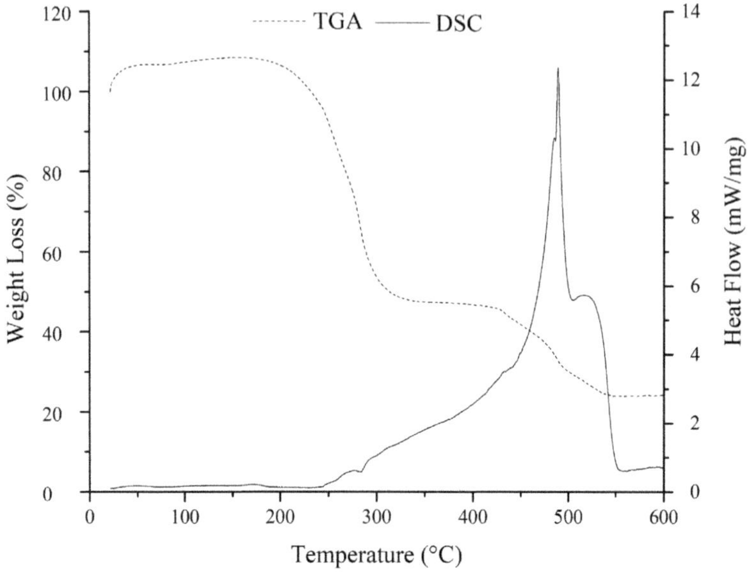

Figure 11.29 Thermogravimetric weight loss curve and differential scanning calorimetry heating curve.

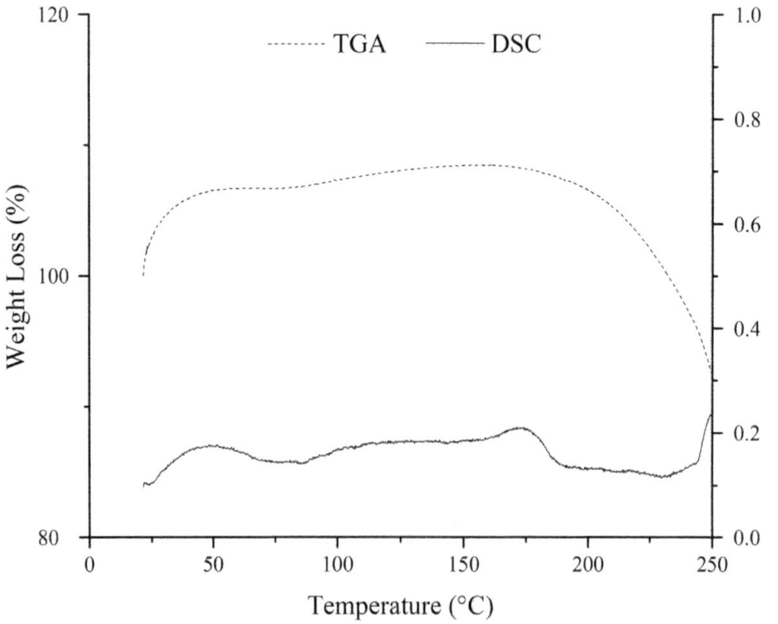

Figure 11.30 Detail of thermogravimetric weight loss curve and differential scanning calorimetry heating curve.

Figure 11.30 shows a detail of the DSC heating curve. Between 50 °C and 200 °C, the sample is seen to exhibit a series of endothermic responses, where energy was taken in by the sample. These correspond to the melting of various crystalline structures formed during the initial processing of the PVC film and a series of underlying chemical alterations.[49]

Above 200 °C the sample is shown to fully decompose. These temperature responses are of particular interest when considering the pyrography techniques used by Peeters in producing *Burn Hole*. When PVC is exposed to heat the polymer undergoes a number of chemical alterations resulting in the release of hydrogen chloride gas, a process called dehydrochlorination. The evolution of this gas poses risks to the environment and health, which is why PVC materials are not suitable for temperature dependent recycling techniques.[50] The result of this initial decomposition is a highly coloured material, which ranges from yellow to orange to brown to black depending on the extent of decomposition.[26,51] These temperature dependent alterations account for the colour changes exhibited across the surface of *Burn Hole*.

It can be seen in Figure 11.31 that the large hole in the centre of the film is surrounded by gradations of brown and black and undulations in the surface of the film, indicating that there were differences in temperature during the scorching and degradation of these regions. The large central region exhibits a melting pattern

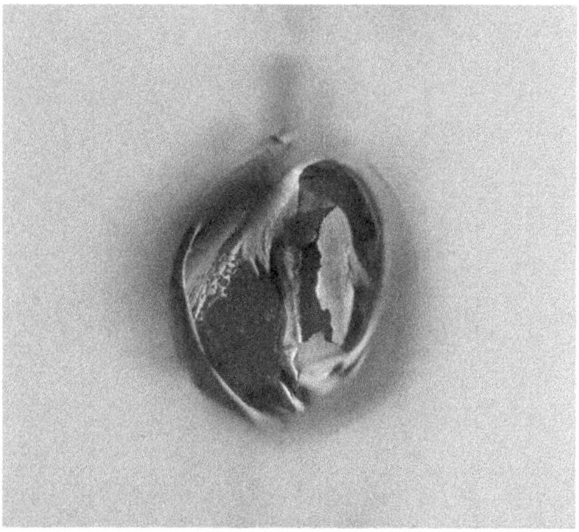

Figure 11.31 Central burn hole in Henk Peeters' *Burn Hole* (1961) taken in raking light.
Reproduced with permission from ©Tate. ©Henk Peeters, c/o Pictoright. DACS 2019.

characteristic of slow heating and distortion of the film due to softening. Although the presence of a hole indicates complete decomposition in a localised area, the regions surrounding this hole show clear deformation of the plastic accompanied by gradations of brown and black.

The nature of this hole suggests that it was created using the heat generated by a candle. Although the internal temperature of a candle flame can reach temperatures between 900–1200 °C,[52] well above the point at which this PVC film shows complete combustion, the regions above a candle flame can produce temperatures sufficient to cause melting, discolouration and the formation of soot. At a vertical distance of 10 cm from the flame of a beeswax candle the air temperature is in the region of 100–150 °C.[52] Such temperatures would be sufficient to soften the PVC film and initiate dehydrochlorination and accompanying colour change. The presence of soot appearing across the work would also indicate the use of a candle, with deposits being formed on the surface of the PVC due to solid particles of unburnt carbon originating from the wax.

The other smaller holes seen across the work exhibit an all-together different form, spherical in shape and showing complete combustion of the polymer, they are reminiscent of cigarette burns (Figure 11.32). The end of a smouldering cigarette is in the region of 600 °C, which quickly rises to temperatures approximating 950 °C as air is drawn into the cigarette.[53] These smaller, highly resolved burn holes could

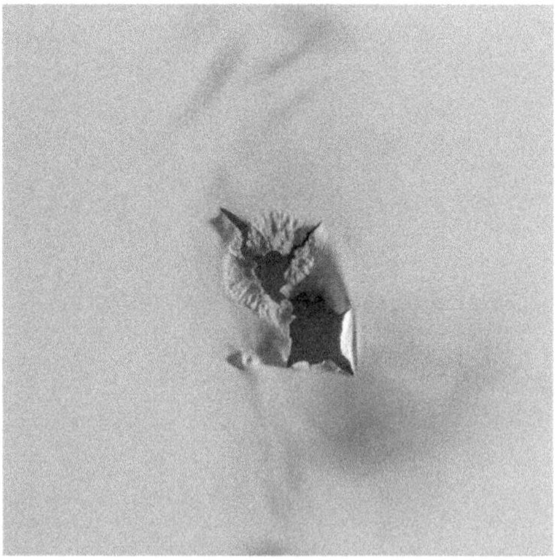

Figure 11.32 The top hole in Henk Peeters' *Burn Hole* (1961). Reproduced with permission from ©Tate. ©Henk Peeters, c/o Pictoright. DACS 2019.

have been easily produced by placing a smouldering cigarette up against the surface of the polymer, taking the temperature well above 250 °C, the temperature of full decomposition.

11.6.1 State of Conservation

Despite the relative instability of plasticised PVC materials, Peeters' *Burn Hole* is in remarkably good condition.[54] The ageing of plasticised PVC is dependent on the composition of the plastic and the degree of plasticisation. The presence of optical brighteners, such as the titanium dioxide pigment found within Henk Peeters' *Burn Hole*, has the capacity to offset some of the colour change caused by a high degree of plasticisation. In addition to aesthetic qualities, titanium dioxide is also reported to provide resistance to light deterioration.[26]

The polymer film is showing signs of cracking at the edges that are wrapped around the wooden frame (Figure 11.33). PVC often experiences the migration of plasticisers, which is seen as an oily exudate on the surface of the material or the formation of crystals. The loss of such material through the polymer matrix causes shrinkage and embrittlement, which in the case of *Burn Hole* may have been exacerbated by the tensile stresses caused by stretching the

Figure 11.33 The facing right top corner of Henk Peeters' *Burn Hole* (1961). Reproduced with permission from ©Tate. ©Henk Peeters, c/o Pictoright. DACS 2019.

film across the work's wooden frame. Not unexpectedly the most fragile sections of the work are those that have been burnt, which are incredibly sensitive to air flow caused by movement and localised stresses. It is for this reason a bespoke conservation backboard was fitted in 2011 to help prevent movement of the film during handling and transport†.[54]

Recommended Reading

J. A. Brydson, *Plastics Materials*, Butterworth-Heinemann, Oxford, 1999.
J. M. G. Cowie, *Polymers: Chemistry and Physics of Modern Materials*, Routledge, Abingdon, 2007.
J. E. Mark, *Physical Properties of Polymers Handbook*, Springer, New York, 2007.
T. van Oosten, *PUR Facts: Conservation of Polyurethane*, RCE Publications, Amserdam, 2014.
Y. Shashoua, *Conservation of Plastics: Materials Science, Degradation and Preservation*, Routledge, Abingdon, 2008.
F. Waentig, *Plastics in Art*, Michael Imhof Verlag, Petersberg, 2009.
D. Walton and P. Lorimer, *Polymers (Oxford Chemistry Primers)*, Oxford University Press, Oxford, 2001.

References

1. B. Keneghan, *Plastics Research in the Victoria and Albert Museum*, The Foundation for the Conservation of Modern Art and the Netherlands Institute for Cultural Heritage, 1999, pp. 356–361.
2. A. Quye and C. Williamson, *Plastics: Collecting and Conserving*, NMS Publishing Ltd., Edinburgh, 1999.
3. R. Barthes, in *Mythologies, Farrar, Straus and Giroux*, 1957, pp. 97–99.
4. R. Mustalish, *Modern Materials: Plastics*, The Metropolitan Museum of Art, 2004.
5. J. Brydson, *Plastics Materials*, Butterworth-Heinemann, Oxford, 7th edn, 1999.
6. J. W. Nicholson, *The Chemistry of Polymers*, The Royal Society of Chemistry, Cambridge, 2nd edn, 1997.
7. R. J. Young and P. A. Lovell, *Introduction to Polymers*, Chapman & Hall, London, 2nd edn, 1991.
8. W.-F. Su, in *Principles of Polymer Design and Synthesis*, Springer Berlin Heidelberg, Berlin, Heidelberg, 2013, https://doi.org/10.1007/978-3-642-38730-2_5, pp. 89–110.
9. J. McMurry, *Organic Chemistry*, Brooks/Cole Publishing Company, California, 2nd edn, 1988.
10. L. Mathias and B. Dizman, *J. Polym. Sci., Part A Polym. Chem.*, 2005, **43**, 5844–5854.

†A Perspex backed inner frame was inserted inside the work's own frame on the reverse by Pamela Steehoulder, Painting Conservation Technician, in 2011.

11. D. Walton and P. Lorimer, *Polymers*, Oxford University Press, Oxford, 2001.
12. N. A. Jones, E. D. T. Atkins, M. J. Hill, S. J. Cooper and L. Franco, *Polymer*, 1997, **38**, 2689–2699.
13. N. S. Murthy, *J. Polym. Sci., Part B: Polym. Phys.*, 2006, 44.
14. Y. Huang, W. Li and D. Yan, *Polym. Bull.*, 2002, **49**, 111–118.
15. J. G. Fatou, *Eur. Polym. J.*, 1971, **7**, 1057–1064.
16. R. Paukkeri and A. Lehtinen, *Polymer*, 1993, **34**, 4075–4082.
17. N. S. Murthy, R. G. Bray, S. T. Correale and R. A. F. Moore, *Polymer*, 1995, **36**, 3863–3873.
18. J. Brydson, *Plastic Materials*, Butterworth-Heinemann, Oxford, 7th edn. 1999.
19. N. S. Murthy, R. G. Bray, S. T. Correale and R. A. F. Moore, *Polymer*, 1995, **36**, 3863–3873.
20. N. Vasanthan, S. B. Ruetsch and D. R. Salem, *J. Polym. Sci., Part B: Polym. Phys.*, 2002, **40**, 1940–1948.
21. D. P. N. Vlasveld, J. Groenewold, H. E. N. Bersee and S. J. Picken, *Polymer*, 2005, **46**, 12567–12576.
22. S. T. McKenna and T. R. Hull, *Fire Sci. Rev.*, 2016, **5**, 3.
23. M. T. Giachet, M. Schilling, K. McCormick, J. Mazurek, E. Richardson, H. Khanjian and T. Learner, *Polym. Degrad. Stab.*, 2014, **107**, 223–230.
24. J. Crook and T. Learner, *The Impact of Modern Paints*, Tate Gallery Publishing Ltd, London, 2000.
25. A. Sawicka, A. Burnstock, F. C. Izzo, K. Keune, J. Boon, K. Kirsch and K. J. Van den Berg, in *Issues in Contemporary Oil Paint*, Springer International, 2014, https://doi.org/10.1007/978-3-319-10100-2_21, pp. 311–332.
26. F. Waentig, *Plastics in Art*, Michael Imhof Verlag, Petersburg, 2009.
27. E. Richardson, G. Martin, P. Wyeth and X. Zhang, *Microchim. Acta*, 2008, **162**, 303–312.
28. R. Rivenc, E. Richardson and T. Learner, presented at ICOM-CC, Lisbon, 2011.
29. C. Paris and C. Coupry, *J. Raman Spectrosc.*, 2005, **36**, 77–82.
30. E. Gillam, *Materials Under Stress*, Butterworth & Co, London, 1969.
31. P. Gabbott, *Principles and Applications of Thermal Analysis*, pp. 87–118, https://doi.org/10.1002/9780470697702.ch3.
32. O. A. Fridman and A. V. Sorokina, *Polym. Sci., Ser. B*, 2006, **48**, 233–236.
33. G. Hohne, W. Hemminger and H.-J. Flammersheim, *Differential Scanning Calorimetry: An Introduction for Practitioners*, Springer-Verlag, Heidelberg, 1996.
34. J. Brandrup, E. H. Immergut and E. A. Grulke, *Polymer Handbook*, Wiley, New York, 4th edn, 1999.
35. A. L. Simal and A. R. Martin, *J. Appl. Polym. Sci.*, 1998, **68**, 441–452.
36. P. N. Thanki, C. Ramesh and R. P. Singh, *Polymer*, 2001, **42**, 535–538.
37. H. B. Chip, *Theories of Modern Art: A Source by Artists and Critics*, University of California Press, Oakland, 1992.
38. B. A. Price, S. Malenka, K. Sutherland, A. Lins and J. H. Carlson, Naum Garbo's *Construction in Space*: *Two Cones*: History and Materials, in *Plastics: Looking at the Future and Learning from the Past, Archetype*, London, 2007, pp. 81–88.
39. E. Richardson, M. Truffa Giachet, M. Schilling and T. Learner, *Polym. Degrad. Stab.*, 2014, **107**, 231–236.
40. E. Rodrigo and L. Beerjens, in *Modern Art: Who Cares*? ed. I. Hummelen and D. Sille, The Foundation for the Conservation of Modern Art and the Netherlands Institute for Cultural Heritage, Amsterdam, 1999, pp. 43–51.
41. *Henk Peeters: Works from the 1960's*, The Mayor Gallery, 2011, https://www.mayorgallery.com/artists/207-henk-peeters/.
42. J. Chen, X. A. Nie, J. C. Jiang and Y. H. Zhou, *IOP Confe. Ser.: Mater. Sci. Eng.*, 2018, **292**, 012008.
43. M. Beltrán, J. C. García and A. Marcilla, *Eur. Polym. J.*, 1997, **33**, 453–462.

44. D. L. Tabb and J. L. Koenig, *Macromolecules*, 1975, **8**, 929–934.
45. I. V. Skornyakov and V. P. Komar, *J. Appl. Spectrosc.*, 1998, **65**, 911–918.
46. Y. Man, L. Mu, Y. Wang, S. Lin, G. L. Rempel and Q. Pan, *Polym. Compos.*, 2015, **36**, 8–16.
47. W. D. Woolley and P. J. Fardell, *Fire Safety J.*, 1977, **1**, 11–21.
48. Y. Soudais, L. Moga, J. Blazek and F. Lemort, *J. Anal. Appl. Pyrolysis*, 2007, **78**, 46–57.
49. P. Gramann, J. Cruz and B. Ralston, *Journal*, 2009, 1–6.
50. Y. Shashoua, *Conservation of Plastics: Materials Science, Degradation and Preservation Butterworth-Heinemann*, Oxford, 2008.
51. W. C. Geddes, *Eur. Polym. J.*, 1967, **3**, 747–765.
52. J. Walker, *Sci. Am.*, 1978, **238**, 154–163.
53. R. R. Baker, *Prog. Energy Combust. Sci.*, 2006, **32**, 373–385.
54. M. Rolfe, *Henk Peeters Conservation File Tate*, London, 2011.

12 3D Printed Material

M. Beecroft

Manchester School of Art, Manchester Metropolitan University, Manchester, UK
Email: m.beecroft@mmu.ac.uk

12.1 Introduction

3D printing, also commonly known as rapid prototyping (RP), is a form of additive manufacturing (AM) whereby 3D objects are created by the building up of layers of material. Originally intended as a prototyping tool, an increasing number of artists and designers have begun to explore the potential of this emerging technology to create bespoke artworks and designs. Objects range from decorative sculptures to functional furniture to wearable fashion pieces. Within the field of textiles there has been a growing interest in exploiting the technology to print flexible textile-based material. An early example of this is Dutch designers Jiri Evenhuis and Janne Kyttanen's series of fashion-based pieces created for Freedom of Creation. These pieces, printed in 2003, make use of traditional chain mail logic and have subsequently become part of New York's Museum of Modern Art's (MoMA) permanent collection.[1] Following this, a host of designers have experimented with 3D printing fashion and textile-based pieces including high profile catwalk pieces by Iris Van Herpen[2,3] and Chanel.[4] With growing interest in these 3D printed pieces, galleries and museums worldwide have begun to include them in their exhibitions and permanent collections. In 2014 MoMA acquired the

Conservation Science: Heritage Materials, 2nd Edition
Edited by Paul Garside and Emma Richardson

Published by the Royal Society of Chemistry, www.rsc.org

world's first fully assembled 3D printed 'Kinematics' dress by Nervous System along with its concept video.[5] Rosenkrantz and Rosenberg of Nervous System have developed a complex computer programme to enable full garments to be printed in one piece by digitally folding and compressing the 3D model. The printed piece consists of thousands of rigid plastic components hinged together to create a fully articulated structure.[6] In 2016 National Galleries Scotland acquired Pringle of Scotland's Autumn/Winter 2014/2015 women's cable knit jumper incorporating 3D-printed nylon elements created by fashion designer Massimo Nicosia in collaboration with architect/designer Richard Beckett.[7] As these pieces become more commonplace in museum and gallery collections a greater understanding of their fabrication techniques, material composition and mechanical properties is required to aid their conservation.

This chapter will categorise the main types of 3D printed textile-based pieces likely to appear in museum and gallery collections. For the purpose of this chapter, 'textile-based' will refer to pieces that follow a textile logic of interlinking, interwoven or interlooping elements that aim to replicate textile properties of strength, flexibility and elasticity.[8] This information will be divided into three sections; the first section will explore the different types of structural geometries used in the development of 3D printed textile-based pieces and their impact on flexibility. The second will discuss the different types of additive manufacturing (AM) processes used to physically fabricate the structures and their suitability to print textile-based material. The third section will discuss the various types of materials used to print with and the mechanical properties they afford the printed pieces. Following this, pieces from Manchester Metropolitan University Special Collections will be used as case studies, focusing on the types of issues conservators and curators may encounter when dealing with 3D printed textile-based pieces as future heritage materials.

12.2 Structure and Geometry

There has been an increased interest in exploring the possibilities of 3D printing flexible textile-based material over the past ten years as demonstrated by the increasing number of research articles published in this field.[9–11] This section will review the most common types of structural geometries that have been used to create flexible 3D printed textile-based pieces.

12.2.1 Multiple Assemblies, Linked Geometries

To date the majority of 3D printed textile-based pieces are made up of discontinuous interlinked structures[12] to create the desired movement and flexibility. These structures are often called multiple assemblies as they consist of separate parts.[13] A simple form of multiple assembly is the chainmail structure. Figure 12.1 shows a traditional chainmail structure 3D printed in the form of interlinking closed loops. The piece is flexible due to the size of the open loops enabling articulation. Rather than stretching like a traditional fabric, the piece has the ability to extend due to movement within the closed loops. As previously mentioned, a notable example of such multiple assembly is Evenhuis and Kyttanen's 3D printed chainmail fashion pieces for Freedom of Creation (purchased by MoMA). These fashion pieces utilise traditional chain mail logic in the form of interlinking closed loops and a variety of other closed interlinking shapes. The weight of the pieces, a result of the thousands of interlinked loops, attributes the textile drapability.

Expanding the chainmail logic, several designers have explored the interlinking closed loop geometry as a flexible base structure to apply surface geometry.[14,15] Notably, Tom Mallinson of London based 3D print bureau Additive.xyz (formerly Digits 2 Widgets) created a series of printed textile-based swatches that utilise various surface shapes including geometric squares and hemispheres alongside more organic petal and leaf forms (see Figure 12.2). These pieces exhibit flexibility and drapability due to the interlocking chain mail base structure and thousands of independently moveable surface shapes.

A variation on the closed interlinking geometries is Bloomfield and Borstrock's 2016 research project Modeclix® (Modeclix® is a registered trade mark of the University of Hertfordshire Higher Education Corporation). Building on the chain mail logic, Modeclix® is made up of interconnecting links, each link consists of four connected spirals. The

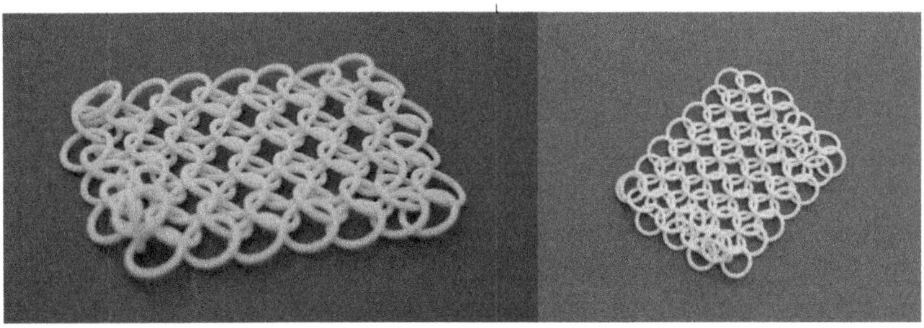

Figure 12.1 3D Printed chainmail structure.
Reproduced with permission from Mark Beecroft, Copyright 2014.

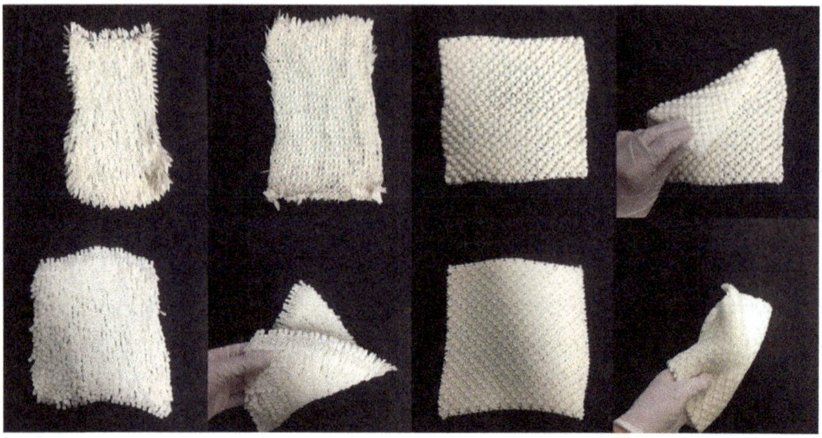

Figure 12.2 3D Printed textile-based structures, MMU Special Collections. Reproduced with permission from Tom Mallinson, Copyright 2019.

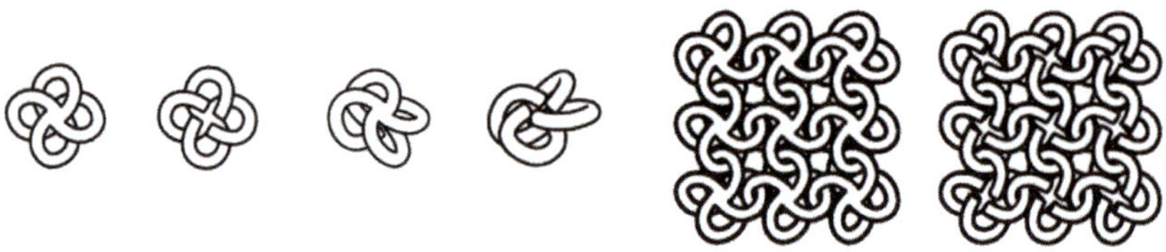

Figure 12.3 Drawing of Modeclix® links. Reproduced with permission from Mark Bloomfield and Shaun Borstrock, Copyright 2016.

links are connected across the top face, but left open on the reverse.[16] As each link is not fully closed this allows them to be disconnected and re-connected as required enabling larger textiles to be created (see Figure 12.3).

A further dimension of this research is the colouration of the pieces enabling customisation. The ability to click together small sections to make larger sheets enables a way around the cost prohibiting scale issue, making it viable to create accessories and garments (see Figure 12.4). Bloomfield describes the research as an adaptable textile which is additively manufactured as linked sheets and then assembled by hand to create customisable clothing.[16]

Alongside the interlinking chainmail type geometry, there are designers exploring hinged geometry to create articulation. A notable example being Bitonti and Schmidt's 2013 dress for performer Dita Von Teese.[17] The dress was made from laser sintered nylon and printed in seventeen sections, which were subsequently hand linked together before being polished, lacquered and embellished with

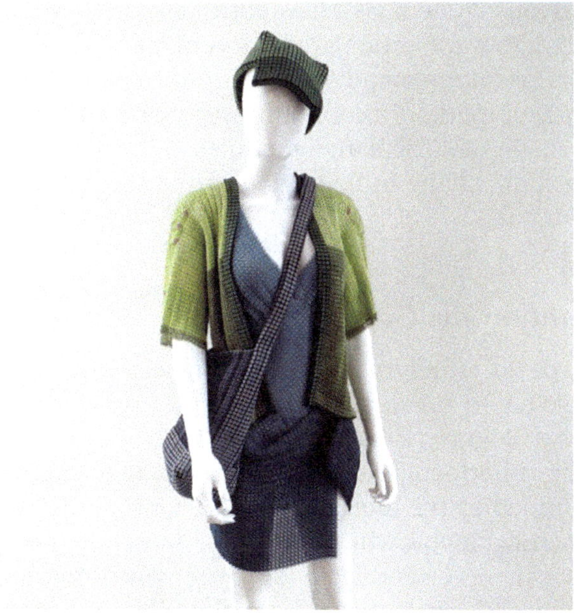

Figure 12.4 Modeclix® fashion pieces.
Reproduced with permission from Mark Bloomfield and Shaun Borstrock, Copyright 2016.

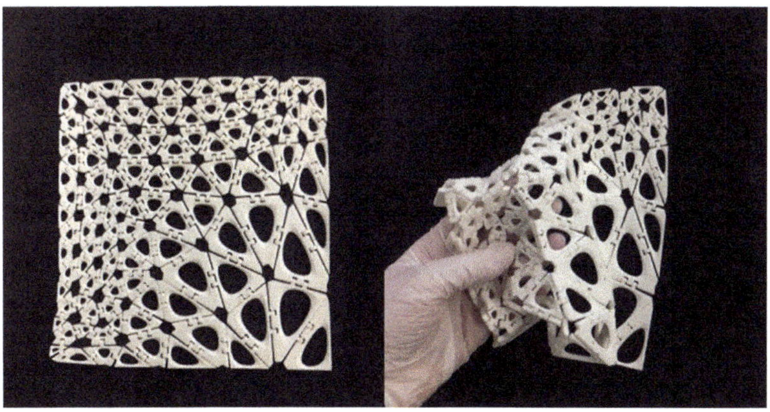

Figure 12.5 3D Printed textile-based structures, MMU Special Collections.
Reproduced with permission from Jessica Rosenkrantz, Nervous System, Copyright 2019.

12 000 Swarovski crystals. Howarth claims that it is the first time a work of fashion conceived and engineered from 3D printed nylon powder is able to move like a textile.[17] The work of Bitonti and Schmidt relates to that of the previously mentioned 'Kinematics'

dress by Nervous System as they both rely on rigid hinged components that allow for articulated movement with the wearer (See Figure 12.5). Rosenkrantz and Rosenberg of Nervous System describe the research as a method for 4D printing that creates complex, foldable forms made of articulated modules. This system enables any three-dimensional shape to be turned into a flexible structure using 3D printing.[6]

12.2.2 Continuous Geometry

The other type of structure used by designers to create flexible 3D printed textile-based material is through the use of continuous fibre type geometry.[18] Figure 12.6 shows an example of a piece printed using the continuous interlooping structure of plain knit. Traditional knitted fabrics are formed by the interlooping of a continuous yarn or fibre to create symmetrical loops, which create the structure (see Chapter 10). These loops can be extended in different directions giving knitted structures their inherent stretch and elasticity. A horizontal row of loops is described as a course, whilst a vertical column of loops is a wale.[19] Due to the geometry of the printed loops and the mechanical properties of the nylon material, the piece can extend longitudinally within the height of the loops as well as extending laterally exhibiting stretch like qualities associated with more traditional textiles.

Other notable examples of continuous geometry include Behnaz Farahi's 2015 piece titled 'Ruff'.[20] This piece explores the continuous geometry of the spiral to create a collar that is able to flex and extend

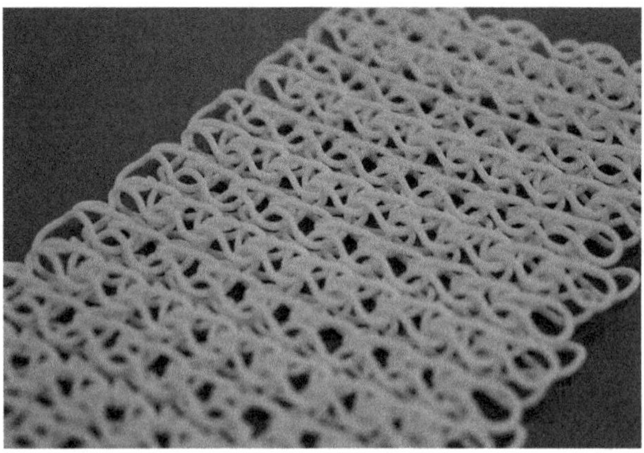

Figure 12.6 3D Printed knit-based structure.
Reproduced with permission from Mark Beecroft, Copyright 2019.

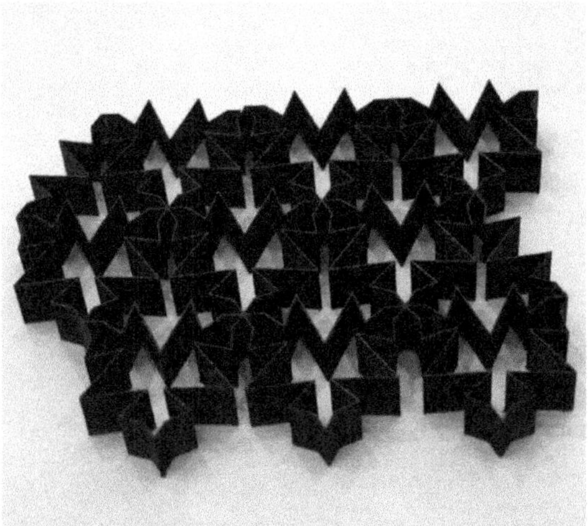

Figure 12.7 3D Printed auxetic structure.
Reproduced with permission from Jack Thomas, Copyright 2019.

around the wearer. Behnaz describes the results of the project as being surprisingly flexible despite being limited to rigid and fragile materials. This can be attributed to the spring like structure of the piece.[21]

Auxetic structures are another type of continuous geometry used to create flexible textile-based material (see Figure 12.7). Auxetic structures are structures or materials that become wider when stretched, perpendicular to the applied force. This occurs due to their particular internal structures, which resemble collapsed or folded structures, and effectively expand laterally when the sample is uniaxially loaded.[22]

12.3 Process

3D printing technologies and the materials available to print with have developed over the past ten years. The two most common types of additive manufacturing (AM) processes that have been used to create textile-based material to date are; fused filament fabrication (FFF), also known as the trademarked term fused deposition modelling (FDM), and selective laser sintering (SLS).

12.3.1 Fused Filament Fabrication (FFF)

FFF is a 3D printing technology that uses a continuous filament of thermoplastic material (see Chapter 11). The filament is a constant

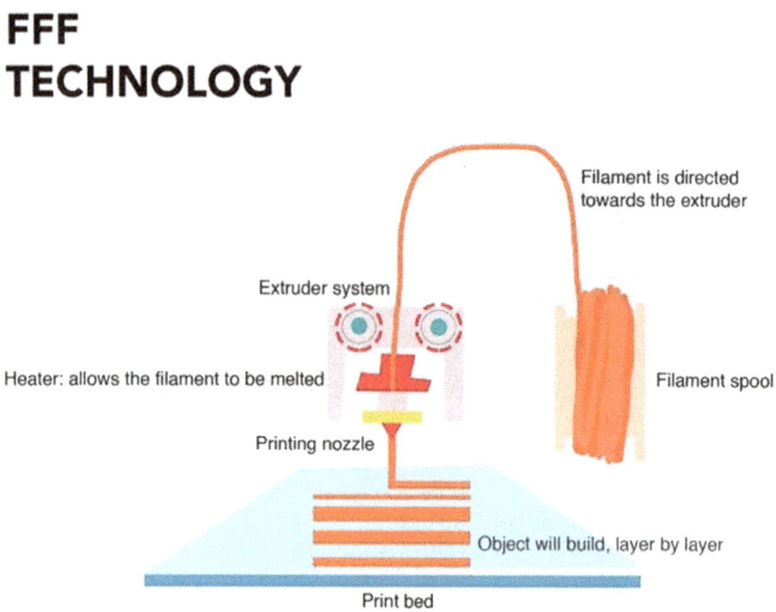

Figure 12.8 Diagram of FFF process.
Reproduced with permission from Georgina Barker Copyright 2019.

width of either 1.75 mm or 2.85 mm fed from a large spool. The FFF process extrudes the filament through a heated nozzle which deposits the melted plastic layer by layer (see Figure 12.8). The print head is moved by computer control in response to the 3D computer aided design (CAD) data. Melted filament from the nozzle is extruded and deposited onto a build plate (see Chapter 11), where it cools and solidifies to form a layer. The bed then moves down by a fraction of a millimetre to start another layer. The process is repeated until the complete object is formed.[23] The advantages of this technology are that it is relatively cheap and simple shapes require no support material. However, complex geometries such as interlinking structures require a second nozzle to deposit a support material that can either be dissolved or cut away after printing.

12.3.2 Selective Laser Sintering (SLS)

In the SLS process a carbon dioxide (CO_2) laser is used to fuse fine powder into solid material. The laser is directed by a computer guided mirror and builds objects in layers of 0.1 mm, the building platform lowers down by this measurement each time allowing the next layer of powder to be rolled onto the surface (see Figure 12.9).

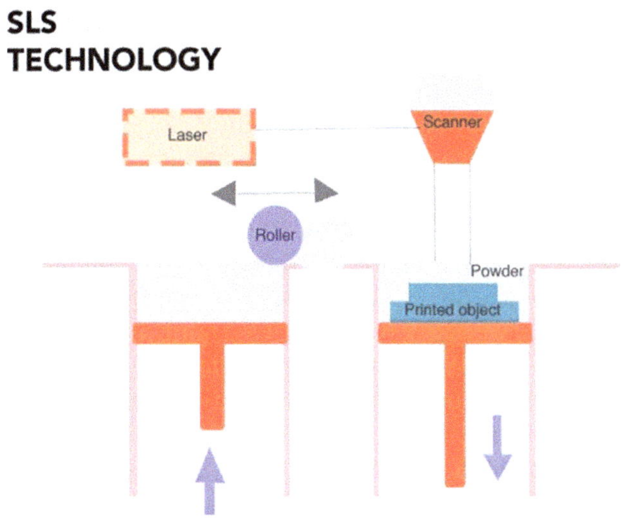

Figure 12.9 Diagram of SLS process.
Reproduced with permission from Georgina Barker Copyright 2019.

The non-sintered powder acts as a support material for the build object. The excess powder is then removed by high pressure after printing. The whole process takes place in an inert nitrogen environment to prevent the nylon oxidising when heated by the laser beam. The temperature inside the building chamber is maintained at 170 °C, just below the melting point of the polymer powder, so that as soon as the laser makes contact with the surface particles, they are instantly fused by the 12 °C rise in temperature.[24] This process has the advantage of being able to print complex geometries without the use of additional support material, making it an ideal method to explore fine interlinking textile-based structures.

12.4 Material

As 3D printing is an emerging technology the range of materials available to print with are constantly evolving, this section reviews the most readily available and commonly used materials to 3D print flexible textile-based material.

12.4.1 FFF Printers

FFF printers use a variety of thermoplastics to print with; the two most commonly used are acrylonitrile butadiene styrene (ABS) and biopolymer polylactic acid (PLA). ABS is a combination of styrene acrylonitrile (SAN) and butadiene rubber (BR) produced as an amorphous blend (see Chapter 11). ABS is suitable for FFF technology as it has an achievable softening point. Once heated to 230 °C it is soft enough for nozzle extrusion and strong enough to bond to the next layer. ABS is around one-third cheaper than other thermoplastics such as polyamide (PA, nylon).[25] Material properties include toughness and ductility. A downside is that ABS gives off harmful fumes when heated so requires extraction. An example of textile-based structures printed using ABS includes Danit Peleg's 2015 fashion collection.[26] Due to ABS having a higher glass transition temperature and a relatively high coefficient of thermal expansion, ABS parts are more sensitive to warping than PLA.[27]

PLA is a bioplastic derived from natural lactic acid. Although more expensive than ABS it does not require extraction, which makes it suitable for use in desktop environments. Melnikova *et al.*[10] have used PLA material to print single-face knit-based structures. Whilst the pieces do exhibit some flexibility there are fine, undesirable connections that remain between the loops, which is a result of the FFF nozzle extrusion.

12.4.2 SLS Printers

The SLS process is compatible with a variety of polyamide (PA) based powders. PA is commonly referred to as 'nylon', the commercial name given to the polymer when first developed by DuPont (see Chapter 10). Nylon's polymer structure is made up of carbon atoms held together by amide groups.[25] Nylon can be formed from either a single monomer unit or from the reaction between two different monomer units (see Chapter 11). This is represented by the numerical nomenclature; in the former case the single monomer only has one functional group capable of forming an amide, whereas in the latter each monomer is capable of forming two amide groups. Nylon is a popular engineering thermoplastic owing to its lightweight, strong and flexible properties, and because of this it has been used by numerous designers to successfully print articulating structures. Nylon 12 (PA12) in particular has outstanding mechanical properties including low moisture absorption, good dimensional stability and superior flexibility. The 'N12' 3D printed bikini developed by Continuum Fashion[28] makes use of thousands of SLS nylon circular components interconnected by thin spring like structures allowing the piece to hold its form whilst still being flexible.

Alongside nylon-based powders the SLS process is compatible with certain polyurethane plastics including thermoplastic polyurethane (TPU). TPU belongs to a group of rubber like plastics called thermoplastic elastomers (TPEs). Unlike thermosetting elastomers, TPEs do not form chemical crosslinks making them compatible with melt forming processes like SLS.[25]

TPU material properties include elasticity and moderate strength. TPU has been used by designers most notably Iris van Herpen and Julia Koerner[29] to 3D print a dress that explores flexibility through interwoven thread like structures. Whilst TPU exhibits flexibility and stretch, a disadvantage as stated by Rothenberg is that it is uncomfortable and has a tendency to stick to the wearer.[30]

12.5 Hybrid 3D Printed Textiles

Alongside fully 3D printed textile-based pieces, there are pieces exploring 3D printed elements combined with traditional textiles. For the purpose of this chapter 'textile' is defined as a planar structure consisting of yarns or fibres.[31] A notable example is architect/designer Richard Beckett's collaboration with Pringle of Scotland creating garments combining 3D printing technology with traditional textile techniques.[7] Figure 12.10 illustrates 3D printed SLS nylon knit-based structures attached to a machine knitted lycra base. The 3D printed elements are manually attached and knitted into the lycra fabric post printing. When the lycra is stretched the 3D printed elements stretch

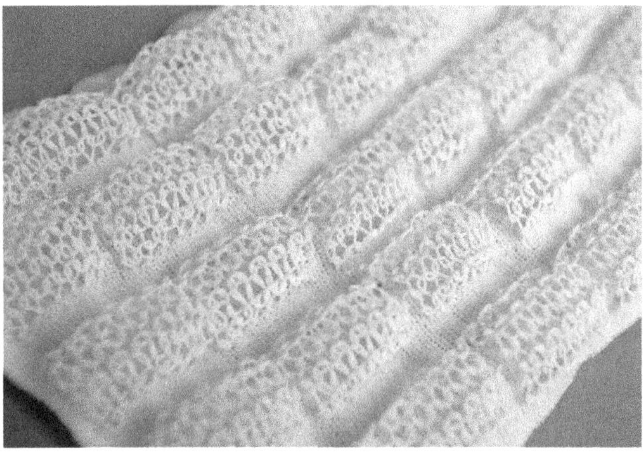

Figure 12.10 3D printed elements combined with machine knitted fabric. Reproduced with permission from Mark Beecroft and Laura McPherson, Copyright 2014.

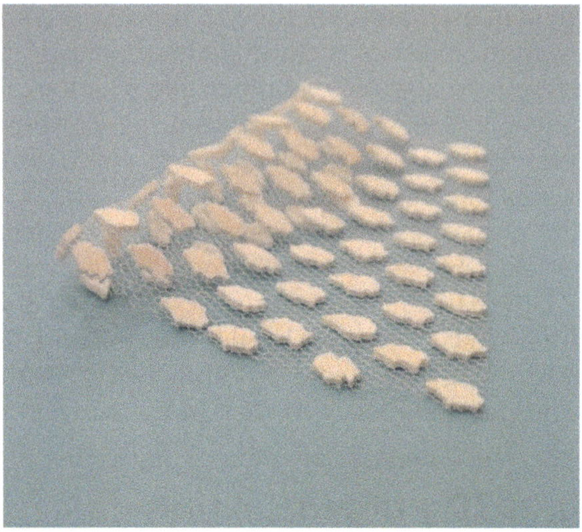

Figure 12.11 ABS shapes printed onto polyester mesh.
Reproduced with permission from Georgina Barker, Copyright 2019.

with the piece. In this example the 3D printed components act as a decorative element rather than altering the mechanical properties of the fabric.

12.5.1 3D Printing Direct onto Fabric Substrates

The FFF process allows for experimentation printing directly onto fabric substrates.[32] To achieve this the fabric is secured to the printer bed before starting the filament extrusion. Figure 12.11 illustrates ABS plastic printed directly onto a polyester mesh base material. The roughened surface of the polyester fibres ensures good adhesion of the printed polymer to the top of the textile. The open mesh structure of the base fabric ensures good adhesion properties inside the textile structure as it offers sufficiently open areas for the molten polymer to penetrate inside.[33]

12.6 Colouration

The FFF process allows for many coloured filaments to be used, designer Georgina Barker interchanges PLA filaments during the printing process to add coloured stripes to the edges of the pieces (see Figure 12.12). Printing directly onto coloured fabric substrates adds to

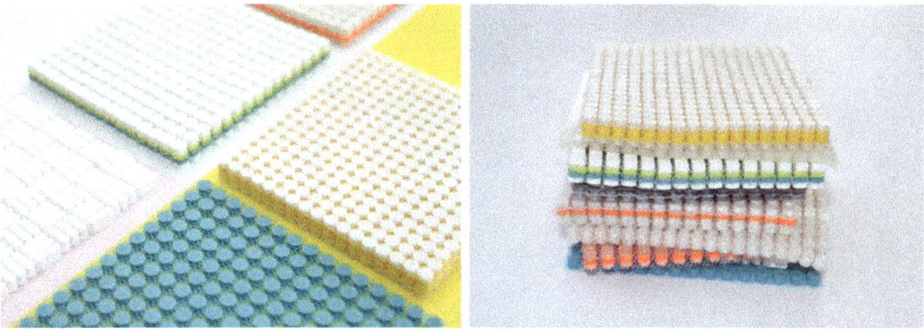

Figure 12.12 Coloured PLA printed onto polyester mesh. Reproduced with permission from Georgina Barker, Copyright 2019.

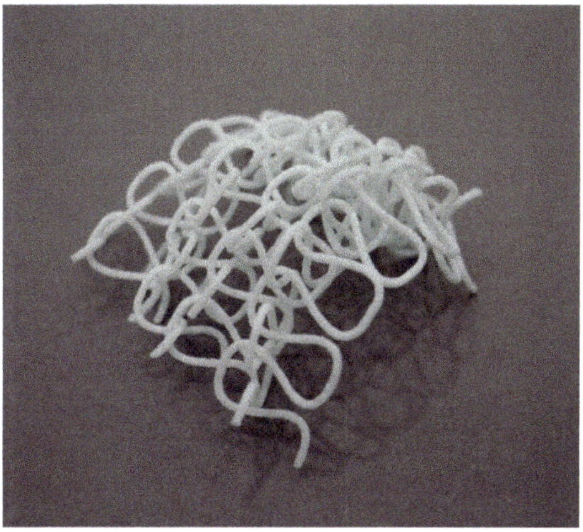

Figure 12.13 Disperse dyed knit-based ABS structure. Reproduced with permission from Mark Beecroft, Copyright 2014.

the aesthetic appeal. Barker explores the limitation of height and its impact on the flexibility of the pieces.

Unlike the FFF process, SLS does not allow for interchangeable colours during the process. As SLS works by laser sintering a bulk of powder, the colour of the piece is determined by the colour of the powder. In the case of SLS nylon the powder is usually white, with colour being added by dyeing the pieces post printing.

12.6.1 Post Processing Dyeing

It is possible to add colour to 3D printed pieces post printing; Figure 12.13 illustrates an ABS knit-based structure printed using FFF

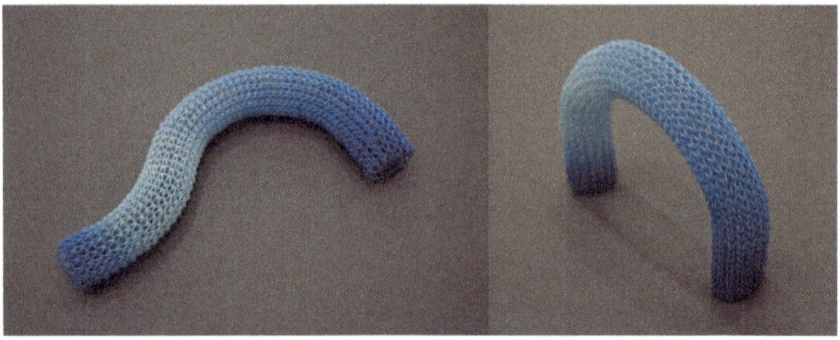

Figure 12.14 Disperse dyed knit-based SLS structure. Reproduced with permission from Mark Beecroft, Copyright 2016.

technology. The piece has been dyed post printing using disperse dyes (see Chapter 8). Disperse dyes were formulated and introduced to permit dyeing of hydrophobic thermoplastic fibres including nylon, polyester, acrylic, and other synthetics.[34] Due to the heat process required to activate the disperse dyes and the relatively low softening point of ABS plastic the shape of the piece has deformed considerably.

Dip dyeing over controlled times allows a colour gradient to be created (see Figure 12.14). Dyeing SLS nylon with disperse dyes is more successful than ABS as the piece does not deform due to nylon's higher melting temperature. Also, SLS pieces are more porous than FFF pieces, a typical SLS part is about 30% porous compared to about 20% for a typical FFF part.[35] Porosity gives SLS pieces their characteristic grainy surface finish.[36]

12.7 Manchester Metropolitan University Special Collections

In 1898 Manchester School of Art opened its Arts and Crafts Museum to showcase the best in art, craft and design. This museum collection became part of Manchester Metropolitan University Special Collections in 2005. Special Collections is a museum accredited by Arts Council England. Since 2013 Manchester Metropolitan University Special Collections has housed Manchester School of Art: the Contemporary Craft and Design (Material and Process Innovation) collection. This collection began as a result of the exhibition, *The Language of Process: how new materials and technologies are changing product design*, initiated by design lecturer David Grimshaw and curator Stephanie Boydell. The aim of the collection is to build a museum quality collection, comprised of objects that are cutting edge in terms of their material and process-led approaches to making, manufacture and distribution whilst also challenging the conservatism of museum collecting by taking innovative objects of untested materials.[37]

12.7.1 Collecting Policy

Taken from the Manchester Metropolitan University Special Collections Development Policy Appendices[38] the themes for collecting objects are:

1. Items can be craft or design, a one-off or mass-produced piece;
2. Items can be any media (including items made from untested, new materials that may deteriorate, and digital media);
3. Items can be international in origin;
4. Items must have been made or designed within the previous 5 years;
5. Items that embody contemporary/cutting edge concepts and practices;
6. Items must reflect current teaching and research concerns of the Manchester School of Art and/or MIRIAD;
7. Items that reflect contemporary practice beyond the Higher Education sector;
8. Items that build a public collection of contemporary craft and design that is unique and unmatched in the region.

12.7.2 Access to the Collection

The Manchester School of Art: the Contemporary Craft and Design (Material and Process Innovation) collection is open access to staff and students from the university as well as a space for hosting public exhibitions. A key feature of the collection is that it is used as a handling collection and teaching resource. This raises issues relating to the conservation of the objects that will be discussed further in the case study. A selection of objects from the collection were exhibited as part of the Crafts Council's 2016 Make: Shift conference[39] where visitors were encouraged to handle the objects to explore their material properties (Figure 12.15). Part of the collection was also exhibited at The International Association of Societies of Design Research (IASDR) 2019 *Design Revolutions* conference hosted by the Manchester School of Art, Manchester Metropolitan University.[40]

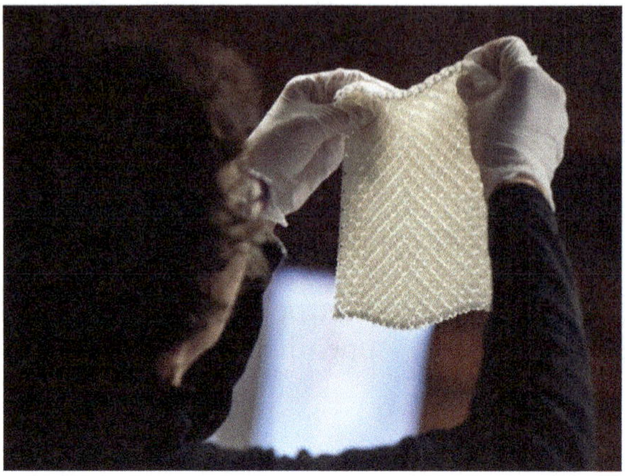

Figure 12.15 Handling session at Make: Shift Conference. Reproduced with permission from the Crafts Council. Photo credit Simon Webb, Copyright 2016.

12.7.3 Conservation Strategy

The Manchester School of Art: the Contemporary Craft and Design (Material and Process Innovation) collection includes examples of 3D printed furniture and sculpture as well as an expanding collection of 3D printed textile-based materials. To date, the 3D printed textile-based pieces include a series of four chainmail based pieces by Tom Mallinson, a sample piece of the 'Kinematics' dress by Nervous System, the 2014/15 Autumn/Winter Pringle of Scotland in collaboration with Richard Beckett jumper as well as pieces of the author's practice-based research: *Digital Interlooping*. The current storage and conservation of the 3D printed pieces constitutes:

- Pieces wrapped in acid free tissue paper;
- Stored in archival boxes;
- Stored in the dark;
- Stored at a temperature between 19 and 23 °C and relative humidity between 40 and 60%;
- Handling of the objects is permitted under supervision and disposable nitrile gloves must be worn.

12.8 Case Study – Digital Interlooping

This section will focus on the author's practice-based research into 3D printed textile-based material, firstly, by discussing how the pieces were created in terms of computer aided design (CAD), manufacturing process, material, structure/geometry. Following this, drawing on current research, the discussion will focus on the conservation issues raised by such pieces.

12.8.1 3D CAD Development

In order to 3D print knit-based structures, a 3D CAD drawing is required. This CAD drawing needs to consider the pipe wall thickness and distance between individual objects, particularly when creating interlooping structures such as knit. To begin the process a CAD drawing of the plain knit structure was created. Digits2Widgets' in house CAD designer Tom Mallinson was identified as having the required CAD modelling skills to translate the knit patterns into three dimensions and then manipulate them to the configurations of interest. His previous experience producing Digits2Widgets' own chainmail based 'fabrics' meant that he had the expertise to understand the material limitations and tolerances that these experiments could be successfully taken to.

McNeel Rhinoceros (Rhino) software was chosen to generate the 3D CAD models due to its powerful modelling capabilities. A single knit loop unit was initially drawn as curves. Once created this loop unit was repeated and arranged using the array command to create desired width (courses) and length (wales). To create the tubular structures the flow along curve command in Rhino was used specifying the diameter of the tube and the length required to flow along. This line drawn tube was then piped with a specified wall thickness. Once piped, these designs were run through the Magics 3D printing software to make sure all neighbouring and intersecting parts have a distance of at least 0.4 mm.

12.8.2 Process

As previously discussed, the SLS process enables fine interlinking structures to be printed without the complication of support material. As such, SLS was identified as the most suitable technology for the *Digital Interlooping* research project to provide the highest print resolution. The 3D printed knit-based structures were printed using an EOS Formiga P1 machine. These machines are calibrated up to $\pm 0.15\%$ on the X- and Y-axis and build with 0.1 mm layers on the Z-axis.[41] As the machine builds using fine layers of powder the software EOS RP tools was used to slice the 3D CAD data into layers.

Following printing the pieces are post-processed using pressurised air to remove any excess powder.

12.8.3 Material

Nylon (PA12) was selected as the material for the research as it requires no support structures and has inherent material properties including good dimensional stability and superior flexibility. These attributes are key when testing stretch behaviour and flexibility. A critical issue relating to the quality of nylon SLS is the powder ratio used. The material for commercial nylon SLS prints is a blend of new and recycled powder. This is due to maintaining cost efficiency. Machine manufacturers EOS recommend a powder ratio of 1 : 1 to achieve the best part quality and material performance. London based 3D print bureau Digits2Widgets were chosen to print the pieces due to their commitment to producing nylon SLS at its optimum quality whilst open to exploring the boundaries of the technology's capabilities. Digits2Widgets guarantee the optimum 1 : 1 new to used powder by housing a proprietary mixing unit that simultaneously mixes and transports the powder directly to the machines.

12.8.4 Structure (Mechanical Behaviour)

Digital Interlooping is a body of practice-based research exploring the primary structures of knit to create flexible 3D printed textile-based material. As previously mentioned, these pieces make use of continuous fibre type geometry[9] in the form of the continuous interlooping structure of the knit loop. In the digital environment, which can be scale-less and weightless, materiality and form can be explored in new physically challenging ways. To date, the unitised loop has been applied to both singe-face (plain) and double-face (interlock) knit-based structures to create tubular seamless forms at various scales (Figure 12.16). However, on analysing the pieces post fabrication, it becomes evident that as well as embedding knit's characteristics of inherent stretch and flexibility, the digital environment in which they were modelled affords the re-fabricated pieces new behaviours in material form. For example, the tubular piece in Figure 12.17 was digitally modelled by sweeping the tube along an assigned pathway. As a result, the piece is 'fixed' in a curved position. The piece when stretched elongates vertically through the wales as well as stretching horizontally across the courses. However, possibly more interesting is the

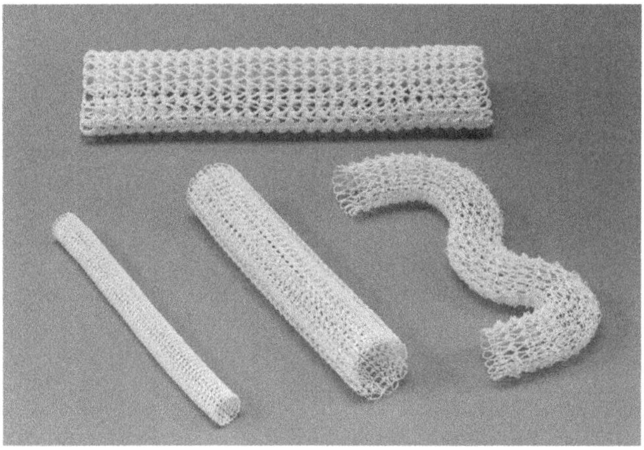

Figure 12.16 Series of tubular knit-based SLS structures.
Reproduced with permission from Mark Beecroft, Copyright 2016.

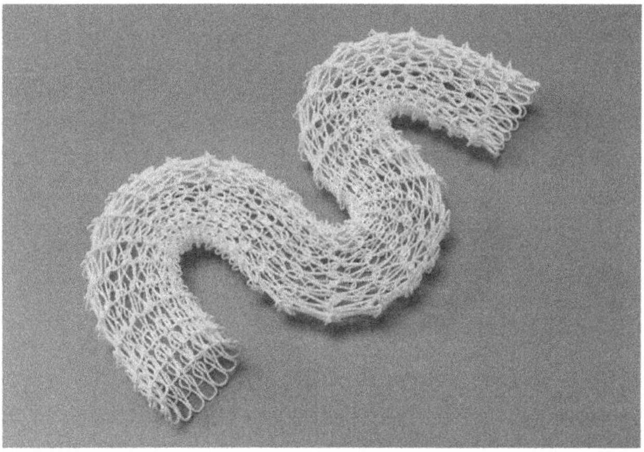

Figure 12.17 Tubular knit-based SLS structure.
Reproduced with permission from Mark Beecroft, Copyright 2016.

articulation of the form at the section where the loops have fused together – this creates restricted movement in these areas. At other sections the loops have been digitally stretched causing fixed elongation this adds a hinged articulation in these areas. Further iterations have explored how applying ply to the 'yarn' by digitally twisting impacts the material behaviour of the re-fabricated piece (see Figure 12.18). As well as further emulating traditional knitted fabrics in appearance, the twisted 'yarn' once printed affords the piece greater porosity and a greater degree of stretch and flexibility.

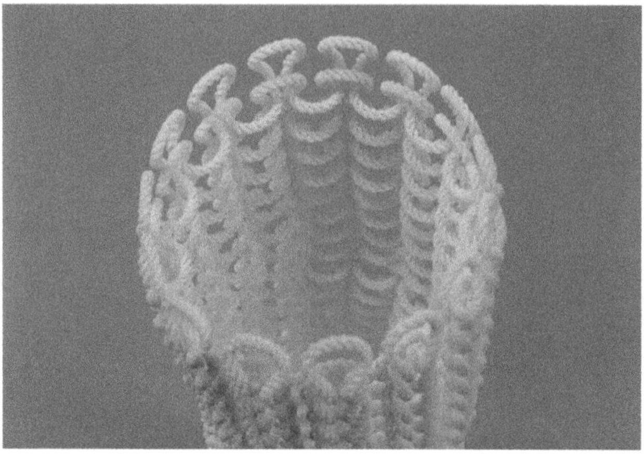

Figure 12.18 Detail of tubular knit-based SLS structure. Reproduced with permission from Mark Beecroft, Copyright 2016.

12.9 Conservation Considerations

Drawing on current research and using the *Digital Interlooping* pieces as examples, this section will discuss some of the key conservation issues and implications for conservators and curators of 3D printed textile-based material. It will also suggest some possible strategies and processes on how these issues may be overcome.

12.9.1 Material Stability

During the SLS process, the machine preheats the bulk powder material in the powder bed to just below its melting point, this makes it easier for the laser to raise the temperature of the selected regions the remaining temperature gap to the melting point.[42] As all of the powder is exposed to long heating and cooling periods, very close to its melting temperature, this causes physical and chemical changes in the mechanical and thermal properties.[43] The polymer chains of SLS thermoplastics are shorter compared to those of other thermoplastic melting processes *i.e.* injection moulding resulting in a less distinctive density of material structure.[44]

As the SLS process is expensive the un-sintered support powder is often reused by 3D print bureaus to reduce costs. The percentage of material that can be successfully reused is dependent on the specific polymer and is often referred to as the 'refresh rate'.[45] Repeated reprocessing of the powder causes further degradation resulting in printed objects of poorer quality.[46] Research has suggested that after approximately 7–8 builds the quality of the powder has degraded so much that it should be discarded.[43] The term 'orange peel'[45] has been used to describe the rough surface of objects printed using over processed powder.

The issue the SLS process may cause in regards to the conservation of 3D printed textile-based material is that the materials may not be completely homogenous.[45] Using the *Digital Interlooping* pieces as examples, it is known that they were printed at Digits2Widgets 3D bureau using the optimum 1:1 ratio of recycled to virgin powder. However, without this depth of knowledge conservators will be unaware of how many times the objects powder has been re-processed this could potentially influence the ageing stability of the finished piece. Lifetime prediction is always challenging, but this is especially the case where the objects will each have very different thermal histories.

12.9.2 Structural Deformation

The physical storage of 3D printed textile-based material is a further concern for conservators. As previously mentioned, objects in Manchester Metropolitan University Special Collection are wrapped in acid free tissue paper and stored in archival boxes. Whilst this is suitable for the flat printed pieces, Figure 12.19 illustrates how one of the large tubular forms has become deformed over time. The tubular object has been stored flat, without support, with the deformation of the structure due to its own weight and the effect of gravity on the piece. Due to both the weight and nature of the interlooping

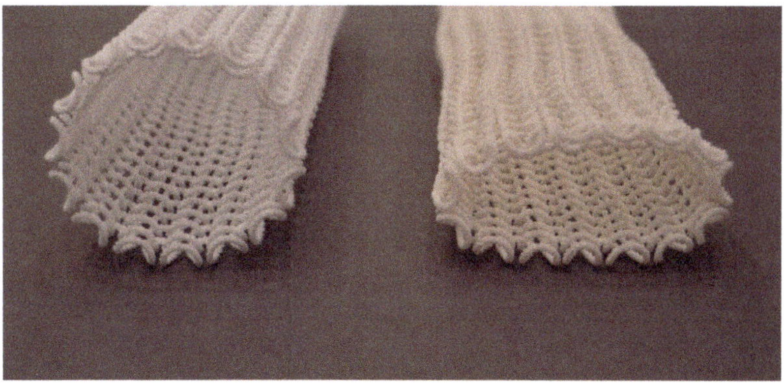

Figure 12.19 Image showing structural distortion of tubular knit-based SLS structure.
Reproduced with permission from Mark Beecroft, Copyright 2016.

structure, it is important to avoid compression on the object as the material has a limit to which it will recover its original form. A solution to this would be to create internal supports for the tube. However, careful consideration of what materials these supports are made from would be required to avoid any chemical reaction between the support and the nylon of the piece. It has been suggested that printing the negative form of the object may prove the best supporting structure.[44] Whilst this may be true in the case of these objects this would require further expensive SLS printing. A hollow printed tubular insert would reduce this printing cost and printed from the same material would limit the possibility of future chemical incompatibility and interaction.

12.9.3 Colour Instability

Colour instability is a key concern for the conservation of 3D printed material. The SLS process itself can initiate pre-oxidation due to the impact of the heat on the nylon particles[47] and this discolouration will continue over time due to ultraviolet (UV) radiation. In the presence of UV, the polymer reacts to form cross linkages between molecules and also the polymer chains are broken down to smaller molecules that absorb visible wavelengths and therefore shift the colour of the polyamide. Figure 12.20 clearly shows a visible change in colour occurring over a relatively short time period of time (24 months). This has occurred despite the object being stored in the dark for the vast majority of the time. The piece has been exhibited several times and has been handled, this has undoubtedly exacerbated the problem by exposing the piece to other pollutants. Research has found that this ageing phenomenon is found to be more acceptable by artists than designers who see the yellowing as a form of 'patina'.[46] This is further echoed by Iverson who describes the sculpture as acquiring a patina that is both physical and historical, linking the piece with the passage of time.[48] A way to combat this ageing effect may be to coat the object with a UV absorber, but this would inevitably alter some of the physical properties of the pieces and would not be reversible.

12.9.4 Repair and Replication

Due to the 'born-digital'[49] nature of 3D printed objects, it could be argued that they should remain perfect and in pristine condition. Recent research found that 84% of artists and designers surveyed

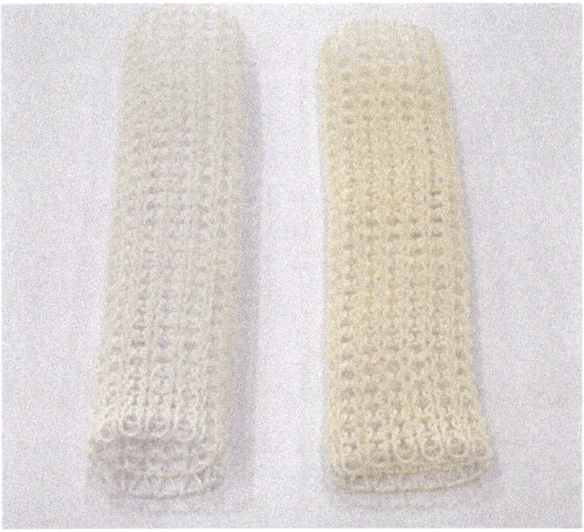

Figure 12.20 Image showing discolouration of tubular knit-based SLS structure.
Reproduced with permission from Mark Beecroft, Copyright 2016.

would be happy for their work to be reprinted if it became damaged or broken.[46] This raises ethical questions regarding the originality and authenticity of a piece. Design Museum London's director Deyan Sudjic states that authenticity and originality are having their meanings redefined in the midst of the current digital explosion.[50] Rolf Fehlbaum, chairman emeritus of VITRA, believes an object to be authentic when its promises are matched by its reality and its appearance and reality correspond. He further suggests that if someone makes a copy of an object and declares it to be so then it is an 'authentic copy' as the promises match the reality.[50] In terms of reprinting an object, this is a complex issue which is discussed later in the chapter. Figure 12.21 shows an example of a 3D printed textile-based piece which has sustained mechanical damage. The fragility of the piece due to printing at such fine resolution has caused it to fracture in places. Should this piece be reprinted? Unlike traditional knitted textiles the 3D printed knit-based structure is not going to unravel like yarn and, as such, is relatively stable. The decision to reprint would be an aesthetic one. It could be argued that in its current state the piece highlights the current limitations of the SLS technology to print at such fineness and as such should be preserved as a piece in the history of the technology. This train of thought echoes Riegl's as he states the historical value of an object is linked to the stage it represents in the development of human activity in a certain

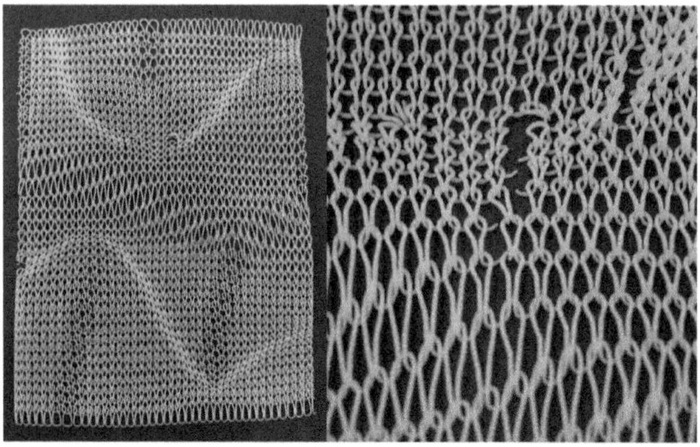

Figure 12.21 Image showing mechanical damage to knit-based SLS piece. Reproduced with permission from Mark Beecroft, Copyright 2016.

field. Suggesting the more faithfully an object's original state is preserved the more significant its historical value.[51] In contrast, as 3D printing technologies advance there may be the temptation to remake the object somehow 'better' than the original making use of the most current state of the art.[51] This is echoed by designer Joni Steiner who describes digital design as moving away from singular outcomes towards multi-authors and iterations.[50]

12.9.5 Re-printability

There is an assumption that as 3D printed objects are created from digital data that to re-print an object is a relatively straightforward process. In a museum or gallery conservation context this poses several issues. The first is related to the originality of the piece purchased by the gallery or museum. Should they prioritise the preservation of the original piece rather than seek to simply re-print as the original deteriorates? In regards to originality, Fehlbaum believes this can only be the case when the object is produced by the legitimate manufacturer in coordination with the author.[50] Another major factor is access to the digital files – unless a gallery or museum obtains or purchases these with the original piece, re-printing is not an option. Even if the gallery or museum does have access to the original digital file then it requires someone who has an understanding of how to work with these files.[52] As time passes and 3D printing software and technology advances, the compatibility of the original files may become an issue. Even with access to the original digital files in a compatible format the range of printers and materials available to

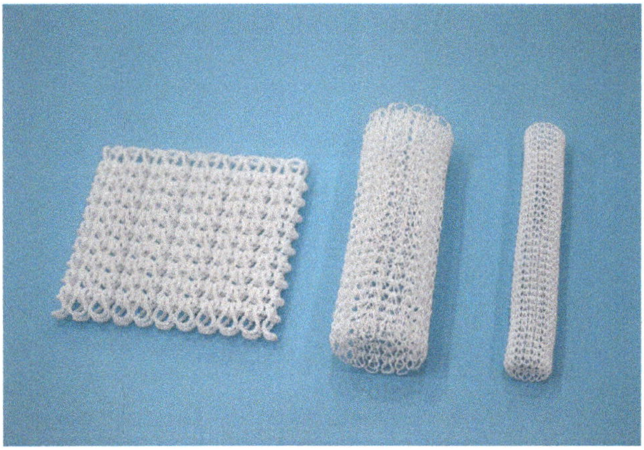

Figure 12.22 Series of SLS knit-based structures.
Reproduced with permission from Mark Beecroft, Copyright 2019.

print will inevitably cause further problems with re-printing. To illustrate this issue a series of re-print tests was carried out at Manchester Metropolitan University Print City.[53] The original *Digital Interlooping* samples (see Figure 12.22) were printed using a high-end industry standard EOS Formiga printer costing in the region of £250 000. As SLS is an expensive process FFF technology was chosen for the re-print test as this is a more readily available technology. The printer used for the re-print tests was an Ultimaker S5, a mid-level printer costing around £6000. The printers have a dual nozzle to allow for a support material to be printed making it possible to print the complex interlooping structures. The original 3D CAD files were not altered in any way.

12.9.5.1 Test 1 Results

Once printed the white PLA piece is encased in the dissolvable PVA support material (see Figure 12.23). This requires post processing by dissolution in water for 48 h. Once the support material has been dissolved the PLA structure is revealed. On initial inspection (see Figure 12.24) the pieces are visually similar to the original in terms of their colour and structure. However, on closer inspection the print quality is significantly coarser with scorch marks visible in a couple of areas. In terms of handling, the flat knit piece is much more fragile than the original SLS piece and as a result easily fractures in places. The piece has some limited lateral extension within the height of the loops, but no stretch capabilities longitudinally. The tubular pieces

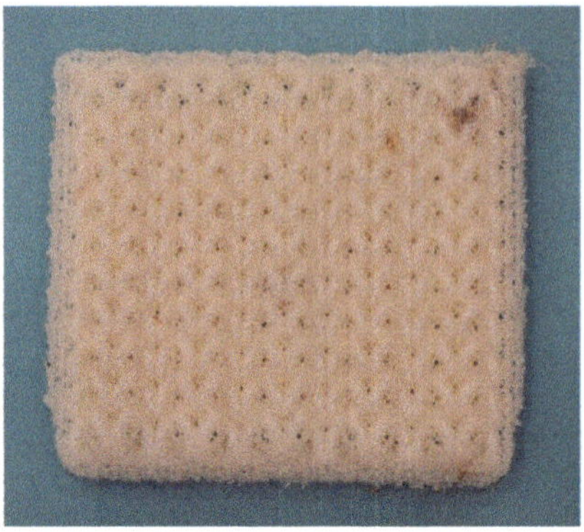

Figure 12.23 PLA knit-based structure encased in dissolvable PVA support material.
Reproduced with permission from Mark Beecroft, Copyright 2019.

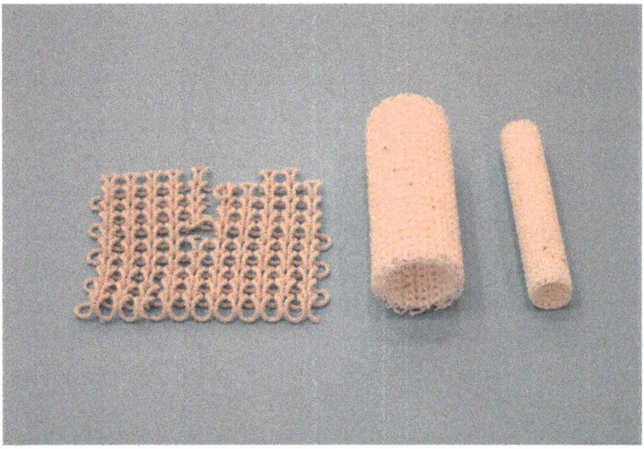

Figure 12.24 Series of FFF knit-based structures in PLA.
Reproduced with permission from Mark Beecroft, Copyright 2019.

are stronger, with the larger tube withstanding compression along its length. The smaller tube is completely rigid as a result of the printed loops become fused together during the printing process (Table 12.1).

12.9.5.2 Test 2 Results

As with Test 1, once printed the nylon pieces were encased in the PVA support material they required dissolution in water for 48 h. On first inspection, the pieces are visually different from the original (see Figure 12.25) in that they are black (white nylon is not currently

Table 12.1 Printer settings and material for Test 1. Reproduced with permission from Mark Beecroft, Copyright 2019.

Test 1	
Nozzle size	0.4 mm
Resolution setting	High
Print speed	40 mm s^{-1}
Print head temperature	210 degrees
Print material	Filamentive recycled white PLA
Support material	PVA

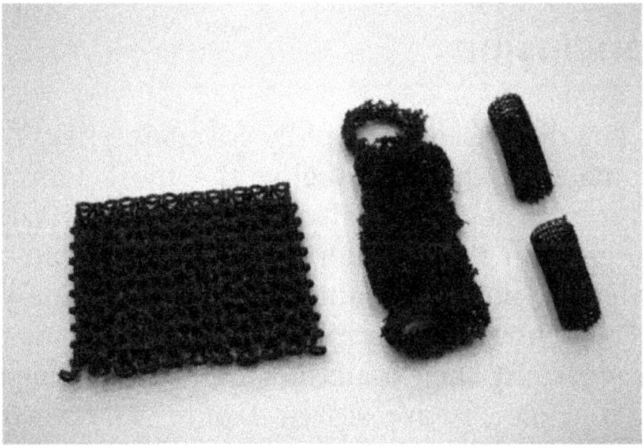

Figure 12.25 Series of FFF knit-based structures in nylon. Reproduced with permission from Mark Beecroft, Copyright 2019.

Table 12.2 Printer settings and material for Test 2. Reproduced with permission from Mark Beecroft, Copyright 2019.

Test 2	
Nozzle size	0.4 mm
Resolution setting	High
Print speed	35 mm s^{-1}
Print head temperature	205 degrees
Print material	Ultimaker Black Nylon
Support material	PVA

available from Ultimaker). The flat knitted piece exhibits some extension and stretch both laterally and longitudinally making it closer to the mechanical properties of the original SLS nylon. However, the two tubular pieces have failed to print successfully. The smaller tube, although having flexibility, has split in half during the dissolving of the support material and the larger tube has split into several pieces (Table 12.2).

These initial tests illustrate that the possibility of re-printing objects is a complex issue. Even with access to the original digital data in the original format, unless the original printer, printer settings and material are known then the printed results may differ greatly from the original object. This was echoed by recent research from the Cooper Hewitt's Digital Collections Management Project, which describes it as a near impossibility to replicate an original artist or designer's print without meticulous documentation of the original settings.[54]

12.10 Conclusion

This chapter has discussed the rapidly advancing field of 3D printed textile-based materials by introducing the most common types of structures, processes and materials used in their fabrication. With an increasing interest in these materials, such pieces are increasingly being acquired by museums and galleries into their permanent collections. These novel pieces raise several issues for conservators and curators regarding their material stability. The chapter has used the author's practice-based research *Digital Interlooping* as a case study to discuss some of the potential conservation issues. It is clear that their conservation is a complex issue. On a material level, the exact composition and structure may have an impact on the ageing stability of the piece, whilst the digital process of fabrication may actually accelerate this deterioration.

Ethical questions are raised around the authenticity and originality of a piece when discussing the option to re-print pieces that are deteriorating or become damaged. As highlighted by the re-print tests, having access to the digital file in the correct format is only one element. The various types of printers and materials available pose further questions for the conservator. Unless the exact printer, printer settings and material of the original are known then re-printing may result in pieces that differ greatly in terms of appearance, handle and material properties. It is also important to consider the artist/designer's original intention for the work and to understand their position in regards to the acceptance of the ageing process.[55]

As 3D printing technologies advance and due to the born-digital nature of the objects produced, there may be the temptation to remake improved versions of the piece. Again this raises questions around ideas of originality and authenticity.

Ultimately, as 3D printing was originally developed for creating prototypes that were not intended to last for many years, the conservation of 3D printed objects in gallery and museum collections will continue to pose questions as they become future heritage materials.

Recommended Reading

Future Talks 011: Technology and Conservation of Modern Materials, ed. T. Bechthold, Die Neue Sammlung: The International Design Museum Munich, Munich, 2013.

F. Bitonti, *3D Printing Design: Additive Manufacturing and the Materials Revolution*, Bloomsbury Publishing, London, 2019.

S. Braddock Clarke and J. Harris, *Digital Visions for Fashion + Textiles: Made in Code*, Thames & Hudson, London, 2012.

Print Shift: How 3D Printing Is Changing Everything (Dezeen), ed. M. Fairs, Blurb, 2013. (A print on demand publication available from: https://www.blurb.co.uk/b/4176869-print-shift)

S. Hoskins, *3D Printing for Artists, Designers and Makers*, 2nd edn, Bloomsbury Publishing, London. 2018.

N. Leach and B. Farahi, *3D-Printed Body Architecture (Architectural Design)*, Wiley & Sons, Oxford, 2017.

H. Lipson and M. Kurman, *Fabricated: The New World of 3D Printing*, Wiley, Indianapolis, 2013.

C. Warnier, D. Verbruggen, S. Ehmann and R Klanten, *Printing Things: Visions and Essentials for 3D Printing*, Gestalten, Berlin, 2014.

References

1. 3D Systems, *FOC Textiles in permanent collection at MOMA*, https://www.3dsystems.com/blog/foc/foc-textiles-to-permanent-collection-at-moma, (accessed July 2019).
2. Dezeen, *Crystallization by Iris van Herpen, Daniel Widrig and .MGX by Materialise*, https://www.dezeen.com/2010/08/11/crystallization-by-iris-van-herpen-daniel-wright-and-mgx-by-materialise/, (accessed July 2019).
3. Dezeen, *Escapism by Daniel Widrig, Iris van Herpen and .MGX by Materialise*, https://www.dezeen.com/2011/03/04/escapism-by-daniel-widrig-iris-van-herpen-and-mgx-by-materialise/ (accessed July 2019).
4. 3D Printing Industry, https://3dprintingindustry.com/news/karl-lagerfeld-showcases-3d-printed-chanel-fashion-at-paris-fashion-week-53000/ (accessed July 2019).
5. Vice, *MoMA Just Acquired A 3D-Printed Dress*, https://www.vice.com/en_uk/search?q=moma%203d%20printed%20dress, (accessed July 2019).
6. Nervous, *System Kinematics*, https://n-e-r-v-o-u-s.com/projects/sets/kinematics/ (accessed July 2019).

7. Dezeen, *3D-printed fabrics by Richard Beckett woven into Pringle of Scotland's ready to wear garments*, https://www.dezeen.com/2014/02/17/3d-printed-fabrics-by-richard-beckett-woven-into-pringle-of-scotlands-ready-to-wear-garments/ (accessed July 2019).
8. S. Grishanov, Structure and properties of textile materials, in *Handbook of Textile and Industrial Dyeing, Volume 1- Principles, Processes and Types of Dyes*, ed. M. Clark, Woodhead Publishing, 2011, ch. 2. Available from: https://app.knovel.com/hotlink/pdf/id:kt0097UD12/handbook-textile-industrial-2/structure-properties-introduction.
9. G. Bingham, R. Hague, C. Tuck, A. Long, J. Crookston and M. Sherburn, Rapid manufactured textiles, *Int. J. Comp. Integr. Manufact.*, 2007, **20**(1), 96–105.
10. R. Melnikova, A. Ehrmann and K. Finsterbusch, 3D printing of textile-based structures by Fused Deposition Modelling (FDM) with different polymer materials, *IOP Conf. Series: Mater. Sci. Eng.*, 2014, **62**, 012018.
11. F. Davis, 3D Printed Textiles from Textile Code: Structural Form and Material Operations, *SIGraDI Proceedings of the 16th Iberoamerican Congress of Digital Graphics*, 2012, pp. 392–396. Available from: http://papers.cumincad.org/data/works/att/sigradi2012_191.content.pdf.
12. J. Crookston, A. Long, G. Bingham and R. Hague, Finite-element modelling of mechanical behaviour of rapid manufactured textiles, *J. Mater.: Design Appl.*, 2008, **222**(1), 29–36.
13. N. Hopkinson, R. Hague and P. Dickens *Rapid Manufacturing: An Industrial Revolution for the Digital Age*, Wiley-Blackwell, 2005.
14. Loughborough University, Revolutionary 3D printed fashion to be developed at Loughborough University, 2016 April 18 [cited 2017 Feb 1]; [news and events]. Available from: https://www.lboro.ac.uk/news-events/news/2016/april/3d-fashion.html.
15. 3 Printr, *Designer Bradley Rothenberg Is Working on "Cellular Textiles"*, 2014 Oct 9 [cited 2018 Sep 3]; [applications fashion]. Available from: https://www.3printr.com/designer-bradley-rothenberg-working-cellular-textiles-0321700/.
16. M. Bloomfield and S. Borstrock, Modeclix. The additively manufactured adaptable textile, *Mater. Today Commun.*, 2018, **16**, 212–216.
17. Dezeen, *3D-printed Dress for Dita Von Teese by Michael Schmidt and Francis Bitonti*, 2013 Mar 7 [cited 2018 Sep 3]; [magazine]. Available from: https://www.dezeen.com/2013/03/07/3d-printed-dress-dita-von-teese-michael-schmidt-francis-bitonti/.
18. G. Bingham, R. Hague, C. Tuck, A. Long, J. Crookston and M. Sherburn, Rapid manufactured textiles, *Int. J. Comput. Integr. Manufact.*, 2007, **20**(1), 96–105.
19. D. Spencer, *Knitting Technology: A Comprehensive Handbook and Practical Guide*, Woodhead Publishing, Cambridge, 3rd edn, 2001, p. 18.
20. B. Farahi, Material Behaviours in 3D Printing, in *Digital Fabrication*, ed. Yuan P, Menges A and Leach N., Tongji University Press, 2017, pp. 211–221.
21. B. Farahi, Ruff, 2015 [cited 2019 Jul 24]. Available from: http://behnazfarahi.com/ruff/.
22. Revolvy, *Auxetics*, 2018 [cited 2018 Sep 3]. Available from: https://www.revolvy.com/page/Auxetics.
23. Livescience, *Fused Deposition Modelling: Most Common 3D Printing Method*, 2013 [cited 2019 Jun 18]. Available from: https://www.livescience.com/39810-fused-deposition-modeling.html.
24. 3D Systems, *Selective Laser Sintering*, [date unknown] [cited 2019 Jun 18]. Available from: https://uk.3dsystems.com/on-demand-manufacturing/selective-laser-sintering?utm_source=crdm.co.uk&utm_medium=301.
25. R. Thompson and M. Thompson, *The Materials Sourcebook for Design Professionals*, Thames & Hudson, 2017, p. 138.

26. D. Peleg, *Liberty Leading the People*, [date unknown] [cited 2019 Jun 18] [collections]. Available from: https://danitpeleg.com/liberty-leading-the-people-2/.
27. B. Msallem, N. Sharma, S. Cao, S. Halbeisen, F. Zeilhofer and M. Thieringer, Evaluation of the Dimensional Accuracy of 3D-Printed Anatomical Mandibular Models Using FFF, SLA, SLS, MJ, and BJ Printing Technology, *J. Clin. Med.*, 2020, **9**, 817.
28. Continuum, *N12*, [date unknown] [cited 2019 Jun 18] [projects]. Available from: https://continuumfashion.com/N12.php.
29. Materialise, *Flexible 3D Printed Fashion Hits the Catwalk with Iris van Herpen, Julia Koerner and Materialise*, 2013 Feb 15 [cited 2019 Jun 18] [blog]. Available from: https://i.materialise.com/blog/en/flexible-3d-printed-fashion-hits-the-catwalk-with-iris-van-herpen-julia-koerner-and-materialise/.
30. WIRED, *The Shattering Truth of 3D-Printed Clothing*, 2017 May 12 [cited 2017 Jul 17]. Available from: https://www.wired.com/2017/05/the-shattering-truth-of-3d-printed-clothing/.
31. M. Kolanjikombil, *Encyclopaedic Dictionary of Textile Terms Volume 1*, Woodhead Publishing, India, 2018, p. 561.
32. L. Sabantina, F. Kinzel, A. Ehrmann and K. Finsterbusch, Combining 3D printed forms with textile structures- mechanical and geometrical properties of mulit-material systems, *IOP Conf. Series: Mater. Sci. Eng.*, 2015, **87**, 012005.
33. M. Korger, J. Bergschneider, M. Lutz, B. Mahltig, K. Finsterbusch and M. Rabe, Possible Applications of 3D Printing Technology on Textile Substrates, *IOP Conf. Series: Mater. Sci. Eng.*, 2016, **141**, 012011.
34. V. Moody and H. Needles, *Tufted Carpet Textile Fibers, Dyes, Finishes, and Processes*, William Andrew, 2004, p. 166.
35. S. Majid, M. Alkahari, F. Ramli, S. Maidin, T. Fai and M. Sudin, Influence of integrated pressing during Fused Filament Fabrication on tensile strength and porosity, *J. Mech. Eng.*, 2017, **2**, 185–197.
36. 3D Hubs, *Introduction to SLS 3D Printing*, [date unknown] [cited 2020 Apr 9]; [knowledge base]. Available from: https://www.3dhubs.com/knowledge-base/introduction-sls-3d-printing/.
37. S. Boydell and D. Grimshaw, Object Lessons: what is the value of engaging with the physical object within design research and education, evaluated through the application of the "Material & Process Innovation Collection", at Manchester Metropolitan University Special Collections, *Int. Assoc. Soc. Design Res.*, 2017 [cited 2019 Jun 18]. Available from: scholar.uc.edu/concern/articles/7s75dc377?locale = en.
38. Manchester Metropolitan University Special Collection Collections Development Policy, Appendix 9, Themes and priorities for future collecting, 2014.
39. Crafts Council, Make: Shift, 2016 [cited 2020 Apr 9]; [what we do]. Available from: https://www.craftscouncil.org.uk/what-we-do/makeshift/.
40. IASDR, IASDR Conference, 2019 [cited 2020 Apr 9]; [conference]. Available from: https://iasdr2019.org.
41. Digits 2 Widgets, Nylon SLS Design Guide [2015] [cited 2019 Jun 15]. Available from: www.digits2widgets.com/3d-printing-cad-services/nylon-sls_3d_printing/).
42. Form labs, Guide to Selective Laser Sintering (SLS) 3D Printing [date unknown] [cited 2019 Jun 15] [guides]. Available from: https://formlabs.com/uk/blog/what-is-selective-laser-sintering/.
43. W. Yusoff, H. Ani, D. Pham and K. Dotchev, Influence of Molecular Weight Average, Degree of Crystallinity, and Viscosity of Different Polyamide PA12 Powder Grades on the Microstructures of Laser Sintered Part, *MATEC Web Conf.*, 2015, **26**, 03005.
44. T. Bechtold, Rapid prototyping in museum collections. Technology—applications—durability, in *Future Talks 011 Technology and Conservation of Modern Materials in Design*, ed. T. Bechtold, Die Neue Sammlung, The International Design Museum, Munich, 2011, pp. 47–55.

45. K. Dotchev and W. Yusoff, Recycling of polyamide 12 based powders in the laser sintering process, *Rapid Prototyping J.*, 2009, **15**(3), 192–203.
46. C. Coon, B. Pretzel, T. Lomax and M. Strlic, Preserving rapid prototypes: a review, *Heritage Sci*, 2016, **4**, 40.
47. D. Madsac, Rapid prototyping–rapid ageing? Technology and ageing properties of contemporary art and design objects made by rapid prototyping technologies, in *Future Talks 011 Technology and Conservation of Modern Materials in Design*, ed. T. Bechtold, Die Neue Sammlung, The International Design Museum, Munich, 2011, pp. 57–65.
48. M. Iversen, Resistance to Replication. *Tate Papers*. 2007 [cited 2020 Apr 13]; 8. Available from: https://www.tate.org.uk/research/publications/tate-papers/08/resistance-to-replication.
49. National Archives, https://www.nationalarchives.gov.uk/information-management/manage-information/digital-records-transfer/what-are-born-digital-records/, (accessed April 2020).
50. Designboom, Defining authentic design within today's digitally-driven age, 2017 [cited 2020 Apr 14]; [magazine]. Available from: https://www.designboom.com/design/museum-london-defining-authentic-design-10-01-2017/.
51. H. Cooper, Thoughts on Thoughts on Replication. *Tate Papers*. 2007 [cited 2020 Apr 13];**8**. Available from: https://www.tate.org.uk/research/publications/tate-papers/08/thoughts-on-thoughts-on-replication.
52. T. Bechtold, T. Van Oosten, R. Griffith, T. Learner and J. Levin, The Material of Many Possibilities: A Discussion about the Conservation of Plastics, *The GCI Newsletter*, 2014, **29**(1), 18–23.
53. PrintCity, *Manchester Metropolitan University*, About print city; [cited 2019 Jun 15]. Available from: https://printcity.mmu.ac.uk/about/.
54. C. Hewitt, *Preserving 3D Data and 3D Prints*; 2019 May [2019 Jul 5]. Available from: https://www.cooperhewitt.org/2019/05/14/preserving-3d-data-and-3d-prints/.
55. D. Mad Sack, Tea. Earl Grey. Hot-about Materials in 3D Printing, *WAAC Newsl.*, 2014, **36**(3), 14–18.

Subject Index

Page numbers in **bold** refer to tables; *italic* to figures; glossary entries as ***bold italic***